Brassica Oilseeds

Breeding and Management

Brassica Oilseeds

Breeding and Management

Edited by

Arvind Kumar

Vice Chancellor, Rani Lakshmi Bai Central Agricultural University, Jhansi-284003, Uttar Pradesh, India

S.S. Banga

ICAR-National Professor, Punjab Agricultural University, Ludhiana, Punjab, India

P.D. Meena

Principal Scientist (Plant Pathology), Directorate of Rapeseed-Mustard Research (ICAR), Bharatpur, India

P.R. Kumar

Ex-Assistant Director General, Indian Council of Agricultural Research, New Delhi, India

www.cabi.org

CABI is a trading name of CAB International

CABI
Nosworthy Way
Wallingford
Oxfordshire OX10 8DE
UK

Tel: +44 (0)1491 832111
Fax: +44 (0)1491 833508
E-mail: info@cabi.org
Website: www.cabi.org

CABI
745 Atlantic Avenue
8th Floor
Boston, MA 02111
USA

Tel: +1 617 682 9015
E-mail: cabi-nao@cabi.org

A catalogue record for this book is available from the British Library, London, UK.

Library of Congress Cataloging-in-Publication Data

Brassica Oilseeds : breeding and management / edited by Arvind Kumar, S.S. Banga, P.D. Meena, P.R. Kumar.
     p. ; cm.
  Includes bibliographical references and index.
  ISBN 978-1-78064-483-7 (hbk : alk. paper) 1. Brassica--Breeding.
2. Brassica--Diseases and pests. I. Kumar, Arvind, 1952- , editor. II. Banga, S. S. (Surinder S.), editor. III. Meena, P. D. (Prabhu Dayal), editor.
IV. Kumar, P. R. (Priya Ranjan), 1942- , editor.
  [DNLM: 1. Brassica--genetics. 2. Breeding. 3. Plant Diseases--prevention & control. 4. Stress, Physiological. SB 317.B65]
  SB299.B7B73 2015
  635'.34--dc23
                            2015021754

ISBN-13: 978 1 78064 483 7

Commissioning editor: Sreepat Jain
Editorial assistant: Emma McCann
Production editors: Shankari Wilford and Emma Ross

Typeset by SPi, Pondicherry, India.
Printed and bound in the UK by CPI Group (UK) Ltd, Croydon, CR0 4YY.

Contents

_____

Contributors

Abha Agnihotri, Head, Centre for Agricultural Biotechnology, Amity Institute of Microbial Technology, Block E-3, 4th Floor, Amity University Uttar Pradesh, Sector 125, Noida-201303 (NCR), India. E-mail: agnihotri.abha@gmail.com

Shashi Banga, Professor (Plant Breeding), Punjab Agricultural University, Ludhiana, Punjab-141027, India. E-mail: nppbg@pau.edu

S.S. Banga, ICAR-National Professor, Oilseed Section, Punjab Agricultural University, Ludhiana, Punjab-141027, India. E-mail: nppbg@pau.edu; surin11@rediffmail.com

M.J. Barbetti, School of Plant Biology, The University of Western Australia (M084), 35 Stirling Highway, Crawley, WA 6009, Australia. E-mail: martin.barbetti@uwa.edu.au

Ram Bhajan, Professor, Department of Plant Breeding & Genetics, College of Agriculture, G.B. Pant University of Agriculture & Technology, Pantnagar-236145, Uttrakhand, India. E-mail: rbhajan@rediffmail.com

Venkatesh Bollina, Saskatoon Research Centre, Agriculture and Agri-Food Canada, 107 Science Place, Saskatoon, Saskatchewan, Canada S7N 0X2. E-mail: Venkatesh.Bollina@nrc-cnrc.gc.ca

M. Hossein Borhan, Research Scientist, 107 Science Place, Saskatoon, Saskatchewan S7N 0X2, Canada. E-mail: hossein.borhan@agr.gc.ca

C. Chattopadhyay, Director, ICAR-National Centre on Integrated Pest Management, LBS Building, Pusa Campus, New Delhi-110012, India. E-mail: chirantan_cha@hotmail.com

Wayne E. Clarke, Agriculture and Agri-Food Canada, 107 Science Place, Saskatoon S7N 0X2, Canada. E-mail: Wayne.Clarke@AGR.GC.CA

Donghui Fu, Key Laboratory of Crop Physiology, Ecology and Genetic Breeding, Ministry of Education, Agronomy College, Jiangxi Agricultural University, Nanchang 330045, China. E-mail: fudhui@163.com

A.M. Gurung, Faculty of Land and Food Resources, The University of Melbourne, Parkville 3010, Australia. E-mail: a.gurung@unimelb.edu.au

Sanjay J. Jambhulkar, Scientific officer 'G', Nuclear Agriculture & Biotechnology Division, Bhabha Atomic Research Centre, Mumbai-400085, India. E-mail: sjj@barc.gov.in

Yogendra Khedikar, Saskatoon Research Centre, Agriculture and Agri-Food Canada, 107 Science Place, Saskatoon, Saskatchewan, Canada S7N 0X2.

S.J. Kolte, Ex-Professor (Plant Pathology), Flat No. 8, Anuprita Apartment, S.No. 78, Plot 160, Left Bhusari Colony, Near Telephone Exchange, Paud Road, Kothrud, Pune, India. E-mail: koltesj@rediffmail.com

Arvind Kumar, Vice Chancellor, Rani Lakshmi Bai Central Agricultural University, Jhansi-284003, Uttar Pradesh, India. E-mail: vcrlbcau@gmail.com

D. Kumar, Emeritus Scientist, ICAR-Central Arid Zone Research Institute, Jodhpur-342003, Rajasthan, India. E-mail: dkumarcazri@gmail.com

P.R. Kumar, Ex-Assistant Director General, Indian Council of Agricultural Research, Krishi Bhawan, New Delhi-110001, India. E-mail: priya_ranjankumar@yahoo.com

Sarwan Kumar, Assistant Professor (Entomology), Department of Plant Breeding and Genetics, Punjab Agricultural University, Ludhiana, Punjab-141004, India. E-mail: Sarwanent@pau.edu

Lisong Ma, Principal Research Scientist, 107 Science Place, Saskatoon, Saskatchewan S7N 0X2, Canada.

P.D. Meena, Principal Scientist (Plant Pathology), ICAR-Directorate of Rapeseed-Mustard Research, Bharatpur-321303, India. E-mail: pdmeena@gmail.com

Isobel A.P. Parkin, Agriculture and Agri-Food Canada, 107 Science Place, Saskatoon S7N 0X2, Canada. E-mail: Isobel.Parkin@agr.gc.ca

G.S. Saharan, Ex-Professor and Head, Department of Plant Pathology, CCS Haryana Agricultural University, Hisar-125004, Haryana, India. E-mail: gssaharan675@gmail.com

P.A. Salisbury, Senior Plant Breeder and Lecturer, University of Melbourne, Institute of Land and Food Resources, University of Melbourne, Parkville, Victoria 3052 and Victorian Institute for Dryland Agriculture, Horsham, Victoria 3402, Australia. E-mail: psalisbury@optusnet.com.au

D.K. Sharma, Director, ICAR-Central Soil Salinity Research Institute, Karnal-132001, Haryana, India. E-mail: director@cssri.ernet.in

P.C. Sharma, Principal Scientist (Agronomy), ICAR-Central Soil Salinity Research Institute, Karnal-132001, Haryana, India. E-mail: pcsharma.knl@gmail.com

Dhiraj Singh, Director, (ICAR) Directorate of Rapeseed-Mustard Research, Sewar, Bharatpur-321303, Rajasthan, India. E-mail: director.drmr@gmail.com; dhirajmustard@gmail.com

Y.P. Singh, Principal Scientist (Entomology), ICAR-Directorate of Rapeseed Mustard Research, Bharatpur-321303, Rajasthan, India. E-mail: ypsingh1777@gmail.com

Gohar Taj, Assistant Professor, Molecular Biology & Genetic Engineering, G.B. Pant University of Agriculture & Technology, Pantnagar, Uttrakhand, India. E-mail: gohartajkhan@rediffmail.com

P.R. Verma, Ex-Senior Plant Pathologist, Agriculture & Agri-Food Canada, Saskatoon, Saskatchewan S7N 0X2, Canada.

Meili Xiao, Chongqing Engineering Research Center for Rapeseed, College of Agronomy and Biotechnology, Southwest University, Chongqing 400716, China. xml846@163.com

Foreword

Oilseed brassicas, also known by their trade name of rapeseed-mustard, are among the largest traded agricultural commodities. The combined volume of exports of rapeseed and mustard-seed oils was almost 10.07% of the total vegetable oilseeds. Growing importance and consequent research investments have led to dramatic breakthroughs in yield and seed quality. Modification of the seed oil and reduction of meal glucosinolate content by Canadian breeders in rapeseed during the early 1970s was a plant breeding feat that stands next only in importance to the architectural modification in cereals. Since then, enormous strides have been made. Hybrids have been commercialized, and all crop brassica genomes sequenced. World output of rapeseed and mustard crops rose from about 36 million tons in 2001/02 to 70 million tons in 2013/14. However, the demand for rapeseed-mustard oils continues to escalate steeply due to increasing consumption and diversion for bioenergy use. The consumption growth rates of the developing countries now almost double those of developed countries. With increasing diversion of competing feed crops for bioenergy usage, demand for rapeseed-mustard as a feedstock is also expected to grow further. Meeting enormous demands requires continuous ameliorations for enhanced productivity and biological value seed storage products. This requires efforts towards sustainable germplasm enhancement, complementary crop management methodologies and effective research feedback to meet challenges of rising needs, cultivation uncertainties and climate change.

This book, *Brassica Oilseeds: Breeding and Management*, is an excellent compilation of researches on oilseed brassicas, especially those relating to biological origin, commercial relevance, germplasm enhancement, genomics and crop management. The editors, Drs Arvind Kumar, Surinder Banga, Prabhu Dayal Meena and Priya Ranjan Kumar, have selected 12 prominent topics as the key subjects to provide a broader perspective of brassica oilseeds. This book is an excellent contribution with chapters written by leading scientists from across the major rapeseed-mustard producing countries. The book interweaves these aspects in an attractive way. Although the emphasis of this book is on germplasm enhancement, it contains much that will be of interest to all stakeholders, indeed to everyone interested in brassica oilseeds.

Fu Ting-Dong

Prof. Ting-Dong Fu
Huazhong Agricultural University
Institute of Crop Genetics & Breeding
Wuhan, China

Preface

It is now a little over 30 years since the research field that this book described was founded. The book's authors, who have had the fortune to be working in the field almost since beginning, felt for some time that a comprehensive book might be useful, mainly as a mean of consolidating what has been achieved so far and making it more accessible.

Oilseed *Brassica* species are grown in around 50 different countries. This diversity of production is reflected in differences in species used, cultivar types, sowing and harvest times, length of growing seasons, oil quality types, yields, use of GM (genetically modified technology) and the major insect and disease challenges faced. *Brassica napus*, *Brassica rapa* and *Brassica juncea* have been widely used for oilseed production over many years. *Brassica carinata* (Ethiopian mustard) has predominantly been grown in Ethiopia. Numerous publications with very useful information have been published, which has encouraged us to compile the data in the form of the present book.

The present book 'Brassica Oilseeds: Breeding and Management' deals with the aspects on 'Breeding' i.e. genetics and breeding, intersubgenomic heterosis, induced mutagenesis and allele mining, seed quality modifications and genomics, and 'the stress management', i.e. diseases, *Albugo candida*, pathogenesis of *Alternaria* species, insect pests and abiotic stresses.

The subject matter is widely illustrated with photographs, graphs, figures, histograms and tables, etc. for stimulating, effective and easy reading and understanding.

Inclusion of important references will be helpful to the researchers, teachers and students in locating the original consultations.

We are sure that this comprehensive treatise on 'Brassica Oilseeds' will be of immense use to researchers, teachers, students and all others who are interested in the breeding and management of brassica oilseed crops worldwide.

Arvind Kumar
S.S. Banga
P.D. Meena
P.R. Kumar

Acknowledgements

———————————

The authors are highly grateful to Dr R.K. Downey, Dr S. Ayyappan, Dr S.K. Datta, Dr P.R. Verma and Prof. Ting-Dong Fu, and to publishers/societies/journals/institutes/websites and all others whose valuable materials such as photographs (macroscopic, microscopic, electron micrographs, scanning electron micrographs), drawings, figures, histograms, graphs, tables and flow charts, etc. have been used through reproduction in the present document.

Abbreviations

Only abbreviations that have been widely used are included in this list.

AB   Alternaria blight
AC   *Albugo candida*
AFLP   Amplified fragment length polymorphism
AICRP-RM   All Indian Coordinated Research Project on Rapeseed-Mustard
AOS   Active oxygen species
ATP   *Arabidopsis* TILLING Project
AUDPC   Area under disease progress curve
BAC   Bacterial artificial chromosome
BDS   Butane diol succinate
BIG   Brassica Improvement Group
BRAD   Brassica Database
Bt   *Bacillus thuringiensis*
CAPS   Cleaved amplified polymorphic sequence
CC   Coiled-Coil
cp-DNA   Chloroplast-deoxyribonucleic acid
CRW   China Rose Winter
CSSRI   Central Soil Salinity Research Institute
cv.   Cultivar
DAS   Days after sowing
DDT   Dichloro Diphenyl Trichloroethane
DEGS   Diethylene glycol succinate
DH   Doubled haploid
DHN   Dihydroxy Naphthalene
DOPA   Dihydroxy Phenylalanine
DRMR   Directorate of Rapeseed-Mustard Research
ECD   European Clubroot Differential
ELISA   Enzyme-linked immune absorbent assay
EPG   Electrical penetration graph
ETI   Effector-triggered immunity
FAE1   Fatty acid elongase 1
FID   Flame ionization detector

FIL	Final intensity of rust on leaf
FIP	Final intensity of rust on plant
GA	Gibberellic acid
GBS	Genotyping by sequencing
GC	Gas chromatography
GEBV	Genomic estimated breeding values
GLC	Gas liquid chromatography
GM	Genetically modified technology
GS	Genomic selection
GSS	Glucosinolate sulfatase enzyme
GWAS	Genome-wide association studies
HIS	Herbivore-induced synomones
HPLC	High performance liquid chromatography
HR	Hypersensitive response
ICAR	Indian Council of Agricultural Research
IPM	Integrated Pest Management
LD	Linkage disequilibrium
LDL	Low density lipoprotein
LEA	Low erucic acid
LRR	Leucine-rich repeat
MAGIC	Multiparent advanced generation inter-cross
MAMPs	Microbe-associated molecular patterns
MAS	Marker assisted selection/breeding
MNU	Methyl nitrosourea
MS	Mass spectrometer
MUFA	Mono-unsaturated Fatty Acids
NAM	Nested association mapping
NB	Nucleotide-binding domain
NBS	Nucleotide-binding sites
NGS	Next generation sequencing
NIRS	Near infrared reflectance spectroscopy
NMR	Nuclear magnetic resonance
NSKE	Neem seed kernel extract
PAMPs	Pathogen-associated molecular patterns
PAU	Punjab Agricultural University
PMG	Polymethyl galacturonase
PR	Pathogenesis related
PRRs	Pattern-recognition receptors
PUFA	Polyunsaturated fatty acids
QTL	Quantitative trait loci
R	Resistant
RAD	Restriction site associated DNA marker
RAPD	Random amplified polymorphic DNA
RES	Round black Spanish
RFLP	Restriction fragment length polymorphisms
RIPs	Ribosome-inhibiting proteins
RLK	Receptor-like kinase
ROS	Reactive oxygen species
ros	Rosette
S	Susceptible
SA	Salicylic acid

SAR	Systemic acquired resistance
SAVERNET	South Asian Vegetable Research Network
SD	Solanaceae domain
SFA	Saturated fatty acids
SMD	Soil moisture deficit
SNP	Single nucleotide polymorphism
SR	Sclerotinia rot
SSRs	Simple sequence repeats
TAL	Transcription activator-like
TD	Transmembrane domain
TILLING	Targeting Induced Local Lesions in Genomes
TIR	Toll-Interleukin-1 receptor
TLC	Thin layer chromatography
TM	Trombay mustard
TMS	Trimethyl silyl
TuMV	Turnip mosaic virus
UPGMA	Unweighted pair group method with arithmetic mean
UPLC	Ultra performance liquid chromatography
VLCFA	Very long chain fatty acids
VLCMFAs	Very long chain mono-unsaturated fatty acids
WGT	Whole genome triplication
WR	White rust
YSM	Yellow seed coat mutants

Symbols and Units

%	per cent
>	more than
μ	micron
μg	microgram
a.i.	active ingredient
cm	centimetre
E	transpiration rate
gs	stomatal conductance
h	hours
ha	hectare
kg	kilogram
Mha	million hectare
Mt	million tonne
°C	degree celcius
P_n	assimilation rate
RH	relative humidity
WP	wettable powder
Y	yield
μm	micrometre

1 Importance and Origin

Arvind Kumar,[1]* P.A. Salisbury,[2,3] A.M. Gurung[4] and M.J. Barbetti[5]

[1]Vice Chancellor, Rani Lakshmi Bai Central Agricultural University, Jhansi-284003, Uttar Pradesh, India; [2]University of Melbourne, Institute of Land and Food Resources, University of Melbourne, Parkville, Victoria, Australia; [3]Victorian Institute for Dryland Agriculture, Horsham, Victoria, Australia; [4]Faculty of Land and Food Resources, The University of Melbourne, Parkville, Australia; [5]School of Plant Biology, The University of Western Australia, Crawley, Australia

Introduction

Oilseed brassicas, also known by their trade name of rapeseed-mustard, include *Brassica napus*, *B. juncea*, *B. carinata* and three ecotypes of *B. rapa*. In 2012/13 global production of these crops exceeded 63.76 Mt, making them the second most valuable source of vegetable oil in the world. The leading oilseed-brassica producers in the world are the European Union, China, Canada and India (USDA, 2015). Different forms of oilseed brassicas are cultivated throughout the world. Winter-type *B. napus* predominates in Europe, parts of China and eastern USA, while spring-type *B. napus* is cultivated in Canada, Australia and China. Spring forms of *B. rapa* are now mainly grown in the Indian subcontinent. Winter-type *B. rapa* has largely been replaced by more higher yielding winter-type *B. napus* and spring crops in its traditional production zones. Only spring types of *B. juncea* are cultivated in the Indian subcontinent, and is now been actively considered as an option in drier areas of Canada, Australia and even in the northern USA. In India, *B. juncea* predominates and is grown on over 90% of the area under rapeseed-mustard crops. The conventional crop improvement objectives focus largely on attempts to produce, protect and tailor the biomass to suit the main requirement as an edible oil. In some countries of the world, Brassica oil is being used as a biodiesel. The goal of developing canola forms has been accomplished for *B. rapa*, *B. napus* and *B. juncea* but remains an important objective in *B. carinata*. The transition of these crops from high erucic to low erucic rapeseed, and the consequent explosive growth as an oilseed crop, began from Canada in 1968 with the commercial release of single low cv. Oro, followed by several other single low cultivars, and the first canola cv. Tower in 1974. Almost all rapeseed produced in Australia, Canada and Europe and to a very large extent in China, is now canola. The cultivation of canola rapeseed-mustard has just begun in India (Chauhan *et al.*, 2010). Further modifications in the acid composition for specialized product applications such as biofuels are being sought.

Brassicas are confronted with several biotic stresses such as diseases (blackleg, Sclerotinia rot, Alternaria blight, white rust, etc.) and pests (aphids, beetles, etc.). As these crops are grown in a wide array of climate and cropping systems, these require general or specific adaptation to

---

*Corresponding author, e-mail: ddgedn@gmail.com

specific situations. Varieties with varying maturity duration are required to escape frost (Canada) or late-season drought (southern Australia) or to fit in multiple cropping sequences (India, China). Breeding programmes are also concerned with a cultivar's suitability for existing or emerging management practices, e.g. herbicide resistance or mechanical harvesting (resistance to pod shattering).

Diversity in Global Production Systems

Oilseed brassica species are grown in around 50 different countries. This diversity of production is reflected in differences in species used, cultivar types, sowing and harvest times, length of growing seasons, oil quality types, yields, use of GM (genetically modified technology) and the major insect and disease challenges faced. *Brassica napus*, *B. rapa* and *B. juncea* have been widely used for oilseed production over many years. *Brassica carinata* (Ethiopian mustard) has predominantly been grown in Ethiopia. Attempts to convert *B. carinata* to a canola quality type have not been successful, but it is finding a niche as industrial products such as biodiesel. Depending on location, typical spring cultivars with no vernalization requirement require a growing season from around 120 to 190 days. This contrasts to over 300 days for European winter lines with a strong vernalization requirement (Table 1.1).

There are both regional and country differences in the relative use of open-pollinated cultivars compared with hybrids. In general, the increased involvement of private companies in canola breeding has led to an increased focus on the use of hybrids. In addition, the canola industry has also seen the introduction of GM cultivars.

Some major diseases and insects are common across different countries. This includes diseases such as blackleg (*Leptosphaeria maculans* (anamorph *Phoma lingam*)), Sclerotinia stem rot (*Sclerotinia sclerotiorum*) and Alternaria blight (*Alternaria* spp.) and insects such as aphids. Other diseases may be more localized. The same disease can be a major challenge to production in one area but of little consequence in another. Some pathogens are a significant problem across several *Brassica* species, while others are more of a major problem in one specific *Brassica* species but cause little or no damage on another species. White rust (*Albugo candida*) is one such species. While some insects are widespread across wide regions and even different countries, many insect problems can be more localized. For example, red-legged earth mite is a widespread and significant establishment pest in Australia but it is not an issue elsewhere. The relative importance of different diseases and insects can change dramatically with changes in race structures or environmental conditions. Changes in tillage practices (e.g. the introduction of minimum tillage) can also influence the severity of different insects and diseases. Several establishment pests have become a bigger problem with the introduction of minimum tillage practices such as stubble retention. The priorities of oilseed brassica breeders do not remain static, but need to be regularly adjusted to meet the challenges brought on by disease and insect variability, changing environmental conditions and other external pressures.

Status of oilseed brassica production in various countries is presented in Table 1.2. The table highlights the similarities and differences between countries in variety types, growing season, area of production, yield and pests. In the case of India, rapeseed-mustard crops are grown in diverse agroclimatic conditions ranging from north-eastern/northwestern hills to down south under irrigated/rainfed, timely/late sown, saline soils and mixed cropping. *Brassica rapa* var. Brown Sarson, which once dominated the entire rapeseed-mustard growing region, has now been largely replaced by Indian mustard. Indian mustard accounts for about 75–80% of the 6.6 Mha under rapeseed-mustard in the country during 2013/14, contributing about 40% to the country's total edible oil supplies. Rajasthan, Haryana, Madhya Pradesh and Uttar Pradesh are the major rapeseed-mustard growing states with 47.2%, 12.6%, 11.0% and 10.3% contribution, respectively, to the national hectarage. The rapeseed-mustard production trends represent a fluctuating scenario with an all-time high production of 8.3 (Mt) from

Table 1.1. Sowing and harvest details for *Brassica* oilseed crops around the world.

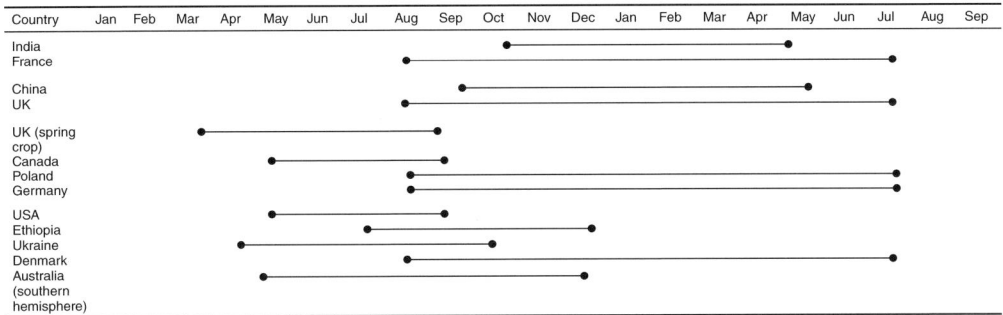

6.9 (Mha) during 2010/11. The yield levels also have been variable, ranging from 854 (2002/03) to 1250 kg/ha (2013/14). Fluctuating and low productivity levels in India are a reflection of the low number of crop growing days, shorter day-length conditions during the bulk of growing season, stressed ecologies of cultivation and low productivity levels.

Taxonomy

Crop brassicas are the economically most important members of family *Brassicaceae* and the sub-tribe *Brassicinae*. A number of taxonomic treatises of the family *Brassicaceae* are available. These include Linnaeus (1753), De Candolle (1821), Baillon (1871), Prantl (1891), Schulz (1919, 1936) and Beilstein *et al.* (2006). Schulz's has so far been most comprehensive and authoritative description. A molecular account of the family has been recently provided by Beilstein *et al.* (2006). *Brassiceae* is among 19 tribes recognized by Schulz in the family *Brassicaceae* and is further subdivided into seven to nine sub-tribes (Gómez-Campo, 1980, 1999). *Brassica* is the major genus in this sub-tribe. Several members of related sub-tribes, such as *Raphaninae* and *Moricandiinae*, also show strong affinity with *Brassica*. This is despite the demonstrated distinctness of these species on the basis of chloroplast-DNA (cp-DNA) and restriction sites (Warwick and Black, 1991; Pradhan *et al.*, 1992; Warwick *et al.*, 1992). Cytogenetic investigations (Prakash and Hinata, 1980) have demonstrated that the three digenomic species, *B. carinata* (2n=34: BBCC), *B. juncea* (2n=36; AABB) and *B. napus* (2n=38: AACC), have originated through crosses between any two of the three elementary species *B. nigra* (2n=16; BB), *B. oleracea* (2n=18; CC) and *B. rapa* (2n=20; AA) (Fig. 1.1).

It is also now known that *B. nigra*, *B. rapa* and *B. oleracea/B. rapa* are the cytoplasm donor species for *B. carinata*, *B. juncea* and *B. napus*, respectively (Banga *et al.*, 1983; Erickson *et al.*, 1983). However, there are now indications that both *B. oleracea* and *B. nigra* might have also been cytoplasm donors for *B. napus* and *B. juncea*, respectively, in independent hybridization events during evolution of these species (Banga, personal communication). Major taxonomic forms are presented in Table 1.3.

Crop History and Ecology

Brassicas were among the earliest crops brought under domestication. Both *B. rapa* and *B. juncea* find mention in ancient Sanskrit literature dating back to 1500 BCE. These are also indicated in Greek, Roman and Chinese writings of 500 to 200 BCE (Downey and Röbbelen, 1989). The Chinese language equivalent of rapeseed was first recorded ca. 2500 years ago, and the oldest archaeological finds may date back as far as ca. 5000 BCE (Yan, 1990). In India, seeds of *B. juncea* found in archaeological sites have been dated to 2300 BCE (Parkash and Hinata, 1980). In Europe, domestication is believed to have occurred in the early middle ages and commercial plantings of rapeseed were carried out in the Netherlands during the 16th century.

Table 1.2. Summary of *Brassica* crop production in various countries.

Character	Australia	Canada	USA	China	India	France	Poland	Germany	UK	Ethiopia	Ukraine	Denmark
Sowing time	May–June (Autumn)	May	May (spring types) September (winter types)	September–October (Autumn)	September–November (winter)	August–September	August–September	August–early September	August–September	July	April–May (spring crop)	August–September (winter crop) March/April (spring crop)
Harvest time	November–December (Summer)	August–September	August–September (spring types) Late June–July (winter types)	May (Summer)	February–May (Summer)	July	July	July	July	December	September–October (spring crop)	July (winter crop)
Length of growing season	150–190 days (spring varieties)	95–125 days	110–125 days (spring types) 270–300 (winter types)	210–230 days (semi-winter varieties)	130–150 days Some very early lines (110 days)	330 days	330 days	330 days	330 days Some spring sowing (180 days)	180 days	150–180 days Some winter sowing	330 days (winter crop)
Species used	Mainly B. napus Small amount of B. juncea B. rapa was also grown in the 1980s, but was discontinued	Mainly B. napus Limited B. rapa (decreasing) Limited B. juncea Some B. carinata (biodiesel production)	B. napus B. rapa	Mostly B. napus Small amount of B. juncea B. carinata	B. juncea B. rapa Limited areas of B. napus B. carinata	B. napus	B. napus	B. napus	B. napus	B. carinata	B. napus	B. napus

Oil quality	Canola quality	Canola quality High oleic, low linolenic acid (HOLL) quality	High oleic, low linolenic acid (HOLL) quality	Mainly canola quality Limited high erucic acid	Mainly high glucosinolates and high erucic acid Some canola-quality *B. napus*	Mainly canola quality Limited high erucic acid rapeseed (HEAR) production	Canola quality	Mainly canola quality Limited high erucic acid rapeseed (HEAR) production	Canola quality Limited high erucic acid rapeseed (HEAR) and high oleic, low linolenic acid (HOLL) production	High glucosinolates and erucic acid	Canola quality	Canola quality
Variety types	Hybrids Open-pollinated types	Mainly hybrids	Hybrids (spring) Open-pollinated types (winter)	Hybrids Open-pollinated types	Mainly open-pollinated types Some hybrids have been developed	Mainly hybrids	Hybrids and open-pollinated types	Predominantly hybrids	Hybrids and open-pollinated types Emphasis on hybrids	Open-pollinated types	Hybrids (winter) Open-pollinated types (spring)	Hybrids, Open-pollinated types
Major diseases	Blackleg Sclerotinia	Blackleg Sclerotinia Brown girdling root rot in (*B. juncea*)	Blackleg Sclerotinia	Sclerotinia Pre-emptive breeding against blackleg White rust (in *B. juncea*)	Sclerotinia Alternaria White rust (in *B. juncea*)	Blackleg Sclerotinia Alternaria Rhizoctonia	Blackleg	Blackleg Sclerotinia White leaf spot Alternaria	Blackleg Light leaf spot Sclerotinia Alternaria	Alternaria		Light leaf spot Sclerotinia Blackleg Alternaria
Major insect problems	Red-legged earth mite Diamondback moth	Flea beetles Bertha armyworm Aphids Diamondback moth	Aphids Harlequin bugs Diamondback moths Southern cabbage worm	Aphids	Aphids	Bertha armyworm Harlequin bug Flea beetle Lygus bug		Bertha armyworm Harlequin bug Flea beetle Lygus bug	Flea beetles Cabbage stem beetles Aphids Pollen beetles Seed weevils	Flea beetle		Flea beetles Pollen beetles

Continued

Table 1.2. Continued.

Character	Australia	Canada	USA	China	India	France	Poland	Germany	UK	Ethiopia	Ukraine	Denmark
Average yield (t/ha)	1.57	1.87	1.72	1.78	0.91	3.34	2.68	3.65	3.43	0.88	1.70	3.54
Average area (Mha)	1,359	6,712	441	6,770	6,582	1,508	798	1,436	639	>40	1,029	166
Average production (Mt)	2,127	12,568	759	12,055	5,990	5,040	2,140	5,248	2,192	>35	1,746	588

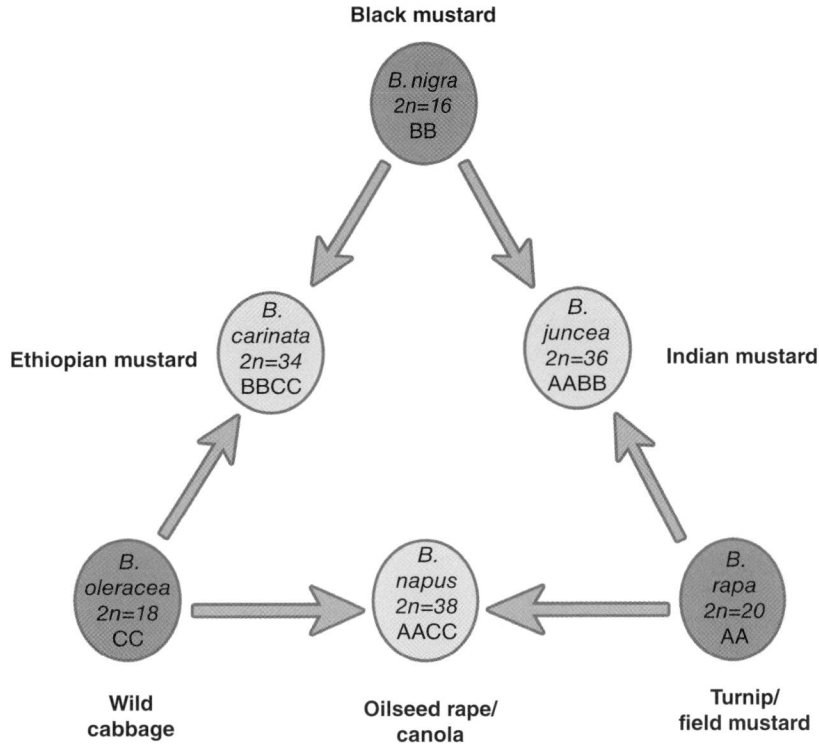

Fig. 1.1. U's triangle showing species relationships among different Brassica species (U, 1935).

At that time, rapeseed oil was used primarily as oil for lamps. Later, it became used as a lubricant for steam engines. Historically, *B. rapa* appears to have the widest distribution among brassica oilseeds. At least 2000 years ago it was distributed from northern Europe to China and Korea, with the primary centre of diversity in the Himalayan region (Hedge, 1976).

Brassica rapa

Wild *B. rapa* subsp. *oleifera* (inc. var. *sylvestris*) is arguably the originator species of var. *rapa* L. (cultivated turnip) and var. *sylvestris* (Lam.) Briggs (turnip-rape). It is native throughout Europe, Russia, Central Asia and the Near East (Prakash and Hinata, 1980), with Europe proposed as one centre of origin. The Asian and Near Eastern type may have originated from an independent centre of origin in Afghanistan, which then moved eastward following domestication. Prakash and Hinata (1980) believed that oleiferous *B. rapa* developed in two places, giving rise to two different races, one European and the other Asian. Recent evidence (through analyses of chloroplast and mitochondrial DNA) suggests that *Brassica montana* (n=9) might be closely related to the prototype that gave rise to both cytoplasms of *B. rapa* and *B. oleracea* (Song and Osborn, 1992). Large morphological variation observed in *B. rapa* might have resulted due to the long history of breeding for different traits along with natural selection for adaptation in different geographical regions. Oleiferous and turnip forms developed in Europe, while in eastern Asia and West Asia these evolved into leafy and oleiferous forms. Various types of leafy vegetables of *B. rapa*, such as Chinese cabbage (*B. rapa* subsp. *pekinensis*) and non-heading pakchoi (*B. rapa* subsp. *chinensis*) and mizuna, are found widely in

Table 1.3. Taxonomy of crop *Brassica* species.

Botanical name	Common name	Usage
B. nigra	Black mustard	Condiment (seed)
B. oleracea		Vegetable fodder (leaves)
var. *acephala*	Kale	Vegetable (head)
var. *capitata*	Cabbage	Vegetable (terminal buds)
var. *sabauda*	Savoy cabbage	Vegetable (head)
var. *gemmifera*	Brussels sprouts	Vegetable, fodder (stem)
var. *botrytis*	Cauliflower	Vegetable (inflorescence)
var. *gongylodes*	Kohlrabi	Vegetable, fodder (stem)
var. *italica*	Broccoli	Vegetable (inflorescence)
var. *fruticosa*	Branching bush kale	Fodder (leaves)
var. *alboglabra*	Chinese kale	Vegetable (stem, leaves)
B. rapa		
subsp. *oleifera*	Turnip rape	Oilseed
var. Brown Sarson	Brown sarson	Oilseed
var. Yellow Sarson	Yellow sarson	Oilseed
var. Toria	Toria	Oilseed
subsp. *rapifera*	Turnip	Fodder, vegetable (root)
subsp. *chinensis*	Bok choi	Vegetable (leaves)
subsp. *pekinensis*	Chinese cabbage	Vegetable, fodder (head)
subsp. *nipposinica*	–	Vegetable (leaves)
subsp. *pamchinensis*	–	Vegetable (leaves)
B. carinata	Ethiopian mustard	Vegetable oilseed
B. juncea	Mustard	Oilseed, vegetable
B. napus		
subsp. *oleifera*	Rapeseed	Oilseed
subsp. *rapifera*	Rutabaga, swede	Fodder

East Asia, particularly in China, Korea and Japan. The oleiferous form of *B. rapa* (*B. rapa* subsp. *oleifera*) has been historically cultivated for production of vegetable oils in China, Canada, India and in northern Europe and is the third brassica oilseed crop. The Indian form developed into a mainly oleiferous form such as the *B. rapa* vars Yellow Sarson, Brown Sarson and Toria that are grown over large areas in eastern and north-eastern parts of India during the winter season (Singh, 1958).

Brassica oleracea

This crop was domesticated as early as 2000 BCE initially as primitive kales or cabbages. These are now believed to be the first cultivated forms of *B. oleracea* (Chiang *et al.*, 1993), especially by Celts whose name 'Bresic' for cabbage may be the likely progenitor of '*Brassica*'. Heading cabbages and other leafy types probably had a common origin from the ancestral kales and non-heading cabbages (Herve, 2003), and they were further developed in Portugal, Spain and France. Wild forms of *B. oleracea* are found on the Atlantic coasts of Europe, northern France and England. Related wild species are endemic to the Mediterranean basin. Because the cultivated *B. oleracea* crops were grown in close proximity to wild relatives, the flow of genes from wild to cultivated types probably occurred. This coupled with mutation, human selection and adaptation must have contributed to current morphogenetic variation in this crop.

Brassica nigra

Brassica nigra or black mustard was originally domesticated in Asia Minor or Iran, and has been cultivated since ancient times. It was the primary condiment mustard all over Europe but subsequently gave way to brown mustard or *B. juncea*. It is native to the Mediterranean region, and archaeological finds show that it has occurred as a weed in association with the cultivation of wheat and barley during domestication. Historically, black mustard appeared to have been harvested mainly from wild populations with only sporadic commercial cultivation.

Brassica carinata

This is an amphiploid derived from natural hybridization between *B. oleracea* and *B. nigra* in Ethiopian highlands. Despite significant efforts, it has not been possible to locate any wild forms (Mizushima and Tsunoda, 1967). However, both the species are commonly observed growing close to each other in cultivation or as escapes. This crop has been domesticated for multiple uses such as oil, condiments, medicines and vegetables (Astley, 1982; Riley and Belayneh, 1982).

Brassica juncea

This is a natural amphiploid derived from crosses between *B. rapa* and *B. nigra*. The origins of *B. juncea* are unclear, but it must have originated in areas like the Middle East and neighbouring regions, where distributions of *B. nigra* and *B. rapa* overlap (Prakash and Hinata, 1980). The concept of a separate origin of foliage types (Chinese) and oleiferous types (Indian) was mainly based upon the investigations of Vaughan *et al.* (1963) and Vaughan (1977) on seed glucosinolate profiles. Vaughan (1977) also suggested that the Indian race was closer to the *B. rapa* progenitor, and the oriental race was closer to the *B. nigra* progenitor, and perhaps this indicates that *B. juncea* evolved into more than one region. Spect and Diederichsen (2001) considered the place of origin somewhere between Eastern Europe and China, where there is parental sympatry. *Brassica juncea* has been divided into four subspecies, with different morphology, quality characteristics and uses (Spect and Diederichsen, 2001). These are: (i) subsp. *integrifolia*, used as a leaf vegetable in Asia; (ii) subsp. *juncea*, cultivated mainly for its seed, occasionally as fodder; (iii) subsp. *napiformis*, used as a root-tuber vegetable; and (iv) subsp. *taisai*, stalks and leafs of which are used as vegetables in China.

Brassica napus

Brassica napus is thought to have multiple origins resulting from independent natural hybridization events between *B. oleracea* and *B. rapa*. In Europe, the winter form has predominantly become a common yellow crucifer of wild areas. In the British Isles, for example, it has been naturalized wherever oilseed rape is grown, and it is a relatively recent introduction into Canada, the USA, China and India.

References

Astley, D. (1982) Collecting in Ethiopia. *Cruciferae Newsletter* 7, 3–4.

Baillon, H.E. (1871) *Historie des plantes – Cruciferes* 3, 188–195, Paris.

Banga, S.S., Banga, S.K. and Labana, K.S. (1983) Nucleo cytoplasmic interactions in *Brassica*. *Proceedings of the 6th International Rapeseed Congress*, Paris, 2, 602–606.

Beilstein, M.A., Al-Shehbaz, I.A. and Kellogg, E.A. (2006) Brassicaceae phylogeny and trichome evolution. *American Journal of Botany* 93, 607–619.

Chauhan, J.S., Singh, K.H., Singh, V.V. and Kumar, S. (2010) Hundred years of rapeseed-mustard breeding in India: accomplishments and future strategies. *Indian Journal of Agricultural Sciences* 81, 1093–1109.

Chiang, M.S., Chong, C., Landry, R.S. and Crete, R. (1993) Cabbage *Brassica oleracea* subsp. Capitata. In: Kalloo, G. and Bergh, B.O. (eds) *Genetic Improvement of Vegetable Crops*. Pergamon Press, Oxford, UK, pp. 113–155.

De Candolle, A.P. (1821) Cruciferae. *Systema Naturale* 2, 139–700.

Downey, R.K. and Röbbelen, G. (1989) *Oil Crops of the World*. Mc Graw-Hill Publishing Company, Baltimore, Maryland, pp. 339–362.

Erickson, I.R., Straus N.A. and Beversdrof, W.D. (1983) Restriction patterns reveal origins of chloroplast genomes in *Brassica* amphiploids. *Theoretical and Applied Genetics* 65, 202–206.

Gómez-Campo, C. (1980) Morphology and morphotaxonomy of the tribe Brassiceae. In: Tsunoda, S., Hinata, K. and Gómez-Campo, C. (eds) *Brassica Crops and Wild Allies: Biology and Breeding*. Japan Scientific Soc. Press, Tokyo, pp. 3–31.

Gómez-Campo, C. (1999) Seedless and seeded beaks in the tribe *Brassiceae* (*Cruciferae*). *Cruciferae Newsletter* 21, 11–12.

Hedge, I.C. (1976) A systematic and geographical survey of the Old World Cruciferae. In: Vaughan, J.G., MacLeod, A.J. and Jones, M.G. (eds) *The Biology and Chemistry of the Cruciferae*. Academic Press, London, pp. 1–45.

Herve, Y. (2003) Choux. In: Pitrat, M. and Foury, C. (eds) *History de legumes, des origins al'oree du XXI siecle*. INRA, Paris, pp. 222–234.

Linnaeus, C. (1753) *Species Plantarum* II, 561. Stockholm.

Mizushima, U. and Tsunoda, S. (1967) A plant exploration in *Brassica* and allied genera. *Tohoku Journal of Agricultural Research* 17, 249–276.

Pradhan, A.K., Prakash, S., Mukhopadhyay, A. and Pental, D. (1992) Phylogeny of *Brassica* and allied genera based on variation in chloroplast and mitochondrial DNA patterns: Molecular and taxonomical classifications are incongruous. *Theoretical and Applied Genetics* 85, 331–340.

Prakash, S. and Hinata, K. (1980) Taxonomy, cytogenetics and origin of crop brassicas, a review. *Opera Botanica* 55, 1–57.

Prantl, K. (1891) Cruciferae. In: Engler, A. and Prantl, K. (eds) *Die Naturlichen Pflanzenfamilien*. Wilhelm Englmann, Leipzig, Germany, pp. 145–208.

Riley, K.W. and Belayneh, H. (1982) Report from an oil crop collection trip in Ethiopia. *Cruciferae Newsletter* 7, 5–6.

Schulz, O.E. (1919) Cruciferae-Brassiceae. Part I: Brassicinae and Raphaninae. In: Engler, A. (ed.) *Das Pflanzenreich*. Wilhelm Engelmann, Leipzig, Heft 68–70, pp. 1–290.

Schulz, O.E. (1936) Cruciferae. In: Engler, A. and Prantl, P. (eds) *Die Naturlichen Pflanzenfamilien*. Wilhelm Engelmann, Leipzig, pp. 17b, 227–658.

Singh, D. (1958) *Rape and Mustard*. Indian Central Oilseeds Committee Examiner Press, Bombay, India.

Song, K.M. and Osborn, T.C. (1992) Polyphyletic origins of *Brassica napus*: new evidence based on organelle and nuclear RFLP analyses. *Genome* 35, 992–1001.

Spect, C.E. and Diederichsen, A. (2001) Brassica. In: Hanelt, P. (ed.) *Mansfeld's Encyclopedia of Agricultural and Horticultural Crops*, vol.3. Springer-Verlag, Berlin, pp. 1453–1456.

U, N. (1935) Genomic analysis in *Brassica* with special reference to the experimental formation of *B. napus* and peculiar mode of fertilization. *Japan Journal of Botany* 7, 389–452.

USDA (United States Department of Agriculture) (2015) *Oilseeds: World Markets and Trade*. Foreign Agricultural Service, Office of Global Analysis, February 2015.

Vaughan, J.G. (1977) A multidisciplinary study of the taxonomy and origin of *Brassica* crops. *Bio Science* 27, 35–40.

Vaughan, J.G., Hemingway, J.S. and Schofield, H.J. (1963) Contributions to a study of variations in *Brassica juncea* (L.) Coss & Czern. *Botany Journal of the Linnaean Society* 58, 435–447.

Warwick, S.I. and Black, L.D. (1991) Molecular systematics of *Brassica* and allied genera (subtribe *Brassicinae, Brassiceae*) – chloroplast genome and cytodeme congruence. *Theoretical and Applied Genetics* 82, 81–92.

Warwick, S.I., Black, L.D. and Aguinagalde, I. (1992) Molecular systematics of *Brassica* and allied genera (subtribe Brassicinae, Brassiceae) – chloroplast DNA variation in the genus *Diplotaxis*. *Theoretical and Applied Genetics* 83, 839–850.

Yan, Z. (1990) Overview of rapeseed production and research in China. *Proceedings of the International Canola Conference Potash and Phosphate Institute*, Atlanta, Georgia, pp. 29–35.

2 Genetics and Breeding

Shashi Banga,[1]* P.R. Kumar,[2] Ram Bhajan,[3] Dhiraj Singh[4] and S.S. Banga[1]

[1]*Punjab Agricultural University, Ludhiana, Punjab;* [2]*Indian Council of Agricultural Research, Krishi Bhawan, New Delhi;* [3]*Department of Plant Breeding & Genetics, College of Agriculture, G.B. Pant University of Agriculture & Technology, Pantnagar, Uttrakhand;* [4]*(ICAR), Directorate of Rapeseed-Mustard Research, Sewar, Bharatpur, Rajasthan, India*

Introduction

Oilseed brassicas are critical for the edible oilseeds economy of the world. Therefore, these crops have received much attention from cytogeneticists, taxonomists, evolutionary biologists, crop breeders and biotechnologists. Notwithstanding unsolved problems of susceptibility to pests and diseases, significant progress has been made towards enhanced productivity and seed quality modifications. A great deal is also understood about the origin and diversification of the genus. Modern molecular technologies have helped to answer many critical questions about the origin of the polyploid brassicas, the relationships among species and species groups, and the genesis of the domesticated forms from their wild progenitors. Also well characterized are wild crucifers and their relatedness with cultivated species. Investigators also focused on introgressing alien genetic resources in crop brassicas. Hybrids are now available in otherwise self-pollinated di-genomic species, *Brassica juncea* (mustard) and *Brassica napus* (rapeseed), thanks to alloplasmic male sterility systems. Genome sequencing has been completed in all three monogenomics and *B. napus*. Complete genome sequences of *B. juncea* and *Brassica carinata*,

though unpublished, are also available. There are also international efforts ongoing to sequence 150 wild crucifer genomes under the Brassicales Map Alignment Project (BMAP). BMAP is analogous to a similar initiative that has been established for *Oryza* (OMAP). This will probably extend the definition of the available *Brassica* gene pool beyond strict taxonomic domains of the cultivated brassicas by emphasizing genetic compatibility. This chapter brings together the genetic and breeding activities in oilseed brassicas.

Cytogenetics

Karyotype studies were initiated as early as 1934 by Catcheside and subsequently by Alam (1936), Richharia (1937) and Sikka (1940). These investigations were primarily focused on the chromosome types, number of satellites, and nucleoli in different genomes. Chromosomes were distinguished as long, medium, small and very small with the median, sub-terminal and terminal constrictions. Sikka (1940) used secondary bivalent associations in three species, *B. monensis* (syn. *Coincya monensis* 2n=24), *B. sinapistrum* (syn. *Sinapis arvensis*, 2n=18) and *B. nigra*, to hypothesize the basic chromosome number as n=5. Venkateswarlu and Kamala

---

*Corresponding author, e-mail: nppbg@pau.edu

(1971) identified six basic types of chromosomes based on pachytene analysis. Subsequent confirmation of this inference (Röbellen, 1960) led to the suggestion of n=6 as the basic chromosome number. This was also supported by meiotic chromosome pairing in haploids of *B. nigra* and *B. tournefortii* (Prakash, 1974a, b). Further information through use of molecular markers firmly suggested that *B. oleracea/B. rapa* originated from one archetype while *B. nigra* originated from the other. These inferences also confirmed earlier cytogenetic studies suggesting genetic divergence between *B. nigra* and *B. oleracea/B. rapa* (Prakash and Hinata, 1980). Genomic studies later confirmed that diploid species themselves evolved from a common ancestor after ancient genome triplication followed by numerical and structural chromosomal changes that took place 7.9–14.6 million years ago (Mya) (Lysak *et al.*, 2005). Recognition of *B. rapa*, *B. nigra* and *B. oleracea* as cytoplasm donor parents for *B. juncea*, *B. carinata* and *B. napus*, respectively, was also reported (Banga *et al.*, 1983; Erickson *et al.*, 1983). Recent research, however, reveals that both the diploid genome donors (e.g. *B. rapa* and *B. oleracea* for *B. napus*; *B. rapa* and *B. nigra* for *B. juncea*) were involved as respective cytoplasm donor parents in independent hybridization events (Allender and King, 2010; Kaur *et al.*, 2014). Synthetic *B. juncea* with *B. nigra* cytoplasm, *B. napus* and *B. carinata* with *B. oleracea* cytoplasm are difficult to obtain but have been bred by Indian scientists. In general, re-synthesis of *B. juncea/B. carinata* was easier than that of *B. napus* (Prakash and Chopra, 1991; Srivastava *et al.*, 2004; Bansal *et al.*, 2009, 2012). These resynthesized genotypes are now being used extensively to improve genetic diversity and enhance heterosis in both *B. napus* and *B. juncea*. A new approach is now available that allows re-synthesis of an amphiploid species through hybridization between the non-parental *Brassica* di-genomic species. This method promises to provide novel sources of genetic diversity (Gupta et al., 2015). This method has the advantage of utilizing intensively bred di-genomics in contrast to the use of unselected diploid progenitors in the re-synthesis route. Extensive and exploitable primary, secondary and tertiary gene pools are available (Prakash *et al.*, 2009). Immunosuppressants, hormones, *in vitro* fertilization, embryo culture and protoplast fusion have helped in overcoming sexual barriers. Chromosome addition lines such as *B. rapa–B. oleracea/B. rapa–B. nigra/B. napus–Diplotaxis erucoides/B. napus–B. nigra/B. rapa–B. alboglabra/B. rapa–B. oxyrhhina* have been produced (Chen *et al.*, 1997; Dhillon *et al.*, 2000; Prakash *et al.*, 2009; Kapoor *et al.*, 2011). C-genome chromosome substitution lines in *B. juncea* have been reported (Banga, 1988). Analysis of meiotic pairing in a very large number of hybrids involving crop *Brassica* species and wild crucifers have helped in establishing genomic relatedness between crop *Brassica* genomes and wild genomes (Kirti *et al.*, 1991, 1992a, b; Bhaskar *et al.*, 2002; Ahuja *et al.*, 2003; Banga *et al.*, 2003a, b; Chandra *et al.*, 2004a, b; Garg *et al.*, 2007; Prakash *et al.*, 2009). It was also possible to introgress wild alleles for tolerance to various stresses in the cultivated brassicas (Garg *et al.*, 2010; Warwick, 2010; Kumar *et al.*, 2011; Atri *et al.*, 2012). Introgression of gene(s) for higher productivity in *B. juncea* has also been achieved from *Erucastrum abyssinicum* (Banga, unpublished) in *B. juncea*. Genetic information for pod shatter resistance from *B. carinata* has been introgressed into *B. napus* (Prakash *et al.*, 2009). Exploitation of alien cytoplasmic variability was very effective in the development of a large number of cytoplasmic male sterility (CMS) lines (Prakash *et al.*, 2009). Besides these known cases of alien genetic introgressions, the genes for fertility restoration (*Rf* genes) for all the CMS systems could be introgressed from cytoplasm donor species (Prakash *et al.*, 2009). Many of these *Rf* genes have been tagged (Janeja *et al.*, 2003a, b).

Genetic Architecture

A large number of genes influencing gross morphological characters of oilseed brassicas have been characterized (Banga, 1996; Séguin-Swartz *et al.*, 1997; Prakash *et al.*, 2004). The majority of the morphological variants showed monogenic recessive inheritance with some role of modifiers. Seed colour was almost always associated with two or more genes. Amplified fragment length polymorphism (AFLP) (Sabharwal *et al.*, 2004) and simple sequence repeat (SSR) (Padmaja *et al.*, 2005) markers associated with gene(s) for seed coat colour were also identified. Due to the wide adaptation of the crop, the genes for vernalization and photo-periodic response have been widely investigated. Spring habit was dominant over winter habit and requirement for vernalization was controlled by a major and minor gene in *B. napus*. Acclimatization ability has been shown to be regulated by genes with very small additive effects. Primacy of additive genetic variance with some dominance and occasional epistasis has been demonstrated in a large number of studies (Singh and Singh, 1972; Lefort-Buson and Datté, 1985). Despite numerous studies, genetic architecture and determinants of yield continue to remain ambiguous (Banga and Banga, 2009). Genotype × environment (G×E) interactions remain a major issue. In an elaborate study, Shi *et al.* (2009) identified 85 quantitative trait loci (QTL) for seed yield along with 785 QTLs for eight yield-associated traits from ten natural environments and two rapeseed populations. A trait-by-trait meta-analysis recognized 401 consensus QTLs, of which 82.5% were clustered closely, 47 of which were relevant for seed yield. The idea of identifying indicator QTLs for yield and underlying candidate genes constitutes an advance in methodology for complex traits. Experiments with different species have not revealed any predictors of heterosis based both on methods of molecular as well as genetic diversity (Bansal *et al.*, 2012). Generalizing such inferences has not been useful due to polyploidy, high genotypic specificity, heterogeneity of parental populations, lack of distinct gene pools with diverse gene frequencies and high G×E interactions. Despite intensive focus, the genetic architecture of heterosis is still to be fully understood. Study of F_2 populations developed following random intercrossing between 202 lines from a double haploid population in rapeseed (*B. napus* L.) led to the identification of a large number of QTLs and epistatic interactions for 15 yield-correlated traits (Shi *et al.*, 2011). These involved variable loci with moderate effect with genome-wide distribution. QTL hotspots were also identified. Varied modes-of-inheritance of QTLs (additive, A; partial-dominant, PD; full-dominant, D; over-dominant, OD) and epistatic interactions (additive × additive, AA; additive × dominant/dominant × additive, AD/DA; dominant × dominant, DD) were observed. AA epistasis was considered the foremost genetic basis of heterosis in rapeseed (Shi *et al.*, 2011).

Mode of Pollination

The crop brassicas harbour large intraspecific variability in the extent of cross-pollination. Outcrossing rates in predominantly self-pollinating crops of *B. juncea* (Labana and Banga, 1984) and *B. napus* vary from 7.6 to 18.1% and up to 16.1%, respectively, while *B. rapa* var. Toria and *B. rapa* var. Brown Sarson are largely cross-pollinating due to self-incompatibility. However, *B. rapa* var. Yellow Sarson is self-pollinating.

Breeding Objectives

The conventional crop improvement objectives focus largely on attempts to produce, protect and tailor the biomass for various edible and industrial uses. The goal of developing canola forms has been accomplished for *B. rapa*, *B. napus* and *B. juncea* but remains an important objective in *B. carinata*. In an Indian context, canola mustard remains to be commercialized. Further modifications in the acid composition for specialized product applications such as oleochemicals are currently being sought. Brassicas are also confronted with several biotic stresses such as diseases (Sclerotinia rot, Alternaria blight, white rust, etc.) and pests (aphids, beetles, etc.). As these crops are grown in a wide array of climate and cropping systems, they all require general or specific adaptation to specific situations. Varieties with varying maturity duration are required to escape frost (northwest India) or late season high temperature and drought (central India) or to fit in multiple-cropping sequences. Breeding programmes are also concerned with a cultivar's suitability for existing or emerging management practices, e.g. herbicide resistance or mechanical harvesting (resistance to pod shattering).

Yield enhancement

Success to improve yields depends upon the ability to substitute bottle-neck loci with more effective alleles (Banga and Banga, 2009). In the self-pollinating brassicas the yield improvements in the conventional programmes were facilitated by different types of pedigree methods (Downey and Rimmer, 1993). Handling of segregating populations from bi-parental or multiple crosses as bulk populations before extraction of pure lines is also a routine practice. Visual selection in F_2 on a single plant basis for seed yield and oil content has proven ineffective. The inefficiency of selection was attributed to large G×E interaction rather than a preponderance of dominance genetic variance (Brown, 1995). Frequently, it is desirable to delay first backcross to F_2 generation as a cycle of recombination provides the possibility of breaking certain undesirable linkages as well as for improving yield and other agronomic traits (Röbbelen and Nitsch, 1975). Delaying selection until F_3 or later generations or through manipulation of simple characters or combination of characters has been more effective. However, the influence of environment on yield components (Thurling, 1993) modulates the response to selection. The dihaploid technique has helped to shorten the breeding cycle in *B. napus*, but this is not successful in *B. rapa* due to inbreeding depression. Efforts are being made to integrate it with mustard breeding in India. Incompatible types of *B. rapa* cultivars/populations are genetically heterogeneous and yield improvement could be achieved through mass selection. Effectiveness of recurrent selection to improve various traits has been well documented in several cross-pollinating crops but no systematic attempt has been made to exploit this technique in *B. rapa*. Recurrent selection in self-pollinated brassicas can be facilitated by genetic male sterility (GMS). To maximize genetic advance, selection must be carried out in the target environment using criteria that combine yield with stability. Several of the advance breeding programmes use molecular markers to increase selection efficiency. Genome sequencing of *Arabidopsis thaliana*, *B. rapa*, *B. oleracea* and *B. napus* and the development of saturated linkage maps and comparative genomics in brassicas and *Arabidopsis* has the enormous potential for application in gene identification and breeding (http://www.brassica.info). This information is likely to improve selection efficiency. In addition, marker-assisted selection during backcrossing can help to eliminate linkage drag and hasten the recovery of a recurrent genotype.

Biotic stresses

Rapeseed-mustard is afflicted by many biotic stresses. These include Alternaria blight caused by *Alternaria brassicae* and *A. brassicicola*, white rust (*Albugo candida*), downy mildew (*Hyaloperonospora brassicae*) and stem rot (*Sclerotinia sclerotiorum*). Aphids (*Lipaphis erysimi*), mustard saw fly (*Athalia proxima*) and painted bug (*Bagrada hilaris*) are major insect pests. Details of these diseases are available elsewhere in this volume. Other than for white rust, genotypic (host) variability for resistance against the other major diseases and insects pests is limited. Stem rot (*S. sclerotiorum*) and Rhizoctonia rot, which were of minor significance during the past, are now assuming great importance. For white rust, monogenic dominant resistance to *A. candida* (race 2) was reported in *B. nigra*, *B. rapa*, *B. juncea* and *B. carinata*. It was also possible to map locus AcB1-A4.1 to a linkage group A4, and locus AcB1-A5.1 to linkage group A5 in *B. juncea*. In both instances, closely linked flanking markers were identified based on synteny between *Arabidopsis* and *B. juncea* (Panjabi *et al.*, 2010). Two or possibly three independent dominant genes have been suggested for inheritance of white rust resistance in *B. napus*. High heritability and nuclear gene control are suggested for *Sclerotinia* resistance in *B. napus*, while only variation for Alternaria blight resistance has been demonstrated. *Brassica* species with the B-genome appear to show greater tolerance to various insects attacking oilseed brassicas. Newly developed sources for Sclerotinia stem rot and mustard aphid have been discussed under alien introgressions. Development of the seedling-stage screening technique for Sclerotinia stem rot (Garg *et al.*, 2008) and framework differential set (Ge *et al.*, 2012) have paved the way for more aggressive breeding strategies for *Sclerotinia* resistance. Marker development and mapping genes associated with resistance to biotic stress factors represent serious goals for brassica breeding. Map-based cloning of genes contributing to blackleg resistance is an important achievement (Rimmer, 2006). Fine mapping led to the identification of two loci governing resistance to the seedling-stage blackleg in canola cultivars (Mayerhofer *et al.*, 2005). Both loci localized to the same position on *B. napus* chromosome N 7. A collinear chromosome region could be identified in *Arabidopsis*. Candidate genes for blackleg resistance are known in *Arabidopsis*. QTL-linked markers associated with resistance against *Verticillium longisporum* were introduced from *B. oleracea*. QTL involved in resistance against club-root disease have been mapped (Werner *et al.*, 2008). Cultivars having resistance to blackleg have been bred in *B. napus* using indigenous variability or through introgression from *B. juncea* or *B. sylvestris*. A simple backcross method has been used to incorporate genes for white rust resistance in *B. juncea*. The introduction of chitinase gene into *B. napus* (Grison *et al.*, 1996) under the control of a constitutive promoter was shown to increase field resistance to three fungal pathogens.

Abiotic stresses

Drought, salt, frost and heat are the major abiotic stresses, which determine the productivity of rapeseed-mustard, with about 25% of the total area sown under rainfed conditions. The yield reduction under moisture stress (drought) in Indian mustard ranged from 17 to 94% (Mohammad *et al.*, 1990; Chauhan *et al.*, 2007) due to considerable effects on yield components (Chauhan *et al.*, 2007; Sharma and Pannu, 2007). Singh *et al.* (2003) opined that mustard genotypes with thicker leaves had greater water-use efficiency. The negative effects of drought during seed development may be mitigated by increasing harvest index. Several physiological parameters such as osmotic adjustment (Singh *et al.*, 1996), transpiration cooling (Chaudhary *et al.*, 1989), epicuticular wax on the leaves (Kumar, 1990), difference between canopy and air temperature (Kumar and Singh, 1998) and drought susceptibility index (Chauhan *et al.*, 2007; Sharma *et al.*, 2011) were reported to be associated with drought tolerance. Although some germplasm lines were found promising, these await genetic characterization. In India, several varieties, including Aravali, PBR 97, Pusa Bahar, Pusa Bold, RH 781, RH 819, RGN 48, RB 50, Shivani, Vaibhav and PBR 378, have been released for

cultivation under rainfed conditions. High temperature, especially terminal heat stress, is the second most important stress next to drought. Two genetic stocks, BPR 541-4 and BPR 543-2, have been registered with the National Bureau of Plant Genetic Resources (NBPGR) for high thermo-tolerance at terminal and juvenile stages, respectively (Chauhan *et al.*, 2010). Several varieties of Indian mustard, viz. Kanti, Pusa Agrani, Urvashi, NRCDR 02, Pusa Mustard 25 and Pusa Mustard 27, showing good thermo-tolerance during crop establishment stage have been commercialized. Oilseed brassicas also face the problem of saline soils and poor-quality brackish irrigation water. RH 8814 and BPR 541-4 have been registered as salinity-tolerant genetic stocks. Varieties CS 52, CS 54 and Narendra Rai 1 that are tolerant to saline soils are now available.

Seed coat colour

A uniform seed colour is an especially desired trait due to its commercial importance. Seed colour variations are very common in oilseed brassicas. The colour ranges from dark brown to reddish brown and from yellow-brown to yellow. In India, most of the commercial cultivars are brown-seeded, while most of the eastern European types of *B. juncea* lines are yellow-seeded. Some exotic cultivars such as Skorospieka, Donskaja and Zem exist in the germplasm collection as potential donors of the yellow-seeded trait in breeding programmes. Inherently, yellow-seeded lines show higher oil content in the seed as compared to their brown-seeded counterparts (Stringam *et al.*, 1974). They also possess higher protein content in the seed meal and lower crude fibre in the seed hull as compared to brown-seeded lines (Simbaya *et al.*, 1995; Slominski, 1997). In addition, it is easier to determine the degree of physiological maturity in yellow seeds, as the occurrence of chlorophyll is not masked by the dark seed-coat colour. Furthermore, yellow-seeded lines have better market value as yellow seeds produce bright yellow-coloured oil, which is rated superior to the dull-coloured oil obtained from brown-seeded lines. Improvement in the quality of seed oil and meal is one of the major import-ant breeding objectives in Indian mustard and these quality parameters of mustard lines can be improved through the development of yellow-seeded mustard cultivars. The inherit-ance of seed coat colour in different *Brassica* species has been extensively studied. The brown seed-coat colour in *B. juncea* is con-trolled by two independently segregating dominant genes with duplicate effect (Vera *et al.*, 1979; Vera and Woods, 1982; Anand *et al.*, 1985; Chauhan and Kumar, 1987). It is hy-pothesized that the genomes of *B. rapa* (AA) and *B. nigra* (BB) each harbour one of the two genes for seed coat colour of *B. juncea*. Chen and Heneen (1992) underlined the role of the maternal genotype and the interaction between the maternal endosperm and/or embryonic genotypes. The influence of environment on the seed colour has also been reported (Van Deynze *et al.*, 1993). Thus, yellow seeds are produced when both the loci are in a homozy-gous recessive condition, and the maternal genotype influences the expression of the trait. Due to its recessive nature coupled with ma-ternal influence, the transfer of yellow seed-coat colour is difficult, since the trait has to be mobilized into Indian cultivars from un-adapted exotic sources of eastern Euro-pean origin through backcross breeding.

Increasing oil content

There has always been an emphasis on im-proving oil content, mostly at the cost of pro-tein. This has helped in raising oil content to 37–42% in current cultivars. Additional increase can accrue from the changeover to yellow-seeded types. Yellow-seeded forms are avail-able in *B. rapa*, *B. juncea* and *B. carinata*. Stable, pure yellow-seeded cultivars are still not avail-able in *B. napus*. Oil content has a complex heredity. It is controlled by a large number of genes with an additive effect besides mater-nal and environment influences (Banga and Banga, 2009; Weselake *et al.*, 2009; Wang *et al.*, 2010). Increasing market demands for edible usage as well as renewable fuels mandates further efforts to increase oil content. Clas-sical plant breeding, molecular breeding and

recombinant DNA techniques are being used to enhance oil content. Analysis of germplasm collections suggest a wide variation (35–46%) for the trait. However, the genes controlling oil biosynthesis are strongly influenced by the environmental and agronomic factors. A large number of QTLs for seed oil content have been identified in different mapping populations of *B. napus* (Delourme *et al.*, 2006; Qiu *et al.*, 2006; Zhao *et al.*, 2006). Of these, three QTLs, located in linkage groups A1, A8 and C3, were consistent across populations (Delourme *et al.*, 2006). The QTL located on linkage groups A8 and C3 also coincide with major QTLs for erucic acid (Howell *et al.*, 1996; Thormann *et al.*, 1996; Burns *et al.*, 2003). There are many potential targets for genetic-ally engineering improved oil content, which may include all the genes involved in increased photosynthesis (Reynolds *et al.*, 2009). From the bioenergetics perspective, leaves and pods are the principal photosynthate sources. Leaves control sink potential in terms of pod number and seeds per pod. During pod fill-ing, the contribution of the leaves declines (Allen *et al.*, 1971) and pod walls supply most of the assimilates to developing seeds (Tayo and Morgan, 1979). Site and tissue-specific photosynthetic activity in the silique wall (a maternal tissue) is responsible for seed oil contents as confirmed by *in planta* manipula-tion of silique-wall photosynthesis. Photosyn-thetic activity in the silique can be a useful breeding tool for the development of varieties with high oil content (Hua *et al.*, 2011). The apetalous and erectophile pods arguably at-tract more photosynthetically active radiation (Fray *et al.*, 1996). Photosynthesis-related genes are over-represented in the expressed genes in the silique wall (Wei *et al.*, 2011). Expression of lipid synthesis regulatory gene *WRINKLED1* was associated with silique-wall photosyn-thetic activity. Oil content also associated with expression of BnRBCS1A levels and Rubisco activities in the silique wall, but not in leaves (Wei *et al.*, 2011). *WRINKLED1* participates in the interaction between the silique wall and the seed. Sucrose from siliquae walls is mobil-ized into developing seeds, and converted into hexose UDP-Glc and Fru. Hexose is then converted into acetyl-CoA, a precursor of fatty acid biosynthesis, in the cytosol and the plastid of the embryo cells (Baud *et al.*, 2002; Hill *et al.*, 2003). A majority of the genes encoding various steps of the fatty acid biosynthesis have been identified and annotated. The precise contribution of individual activities through the pathways and final levels of triacylglycerol remains to be understood (Napier and Grahm, 2010). Efforts have been made to manipulate key fatty acid synthetic genes using transgenic technology. Incorporation of transgenes encoding key enzymes or enzyme subunits resulted in varying levels of lipids (Zou *et al.*, 1997; Jako *et al.*, 2001). In some cases, the oil content actually declined (Dehesh *et al.*, 2001). As oil biosynthesis is an intricately coordinated process involving not only the fatty acid synthesis genes but also genes associated with carbon metabolism (Cahoon *et al.*, 2007; Weselake *et al.*, 2009), genetic manipulations of fatty acid transcription factors or protein kinases may help to overcome the identified bottlenecks (Ruuska *et al.*, 2002; Cahoon *et al.*, 2007; Mu *et al.*, 2008). Incorporation of yeast gene coding for cytosolic glycerol-3-phosphate dehydrogenase (gpd1) with a seed-specific napin promoter precipitated a twofold increase in glycerol-3-phosphate dehydrogenase activity and a three- to four-fold increase in the level of glycerol-3-phosphate in developing seeds of transgenic oilseed rape (Vigeolas *et al.*, 2007). Final lipid content increased by 40%. While the protein content remained substantially unchanged, glycolytic intermediate dihydroxyacetone phosphate, a direct precursor of glycerol-3-phosphate dehydrogenase, declined. The levels of sucrose and other metabolites in the pathway from sucrose to fatty acids also remained unaltered. These results suggest that the glycerol-3-phosphate supply is a limiting factor for oil accumulation in developing seeds. MicroRNAs and siRNAs are considered to be significant regulators of plant development and seed formation. A comparison of small RNA expression profiles at early embryonic developmental stages in high oil content and low oil content *B. napus* cultivars helped to identify 50 conserved miRNAs and 11 new miRNAs, together with new miRNA targets (Zhao *et al.*, 2012). It is likely that some miRNAs that may be involved in the regulation of *B. napus* seed-oil production. In *Arabidopsis*, several genes encoding embryo-specific fatty acid transcription factors LEAFY COTYLEDON1 (AtLEC1), LEC1-LIKE (AtL1L) were found to play critical roles in the fatty acid biosynthesis.

Fatty acid modifications

Oil quality in brassicas has been successfully modified using selection, hybridization, induced mutagenesis and transformation. Half-seed technique is routinely applied in the backcross programme to incorporate gene-preventing synthesis of erucic acid (Morice, 1974). Mutagenesis or selective hybridization has helped in reducing linolenic acid content (<3%) and enhancing oleic acid levels (≈80%). Plants with genetic blocks in the chain elongation step towards eicosenoic acid and erucic acid were initially identified in summer rape (Stefansson *et al.*, 1961), summer turnip rape (Downey, 1964) and later in *B. juncea* (Kirk and Oram, 1981). Occurrence of a single gene controlling high erucic acid content has been reported in *B. rapa* (Davik and Heneen, 1996). Studies of *B. juncea* and *B. carinata* showed a good fit to two locus model (Kirk and Hurlstone, 1983: Fernandez-Escobar *et al.*, 1988). Higher expression of erucic acid and linoleic acid appear to be under the control of two genes with additive effects in *B. napus* and *B. juncea* (Jourdren *et al.*, 1996). Maternal influence, realized by interaction between maternal genotype and nuclear genes of the embryo, is also indicated. A number of investigators have attempted to modify the C18 fatty acid profile in *B. rapa* and *B. napus*, utilizing standard plant breeding techniques of induced mutagenesis and selection. Aside from the traditional crop improvement strategies, many successful attempts leading to cloning of the genes controlling different steps of fatty acid biosynthesis have been made using gene technologies. Novel genotypes with a desired level of specified fatty acid have been achieved in brassica oilseeds either by harnessing natural variability within primary and secondary gene pools, induced mutagenesis or genetic engineering. Identification of a single turnip rape plant with high level of C18:1 (69%) from a population with mean value of 58%

spawned the earlier breeding efforts for high C18:1 (Vikki and Tanhuanpää, 1995) in rapeseed-mustard crops. Recurrent selection for six generations for high C18:1 led to greatly enhanced C18:1 (85–90%) in combination with low C18:2 (1–3%) and C18:3 (3–6%) fatty acids. A similar kind of continuous response to selection experiments over 3 years was successful in decreasing C18:3 levels by fourfold. Modified fatty acid lines with high C18:1 and lower C18:2 than normal fatty acid have been bred in *B. napus* (Scarth, 1995; Potts and Males, 1999) through simple selection. Limited genetic variability available in *Brassica* germplasm for C18 fatty acid composition also prompted breeders to exploit induced variations using both physical and chemical mutagens. Chemical mutagens, such as ethyl methane sulfonate (EMS) that creates point mutations, have been the preferred mutagens in such studies. In many instances, combinations of physical and chemical mutagens were used (Rakow, 1973; Röbbelen and Nitsch, 1975). Most of the described mutations were recessive and obtained after screening several thousand M_2 or M_3 plants. Because of their relatively low frequency, few mutations in the fatty acid biosynthesis have been identified. Repeated applications of EMS followed by selection in M_2 and subsequent generations facilitated isolation of low C18:3 lines in *B. napus* (Rakow, 1973). Combinations of X-rays and chemical mutagens helped to develop mutants of *B. napus* with low C18:3 and elevated levels of C18:2 (Röbbelen and Thies, 1980; Rakow *et al.*, 1987). Some of these identified low C18:3 mutants were later used in hybridization programmes to develop low C18:3 cultivars of *B. napus* such as 'Stellar' (Scarth *et al.*, 1988) and 'Apollo' (Scarth *et al.*, 1995). Oil from these low C18:3 cultivars improved storage and frying stability and shortened hydrogenation time compared to standard canola oil (Eskin *et al.*, 1989; Przybylski *et al.*, 1993). Genotypes having >88% C18:1 are now available in *B. napus* (Auld *et al.*, 1992). The general increase in the amounts of C18:2 and C18:3 following termination of the chain elongation step in *B. juncea* suggested the presence of strong desaturase promotor gene(s) possibly associated with its B genome. In contrast, low C22:1 strains of *B. napus* possess

C18:1 as the major fatty acid, which is attributed to the desaturase suppressor gene(s) located on C genome chromosomes (Raney *et al.*, 1995a). Attempts have been made towards the transfer of desaturase suppresser gene(s) from C genome of *B. napus* to B genome of *B. juncea*. Raney *et al.* (1995a) crossed *B. juncea* line J90-4253 with the low C18:3 *B. napus* line C92-0226. The interspecific plants were backcrossed to *B. juncea* with the objective to develop *B. juncea* with lower C18:2 and C18:3 contents. Fatty acid analysis of single seeds of the BC_1F_2 generation allowed identification of plants with increased C18:1 and decreased C18:2 and C18:3 contents. Continuous selection for C18:1 and C18:2 in segregating material of BC_1F_3 and BC_1F_4 generations permitted the development of lines having low C18:3 (5.5–6.2%) coupled with high C18:1 (53.7%) or high C18:2 (47.2%). *B. napus* lines with high C18:2 and low C18:3 contents and *B. carinata* lines with high C18:1 were also developed by exploiting the interspecific variability that existed between the two species. Low C22:1 and high C18:2 *B. carinata* were crossed with low C18:3 *B. napus* and plants were backcrossed to the parental species to reconstitute both parental genotypes with introgression of genes from respective donors. Continuous selection of particular species for high C18:1 and low C18:3 yielded lines with desirable content for fatty acid composition (Raney *et al.*, 1995b). Markers associated with linolenic acid and erucic acid (Jourdren *et al.*, 1996) have been identified. These are likely to improve selection efficiency. In addition, marker-assisted selection during backcrossing can help to eliminate linkage drag and hasten the recovery of a recurrent genotype. Work on breeding of zero erucic acid mustard was initiated following the identification of low erucic acid lines Zem 1 and Zem 2 by Kirk and Oram (1981). This helped in the development of genotypes with low erucic and high oleic acids (Banga *et al.*, 1998). The high erucic acid trait in *B. juncea* is controlled by two genes with additive effects, zero erucic being recessive in expression (Bhat *et al.*, 2002; Gupta *et al.*, 2004). Further refinement in the genetic analysis was undertaken by Gupta *et al.* (2004) with an aim to develop markers tagged to erucic acid for use in the marker-assisted selection of low/high erucic acid lines in a backcross breeding programme. Two *FAE1* genes, *FAE1.1* and *FAE1.2*, cloned from low and high erucic acid mustard lines, revealed the presence of four and three single nucleotide polymorphisms (SNPs) in *FAE1.1* and *FAE1.2*, respectively. Using these SNPs, the *FAE1.1* and *FAE1.2* genes were mapped to the *B. juncea* map described by Pradhan *et al*. (2003). Their map positions coincided with the two QTLs for erucic acid located on two different linkage groups, one belonging to the A genome and the other belonging to the B genome. The QTL in the A genome explained 60% and the QTL in the B genome explained 38% of the phenotypic variance. Several low erucic acid strains have been commercialized for mustard in India.

Reducing meal glucosinolates

High levels of glucosinolates in rapeseed meal impair palatability and acceptance by animals and can also lead to goitrogenic hypertrophy. One spring cultivar ('Bronowski') was identified as a low glucosinolate form and together with 'Liho' provided the alleles to create the first double zero (canola quality) spring variety (Tower) exhibiting low erucic acid and low glucosinolate content (8–15 µmol/g) in seeds. Analysis of the segregating populations from various crosses has indicated the involvement of as many as 11 recessive alleles (Kondra and Stefansson, 1970) conditioning the low values of 3-butenyl-4-pentenyl and 2-hydroxy-3-butenyl glucosinolates. Studies in *B. juncea* (Stringam and Thiagarajah, 1995) also show a comparable pattern with involvement of five to eight recessive alleles conditioning lower values for total alkenyl glucosinolates. Maternal influence on the heredity of glucosinolates complicates breeding for low glucosinolates using the backcross method. Four seed glucosinolate content QTLs located on chromosomes A9, C2, C7 and C9 have been recurrently mapped in different *B. napus* studies (Uzunova *et al*., 1995; Howell *et al*., 2003; Sharpe and Lydiate, 2003; Zhao and Meng, 2003; Basunanda *et al*., 2007). Breeding efforts to further reduce glucosinolate content are ongoing and targeted to reduce specific types of glucosinolates. A recent study used comparative

genomics with *Arabidopsis* to identify gene-linked SSR markers to seed glucosinolate content in *B. napus* (Hasan *et al*., 2008). It is, therefore, possible to apply the vast biochemical knowledge developed in this model species to select molecular markers with putative linkage to genes active on specific branches of the glucosinolate biosynthetic pathway and thus facilitate targeting the reduction of specific glucosinolate types (Hasan *et al*., 2008).

Genetic variations with <2% erucic acid in the seed oil and <20 µmol of glucosinolate in seed meal are the basic requirements for the development of canola-quality *B. juncea*. The first low glucosinolate *B. juncea* line BJ 1058 was developed in Canada from an interspecific cross between an Indian *B. juncea* line having non-allyl, 3-butenyl glucosinolate with low glucosinolate *B. rapa*, followed by one backcross to the non-allyl, 3-butenyl glucosinolate *B. juncea* parent and selfing of backcrossed plants. Studies conducted by Sodhi *et al*. (2002) later revealed that the aliphatic glucosinolate profiles of east European mustard lines are less complex than the glucosinolate profiles of the Indian types. The former contains mostly 2-propenyl glucosinolate while the latter contain both 2-propenyl and 3-butenyl glucosinolates. The difference in the glucosinolate profile was also reflected at the genetic level; aliphatic glucosinolate in the east European lines was controlled by two loci while the crosses involving low glucosinolate east European and high glucosinolate Indian lines revealed the contribution of six to eight loci to the glucosinolate content. Attempts at mapping loci involved in the biosynthesis of aliphatic glucosinolates were reported recently in *B. juncea* (Ramchiary *et al*., 2007a; Bisht *et al*., 2009). This was accomplished through the use of DNA markers based on the functional genomics information available from allied *Brassica* species and *A. thaliana*. Ramchiary *et al*. (2007a) helped in the identification of six 'true' QTLs, four major (J2Gsl1, J3Gsl2, J9Gsl3 and J16Gsl4) and two minor QTL (J17Gsl5 and J3Gsl6). Suitability of these markers in the marker-assisted introgression to the Indian germplasm has since been validated. In a subsequent study, Bisht *et al*. (2009) undertook fine mapping of the QTL identified earlier by Ramchiary *et al*. (2007b) for efficient marker-assisted introgression of the low gluco-sinolate trait into the germplasm of the Indian gene pool. Candidate genes were cloned from high and low glucosinolate *B. juncea* lines Varuna and Heera, respectively. Seventeen paralogues belonging to six gene families were mapped in *B. juncea*. Despite elaborate efforts, it has not been possible to commercialize canola mustard in India due to the comparatively low productivity and oil content. This has been attributed to a negative pleiotropic effect of low erucic acid alleles on the oil biosynthesis.

Double Haploidy

The term doubled haploid or di-haploid (DH) refers to the two exact copies of the nuclear genome, present in the haploid plant and is produced by spontaneous or artificial doubling of the chromosomes. The doubled haploid technique offers numerous advantages over conventional breeding in *Brassica* improvement programmes, especially production of homozygous plants in a single generation. These could be utilized in different ways such as development of pure-line variety, mapping population, diversification of restorer lines for hybrid breeding, recovery of recessive mutants and *in vitro* selection for disease resistance etc. In natural conditions, spontaneously occurring haploid plants were discovered in the 1920s (Blakeslee *et al.*, 1922), but utilization of haploid plants was not a practical technique until the discovery of colchicine and methods for controlled production of haploid plants were developed. Guha and Maheshwari (1964) were the first to demonstrate that anthers of *Datura inoxia* cultured *in vitro* may produce embryos that originate from immature pollen grains or microspores. This finding was a landmark in plant tissue culture research, which resulted in the production of double haploids in most of the economically important plant species that play a crucial role in human nutritional security. In *Brassica* species, DH plants are mainly produced either through anther culture or microspore embryogenesis. The first anther culture in *Brassica* was reported by Canadian scientists in *B. napus* during the 1970s and later on

several researchers extended the anther culture techniques in other *Brassica* species. The major constraints in anther culture technique are the very low frequency of embryogenesis. Therefore; the routine use of anther culture has not been reported for any *Brassica* species. However, this limitation can be circumvented to a large extent by removing microspores from the anthers and culturing them separately. On the other hand, Lichter (1982) for the first time isolated microspore culture in *B. napus*, a plant species not belonging to the family *Solanaceae*. Since 1982, several researchers have developed and standardized the protocols for isolated microspore culture in different *Brassica* species, including Indian mustard (Thiagarajah and Stringam, 1990; Hiramatsu *et al.*, 1995; Rawat and Rawat, 1997; Lionneton *et al.*, 2001; Prem *et al.*, 2005). In general, successful implementation of the DH technology using microspore culture technique relies upon the availability of an efficient microspore embryo induction and regeneration protocol that is affected by a number of factors such as the donor plant genotype, its growth condition, stage of microspore development, microspore density in culture, media composition and addition of growth regulators.

Linkage Maps

The linkage map of *B. juncea* based primarily on AFLP markers was first developed by Pradhan *et al.* (2003); it was subsequently enriched with more AFLP, restriction fragment length polymorphisms (RFLP), microsatellites and gene-based markers by Ramchiary *et al.* (2007b). This *B. juncea* map covered a total length of 1840.1 centimorgan (cM). Panjabi *et al.* (2008) developed a comparative linkage map based on polymorphism in the intronic regions. A total of 486 introns polymorphic (IP) loci of *A. thaliana* covering all the five chromosomes were mapped relative to the earlier framework map of *B. juncea* (Pradhan *et al.*, 2003). The markers were distributed to all the 18 linkage groups of *B. juncea* covering 1992.2 cM. The comparative map of Panjabi *et al.* (2008) was used to study homoeologous relationships, diversification and evolution of the A, B and C genomes of *Brassica* species.

Gene-based markers allowed comparative mapping between *B. juncea* and the other allotetraploid and diploid *Brassica* species of U's triangle. The comparison of A, B and C genomes showed a high level of conserved macro-level collinearity between the A genome of *B. juncea* and *B. napus* and between the B genome of *B. juncea* and *B. nigra*, signifying the absence of large-scale perturbation during the formation of the allopolyploid species. It also allowed the prediction of possible macro-level chromosomal changes that led to the diversification of the three diploid *Brassica* genomes. Linkage maps based in SSR primers are also available for *B. rapa* (Kapoor *et al.*, 2009) and *B. carinata* (Priyamedha *et al.*, 2012). Detailed QTL analysis of yield-associated traits in *B. juncea* from a DH population (Ramchiary *et al.*, 2007b) showed that the contribution of the A genome toward agronomically important traits in *B. juncea* was more than that of the B genome, and many QTL were found to be clustered in a few linkage groups (LGs). A high density consensus (integrated) map in *B. juncea* was developed using two different maps, Varuna × Heera map (Pradhan *et al.*, 2003; Ramchiary *et al.*, 2007b; Panjabi *et al.*, 2008) and TM-4 × Donskaja-IV. Comparative QTL analysis between the two QTL maps identified many consistent QTL regions and also population-specific QTL regions in *B. juncea* (Yadava *et al.*, 2012).

Molecular Markers

Plant breeding techniques became precise with the advent of DNA-based genetic markers in the 1980s (Paterson *et al.*, 1988). DNA markers are sections of the genome that may either be located within the gene of interest or be linked to a gene determining a trait of interest. Therefore; it is now possible to select desirable traits more directly based on genotype using associated markers rather than the phenotype of the trait (Foolad and Sharma, 2005). Marker-assisted selection (MAS) is an important tool for plant breeders to increase the pace and efficiency of a breeding programme, especially for traits with multi-genic inheritance. Cheung *et al.* (1997) constructed

the first linkage map in *B. juncea* based on RFLP markers to resolve white rust resistance into discrete Mendelian factors (Cheung *et al.*, 1998). Since 1998, various types of molecular markers associated with different economically important traits have been identified in brassica oilseeds (Table 2.1).

Transgenics

Male sterility, herbicide tolerance and resistance to biotic and abiotic stresses have been the targets for transgenic research (Table 2.2). A glyphosate-resistant canola was first developed from *B. napus* cv. Westar by transgenic insertion of one gene associated with a mutant form of endogenous 5-enolypyruvyl shikimate-3-phosphate synthase (EPSPS) and another gene associated with a bacterial enzyme linked to the degradation of glyphosate to aminomethyl-phosphonic acid (AMPA) and glyoxalate. Glyphosate explicitly binds to and inactivates EPSPS, a key enzyme involved in the biosynthesis of the aromatic amino acids tyrosine, phenylalanine and tryptophan.

These genetically modified lines are now widely grown in Canada, the USA and Australia, permitting the use of glyphosate-containing herbicides, such as Roundup®, for weed-free canola cultivation. Reports also suggest successful development of transgenic mustard with resistance to 2-4D (Bisht *et al.*, 2004) and phosphinothricin (Mehra *et al.*, 2000). Another important use of transgenic technology has been in developing genetically engineered male sterility. Insertion of a chimeric dominant gene *barnase* (bacterial *RNase*), from *Bacillus amyloliquefaciens*, driven by a tapetum-specific promoter (*TA 29*) in tobacco and rapeseed facilitated development of male sterile lines (Mariani *et al.*, 1990). The sterility was caused by *TA29*-programmed expression of the *barnase* gene, specifically to anther tapetal cells. This caused premature selective ablation of the tapetal cell layer that surrounds the pollen sac, presumably by hydrolysing the tapetal cells. This disrupted the normal pollen formation and caused male sterility. Male-sterile plants formed flowers with narrow petals and reduced anthers carrying sterile pollen grains.

Table 2.1. Tagging of economically important traits with molecular markers.

Trait	Type of marker	References
Morphological and yield component traits		
Plant height, primary branches, secondary branches, main shoot length, siliquae/	RFLP, AFLP and SSR	Ramchiary *et al.*, 2007a
main shoot, siliquae/plant, silique length,	AFLP, SSR and IP	Yadava *et al.*, 2012
seeds/silique, 1000-seed weight, silique density, days to flower, oil content	AFLP	Lionneton *et al.*, 2004
Leaf colour	SRAP	Luo *et al.*, 2011
Seed coat colour	RFLP	Upadhyay *et al.*, 1996
	AFLP and SCAR	Negi *et al.*, 2000
	SNP	Li *et al.*, 2010
	AFLP and SSR	Huang *et al.*, 2011
Oil quality		
Oleic acid	RAPD	Sharma *et al.*, 2002
Erucic acid	SNP	Gupta *et al.*, 2004
Fatty acids	RFLP	Mahmood *et al.*, 2003
	AFLP	Lionneton *et al.*, 2002, 2004; Singh *et al.*, 2012
Seed meal quality		
Protein	RFLP	Mahmood *et al.*, 2006
Glucosinolates	RFLP	Stringam and Thiagarajah, 1995; Good *et al.*, 2003; Mahmood *et al.*, 2003
	ISSR and SCAR	Ripley and Roslinsky, 2005
	AFLP	Lionneton *et al.*, 2004; Ramchiary *et al.*, 2007b
Biotic stress		
White rust	RFLP	Cheung *et al.*, 1998
	RAPD	Prabhu *et al.*, 1998; Mukherjee *et al.*, 2001
	AFLP and CAPS	Varshney *et al.*, 2004
	IP	Panjabi *et al.*, 2010
Fertility restoration		
Fertility restorer gene of CMS *Moricandia arvensis*	AFLP and SCAR	Ashutosh *et al.*, 2007

RFLP, restriction fragment length polymorphism; AFLP, amplified fragment length polymorphism; SSR, simple sequence repeat; IP, PCR-based intron polymorphism; SRAP, sequence-related amplified polymorphism; SCAR, sequence characterized amplified regions; SNP, single nucleotide polymorphism; RAPD, random-amplified polymorphic DNA; ISSR, inter-simple sequence repeat; CAPS, cleaved amplified polymorphic sequences.

To restore fertility, the pollen parents were transformed with *barstar* gene (RNase inhibitor), also from *B. amyloliquefaciens* (Mariani *et al.*, 1992). The F_1 progeny produced by crossing *barnase* male-sterile plants with transgenic fertile plants carrying *TA29-barstar* chimaeric gene was male fertile. These plants also showed co-expression of both the genes in the anthers. The *barstar* gene was dominant to the *barnase* gene, and fertility restoration resulted from the formation of tapetal cell-specific *barnase-barstar* complex inhibiting the expression of *barnase* gene. By coupling the *TA29-barnase* gene to a

dominant herbicide resistance '*Bar*' gene, uniform populations of male sterile plants could be produced. '*Bar*' gene confers resistance to the herbicide bialophos (phosphinothricin or PPT) and is used as a selectable marker for male sterility (Denis *et al.*, 1993). Maintenance progeny showed 50% plants that were male fertile but susceptible to the herbicides. Application of PPT permitted the elimination of male fertile susceptible segregants in the field by herbicidal application and assured 100% production of hybrid seed on male-sterile plants. Canola hybrids based on this technology are

Table 2.2. Name of selected transgenes and their functions.

Name of introduced gene	Source of introduced gene	Promoter	Phenotype	Reference
Abiotic stress				
Bacterial CodA	*Arthrobacter globiformis*	35S	Enhanced salt and cold tolerance	Prasad et al., 2000
NHX1 (vacuolar Na+/ H+ antiporter)	*Pennisetum glaucum*	35S	Enhanced salt tolerance	Rajagopal et al., 2007
Biotic stress				
WGA (Wheat germ agglutinin)	Wheat	35S	Resistance to mustard aphid	Kanrar et al., 2002b
Chitinase	Tobacco	35S	Delay in the onset of Alternaria leaf spot (A. brassicae) as well as reduced lesion number and size	Mondal et al., 2003
ASAL (*Allium sativum* leaf lectin)	*Allium sativum*	35S and RSs-1 (rice sucrose synthase)	Reduce the survival and fecundity of mustard aphid	Dutta et al., 2005
Glucanase	Tomato	35S	Restricted number, size and spread of lesions of Alternaria leaf spot (A. brassicae)	Mondal et al., 2007
Bt genes, namely cry1C, cry1Ac, and cry1Ac + cry1C	*Bacillus thuringiensis*	35S	cry1C plants controlled Cry1A-resistant diamondback moth (DBM) while cry1Ac plants controlled Cry1C-resistant DBM, and the pyramided cry1Ac + cry1C plants effectively controlled all three types of DBM	Cao et al., 2008
Class II chitinase and type I RIP	Barley	35S	Showed resistance through delayed onset of the Alternaria leaf spot and restricted number, size, and expansion of lesions	Chhikara et al., 2012
Heavy metal tolerance				
GR (glutathione reductase)	Bacteria	35S	In the plastidic transgenics (cpGR) plants experienced lower cadmium stress	Elizabeth et al., 2000
CGS (cystathionine-γ-synthase)	*Arabidopsis thaliana*	35S	2 to 3-fold higher Se volatilization rates after supplied with selenate or selenite; transgenic plants contained 20–40% and 50–70% lower shoot and root Se levels, respectively	Huysen et al., 2003

Continued

Table 2.2. Continued.

Name of introduced gene	Source of introduced gene	Promoter	Phenotype	Reference
PCS (phytochelatin synthase)	*Arabidopsis thaliana*	35S	Moderate expression levels of phytochelatin synthase improved tolerance to Cd and Zn	Gasic and Korban, 2007a
PCS1 (phytochelatin synthase)	*Arabidopsis thaliana*	Native promoter	Higher tolerance to Cd and As	Gasic and Korban, 2007b
Nutritional quality				
γ-TMT (γ-tocopherol methyl transferase)	*Arabidopsis thaliana*	35S	More than 6-fold increase in the level of α-tocopherol and enhanced tolerance to the induced stresses	Yusuf *et al.*, 2010
PiD6 (Δ6 desaturases *Pythium irregulare*)	*Pythium irregulare*	Oilseed ('*Brassica napus*) napin	Improved γ-linolenic acid in seed	Hong *et al.*, 2002
Antisense constructs using the sequence fad2	*Brassica rapa*	Truncated napin	Improved C18:1 but decreased C18:2 and C18:3 in comparison to the parental line	Sivaraman *et al.*, 2004
Hybrid breeding				
bar gene encoding phosphinothricin-acetyl-transferase (PAT)	*Bacillus amyloliquefaciens*	Double enhancer version of CaMV35S promoter (35Sde *bar*) or CaMV35S promcter with a leader sequence from RNA4 of alfalfa mosaic virus	Resistant to herbicide phosphinothricin (PPT)	Mehra *et al.*, 2000
Barnase	*Bacillus amyloliquefaciens*	TA29	Fertility	Jagannath *et al.*, 2001
Agronomic traits				
FRUITFULL	*Arabidopsis thaliana*	–	Improved pod shattering	Østergaard *et al.*, 2006

now under cultivation in Canada. Jagannath *et al.* (2001) modified the system by using gene constructs with spacer DNA in between the *barnase* gene and the CaMV 35S promoter-driven *bar* gene. To restore fertility in such male-sterile lines modified *barstar* constructs were developed (Jagannath *et al.*, 2002; Bisht *et al.*, 2004, 2007). To impart aphid resistance, transgenics carrying gene(s) governing constitutive or phloem-specific expression of chitin-binding *lectine* gene from wheat (Kanrar *et al.*, 2002a) and *Allium sativum* leaf lectin (Dutta *et al.*, 2005; Hossain *et al.*, 2006) have been developed. Similarly, transgenics showing expression of a barley class II chitinase and type I ribosome inactivating protein (Chhikara *et al.*, 2012), tomato glucanase (Mondal *et al.*, 2007) and the rubber-tree lectin hevein (Kanrar *et al.*, 2002b) in Indian mustard were found to confer protection against *Alternaria brassicae*. Tocopherols (vitamin E) are lipid soluble antioxidants synthesized by plants and some cyanobacteria. Over-expression of the gamma-tocopherol methyl transferase (gamma-TMT) gene from *A. thaliana* in transgenic *B. juncea* plants resulted in over a six-fold increase in the level of alpha-tocopherol, the most active form of all the tocopherols. Such transgenics also had enhanced tolerance to the stress induced by salt, heavy metal and osmoticum (Yusuf *et al.*, 2010). Sivaraman *et al.* (2004) have reported success in development of high oleic and low linoleic acid transgenics in a zero erucic acid mustard line by antisense suppression of the *fad2* gene. Stable integration and inheritance of transgene(s) have been shown through Southern hybridization and *in planta* bioassays. However, extensive field assessment of these transgenics is still awaited. The *Arabidopsis AtmiR156b* gene was expressed in *B. napus* under the control of the cauliflower mosaic virus (CaMV) 35S promoter and the seed-specific napin promoter. Constitutive expression of *AtmiR156b* in *B. napus* resulted in enhanced levels of seed lutein and β-carotene and a two-fold increase in the number of flowering shoots (Wei *et al.*, 2010). These data suggest that *AtmiR156b* gene expression could be applied in plant-breeding initiatives for enhancing carotenoid production in brassicas. Targeted silencing of BjMYB28 transcription factor gene have been found to be associated

with the development of low glucosinolate lines in oilseed *B. juncea* (Augustine *et al.*, 2013).

Case Studies for Germplasm Enhancement from India

Variation in *Brassica rapa* land races

Brassica rapa L. (syn. *Brassica campestris* L.) has the widest distribution among brassica oilseeds, with the secondary centres of diversity in Europe, western USSR, Central Asia and the Near East (Mizushima and Tsunoda, 1967). The Asian types comprising three ecotypes including *B. rapa* var. Toria, *B. rapa* var. Yellow Sarson and *B. rapa* var. Brown Sarson, are distinct from the European oleiferous types, not only in general morphology but also in chemical composition of the seed. The protein analysis supports a separate origin of Asian and European types. *B. rapa* var. Brown Sarson, including both *lotni* (self-incompatible) and *tora* (self-compatible) types, are the oldest among *B. rapa* ecotypes. Of these two, the *lotni* type is older than *tora*. Tora were derived from hybridization between *B. rapa* var. Yellow Sarson and *lotni* type of *B. rapa* var. Brown Sarson. *B. rapa* var. Toria and *B. rapa* var. Yellow Sarson, are again derivative of *B. rapa* var. Brown Sarson, the former being the result of selection, the latter being the mutant for quality (Hinata and Prakash, 1984). Two landraces from Utttarakhand in India were found to have extra early maturity (<60 days). Further selection for earliness, uniformity, height and other characteristics resulted in extra-early maturing lines of *B. rapa* var. Toria named as PT-141 and PT-145. At maturity, plants of these lines attain 30–35 cm height, bear 3–6 primary and 4–9 secondary branches, 65–85 siliquae/plant, 10–12 seeds/silique and 2.0–2.5 g/1000 seed weight. Further selection for earliness, uniformity, height and other characteristics resulted in extra-early maturing lines of toria named as PT-141 and PT-145. At maturity, plants of these lines attain 30–35 cm in height, bear 3–6 primary and 4–9 secondary branches, 65–85 siliquae/plant, 10–12 seeds/silique and 2.0–2.5 g/1000 seed weight (Fig. 2.1).

In these lines, flowering period was considerably reduced, which increased synchrony

Fig. 2.1. Dwarf and extra early-maturing plants of PT-141, an early-maturing line of *B. rapa* var. Toria.

both within and between racemes of a plant. It is known that earliness is associated with increased photoperiod insensitivity (day neutrality) in different crop plants (Singh and Sharma, 1996). Earliness or day neutrality is also known to increase the geographic adaptability of varieties. With the availability of such extra-early maturing germplasm, it is possible to design a plant type of *B. rapa* var. Toria which is high yielding, possesses short stature and also matures early. The extra-early maturity of these lines makes them more suitable for their use as catch crop and also to regain the *B. rapa* var. Toria area that has declined in the recent past. Climate changes are also likely to pose a bigger challenge in the form of temperature rise, unexpected high or low rainfall, and shifts in the onset of monsoon and also in growing seasons. A slight shift towards earliness in sowing time of major winter crops, including oilseed brassica, or due to the early rise in temperature at the terminal stage of the crop may necessitate the need for extra-early genotypes of *B. rapa* var. Toria to fit in as catch crop in the emerging cropping situations. The short plant height of PT-141 and PT-145 also offers opportunities for re-synthesizing dwarf stature, short-duration genotypes of *B. juncea*, the major brassica oilseed crop grown in India. It has been noticed that PT-141 and PT-145 had satisfactory growth, seed and silique set even under delayed sowing in the month of November, that was comparable to that of a September-sown crop. With increased day neutrality of varieties, it may be possible to introduce

Toria as a *Zaid* season crop (these are crops grown on irrigated lands that do not have to wait for monsoons, mainly from March to June). However, such possibility requires critical analysis, experimentation as well as a selection for specific adaptation from the base collection of such short duration germplasm. Normally, branching in *B. rapa* var. Toria arises from the main stem at a wide angle with a semi-spreading to spreading growth habit, which means that plants tend to spread more in spaced growing environments than at normal spacing intervals. A *B. rapa* var. Toria line, PTHC-11-22, with a profuse base branching trait, has been identified out of local collections adapted to rainfed conditions of the Uttarakhand hills. Apart from this, types having erect growth habit and uniformity in flowering after eliminating undesirable plants before flowering and permitting inter-mating among the selects. The plants of this line had 9–16 primary branches and about 40% of branches were formed within 10 cm height from the base. The primary branches were found to grow as tall as the main raceme, which probably caused relatively more synchronous flowering among racemes on a plant. As such this appears to be a desirable trait which is lacking in existing varieties.

Germplasm enhancement in mustard

The history of brassica improvement has been reviewed by Chauhan *et al.* (2010). Organized crop improvement work was first initiated by

1900 at Pusa (Bihar), by the then Bengal Presidency. Initially, the crop improvement work involved collection of landraces and their purification. The scientific work for varietal improvement of Indian oleiferous *Brassiceae* started at Layallpur (now Faisalabad in Pakistan), then in Punjab of undivided India. This resulted in identification and commercial release of Indian mustard variety RL 18 (Raya Layallpur 18) during 1937 and L1 of *B. rapa* var. Yellow Sarson through selection. Indian mustard strain RT 11 from Uttar Pradesh was found superior to local varieties and released in 1936. However, systematic research efforts were started after independence. Until the mid-1940s, barring sporadic and isolated research efforts, the rapeseed-mustard crops did not receive adequate attention in terms of research and development. The earliest attempt to organize research in oilseeds was made by constituting the Indian Central Oilseeds Committee (ICOC) in 1947 under the Indian Oilseeds Committee Act IX in 1946. The main objective of the ICOC was to increase production of individual oilseeds through ad hoc funding of research programmes carried out by the State Department of Agriculture, Universities and Central Institutes. The Oilseeds Development Council replaced the ICOC in 1966. During the course of about two decades (1947–67), a number of high-yielding varieties of mustard (Laha 101, Varuna, Durgamani, Patan Mustard), *B. rapa* var. Toria (Abohar, BR 23, M 27, T 9, ITSA, T 36, DK 1), *B. rapa* var. Brown Sarson (BSA, BSG, BSH 1, BS 2, BS 65,BS 70), *B. rapa* var. Yellow Sarson (T 151, Patan sarson, YSPb 24, T 42) and taramira (*Eruca sativa*)(ITSA) were developed (Chauhan *et al.*, 2010). In India, the Indian Council of Agricultural Research (ICAR) gave a major fillip to the research programme during April 1967 by launching a comprehensive multidisciplinary research programme on the improvement of oilseeds in the country under All India Coordinated Research Project on Oilseeds (AICORPO), including rapeseed-mustard. ICAR also established the National Research Centre on Rapeseed-Mustard (NRCRM; now re-designated as Directorate of Rapeseed-Mustard Research, DRMR) during 1993 at Bharatpur (Rajasthan) to promote basic, strategic and applied research. The All India Coordinated Research Project on Rapeseed-Mustard with 19 research centres across the country was also brought under its umbrella.

Plant genetic resources

India presents a rich diversity of oilseed brassicas. *Brassica rapa* var. Toria and *B. rapa* var. Brown Sarson, *B. juncea* and *Eruca vesicaria* were considered to be of an Indian gene centre (Arora, 1998). The north-eastern and north-western regions hold promising genetic variability in rapeseed-mustard. The NBPGR, New Delhi, conserved 10,259 germplasm collections until December 2010 (NBPGR, 2011). The NRCRM/ICAR-DRMR and other collaborative centres under the umbrella of the All India Coordinated Research Project on Rapeseed-Mustard held 14,722 germplasm accessions. In addition, a very healthy collection of wild and weedy crucifers is being maintained and evaluated at the National Research Centre (Plant Biotechnology), Punjab Agricultural University and ICAR-DRMR. Two varieties of *B. rapa* var. Brown Sarson and ten varieties of *B. napus* (including GSC 5 and GSC 6) have exotic genotypes in their pedigree. In Indian mustard, two varieties (JM 1 and JM 2) were derived by utilizing white rust-resistant accessions L 4 and L 6 from Canada; all the low erucic acid varieties were derived (Pusa Karishma, LET 17, LET 18, LES 1-27, RLC 1 and RLC 2) using ZEM 1 and ZEM 2 from Australia (Kirk and Oram, 1981) as donors for the low erucic acid trait. Furthermore, BJ 1058, the only source of low glucosinolate in *B. juncea* having Bronowski block, has been widely utilized in the Indian breeding programme to develop canola *B. juncea*. Almost 5605 accessions have been evaluated/characterized for morphological traits at DRMR, Bharatpur up to 2009/10. Of these, 42 accessions have been registered with ICAR-NBPGR (ICAR), New Delhi up to December 2010.

Breeding for improved productivity depends upon the ability to substitute bottleneck loci with more effective alleles (Banga and Banga, 2009). Major breeding objectives were genetic enhancement for oil and seed yield through developing varieties for early, timely and late-sown conditions to cater to the need of diverse agroecological situations of the country, improvement of oil (low erucic acid) and seed meal (low glucosinolate) quality, introgression of resistance/tolerance to major biotic (white rust, Alternaria blight, Sclerotinia rot diseases and aphid and painted bug) and abiotic stresses (drought, high temperature and salinity), etc. Pure-line and mass selection have been the methods of choice precipitating in development of 26 varieties, the first release being in 1936. The number of varieties developed through hybridization consistently increased after 1980, resulting in 22 varieties being released in each of the eighth and ninth decades of the 20th century. This number further increased to 41 during the first decade of the 21st century. Simultaneously, mutation breeding was also pursued, and 12 varieties were developed (Chauhan *et al.*, 2010). Chauhan and Singh (2004) opined that most of the Indian mustard varieties were the pure-line selections derived from a few common ancestors and a limited number of donors was utilized in the breeding programme resulting in a narrow genetic base. Excellent analysis conducted by Chauhan *et al.* (2010) also revealed that a single variety of Indian mustard, Varuna, which is itself a pure-line selection from Banarasi Rai, a local germplasm, has been extensively used in the development of 46 varieties, including important varieties such as Kranti, Rohini, Pusa Jaikisan. The direct and indirect contribution of Varuna to Indian mustard breeding was estimated to about 42.6%. In all, 203 rapeseed-mustard varieties have been identified. The Central Sub-Committee on Crop Standards, Notification and Release of Varieties for Agriculture Crops has notified 144 varieties up to December 2010. The narrow genetic base of mustard germplasm has also been reported through the study of molecular markers (Srivastava *et al.*, 2001; Chen *et al.*, 2011).

Hybrid breeding

A high level (35–100%) of heterosis has been reported in the entire oilseed brassica group (Banga, 1996; Banga and Banga, 2009). Heterosis is generally higher in crosses involving parents with diverse phonologies or geographic

origin. Attempts at the production of F_1 hybrids have been stimulated by the availability of a large number of CMS sources. These are *nap, polima, ogura, tournefortii, axyrrhina, siifolia, catholica, sinapis, trachystoma, moricandia* and *lyratus*. Hybrids based on *polima* or *ogura* and/or transgenic *barnase-barstar* system have been commercialized in *B. napus* and now cover very large areas (>50%) in Canada, China and Europe. In India, hybrids have been commercialized in *B. juncea*. These are based on either *ogura*, INS 126 or *mori* CMS systems. Average yield heterosis over the best inbred parents, however, continues to be below 15%. In Canada, hybrids have averaged 11% higher yields over open-pollinated (OP) varieties during the past decade. Many instances of high level heterosis in rapeseed-mustard have been reported (Banga and Banga, 2009), with heterosis ranging over 100% on the plant basis in Indian mustard. Heterosis was generally higher in crosses involving parents with diverse phenologies or geographic origin (Pradhan *et al.*, 1993; Jain *et al.*, 1994; Sodhi *et al.*, 2006). The concept of fixed heterosis (heterosis due to allopolyploidy) has been elucidated in *B. juncea*. Experiments with different species have not revealed any predictors of heterosis based both on methods of molecular as well as genetic diversity (Bansal *et al.*, 2012). Development of a population structure in mustard (Fig. 2.2) may permit synthesis of heterotic gene pools. Six gene pools are clearly evident.

Recent development of a determinate growth habit in mustard (Banga, unpublished) has opened up new possibilities to enhance heterosis in this critical oilseed crop. Most of the hybrids involving determinate × indeterminate crosses show high heterosis. Initially, however, heterosis could also not be exploited in *B. juncea* and *B. napus* due to their predominantly self-fertilized crops. A suitable pollination control mechanism is required to produce commercial hybrid seed. A CMS fertility restoration system is an excellent potential means to facilitate hybridization because it is easy to maintain. CMS, a maternally inherited inability to produce fertile pollen, is encoded in the mitochondrial genome and can arise spontaneously due to mutation in the genome (autoplasmy) or can

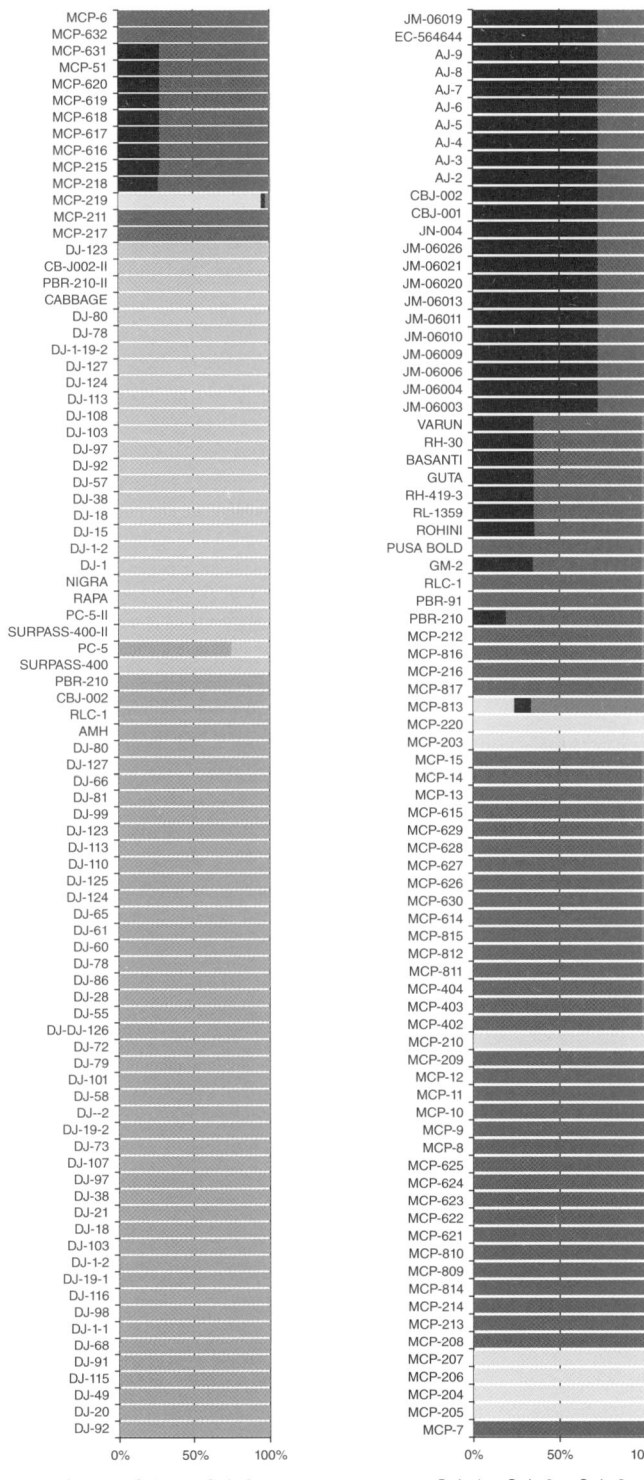

Fig. 2.2. Population structure in *Brassica* germplasm assemblage.

be expressed following cytoplasmic substitutions due to nuclear-mitochondrial incompatibility (alloplasmy). CMS lines could be developed following backcrossing either sexually synthesized allopolyploids or somatic hybrids between wild and crop species (Table 2.3). As expected, CMS originating from sexual hybridizations possess unaltered organelle genomes because of exclusive maternal inheritance. Since organelle assortment and inter-genomic mitochondrial recombinant is of frequent occurrence in *Brassiceae*, the cytoplasmic constitution is entirely different in those originating from somatic hybrids, and different combinations of mitochondrial and chloroplast genomes have been reported in different CMS lines (Prakash *et al.*, 2009). A duplication of *Cox I* gene was associated with male sterility for catholica CMS system (Pathania *et al.*, 2007). Alloplasmic CMS plants, in general, are similar to euplasmic plants in development and morphology. However, many of them exhibited developmental and floral abnormalities as a consequence of altered nucleo-cytoplasmic interactions. Varying degrees of leaf chlorosis was associated with *raphanus/ogu, oxyrrhina, tournefortii, moricandia* and *enarthrocarpus* systems. Floral abnormalities in male-sterile plants included: petaloid anthers (*nigra, muralis, trachystoma, raphanus, tournefortii, canariense*); poor or absent nectarines (*tournefortii* and *raphanus*); crooked style (*tournefortii, raphanus*); thick pistil (*raphanus*); and low seed fertility (*raphanus, tournefortii, enarthrocarpus* and *trachystoma*). Fertility restorers for *moricandia, ogura, catholica, erucoides* and *lyratus* CMS systems could be introgressed from cytoplasm donor wild species (Banga and Banga, 2009; Prakash *et al.*, 2009). Fertility restorer for the *mori* CMS could also restore fertility of *eru* CMS system (Bhat *et al.*, 2005). At present CMS is being used to develop hybrids. Oilseed rape hybrid (PGSH 51) based on the *tournefortii* system was the first India-bred oilseed brassica hybrid to be released in India during 1994 (Banga *et al.*, 1995). A mustard hybrid, NCR Sankar Sarson (NRCHB506) based on *moricandia* CMS was released during 2008 by DRMR. Besides these alloplasmic systems, a novel CMS system 126-1 has also been reported (Sodhi *et al.*, 2006). Although the genetic basis of this CMS is yet to be clearly elucidated, the advantage of this system is that any line can be used as fertility restorer or maintainer of sterility. This obviates the necessity of introgression of *Rf* genes as is the case with alloplasmic CMS systems.

Two hybrids, DMH 1 and NRCHB-506, based on the 126 I and *mori* CMS system, respectively, were released during 2008. In addition, a few private companies have also commercialized *ogura* CMS-based mustard hybrids. The first *ogura* CMS-based hybrid, Coral 432 (PAC 432), was released during 2009. Current hybrids have shown yield advantage of 15–20% across locations over best commercial varieties. Development of a genetically engineered system of male sterility (*barnase-barstar* system) by Jagannath *et al.* (2002) was another milestone in utilizing the hybrid vigour in rapeseed-mustard. Hybrid DMH 11 based on this system is presently undergoing biosafety assessment, being a transgenic.

Table 2.3. Major alloplasmic male sterile systems developed in India.

CMS system	Cytoplasm donor	Reference
ogura	*Raphanus sativus*	Kirti *et al.*, 1995a
oxy	*Brassica oxyrrhina*	Prakash and Chopra, 1988, 1990; Kirti *et al.*, 1993
sii	*Diplotaxis siifolia*	Rao *et al.*, 1994; Rao and Shivanna, 1996
Trachy	*Trachystoma ballii*	Kirti *et al.*, 1995b; Kaur *et al.*, 2004
Mori	*Moricandia arvensis*	Kirti *et al.*, 1998; Prakash *et al.*, 1998
eru	*Diplotaxis erucoides*	Malik *et al.*, 1999; Prakash *et al.*, 2001; Bhat *et al.*, 2006
berth	*Diplotaxis berthauti*	Malik *et al.*, 1999; Bhat *et al.*, 2008
can	*Erucastrum canariense*	Prakash *et al.*, 2001
cath	*Diplotaxis catholica*	Pathania *et al.*, 2007
lyr	*Enarthrocarpus lyratus*	Deol *et al.*, 2003; Janeja *et al.*, 2003a, b

Conclusions

Oilseed-brassica breeding is now confronted with challenges of increased productivity growth under the scenarios of climate change and worsening land and water resources. For breeders, this implies continuous reappraisal of breeding goals, ability to enhance the germplasm base and constantly upgrading breeding skills. The diversity of the *Brassica* germplasm pool is currently narrow. However, there are many sources of diversity available from the primary, secondary, and tertiary gene pools. The phenotypic and genetic consequences (variation bottlenecks) of the domestication need to be investigated in their entirety by characters found in the domestication syndrome that is common to many crop plants. Brassica crops have been extensively utilized to source cytoplasmic variations from wider *Brassicaceae* gene pools. Basic work on the genetic tools, especially study of population structure, is necessary to manipulate the germplasm for continuing selection gains. That alone will help in efficient renewal of available genetic armoury and provide growers with varieties/hybrids adapted to specific needs. Numerous possibilities of lipid engineering to suit specialized food or factory needs will necessitate a focus on traditional oilseed crops despite competition from cheaper alternatives such as palm oil. With technological constraints behind us, it is now time to push frontiers of hybrid breeding in a broader commercial realm, especially in the stressed ecologies of India and Australia. Further improvement in harvest index and growth rate will improve productivity of inbred lines, both for use directly as a cultivar or parents of a hybrid. Determinacy may be a huge way forward. Integration of conventional breeding with marker assisted selection and genetic engineering is important to hasten the pace and scope of varietal development. In spite of several gains, the task of checking yield losses by biotic and abiotic stresses is far from accomplished and deserves greater attention from brassica scientists. There is a need to initiate massive international efforts towards pre-breeding/germplasm enhancement. Brassica breeding will have to become more responsive to the challenges of climate change besides the demands of farmers, millers and consumers alike.

References

Ahuja, I., Bhaskar, P.B., Banga, S.K. and Banga, S.S. (2003) Synthesis and cytogenetic characterization of the intergeneric hybrids of *Diplotaxis siifolia* with *Brassica rapa* and *B. juncea*. *Plant Breeding* 122, 447–449.

Alam, Z. (1936) Cytological studies of some Indian oleiferous Cruciferae III. *Annals of Botany* 50, 85–102.

Allen, E.J., Morgan, D.G. and Ridgman, W.I. (1971) A physiological analysis of the growth of oilseed rape. *The Journal of Agricultural Science Cambridge* 77, 339–341.

Allender, C.J. and King, G.J. (2010) Origins of the amphiploid species *Brassica napus* L. investigated by chloroplast and nuclear molecular markers. *BMC Plant Biology* 10, 54.

Anand, I.J., Reddy, W.R. and Rawat, D.S. (1985) Inheritance of seed colour in mustard. *Indian Journal of Genetics* 45, 34–37.

Arora, R.K. (1998) The Indian gene center – priorities and prospects for collection. In: Paroda, R.S., Arora, R.K. and Chandel, K.P.S. (eds) *Plant Genetic Resources: Indian Perspective*. Indian Society of Plant Genetic Resources, National Bureau of Plant Genetic Resources, New Delhi, pp. 66–75.

Ashutosh Sharma, P.C., Prakash, S. and Bhat, S.R. (2007) Identification of AFLP markers linked to the male fertility restorer gene of CMS (*Moricandia arvensis*) *Brassica juncea* and conversion to SCAR marker. *Theoretical and Applied Genetics* 114, 385–392.

Atri, C., Kumar, B., Kumar, H., Kumar, S., Sharma, S. and Banga, S.S. (2012) Development and characterization of *Brassica juncea - fruticulosa* introgression lines exhibiting resistance to mustard aphid (*Lipaphis erysimi* Kalt). *BMC Genetics* 13, 104.

Augustine, R., Mukhopadhyay, A. and Bisht, N.C. (2013) Targeted silencing of BjMYB28 transcription factor gene directs development of low glucosinolate lines in oilseed *Brassica juncea*. *Plant Biotechnology Journal* 11(7), 855–866.

Auld, D., Heikkinen, M.K., Erickson, D.A., Sernyk, L. and Romero, E. (1992) Rapeseed mutants with reduced levels of polyunsaturated fatty acids and increased levels of oleic acid. *Crop Science* 32, 657–662.

Banga, S.S. (1988) C-genome chromosome substitution lines in *Brassica juncea*. *Genetica* 77, 81–86.

Banga, S.S. (1996) Breeding and Genetics. In: Chopra, V.L. and Shyam Prakash (ed.) *Oilseeds and Vegetable Brassicas: Indian perspective*. IBH, New Delhi, pp. 50–76.

Banga, S.S. and Banga, S.K. (2009) Crop improvement strategies in rapeseed-mustard. In: D.M. Hegde (ed.) *Vegetable Oils Scenario: Approaches to Meet the Growing Demands*. ISOR, Hyderabad, pp. 13–35.

Banga, S.S., Banga, S.K. and Labana, K.S. (1983) Nucleo-cytoplasmic interaction in Brassica. *Proceedings of the 6th International Rapeseed Congress*, 17–19 May 1983, Paris, pp. 602–606.

Banga, S.S., Labana, K.S., Banga, S.K., Sandha, G.S. and Gupta, T.R. (1995) PGSH 51: the first hybrid of gobhi sarson. *PAU Journal of Research* 32, 242.

Banga, S.S., Banga, S.K., Gupta, M.L. and Sandha, G.S. (1998) Synthesis of genotypes with specialized fatty acid composition in Indian mustard [*Brassica juncea* (L.) Czern.& Coss.]. *Crop Improvement* 25, 21–25.

Banga, S.S., Deol, J.S. and Banga, S.K. (2003a) Alloplasmic male sterile *Brassica juncea* with *Enarthrocarpus lyratus* cytoplasm and the introgression of gene(s) for fertility restoration from cytoplasm donor species. *Theoretical and Applied Genetics* 106, 1390–1395.

Banga, S.S., Bhaskar, P.B. and Ahuja, I. (2003b) Synthesis of intergeneric hybrids and establishment of genomic affinity between *Diplotaxis catholica* and crop *Brassica* species. *Theoretical and Applied Genetics* 106, 1244–1246.

Bansal, P., Kaur, P., Banga, S.K. and Banga, S.S. (2009) Augmenting genetic diversity in *Brassica juncea* through its resynthesis using purposely selected diploid progenitors. *International Journal of Plant Breeding* 3, 41–45.

Bansal, P., Banga, S.K. and Banga, S.S. (2012) Heterosis as investigated in terms of polyploidy and genetic diversity using designed *Brassica juncea* amphiploid and its progenitor diploid species. *PLoS One* 7: e29607. doi:10.1371/journal.pone.0029607.

Basunanda, P., Spiller, T.H., Hasan, M., Gehringer, A., Schondelmaier, J., Lühs, W., Friedt, W. and Snowdon, R.J. (2007) Marker-assisted increase of genetic diversity in a double-low seed quality winter oilseed rape genetic background. *Plant Breeding* 126, 581–587.

Baud, S., Boutin, J.P., Miquel, M., Lepiniec, L. and Rochat, C. (2002) An integrated overview of seed development in *Arabidopsis thaliana* ecotype WS. *Plant Physiology and Biochemistry* 40, 151–160.

Bhaskar, P.B., Ahuja, I., Juneja, H.S. and Banga, S.S. (2002) Intergeneric hybridization between *Erucastrum canariense* and *Brassica rapa*. Genetic relatedness between EC and A genomes. *Theoretical and Applied Genetics* 105, 754–758.

Bhat, M.A., Gupta, M.L., Banga, S.K., Raheja, R.K. and Banga, S.S. (2002) Erucic acid heredity in *Brassica juncea* (L.) Czern & Coss. – some additional information. *Plant Breeding* 121, 456–458.

Bhat, S.R., Prakash, S., Kirti, P.B., Dinesh Kumar, V. and Chopra, V.L. (2005) A unique introgression from *Moricandia arvensis* confers male fertility to two different cytoplasmic male sterile lines of *Brassica juncea*. *Plant Breeding* 124, 117–120.

Bhat, S.R., Priya, V., Ashutosh, Dwivedi, K.K., and Prakash, S. (2006) *Diplotaxis erucoides* induced cytoplasmic male sterility in *Brassica juncea* is rescued by the *Moricandia arvensis* restorer: genetic and molecular analyses. *Plant Breeding* 125, 150–155.

Bhat, S.R., Kumar, P. and Prakash, S. (2008) An improved cytoplasmic male sterile (*Diplotaxis berthautii*) *Brassica juncea*: Identification of restorer and molecular characterization. *Euphytica* 159, 145–152.

Bisht, N.C., Jagannath, A., Gupta, V., Burma, P.K. and Pental, D. (2004) A two gene–two promoter system for enhanced expression of a restorer gene (*barstar*) and development of improved fertility restorer lines for hybrid seed production in crop plants. *Molecular Breeding* 14, 129–144.

Bisht, N.C., Jagannath, A., Burma, P.K., Pradhan, A.K. and Pental, D. (2007) Retransformation of a male sterile *barnase* lines with the *barstar* gene as an efficient alternative method to identify male sterile-restorer combinations for heterosis breeding. *Plant Cell Reports* 26, 727–733.

Bisht, N.C., Gupta, V., Ramchiary, N., Sodhi, Y.S., Mukhopadhyay, A., Arumugam, N., Pental, D. and Pradhan, A.K. (2009) Fine mapping of loci involved with glucosinolate biosynthesis in oilseed mustard (*Brassica juncea*) using genomic information from allied species. *Theoretical and Applied Genetics* 118, 413–421.

Blakeslee, A.F., Belling, F., Farnham, M.E. and Bergner, A.D. (1922) A haploid mutant in the Jimson weed, *Dhatura stramonium*. *Science* 55, 646–647.

Brown, J. (1995) Early generation selection in spring canola (*Brassica napus* L.). *Proceedings of 9th International Rapeseed Congress,* Cambridge, UK, pp. 699–701.

Burns, M.J., Barnes, S.R., Bowman, J.G., Clarke, M.H.E., Werner, C.P. and Kearsey, M.J. (2003) QTL analysis of an intervarietal set of substitution lines in *Brassica napus*: (i) Seed oil content and fatty acid composition. *Heredity* 90, 39–48.

Cahoon, E.B., Shockey, J.M., Dietrich, C.R., Gidda, S.K., Mullen, R.T. and Dyer, J.M. (2007) Engineering oil-seeds for sustainable production of industrial and nutritional feedstocks: solving bottlenecks in fatty acid flux. *Current Opinion in Plant Biology* 10, 236–244.

Cao, J., Shelton, A.M. and Earle, E.D. (2008) Sequential transformation to pyramid two Bt genes in vegetable Indian mustard (*Brassica juncea* L.) and its potential for control of diamondback moth larvae. *Plant Cell Reporter* 27, 479–487.

Chandra, A., Gupta, M.L., Ahuja, I., Kaur, G. and Banga, S.S. (2004a) Intergeneric hybridization between *Erucastrum cardaminoides* and two diploid crop *Brassica* species. *Theoretical and Applied Genetics* 108, 1620–1626.

Chandra, A., Gupta, M.L., Banga, S.S. and Banga, S.K. (2004b) Production of an interspecific hybrid between *Brassica fruticulosa* and *B. rapa*. *Plant Breeding* 123, 497–498.

Chaudhary, B.D., Singh, P., Singh, P. and Kumar, P. (1989) Inheritance studies of plant water relations in *Brassica juncea*. *Biologia Plantarum* 31, 202–206.

Chauhan, J.S. and Singh, N.B. (2004) Breeding approaches in rapeseed-mustard varietal improvement. In: Singh, N.B. and Arvind Kumar (eds) *Rapeseed-Mustard Research in India*. National Research Center on Rapeseed-Mustard, Sewar, Bharatpur (Rajasthan), pp. 51–64.

Chauhan, J.S., Tyagi, M.K., Kumar, A., Nashaat, N.I., Singh, M., Singh, N.B., Jakhar, M.L. and Welham, S.J. (2007) Drought effects on seed yield and its components in Indian mustard (*Brassica juncea* L.). *Plant Breeding* 126, 399–402.

Chauhan, J.S., Singh, K.H., Singh, V.V. and Kumar, S. (2010) Hundred years of rapeseed-mustard breeding in India: accomplishments and future strategies. *Indian Journal of Agricultural Sciences* 81, 1093–1109.

Chauhan, Y.S. and Kumar, K. (1987) Genetics of seed colour in mustard (*Brassica juncea* L. Czern and Coss). *Cruciferae Newsletter* 12, 22–23.

Chen, B.Y. and Heneen (1992) Inheritance of seed coat colour in *Brassica campestris* (L.) and breeding for yellow-seeded *B. napus* L. *Euphytica* 59, 157–163.

Chen, B.Y., Jørgensen, R.B., Cheng, B.F. and Heneen, W.K. (1997) Identification and chromosomal assignment of RAPD markers linked with a gene for seed colour in a *Brassica campestris-alboglabra* addition line. *Hereditas* 126, 133–138.

Chen, S., Wan, Z., Nelson, M.N., Chauhan, J.S., Lin, P., Redden, B., Burton, W.A., Banga, S.S., Chen, Y., Salisbury, P.A., Fu, T.D. and Cowling, W.A. (2011) Two distinct genetic diversity groups of oilseed *Brassica juncea* in both China and India. 17th Australian Research Assembly on Brassicas. Available at: http://www.australianoilseeds.com/data/assets/pdf_file/0012/8301/S1-P6-ShengChen.pdf (accessed 25 February 2015).

Cheung, W.Y., Friesen, L., Rakow, G.F.W., Séguin-Swartz, G. and Landry, B.S. (1997) A RFLP-based linkage map of mustard [*Brassica juncea* (L.) Czern. & Coss.]. *Theoretical and Applied Genetics* 94, 841–851.

Cheung, W.Y., Gugel, R.K. and Landry, B.S. (1998) Identification of RFLP markers linked to the white rust resistance gene (*Acr*) in mustard [*Brassica juncea* (L.) Czern.& Coss.]. *Genome* 41, 626–628.

Chhikara, S., Chaudhury, D., Dhankar, O.P. and Pawan, K. (2012) Combined expression of a barley class II chitinase and type I ribosome inactivating protein in transgenic *Brassica juncea* provides protection against *Alternaria brassicae*. *Plant Cell, Tissue and Organ Culture* 108, 83–89.

Davik, J. and Heneen, W. (1996) Fatty acid inheritance in wide reciprocal oilseed cross (*Brassica rapa* and *Brassica napus*). *Plant Soil Sciences* 46, 234–239.

Dehesh, K., Tai, H., Edwards, P., Byrne, J. and Jaworski, J.G. (2001) Overexpression of 3-ketoacyl-acyl-carrier protein synthase IIIs in plants reduces the rate of lipid synthesis. *Plant Physiology* 125, 1103–1114.

Delourme, R., Chevre, A.M., Brun, H., Rouxel, T., Balesdent, M.H., Dias, J.S., Salisbury, P., Renard, M. and Rimmer, S.R. (2006) Major gene and polygenic resistance to *Leptosphaeria maculans* in oilseed rape (*Brassica napus*). *European Journal of Plant Pathology* 114, 41–52.

Denis, M., Delourme, R., Gourrel, J.P., Mariani, C. and Renard, M. (1993) Expression of engineered nuclear male sterility in *Brassica napus*. Genetics, morphology, cytology and sensitivity to temperature. *Plant Physiology* 101, 1295–1304.

Deol, J.S., Shivanna, K.R., Prakash, S. and Banga, S.S. (2003) *Enarthrocarpus lyratus* based cytoplasmic male sterility and fertility restorer system in *Brassica rapa*. *Plant Breeding* 122, 438–440.

Dhillon, S.K., Sandha, G.S. and Banga, S.S. (2000) Fertility and self incompatibility of some 'B' and 'C' genome addition lines of *Brassica rapa* ssp. *Toria*. *Crop Improvement* 27, 198–201.

Downey, R.K. (1964) A selection of *Brassica campestris* containing no erucic acid in its seed oil. *Canadian Journal of Plant Science* 44, 295.

Downey, R.K. and Rimmer, S.R. (1993) Agronomic improvement in oilseed Brassicas. *Advances in Agronomy* 50, 1–65.

Dutta, I., Majumder, P., Saha, P., Ray, K. and Das, S. (2005) Constitutive and phloem specific expression of *Allium sativum* leaf agglutinin (ASAL) to engineer aphid (*Lipaphis erysimi*) resistance in transgenic Indian mustard (*Brassica juncea*). *Plant Science* 169, 996–1007.

Elizabeth, A.H., Zhu, Y.L., Sears, T. and Terry, N. (2000) Overexpression of glutathione reductase in *Brassica juncea*: effects on cadmium accumulation and tolerance. *Physiologia Plantarum* 110, 455–460.

Erickson, I.R., Straus, N.A. and Beversdrof, W.D. (1983) Restriction patterns reveal origins of chloroplast genomes in *Brassica* amphiploids. *Theoretical and Applied Genetics* 65, 202–206.

Eskin, N.A.M., Vaisey-Genser, M., Durance-Todd, S. and Przybylski, R. (1989) Stability of low linolenic acid canola oil to frying temperatures. *Journal of the American Oil Chemists' Society* 66, 1081–1084.

Fernandez-Escobar, J., Dominguez, J., Martin, A. and Fernandez-Martinez, J.M. (1988) Genetics of erucic acid content in interspecific hybrids of Ethiopian mustard (*Brassica carinata* Braun) and rapeseed (*B. napus* L.). *Plant Breeding* 100, 310–315.

Foolad, M.R. and Sharma, A. (2005) Molecular markers as selection tools in tomato breeding. *Acta Horticulturae* 695, 225–240.

Fray, J., Evans, E.J., Lydiate, D.J. and Arthur, A.E. (1996) Physiological assessment of apetalous flowers and erectophile pods in oilseed rape (*Brassica napus*). *The Journal of Agriculture Science* 127, 193–200.

Garg, H., Banga, S.K., Bansal, P., Atri, C. and Banga, S.S. (2007) Hybridizing *Brassic rapa*, with wild crucifers *Diplotaxis erucoides* and *Brassica maurorum*. *Euphytica* 156, 412–424.

Garg, H., Sivasithamparam, B.K., Banga, S.S. and Barbetti, M.J. (2008) Cotyledon assay as a rapid and reliable method of germplasm screening for resistance against *Sclerotinia sclerotiorum* in *Brassica napus* genotypes. *Australasian Plant Pathology* 37, 106–111.

Garg, H., Atri, C., Sandhu, P.S., Kaur, B., Renton, M., Banga, S.K., Singh, H., Singh, C., Barbetti, M.J. and Banga, S.S. (2010) High level of resistance to *Sclerotinia sclerotiorum* in introgression lines derived from hybridization between wild crucifers and the crop *Brassica species B. napus* and *B. juncea*. *Field Crop Research* 117, 51–58.

Gasic, K. and Korban, S.S. (2007a) Transgenic Indian mustard (*Brassica juncea*) plants expressing an *Arabidopsis* phytochelatin synthase (AtPCS1) exhibit enhanced As and Cd tolerance. *Plant Molecular Biology* 64, 361–369.

Gasic, K. and Korban, S.S. (2007b) Expression of *Arabidopsis* phytochelatin synthase in Indian mustard (*Brassica juncea*) plants enhances tolerance for Cd and Zn. *Planta* 225, 1277–1285.

Ge, Z.T., Li, Y.P., Wan, Z.J., You, M.P., Finnegan, P.M., Banga, S.S., Sandhu, P.S., Garg, H., Salisbury, P.A. and Barbetti, M. (2012) Delineation of *Sclerotinia sclerotiorum* pathotypes using differential resistance responses on *Brassica napus* and *Brassica juncea* genotypes enables identification of resistance to prevailing pathotypes. *Field Crop Research* 127, 248–255.

Good, A.G., Stringam, G.R., Mahmood, T., Ekuere, U. and Yeh, F. (2003) Molecular mapping of seed aliphatic glucosinolates in *Brassica juncea*. *Genome* 46, 753–760.

Grison, R., Grezes-Besset, B., Schneider, M., Lucante, N., Olsen, L., Leguay, J.J. and Toppan, A. (1996) Field tolerance to fungal pathotypes of *Brassica napus* constitutively expressing a chimeric chitinase gene. *Nature Biotechnology* 14, 643–646.

Guha, S. and Maheshwari, S.C. (1964) *In vitro* production of embryos from anthers of *Datura*. *Nature* 204, 497.

Gupta, M., Gupta, S., Kumar, H., Kumar, N. and Banga, S.S. (2015) Population structure and breeding value of a new type of *Brassica juncea* created by combining A and B genomes from related allotetraploids. *Theoretical and Applied Genetics* 128, 221–234.

Gupta, V., Mukhopadhyay, A., Arumugam, N., Sodhi, Y.S., Pental, D. and Pradhan, A.K. (2004) Molecular tagging of erucic acid trait in oilseed mustard (*Brassica juncea*) by QTL mapping and SNPs in FAE 1 gene. *Theoretical and Applied Genetics* 108, 743–749.

Hasan, V., Friedt, W., Pons-Kühnemann, J., Freitag, N.M., Link, K. and Snowdon, R.J. (2008) Association of gene-linked SSR markers to seed glucosinolate content in oilseed rape (*Brassica napus* ssp. *napus*). *Theoretical and Applied Genetics* 116, 1035–1049.

Hill, L.M., Morley-Smith, E.R. and Rawsthorne, S. (2003) Metabolism of sugars in the endosperm of developing seeds of oilseed rape. *Plant Physiology* 131, 228–236.

Hinata, K. and Prakash, S. (1984) Ethnobotany and evolutionary origin of Indian oleiferous Brassicae. *Indian Journal of Genetics* 44, 102–112.

Hiramatsu, M., Odahara, K. and Matsue, Y. (1995) A survey of microspore embryogenesis in leaf mustard (*Brassica juncea*). *Acta Horticulturae* 392, 139–145.

Hong, H., Datla, N., Reed, D.W., Covello, P.S., MacKenzie, S.L. and Qiu, X. (2002) High-level production of γ-linolenic acid in *Brassica juncea* using a Δ6 desaturase from *Pythium irregular*. *Plant Physiology* 129, 354–362.

Hossain, M.A., Maiti, M.K., Basu, A., Sen, S., Ghosh, A.K. and Ghosh, S.K. (2006) Transgenic expression of onion leaf lectin gene in Indian mustard offers protection against aphid colonization. *Crop Science* 46, 2022–2032.

Howell, B.C., Barker, G.C., Jones, G.H., Kearsey, M.J., King, G.J., Kop, E.P., Ryder, C.D., Teakle, G.R., Vicente, J.G. and Armstrong, S.J. (2003) Integration of the cytogenetic and genetic linkage maps of *Brassica oleracea*. *Genetics* 161, 1225–1234.

Howell, P.M., Marshall, D.F. and Lydiate, D.J. (1996) Towards developing intervarietal substitution lines in *Brassica napus* using marker-assisted selection. *Genome* 39, 348–358.

Hua, W., Li, R.J., Zhan, G.M., Liu, J., Li, J., Wang, X.F., Liu, G.H. and Wang, H.Z. (2011) Maternal control of seed oil content in *Brassica napus*: the role of silique wall photosynthesis. *Plant Journal* doi: 10.1111/j.1365-313X.2011.04802.x.

Huang, Z., Zhang, Y., Li, H.Q., Yang, L., Ban, Y.Y., Xu, A.X. and Xiao, E.S. (2011) AFLP and SSR markers linked to the yellow seed colour gene in *Brassica juncea* L. *Czech Journal of Genetics and Plant Breeding* 47, 149–155.

Huysen, T.V., Abdel-Ghany, S., Hale, K.L., LeDuc, D., Terry, N. and Pilon-Smits, E.A.H. (2003) Overexpression of cystathionine-γ-synthase enhances selenium volatilization in *Brassica juncea*. *Planta* 218, 71–78.

Jagannath, A., Bandyopadhyay, P., Arumugam, N., Gupta, V., Burma, P.K. and Pental, D. (2001) The use of a Spacer DNA fragment insulates the tissue-specific expression of a cytotoxic gene (*barnase*) and allows the high-frequency generation of transgenic male sterile lines in *Brassica juncea* L. *Molecular Breeding* 8, 11–23.

Jagannath, A., Arumugam, N., Gupta, V., Pradhan, A.K., Burma, P.K. and Pental, D. (2002) Development of transgenic *barstar* lines and identification of a male sterile (*barnase*)/restorer (*barstar*) combination for heterosis breeding in Indian oilseed mustard (*Brassica juncea*). *Current Science* 82, 46–52.

Jain, A., Bhatia, S., Banga, S.S., Prakash, S. and Laxmikumaran, M. (1994) Potential use of random amplified polymorphic DNA (RAPD) technique to study the genetic diversity in Indian mustard (*Brassica juncea*) and its relatedness to heterosis. *Theoretical and Applied Genetics* 88, 116–122.

Jako, C., Kumar, A., Wei, Y., Zou, J., Barton, D.L., Giblin, E.M., Covello, P.S. and Taylor, D.C. (2001) Seed-specific over-expression of an *Arabidopsis* cDNA encoding a diacylglycerol acyltransferase enhances seed oil content and seed weight. *Plant Physiology* 126, 861–874.

Janeja, H.S., Banga, S.K., Bhaskar, P.B. and Banga, S.S. (2003a) Alloplasmic male sterile *Brassica napus* with *Enarthrocarpus lyratus* cytoplasm: Introgression and molecular mapping of lyratus chromosome segment carrying fertility restoring gene. *Genome* 46, 792–797.

Janeja, H.S., Banga, S.S. and Lakshmikumaran, M. (2003b) Identification of AFLP markers linked to fertility restorer genes for tournefortii cytoplasmic male sterility system in *Brassica napus*. *Theoretical and Applied Genetics* 107, 148–154.

Jourdren, C., Barret, P., Horvais, R., Foisset, N., Delourme, R. and Renard, M. (1996) Identification of RAPD markers linked to the loci controlling erucic acid level in rapeseed. *Molecular Breeding* 2, 61–71.

Kanrar, S., Venkateswari, J.C., Kirti, P.B. and Chopra, V.L. (2002a) Transgenic expression of hevein, the rubber tree lectin, in Indian mustard confers protection against *Alternaria brassicae*. *Plant Science* 162, 441–448.

Kanrar, S., Venkateswari, J., Kirti, P.B. and Chopra, V.L. (2002b) Transgenic Indian mustard (*Brassica juncea*) with resistance to the mustard aphid (*Lipaphis erysimi* Kalt.). *Plant Cell Reports* 20, 976–981.

Kapoor, R., Banga, S.S. and Banga, S.K. (2009) A microsatellite (SSR) based linkage map of *Brassica rapa*. *New Biotechnology* 26, 239–243.

Kapoor, R., Kaur, G., Banga, S.K. and Banga, S.S. (2011) Generation of *Brassica rapa*-*Brassica nigra* chromosome addition stocks: cytology and microsatellite (SSR) based characterization. *New Biotechnology* 28, 407–417.

Kaur, G., Banga, S.K., Gogna, K.P.S., Joshi, S. and Banga, S.S. (2004) *Moricandia arvensis* cytoplasm based system of cytoplasmic male sterility in *Brassica juncea*: reappraisal of fertility restoration and agronomic potential. *Euphytica* 138, 271–276.

Kaur, P., Banga, S., Kumar, N., Gupta, S., Akhatar, J. and Banga, S.S. (2014) Polyphyletic origin of *Brassica juncea* with *B. rapa* and *B. nigra* (Brassicaceae) participating as cytoplasm donor parents in independent hybridization events. *American Journal of Botany* 101. Available at: http://www.amjbot.org/cgi/doi/10.3732/ajb.1400232 (accessed 25 February 2015).

Kirk, J.T.O. and Hurlstone, C.G. (1983) Variation and inheritance of erucic acid content in *Brassica juncea*. *Journal of Plant Breeding* 90, 331–338.

Kirk, J.T.O. and Oram, R.N. (1981) Isolation of erucic acid free lines of *Brassica juncea*: Indian mustard now a potential oilseed crop in Australia. *Journal of the Australian Institute of Agricultural Sciences* 47, 51–52.

Kirti, P.B., Prakash, S. and Chopra, V.L. (1991) Interspecific hybridization between *Brassica juncea* and *B. spinescens* through protoplast fusion. *Plant Cell Reports* 9, 639–642.

Kirti, P.B., Narasimhulu, S.B., Prakash, S. and Chopra, V.L. (1992a) Production and characterization of somatic hybrids of *Trachystoma balli* and *Brassica juncea*. *Plant Cell Reports* 11, 90–92.

Kirti, P.B., Narasimhulu, S.B., Prakash, S. and Chopra, V.L. (1992b) Somatic hybridization between *Brassica juncea* and *Moricandia arvensis* by protoplast fusion. *Plant Cell Reports* 11, 318–321.

Kirti, P.B., Narasimhulu, S.B., Mohapatra, T., Prakash, S. and Chopra, V.L. (1993) Correction of chlorophyll deficiency in alloplasmic male sterile *Brassica juncea* through recombination between chloroplast genome. *Genetics Research (Cambridge)* 62, 11–14.

Kirti, P.B., Banga, S.S., Parkash, S. and Chopra, V.L. (1995a) Transfer of Ogu cytoplasmic male sterility to *Brassica juncea* and improvement of the male sterile through somatic cell fusion. *Theoretical and Applied Genetics* 91, 517–521.

Kirti, P.B., Mohapatra, T., Prakash, S. and Chopra, V.L. (1995b) Development of a stable cytoplasmic male sterile line of *Brassica juncea* from somatic hybrid *Trachystoma ballii* and *Brassica juncea*. *Plant Breeding* 114, 434–438.

Kirti, P.B., Prakash, S., Gaikwad, K., Bhat, S.R., Dineshkumar, V. and Chopra, V.L. (1998) Chloroplast substitution overcomes leaf chlorosis in *Moricandia arvensis* based cytoplasmic male sterile *Brassica juncea*. *Theoretical and Applied Genetics* 97, 1179–1182.

Kondra, Z.P. and Stefansson, B.R. (1970) Inheritance of major glucosinolates of rapeseed (*Brassica napus*) meal. *Canadian Journal of Plant Science* 50, 643–647.

Kumar, A. and Singh, D.P. (1998) Use of physiological indices as a screening technique for drought tolerance in oil seeds *Brassica* species. *Annals of Botany* 81, 413–420.

Kumar, C. (1990) Drought management: research on rapeseed and mustard. *Proceedings of Indo-Swedish Symposium*. The Swedish University of Agricultural Sciences, Uppsala, Sweden, 4–6 September 1989, pp. 44–48.

Kumar, S., Atri, C., Sangha, M.K. and Banga, S.S. (2011) Screening of wild crucifers for resistance to mustard aphid, *Lipaphis erysimi* (Kaltenbach) and attempt at introgression of resistance gene(s) from *Brassica fruticulosa* to *Brassica juncea*. *Euphytica* 179, 461–470.

Labana, K.S. and Banga, S.K. (1984) Floral biology in Indian mustard [*Brassica juncea* (L.) Czern.& Coss]. *Genetica Agraria* 38, 131–138.

Lefort-Buson, M. and Datté, Y. (1985) Etude del' heterosis chezle conzla oleagineux d'hiver (*Brassica napus* L.) I. Comparison de deux populations, Fune homozygote et. Fautre heterozygote. *Agronomie* 5, 101–110.

Li, Y.M., Jun, L.X., Yun, G.C. and Li, L.L. (2010) Cloning and SNP analysis of TT1 gene in *Brassica juncea*. *Acta Agronomica Sinica* 36, 1634–1641.

Lichter, R. (1982) Induction of haploid plants from isolated pollen of *Brassica napus*. *Zeitschrift Fur Pflanzenphysiologie* 105, 427–434.

Lionneton, E., Beuret, W., Delaitre, C., Ochatt, S. and Rancillae, M. (2001) Improved microspore culture and doubled haploid plant regeneration in the brown condiment mustard (*B. juncea*). *Plant Cell Reports* 20, 126–130.

Lionneton, E., Ravera, S., Sanchez, L., Aubert, G., Delourme, R. and Ochatt, S. (2002) Development of an AFLP-based linkage map and localization of QTLs for seed fatty acid content in condiment mustard (*Brassica juncea*). *Genome* 45, 1203–1215.

Lionneton, E., Aubert, G., Ochatt, S. and Merah, O. (2004) Genetic analysis of agronomic and quality traits in mustard (*Brassica juncea*). *Theoretical and Applied Genetics* 109, 792–799.

Luo, Y.X., Du, D.Z., Fu, G., Xu, L., Li, X.P., Xing, X.R., Yao, Y.M., Zhang, X.M., Zhao, Z. and Liu, H.D. (2011) Inheritance of leaf color and sequence-related amplified polymorphic (SRAP) molecular markers linked to the leaf color gene in *Brassica juncea*. *African Journal of Biotechnology* 10, 14724–14730.

Lysak, M.A., Koch, M., Pecinka, A. and Schubert, I. (2005) Chromosome triplication found across the tribe Brassiceae. *Genome Research* 15, 516–525.

Mahmood, T., Ekuere, U., Yeh, F., Good, A.G. and Stringam, G.R. (2003) Molecular mapping of seed aliphatic glucosinolates in *Brassica juncea*. *Genome* 46, 753–760.

Mahmood, T., Rahman, M.H., Stringam, G.R., Yeh, F. and Good, A.G. (2006) Identification of quantitative trait loci (QTL) for oil and protein contents and their relationships with other seed quality traits in *Brassica juncea*. *Theoretical and Applied Genetics* 113, 1211–1220.

Malik, M., Vyas, P., Rangaswamy, N.S. and Shivanna, K.R. (1999) Development of two new cytoplasmic male-sterile lines of *Brassica juncea* through wide hybridization. *Plant Breeding* 118, 75–78.

Mariani, C., De Beuckeleer, M., Truettner, J., Leemans, J. and Goldberg, R.B. (1990) Induction of male sterility in plants by a chimaeric ribonuclease gene. *Nature* 347, 737–741.

Mariani, C., Gossele, V., De Beuckeleer, M., De Block, M., Goldberg, R.B., Degreef, R.B.W. and Leemans, J. (1992) A chimaeric ribonuclease inhibitor gene restores fertility to male sterile plants. *Nature* 357, 384–387.

Mayerhofer, R., Wilde, K., Mayerhofer, M., Lydiate, D., Bansal, V., Good, A. and Parkin, I. (2005) Complexities of chromosome landing in a highly duplicated genome: towards map based cloning of a gene controlling blackleg resistance in *Brassica napus*. *Genetics* 171, 1977–1988.

Mehra, S., Pareek, A., Bandyopadhyay, P., Sharma, P., Burma, P.K. and Pental, D. (2000) Development of transgenics in Indian oilseed mustard (*Brassica juncea*) resistant to the herbicide phosphinothricin. *Current Science* 78, 1358–1364.

Mizushima, U. and Tsunoda, S. (1967) A plant exploration in *Brassica* and allied genera. *Tohoku Journal of Agricultural Research* 17, 249–277.

Mohammad, A., Olsson, A.I. and Sevensk, H. (1990) Drought management. In: Olsson, I. and Kumar, P.R. (eds) *Research on Rapeseed and Mustard*. Proceedings of the Indo-Swedish Symposium, Swedish University of Agricultural Sciences, Uppsala, Sweden, 4–6 September 1989, pp. 49–67.

Mondal, K.K., Chatterjee, S.C., Viswakarma, N., Bhattacharya, R.C. and Grover, A. (2003) Chitinase-mediated inhibitory activity of *Brassica* transgenic on growth of *Alternaria brassicae*. *Current Microbiology* 47, 171–173.

Mondal, K.K., Bhattacharya, R.C., Koundal, K.R. and Chatterjee, S.C. (2007) Transgenic Indian mustard (*Brassica juncea*) expressing tomato glucanase leads to arrested growth of *Alternaria brassicae*. *Plant Cell Reports* 26, 247–252.

Morice, J. (1974) Selection d'une variete de colza, sans acids erucique et sans glucosinolates. In: Giestten, W. (ed.) *Proceedings of the 4th International Rapeseed Congress*, Giessen, Germany, pp. 31–47.

Mu, J., Tan, H., Zheng, Q., Fu, F., Liang, Y., Zhang, J., Yang, X., Wang, T., Chong, K., Wang, X.J. and Zuo, J. (2008) LEAFY COTYLEDON1 is a key regulator of fatty acid biosynthesis in *Arabidopsis*. *Plant Physiology* 148, 1042–1054.

Mukherjee, A.K., Mohapatra, T., Varshney, A., Sharma, R. and Sharma, R.P. (2001) Molecular mapping of a locus controlling resistance to *Albugo candida* in Indian mustard. *Plant Breeding* 120, 483–497.

Napier, J.A. and Grahm, I.A. (2010) Tailoring plant lipid composition: designer oilseeds come of age. *Current Opinion in Plant Biology* 13, 330–337.

NBPGR (2011) *Annual Report*. National Bureau of Plant Genetic Resources, New Delhi, India.

Negi, M.S., Devic, M., Delseny, M. and Lakshmikumaran, M. (2000) Identification of AFLP fragments linked to seed coat colour in *Brassica juncea* and conversion to SCAR marker for rapid selection. *Theoretical and Applied Genetics* 101, 146–152.

Østergaard, L., Kempin, S.A., Bies, D., Klee, H.J. and Yanofsky, M.F. (2006) Pod shatter-resistant *Brassica* fruit produced by ectopic expression of the FRUITFULL gene. *Plant Biotechnology Journal* 4, 45–51.

Padmaja, K.L., Arumugam, N., Gupta, V., Mukhopadhyay, A., Sodhi, Y.S., Pental, D. and Pradhan, A.K. (2005) Mapping and tagging of seed coat color and identification of microsatellite markers for marker assisted manipulation of the trait in *Brassica juncea*. *Theoretical and Applied Genetics* 111, 8–14.

Panjabi, P., Jagannath, A., Bisht, N.C., Lakshmi Padmaja, K., Sharma, S., Gupta, V., Pradhan, A.K. and Pental, D. (2008) Comparative mapping of *Brassica juncea* and *Arabidopsis thaliana* using Intron Polymorphism (IP) markers: homoeologous relationships, diversification and evolution of the A, B and C *Brassica* genomes. *BMC Genomics* 9, 113–132.

Panjabi, P., Yadava, S.K., Sharma, P., Kaur, A., Kumar, A., Arumugam, N., Sodhi, Y.S., Mukhopadhyay, A., Gupta, V., Pradhan, A.K. and Pental, D. (2010) Molecular mapping reveals two independent loci conferring resistance to *Albugo candida* in the east European germplasm of oilseed mustard *Brassica juncea*. *Theoretical and Applied Genetics* 121, 137–145.

Paterson, A.H., Lander, E.S., Hewitt, J.D., Peterson, S., Lincoln, S.E. and Tanksley, S.D. (1988) Resolution of quantitative traits into Mendelian factors by using a complete linkage map of restriction fragment length polymorphisms. *Nature* 335, 721–726.

Pathania, A., Kumar, R., Dinesh Kumar, V., Ashutosh, Dwivedi, K.K., Kirti, P.B., Prakash, S., Chopra, V.L. and Bhat, S.R. (2007) A duplication of *Cox I* gene is associated with CMS (*Diplotaxis catholica*) *Brassica juncea* derived from somatic hybridization with *Diplotaxis catholica*. *Journal of Genetics* 86, 93–101.

Potts, D.A. and Males, D.R. (1999) Inheritance of fatty acid composition in *Brassica juncea*. In: Wratten, N. and Salisbury, P.A. (eds) *Proceedings 10th International Rapeseed Congress*, Canberra, Australia.

Prabhu, K.V., Somers, D.J., Rakow, G. and Gugel, R.K. (1998) Molecular markers linked to white rust resistance in mustard *Brassica juncea*. *Theoretical and Applied Genetics* 97, 865–870.

Pradhan, A.K., Sodhi, Y.S., Mukhopadhyay, A. and Pental, D. (1993) Heterosis breeding in Indian mustard [*Brassica juncea* (L.) Czern & Coss]: Analysis of component characters contributing to heterosis for yield. *Euphytica* 69, 219–229.

Pradhan, A.K., Gupta, V., Mukhopadhyay, A., Sodhi, Y.S., Arumugam, N. and Pental, D. (2003) A high-density linkage map in *Brassica juncea* using AFLP and RFLP markers. *Theoretical and Applied Genetics* 106, 607–614.

Prakash, S. (1974a) Haploid meiosis and origin of *Brassica tournefortii* Gouan. *Euphytica* 23, 591–595.

Prakash, S. (1974b) Haploidy in *Brassica nigra* Koch. *Euphytica* 22, 613–614.

Prakash, S. and Chopra, V.L. (1988) Synthesis of alloplasmic *Brassica campestris* as a new source of cytoplasmic male sterility. *Plant Breeding* 101, 235–237.

Prakash, S. and Chopra, V.L. (1990) Male sterility caused by cytoplasm of *Brassica oxyrrhina* in *B. campestris* and *B. juncea. Theoretical and Applied Genetics* 79, 285–287.

Prakash, S. and Chopra, V.L. (1991) Cytogenetics of crop Brassicas and their allies. In: Tsuchyia, T. and Gupta, P.K. (eds) *Chromosome Engineering in Plants: Genetics, breeding, evolution.* Elsevier Science Publishers, Amsterdam, the Netherlands, pp. 161–180.

Prakash, S. and Hinata, K. (1980) Taxonomy, cytogenetics and origin of crop brassicas, a review. *Opera Botanica a Societate Botanice Lundensi* 55, 1–57.

Prakash, S., Kirti, P.B., Bhat, S.R., Gaikwad, K., Dinesh Kumar, V. and Chopra, V.L. (1998) A *Moricandia arvensis*-based cytoplasmic male sterility and fertility restoration system in *Brassica juncea. Theoretical and Applied Genetics* 97, 488–492.

Prakash, S., Ahuja, I., Uprety, H.C., Kumar, V.D., Bhat, S.R., Kirti, P.B. and Chopra, V.L. (2001) Expression of male sterility in alloplasmic *B. juncea* with *Erucastrum canariense* cytoplasm and the development of fertility restoration system. *Plant Breeding* 120, 479–482.

Prakash, S., Bhat, S.R., Kirti, P.B., Banga, S.K., Banga, S.S. and Chopra, V.L. (2004) Oilseed Brassica crops in India: history and improvement. *Brassica* 6, 1–54.

Prakash, S., Bhat, S.R., Quiros, C.R., Kirti, P.B. and Chopra, V.L. (2009) Brassica and its close allies: cytogenetics and evolution. In: Janick, J. (ed.) *Plant Breeding Reviews* vol 31, John Wiley & Sons, Hoboken, New Jersey, pp. 21–161.

Prasad, K.V.S.K., Sharmila, P., Kumar, P.A. and Saradhi, P.P. (2000) Transformation of *Brassica juncea* (L.) with bacterial coda gene enhances its tolerance to salt and cold stress. *Molecular Breeding* 6, 489–499.

Prem, D., Gupta, K. and Agnihotri, A. (2005) The effect of various exogenous and endogenous factors on microspore embryogenesis in Indian mustard (*B. juncea*). *In Vitro Cellular and Developmental Biology – Plant* 41, 266–273.

Priyamedha, Singh, B.K., Kaur, G., Sangha, M.K. and Banga, S.S. (2012) RAPD, ISSR and SSR based integrated linkage map from an F2 hybrid population of resynthesized and natural *Brassica carinata. National Academy Science Letters* 35, 303–308.

Przybylski, R., Malcolmson, L.J., Eskin, N.A.M., Durance-Todd, S., Mickle, J. and Carr, R. (1993) Stability of low linolenic acid canola oil to accelerate storage at 60°C. *Food Science and Technology* 26, 205–209.

Qiu, D., Morgan, C., Shi, J., Long, Y., Liu, J., Li, R., Zhuang, X., Wang, Y., Tan, X., Dietrich, E., Weihmann, T., Everett, C., Vanstraelen, S., Beckett, P., Fraser, F., Trick, M., Barnes, S., Wilmer, J., Schmidt, R., Li, J., Li, D., Meng, J. and Bancroft, I. (2006) A comparative linkage map of oilseed rape and its use for QTL analysis of seed oil and erucic acid content. *Theoretical and Applied Genetics* 114, 67–80.

Rajagopal, D., Agarwal, P., Tyagi, W., Singla-Pareek, S.L., Reddy, M.K. and Sopory, S.K. (2007) *Pennisetum glaucum* Na+/H+ antiporter confers high level of salinity tolerance in transgenic *Brassica juncea. Molecular Breeding* 19, 137–151.

Rakow, G. (1973) Selection of linoleic and linolenic fatty acid in rapeseed with mutation breeding [In German]. *Journal of Plant Breeding* 69, 62–68.

Rakow, G., Stringam, G.R. and McGregor, D.I. (1987) Breeding *Brassica napus* (L.) Canola with improved fatty acid composition, high oil content and high seed yield. In: Piekarczyk K., Czarnik, W. and Pruszynski, S. (eds) *Proceedings of the 7th International Rapeseed Congress*, Paris, pp. 4–33.

Ramchiary, N., Padmaja, K.L., Sharma, S., Gupta, V., Sodhi, Y.S., Mukhopadhyay, A., Arumugam, N., Pental, D. and Pradhan, A.K. (2007a) Mapping of yield influencing QTL in *Brassica juncea*: Implications for breeding of major oilseed crop of dryland areas. *Theoretical and Applied Genetics* 115, 807–817.

Ramchiary, N., Bisht, N.C., Gupta, V., Mukhopadhyay, A., Arumugam, N., Sodhi, Y.S., Pental, D. and Pradhan, A.K. (2007b) QTL analysis reveals context-dependent loci for seed glucosinolate trait in the oilseed *Brassica juncea*: importance of recurrent selection backcross (RSB) scheme for the identification of 'true' QTL. *Theoretical and Applied Genetics* 116, 77–85.

Raney, J.P., Rakow, G. and Olson, T.V. (1995a) Development of low erucic, low linolenic *Brassica juncea* utilizing interspecific crossing. In: Murphy, D. (ed.) *Proceedings of 9th International Rapeseed Congress*, John Innes Centre, Norwich, UK, pp. 413–415.

Raney, J.P., Rakow, G. and Olson, T.V. (1995b) Modification of *Brassica* seed oil fatty acid composition utilizing interspecific crossing. In: Murphy, D. (ed.) *Proceedings of 9th International Rapeseed Congress*, John Innes Centre, Norwich, UK, pp. 410–412.

Rao, G.U. and Shivanna, K.R. (1996) Development of a new alloplasmic CMS *Brassica napus* in the cytoplasmic background of *Diplotaxis siifolia*. *Cruciferae Newsletter* 18, 68–69.

Rao, G.U., Batra, V.S., Prakash, S. and Shivanna, K.R. (1994) Development of a new cytoplasmic male sterile system in *Brassica juncea* through wide hybridization. *Plant Breeding* 112, 171–174.

Rawat, P. and Rawat, S. (1997) Regenerating *Brassica juncea* plants from *in vitro* microspore culture. *Cruciferae Newsletter* 19, 47–48.

Reynolds, M., Foulkes, M.J., Slafer, G.A., Berry, P., Parry, M.A., Snape, J.W. and Angus, W.J. (2009) Raising yield potential in wheat. *Journal of Experimental Botany* 60, 1899–1918.

Richharia, R.H. (1937) Cytological investigations of *Raphanus sativus*, *Brassica oleracea* and their F1 hybrids. *Journal of Genetics* 34, 45–55.

Rimmer, S.R. (2006) Resistance genes to *Leptosphaeria maculans* in *Brassica napus*. *Canadian Journal of Plant Pathology* 28, 288–297.

Ripley, V.L. and Roslinsky, V. (2005) Identification of an ISSR marker for 2-propenyl glucosinolate content in *Brassica juncea* and conversion to SCAR marker. *Molecular Breeding* 16, 57–66.

Röbbelen, G. (1960) Beitrage zur Analyse des *Brassica*-Genomes. *Chromosoma* 11, 205–228.

Röbbelen, G. and Nitsch, A. (1975) Genetical and physiological investigations on mutants for polyenoic fatty acids in rapeseed *B. napus* (L.). *Journal of Plant Breeding* 75, 93–105.

Röbbelen, G. and Thies, W. (1980) Biosynthesis of seed oil and breeding for improved oil quality of rapeseed. In: Tsunoda, S., Hinata, K. and Gomez-Campo, C. (eds) *Brassica Crops and Wild Allies*. Japan Science Society Press, Tokyo, pp. 253–283.

Ruuska, S.A., Girke, T., Benning, C. and Ohlrogge, J.B. (2002) Contrapuntal networks of gene expression during *Arabidopsis* seed filling. *Plant Cell* 14, 1191–1206.

Sabharwal, V., Negi, M.S., Banga, S.S. and Lakshmikumaran, M. (2004) Mapping of AFLP markers linked to seed coat colour loci in *Brassica juncea* (L.) Czern .& Coss. *Theoretical and Applied Genetics* 109, 160–166.

Scarth, R. (1995) Developments in the breeding of edible oil in *Brassica napus* and *Brassica rapa*. In: Murphy, D. (ed.) *Proceedings of 9th International Rapeseed Congress*, John Innes Centre, Norwich, UK, pp. 378–382.

Scarth, R., McVetty, P.B.V., Rimmer, S.R. and Stefansson, B.R. (1988) Stellar low linolenic–high linoleic acid summer rape. *Canadian Journal of Plant Science* 68, 509–510.

Scarth, R., Rimmer, S.R. and McVetty, P.B.E. (1995) Apollo low linolenic summer rape. *Canadian Journal of Plant Science* 75, 203–204.

Séguin-Swartz, G., Warwick, S.I. and Scarth, R. (1997) *Cruciferae: Compendium of Trait Genetics*. Agriculture and Agri-Food Canada, Saskatoon, Canada.

Sharma, K.D. and Pannu, R.K. (2007) Biomass accumulation and its mobilization in Indian mustard, *Brassica juncea* (L.) Czern & Coss. under moisture stress. *Journal of Oilseeds Research* 24, 267–270.

Sharma, P., Sardana, V., Banga, S., Salisbury, P.A. and Banga, S.S. (2011) Morpho-physiological responses of oilseed rape (*Brassica napus* L.) genotypes to drought stress. *Proceedings of 13th International Rapeseed Congress*, 5–9 June 2011, Prague, Czech Republic, pp. 238–239.

Sharma, R., Aggarwal, A.K., Kumar, R., Mohapatra, T. and Sharma, R.P. (2002) Construction of an RAPD linkage map and localization of QTLs for oleic acid level using recombinant inbreds in mustard (*Brassica juncea*). *Genome* 45, 467–472.

Sharpe, A.G. and Lydiate, D.J. (2003) Mapping the mosaic of ancestral genotypes in a cultivar of oilseed rape (*Brassica napus*) selected via pedigree breeding. *Genome* 46, 461–468.

Shi, J., Li, R., Qiu, D., Jiang, C., Long, Y., Morgan, C., Bancroft, I., Zhao, J. and Meng, J. (2009) Unraveling the complex trait of crop yield with quantitative trait loci mapping in *Brassica napus*. *Genetics* 182, 851–861.

Shi, J., Li, R., Zou, J., Long, Y. and Meng, J. (2011) A dynamic and complex network regulates the heterosis of yield-correlated traits in rapeseed (*Brassica napus* L.). *PLoS One* 6(7), e21645.

Sikka, S.M. (1940) Cytogenetics of Brassica hybrids and species. *Journal of Genetics* 40, 441–509.

Simbaya, J., Slominski, B.A., Rakow, G., Campbell, L.D., Downey, R.K. and Bell, J.M. (1995) Quality characteristics of yellow seeded Brassica seed meal: protein, carbohydrates and dietary fibre components. *Journal of Agricultural Food Chemistry* 43, 2062–2066.

Singh, B.B. and Sharma, B. (1996) Restructuring cowpea for higher yield. *Indian Journal of Genetics* 56, 389–405.

Singh, D.P., Sangwan, V.P., Pannu, R.K. and Chaudhary, B.D. (1996) Comparison of osmotic adjustment in leaves and siliquae of oilseed Brassicae. *Indian Journal of Plant Physiology* 1, 284–285.

Singh, M., Chauhan, J.S. and Kumar, P.R. (2003) Response of different rapeseed-mustard varieties for growth, yield and yield component under irrigated and rainfed condition. *Indian Journal of Plant Physiology* 8, 53–59.

Singh, S., Mohapatra, T., Singh, R. and Hussain, Z. (2012) Mapping of QTLs for oil content and fatty acid composition in Indian mustard [*Brassica juncea* (L.) Czern.& Coss.]. *Journal of Plant Biochemistry and Biotechnology* DOI: 10.1007/s13562-012-0113-6.

Singh, S.P. and Singh, D.P. (1972) Inheritance of yield and other agronomic characters in Indian mustard. *Canadian Journal of Genetics and Cytology* 14, 227–233.

Sivaraman, I., Arumugam, N., Sodhi, Y.S., Gupta, V., Mukhopadhyay, A., Pradhan, A.K., Burma, P.K. and Pental, D. (2004) Development of high oleic and low linoleic acid transgenics in a zero erucic acid *Brassica juncea* L. (Indian mustard) line by antisense suppression of the *fad2* gene. *Molecular Breeding* 13, 365–375.

Slominski, B.A. (1997) Development in the breeding of low fibre rapeseed/canola. *Journal of Animal Feed Science* 6, 303–317.

Sodhi, Y.S., Mukhopadhyay, A., Arumugam, N., Verma, J.K., Gupta, V. and Pental, D. (2002) Genetic analysis of total glucosinolate in crosses involving a high glucosinolate Indian variety and a low glucosinolate line of *Brassica juncea*. *Plant Breeding* 121, 508–511.

Sodhi, Y.S., Chandra, A., Verma, J.K., Arumugam, N., Mukhopadhyay, A., Gupta, V., Pental, D. and Pradhan, A.K. (2006) A new cytoplasmic male sterility system for hybrid seed production in Indian oilseed mustard *Brassica juncea*. *Theoretical and Applied Genetics* 114, 93–99.

Srivastava, A., Gupta, V., Pental, D. and Pradhan, A.K. (2001) AFLP-based genetic diversity assessment amongst agronomically important natural and some newly synthesised lines of *Brassica juncea*. *Theoretical and Applied Genetics* 102, 193–199.

Srivastava, A., Mukhopadhyay, A., Arumugam, M., Gupta, V., Verma, J.K., Pental, D. and Pradhan, A.K. (2004) Resynthesis of *Brassica juncea* through interspecific crosses between *B. rapa* and *B. nigra*. *Plant Breeding* 123, 204–206.

Stefansson, B.R., Hougen, F.W. and Downey, R.K. (1961) Note on isolation of rape plants with seed oil free from erucic acid. *Canadian Journal of Plant Science* 41, 218–219.

Stringam, G.R. and Thiagarajah, M.R. (1995) Inheritance of alkenyl glucosinolates in traditional and micro-spore-derived doubled haploid populations of *Brassica juncea* (L.) Czern & Coss. *Proceedings of 9th International Rapeseed Congress UK* 3, 804–806.

Stringam, G.R., McGregor, D.I. and Pawlowski, S.H. (1974) Chemical and morphological characteristics associated with seed coat colour in rapeseed. In: *Proceedings of 4th International Rapeseed Conference*, Giessen, West Germany, pp. 99–108.

Tayo, T.O. and Morgan, D.G. (1979) Factors are influencing flower and pod development in oil-seed rape (*Brassica napus* L.). *Journal of the Agricultural Sciences* 92, 363–373.

Thiagarajah, M.R. and Stringam, G.R. (1990) High frequency embryo induction from microspore culture of *Brassica juncea* (L.) Czern and Coss. In: *Proceedings of the 7th International Congress on Plant Tissue and Cell Culture*, Amsterdam, pp. 190.

Thormann, C.E., Romero, J., Mantet, J. and Osborn, T.C. (1996) Mapping loci controlling the concentrations of erucic and linolenic acids in seed oil of *Brassica napus* (L). *Theoretical and Applied Genetics* 93, 282–286.

Thurling, N. (1993) Physiological constraints and their genetic manipulation. In: Labana, K.S., Banga, S.S. and Banga, S.K. (eds) *Breeding Oilseed Brassicas*. Springer-Verlag, Berlin, pp. 44–68.

Upadhyay, A., Mohapatra, T., Pai, R.A. and Sharma, R.P. (1996) Molecular mapping and character tagging in Indian mustard (*Brassica juncea*) II.RFLP marker association with seed coat colour and quantitative traits. *Journal of Plant Biochemistry and Biotechnology* 5, 17–22.

Uzunova, M., Ecke, W., Weissleder, K. and Röbbelen, G. (1995) Mapping the genome of rapeseed (*Brassica napus* L.). I. Construction of an RFLP linkage map and localization of QTLs for seed glucosinolate content. *Theoretical and Applied Genetics* 90, 194–204.

Van Deynze, A.E., Beversdorf, W.D. and Pauls, K.P. (1993) Temperature effect on seed colour in black and yellow-seeded rapeseed. *Canadian Journal of Plant Science* 73, 383–387.

Varshney, A., Mohapatra, T. and Sharma, R.P. (2004) Development and validation of CAPS and AFLP markers for white rust resistance gene in *Brassica juncea*. *Theoretical and Applied Genetics* 109, 153–159.

Venkateswarlu, J. and Kamala, T. (1971) Pachytene chromosome complements and genome analysis in *Brassica*. *Journal of the Indian Botanical Society* 50, 442–449.

Vera, C.L. and Woods, D.L. (1982) Isolation of independent gene pairs at two loci for seed coat colour in *Brassica juncea*. *Canadian Journal of Plant Science* 62, 47–50.

Vera, C.L., Woods, D.L. and Downey, R.K. (1979) Inheritance of seed coat colour in *Brassica juncea*. *Canadian Journal of Plant Science* 59, 635–637.

Vigeolas, H., Waldeck, P., Zank, T. and Geigenberger, P. (2007) Increasing seed oil content in oil-seed rape (*Brassica napus* L.) by over-expression of a yeast glycerol-3-phosphate dehydrogenase under the control of a seed-specific promoter. *Plant Biotechnology Journal* 5, 431–441.

Vikki, J.P. and Tanhuanpää, P.K. (1995) Breeding of high oleic acid spring turnip rape in Finland. In: Murphy, D. (ed.) *Proceedings of the 9th International Rapeseed Congress*, John Innes Centre, Norwich, UK, pp. 383–388.

Wang, X.F., Liu, G.H., Yang, Q., Hua, W., Liu, J. and Wang, H.Z. (2010) Genetic analysis on oil content in rapeseed (*Brassica napus* L.). *Euphytica* 173, 17–24.

Warwick, S.I. (2010) Brassicaceae in Agriculture. In: Schmidt, R. and Bancroft, I. (eds) *Genetics and Genomics of the Brassicaceae*. Springer Science and Business Media, pp. 1–67.

Wei, H., Rong-Jun, L., Gao-Miao, Z.J.L., Jun, L., Xin-Fa, W., Gui-Hua, L. and Wang, H.Z. (2011) Maternal control of seed oil content in *Brassica napus*: the role of silique wall photosynthesis. *Plant Journal* doi: 10.1111/j.1365-313X.2011.04802.

Wei, S., Yu, B., Gruber, M.Y., Khachatourians, G.G., Hegedus, D.D. and Hannoufa, A. (2010) Enhanced seed carotenoid levels and branching in transgenic *Brassica napus* expressing the *Arabidopsis miR156b* gene. *Journal of Agricultural and Food Chemistry* 58, 9572–9578.

Werner, S., Diederichsen, E., Frauen, M., Schondelmaier, J. and Jung, C. (2008) Genetic mapping of clubroot resistance genes in oilseed rape. *Theoretical and Applied Genetics* 116, 363–372.

Weselake, R.J., Taylor, D.C., Rahman, M.H., Shah, S., Laroche, A., McVetty, P.B.E. and Harwood, J.L. (2009) Increasing the flow of carbon into seed oil. *Biotechnology Advances* 27, 866–878.

Yadava, S.K., Arumugam, N., Mukhopadhyay, A., Sodhi, Y.S., Gupta, V., Pental, D. and Pradhan, A.K. (2012) QTL mapping of yield-associated traits in *Brassica juncea*: meta-analysis and epistatic interactions using two different crosses between east European and Indian gene pool lines. *Theoretical and Applied Genetics* DOI 10.1007/s00122-012-1934-3.

Yusuf, M.A., Kumar, D., Rajwanshi, R., Strasser, R.J., Tsimilli-Michael, M., Govindjee, and Sarin, N.B. (2010) Overexpression of gamma-tocopherol methyl transferase gene in transgenic *Brassica juncea* plants alleviates abiotic stress: physiological and chlorophyll fluorescence measurements. *Biochimica Biophysica Acta* 1797, 1428–1438.

Zhao, J. and Meng, J. (2003) Detection of loci controlling seed glucosinolate content and their association with *Sclerotinia* resistance in *Brassica napus*. *Plant Breeding* 122, 19–23.

Zhao, J., Becker, H., Zhang, D., Zhang, Y. and Ecke, W. (2006) Conditional QTL mapping of oil content in rapeseed with respect to protein content and traits related to plant development and grain yield. *Theoretical and Applied Genetics* 133, 33–38.

Zhao, Y.T., Wang, M., Fu, S.X., Yang, W.C., Qi, C.K. and Wang, X.J. (2012) Small RNA profiling in two *Brassica napus* cultivars identifies microRNAs with oil production and developmental correlated expressions and new small RNA classes. *Plant Physiology* DOI:10.1104/pp.111.187666.

Zou, J., Katavic, V., Giblin, E.M., Barton, D.L., MacKenzie, S.L., Keller, W.A., Hu, X. and Taylor, D.C. (1997) Modification of seed oil content and acyl composition in the Brassicaceae by expression of a yeast sn-2 acyltransferase gene. *Plant Cell* 9, 909–923.

3 Intersubgenomic Heterosis: *Brassica napus* as an Example

Donghui Fu[1]* and Meili Xiao[2]

[1]*Key Laboratory of Crop Physiology, Ecology and Genetic Breeding, Ministry of Education, Agronomy College, Jiangxi Agricultural University, Nanchang;*
[2]*Chongqing Engineering Research Center for Rapeseed, College of Agronomy and Biotechnology, Southwest University, Chongqing, China*

Introduction

There are six predominant *Brassica* species. These include three primary diploids, *B. rapa* (AA, 2n=20), *B. oleracea* (CC, 2n=18), *B. nigra* (BB, 2n=16) and three amphidiploids, *B. napus* (AACC, 2n=38), *B. juncea* (AABB, 2n=36) and *B. carinata* (BBCC, 2n=34). Each pair of diploid species hybridized during evolution has undergone chromosome doubling to generate amphidiploids. Of the three amphiploids, *B. napus* is now a major oilseed crop of the world. Both open-pollinated (OP) varieties and hybrids are cultivated. Hybrids are now more popular as these provide 30% higher yields than OP varieties and also exhibit superior stress tolerance and adaptability. Hybrid cultivars presently occupy over 75% of the area in China, Canada and Europe. Efforts to further increase the level of heterosis have not been very successful; this has been attributed to the narrow genetic base, which hindered further increase in hybrid vigour (Becker *et al.*, 1995; Rana *et al.*, 2004; Cowling, 2007). A narrow genetic base in this crop can be attributed to the relatively short history of domestication and intensive breeding over the years. Fortunately, *B. rapa*, *B. oleracea* and *B. carinata*, species closely related with *B. napus*, possess abundant germplasm and are also compatible with *B. napus*. Although the compatibility is species-specific, these relatives can be used as a resource to broaden the genetic base of *B. napus*. Over the long history of evolution and human selections, diploid genomes (A/C) have accumulated huge genetic differences. For example, the A genomes resident in *B. rapa*, *B. juncea* and *B. napus*, and the C genome(s) from *B. oleracea*, *B. carinata* and *B. napus* show large differences. In order to factor in such differences, a new term 'subgenome' was proposed in Brassicas. Terms A[r], A[n] and A[j] were suggested to mark the respective A genome(s) in *B. rapa*, *B. napus* and *B. juncea*; B[n], B[j] and B[c] represented the B genome(s) of *B. nigra*, *B. juncea* and *B. carinata*; C[o], C[n] and C[c] symbolized the C genome(s) of *B. oleracea*, *B. napus* and *B. carinata* (Li *et al.*, 2004, 2005, 2007; Qian *et al.*, 2005). The introgression in *B. napus* from other species is expected to result in a new-type *B. napus*. Another term 'intersubgenomic heterosis' was proposed for the heterosis emanating from crosses between natural *B. napus* and new-type *B. napus* with introgression of novel genetic components, such as *B. rapa* genomic fragments. The concept

---

*Corresponding author, e-mail: fudhui@163.com

can be extended to other polyploids. Therefore, factually speaking, the intersubgenomic heterosis is the heterosis from a cross between natural polyploids and introgressed polyploids, developed by including the genetic components of other species. Intersubgenomic heterosis exhibits stronger heterosis potential and has shown large application value for *B. napus* breeding. However, there are several associated limitations. This chapter recalls the history of research of intra-species heterosis, sums up the avenues and problems of generating new-type *B. napus*, dissects the research advance of genetics and genomics of intersubgenomic heterosis, and points out the problems and countermeasures of improving intersubgenomic heterosis. This review is beneficial to further improve the potential of intersubgenomic heterosis and clarify the mechanism of intersubgenomic heterosis.

Generating New-type *Brassica napus*

The development of new-type *B. napus* is a prerequisite for breeding intersubgenomic hybrids. High compatibility between *B. napus* and *B. rapa* facilitates the improvement of *B. napus* using *B. rapa* genetic introgressions. *B. rapa* is an oilseed and vegetable crop in Asia and possesses abundant variation and trait options, including early maturity and yellow seed-coat colour (Liu, 2000). The trigenomic hybrids between *B. napus* ($A^nA^nC^nC^n$) and *B. rapa* (A^rA^r) can be easily obtained (Liu *et al.*, 2002; Qian *et al.*, 2005) and self-crossed. The plants with 38 chromosomes can be selected in the progenies as new-type *B. napus*. Alternatively, the trigenomic hybrids between *B. napus* and *B. rapa* are backcrossed with *B. rapa* again and self-pollinated. From the progeny, new-type *B. napus* with 38 chromosomes was selected (Qian *et al.*, 2005). If *B. rapa* belongs to the oilseed type, it is easy to obtain new-type *B. napus*. This approach has several advantages, such as with the abundant and diverse germplasm in *B. rapa* and ease of developing hybrids between *B. napus* and *B. rapa*, excellent new-type *B. napus* can be obtained if oilseed forms of *B. rapa* are used. A limitation to complete exploitation of intersubgenomic heterosis was due to the fact that improvement of *B. napus* was

associated more with the A genome and less with the C genome. This does not help in maximizing intersubgenomic heterosis.

The second method involves re-synthesis of *B. napus* amphiploids through hybridization between *B. oleracea* and *B. rapa* and chromosome doubling. A large number of *B. napus* genotypes were resynthesized through crossing 11 wild *Brassica* species (wild *B. oleracea*, *B. bourgaei*, *B. cretica*, *B. incana*, *B. insularis*, *B. hilarionis*, *B. macrocarpa*, *B. montana*, *B. rupestris*, *B. taurica*, *B. villosa*) and *B. rapa* (Jesske *et al.*, 2013). Similarly, synthetic *B. juncea* and *B. carinata* were obtained from the reciprocal crosses among *B. rapa*, *B. nigra* and *B. oleracea* (Cui *et al.*, 2012). Besides their re-synthesis through interspecific hybridization, fusion between pollen protoplasts of *B. oleracea* and haploid mesophyll protoplasts of *B. rapa* has also been used for developing new amphiploids (Liu *et al.*, 2007). Though some rearrangements can affect genome stability of synthetic *B. napus*, the new-type *B. napus* appeared more stable than those developed through hybridization between *B. napus* and *B. rapa* as resynthesized genotypes were true amphiploids. The limitation with this approach was the difficulty of producing initial hybrids between *B. oleracea* and *B. rapa* due to low seed setting, and since most *B. oleracea* are vegetable type, the synthetic *B. napus* was not found suitable as an oilseed crop. Despite these limitations, re-synthesis is the most preferred approach.

The third method involves synthesis of trigenomic hybrids ($A^rB^cC^c$) by hybridizing *B. carinata* ($B^cB^cC^cC^c$) and *B. rapa* (A^rA^r) followed by chromosome doubling. The hexaploids ($A^rA^rB^cB^cC^cC^c$, 2n=54) thus generated were used as female parents to pollinate the natural *B. napus* ($A^nA^nC^nC^n$) to produce pentaploid ($A^rA^nB^cC^cC^n$) plants with 46 chromosomes. Selfing of the pentaploid hybrids led to the elimination of B^c chromosomes during meiosis, resulting in a new *B. napus* with the euploid chromosome number (Li *et al.*, 2007). The yellow-seeded *B. napus* genotypes were also obtained through interspecific hybridization involving (*B. rapa* × *B. carinata*) × *B. napus* (Meng *et al.*, 1998). This approach allowed simultaneous improvement of the A and C genomes of *B. napus* to exhibit stronger heterosis potential. However, the process of developing stable new-type *B. napus* by this method is long and complex. In another procedural alternative, the trigenomic hybrids between *B. napus* ($A^nA^nC^nC^n$) and *B. oleracea* (C^cC^c) were treated with colchicine to generate the hexaploid ($A^nA^nC^nC^nC^cC^c$). Such hexaploid plants showed good cross ability with *B. rapa* (9.24 seeds per pod) and allowed production of a genetically diverse *B. napus*. These hexaploids displayed large flowers and high normal chromosome segregation, and had high seed set (4.48 seeds per pod on average) and pollen fertility (87.05% on average). This is expected to become a large-scale and highly efficient method to import *B. rapa* genetic components into *B. napus* (Li *et al.*, 2013). In addition, *B. napus* could also be extracted from the progeny of interspecific hybrids between *B. napus* and *B. carinata*/*B. juncea* (Nelson *et al.*, 2009; Navabi *et al.*, 2010). The B genome may be used to improve the phenotypes of *B. napus*, including gene resistance to blackleg, morphological variation (e.g. days to flowering and days to maturity) and seed quality traits (e.g. erucic acid content). The genes for specific traits were also imported into *B. napus* by chromosome translocation. However, how to evaluate these new-type *B. napus* has not been reported. Large-scale application of these approaches can help to construct a heterotic gene pool for *B. napus* hybrid breeding (Xiao *et al.*, 2010). Combined with recurrent selection, a new *B. napus* gene pool can provide a permanent source of novel alleles.

Genetics and Genomics of Intersubgenomic Heterosis

The interspecific hybrids between *B. napus* and *B. rapa* exhibited the average over-mid-parent heterosis for biomass production (approximately 30%). There was also a close relationship between some DNA fragments of A^r and biomass production in trigenomic hybrids ($A^rA^nC^n$) (Liu *et al.*, 2002). *B. napus* type of genotypes obtained through complex combinations of *B. napus* × *B. rapa* and (*B. napus* × *B. rapa*) × *B. rapa* (Qian *et al.*, 2005) demonstrated a very high subgenomic heterosis when crossed with natural *B. napus* to produce

intersubgenomic hybrids. Almost 90% of combinations surpassed their respective tester lines in yield, and about 75% and 25% of hybrid combinations exceeded the productivity levels of two elite Chinese cultivars, respectively (Qian *et al.*, 2005). Some DNA segments from A^r appeared to play a positive role for increasing seed yield of intersubgenomic hybrids ($A^rA^nC^nC^n$) (Qian *et al.*, 2005). This constituted the first report of intersubgenomic heterosis. Seyis *et al.* (2006) also validated the existence of intersubgenomic heterosis with new-type *B. napus* from the hybridization of *B. oleracea* and *B. rapa* and chromosome doubling. Li *et al.* (2007) developed another type of new-type *B. napus* with introgression of *B. carinata* and *B. rapa*. A close relationship was detected between the seed yields of intersubgenomic hybrids of $A^rA^nC^cC^n$ and the genomic contents of A^r, C^c and $A^r + C^c$ in the reconstituted *B. napus* (Li *et al.*, 2006). This study also underlined the role of alien germplasm in intersubgenomic heterosis.

Altered gene expression patterns were also detected among new-type *B. napus*, natural *B. napus* cultivars and their hybrids through the cDNA amplified fragment length polymorphism technique (cDNA-AFLP) (Chen *et al.*, 2008). The result indicated that the gene expression profiles of the new-type *B. napus* lines were sharply different from those of their parents because the new-type *B. napus* lines contained A^r and C^c genome components from *B. rapa* and *B. carinata*. Twenty transcript-derived fragments (TDFs) related to intersubgenomic heterosis were randomly selected and developed into molecular markers. These markers were mapped in the confidence intervals of QTLs for seed yield and yield-related traits by comparative mapping. Although this study aimed at finding the molecular mechanism of intersubgenomic heterosis, no key genes regulating intersubgenomic heterosis could be detected. Zou *et al.* (2010) reported the development of the second generation of resynthesized *B. napus* by pyramiding breeding based on hybridization between the first generation of new-type *B. napus*. The second generation *B. napus* forms higher heterosis potential than the first generation in heterosis potential. In addition, a significant positive correlation was observed between the heterosis potential and the introgressed subgenomic components in the new type *B. napus*. In another study, Xiao *et al.* (2010) created a new population containing a large number of *B. napus* carrying introgressions from 25 cultivars of *B. rapa* and 72 accessions of *B. carinata*. This acted as a sustained heterotic gene pool for *B. napus*. Very large genetic differences were observed between this population and natural *B. napus*. There were also evidences of *de novo* variation as a consequence of polyploidization. A recombinant inbred line population of new type *B. napus* derived from a cross between *B. napus* and *B. rapa* was used to identify QTLs associated with heterosis (Fu *et al.*, 2012). More than half of single-locus QTLs and interacting QTL pairs were associated with the novel alleles produced by the introgression of *B. rapa*. The intersubgenomic heterosis between natural *B. napus* and *B. napus* with introgressions from *B. rapa* was attributed to the single-locus effect and two-loci interactions of both novel loci and the loci directly from the *B. rapa* genome. This study was the first major attempt to understand the genetic mechanism of intersubgenomic heterosis. However, subsequent research about fine mapping of QTLs related to intersubgenomic heterosis has not been reported.

Girke *et al.* (2012) obtained intersubgenomic hybrids from crosses of 44 resynthesized lines with diverse genetic background to two male sterile winter *B. napus* as tester lines. Multiple field trials showed that the yield of resynthesized lines ranged from 85.1 to 44.6% of check cultivar yields while hybrid yields varied between 91.6 and 116.6% of check cultivar. Mid-parent heterosis ranged from −3.5 to 47.2% for yield. There appeared to be no correlation between the genetic distance of parental lines and heterosis or hybrid yield. Jesske *et al.* (2013) observed a slightly negative correlation (r =−0.29) between genetic distance and hybrid yield of resynthesized × natural *B. napus* hybrids. It showed that the content of alien genetic composition in new-type *B. napus* was not a key factor contributing to heterosis. In addition, most of the hybrids with the resynthesized *B. napus* from wild *B. oleracea* displayed lower yield than hybrids with resynthesized *B. napus* from domesticated *B. oleracea*. Elite parents seemed critical to increase heterosis. The whole history of intersubgenomic heterosis incorporates the ideas from 'point' (individual or less combinations) to 'gene pool', and from 'utilization in breeding' to 'theory' (molecular mechanism), which provides a case of the growth process of one new research field. To sum up, these studies prove that the intersubgenomic heterosis exists and can be utilized in *B. napus* breeding. However, there is a long way to go before intersubgenomic heterosis is utilized on a large scale.

Limitations of Intersubgenomic Heterosis

A key problem in the utilization of intersubgenomic heterosis is stability of genomes of new-type *B. napus*. New-type *B. napus* originates from the offspring of interspecific hybridization and contains more genetic components of alien species. Interspecific hybridization and polyploidization can result in 'genome shock', including chromosomal rearrangements, sequence elimination, emergence of novel sequences, transposon activation and epigenetic modifications. The incompatible relationship between host genome and alien genome is expected to continue for several generations, which delays breeding. Therefore, it is necessary to investigate the types and mechanism of different levels of genomic variations in new-type *B. napus*.

Changes of DNA Sequences

Allopolyploidization, including two major events, interspecific hybridization and genome doubling, is a prominent evolutionary force. During the process of generation of resynthesized *B. napus*, chromosome number may vary for several generations. For example, in multiple generations of resynthesized *B. napus*, the average chromosome number was up to 38 with a range from 36 to 42. There were evidences of aneuploidy, chromosome breakage and fusion, inter- and intra-genomic rearrangements, rDNA variation, and the disappearance of repeat sequences (Xiong *et al.*, 2011).

Likewise, there were the loss and/or gain of parental restriction fragments and appearance of novel fragments in each generation from F_2 to F_5 (Song *et al.*, 1995). By the AFLP approach, 1.17% of the loci exhibited genetic changes in the resynthesized *B. napus* in contrast to parents, *B. rapa* and *B. oleracea* (Xu *et al.*, 2012). All these studies suggest that the genomic variation is often found in synthetic *B. napus*. There was no significant difference for genetic changes in the A and C genomes (Xu *et al.*, 2012). In addition, the frequency of genome variation was correlated with the divergence degree of the diploid parental genomes (Song *et al.*, 1995). Combinations which showed greater divergence of the diploid parental genomes can accumulate more alleles in new type *B. napus* but also carry increased risk of genome instability. Such genomic instabilities in resynthesized *B. napus* can be the outcome of homologous recombination between A and C genomes that destabilizes the karyotype, leading to aberrant meiotic behaviour, and generate novel gene combinations and phenotypes (e.g. reduced fertility) (Gaeta and Pires, 2010). The extensive homology of chromosome sets exhibited the highest instability. The homologous chromosome replacement and compensation frequently cause loss and gain of chromosomes (Xiong *et al.*, 2011). In fact, the efficiency of interspecific hybridization is dependent on the similarity between the parental genomes because a high similarity can promote accurate chromosome pairing and genetic recombination (Leflon *et al.*, 2006). Therefore, the exchanges among homologous chromosomes are a key reason to generate novel allele combinations and phenotypic variation in synthetic *B. napus* (Gaeta *et al.*, 2007). Furthermore, increased aneuploidy was negatively associated with pollen viability and seed yield (Xiong *et al.*, 2011). It may also be pertinent to consider the effect of transposons in genome instability of new-type *B. napus*. In a recombinant inbred line population consisting of a new-type *B. napus* lines introgressed by *B. rapa*, considerable novel genomic alterations were found. These included retrotransposon activations, SSR mutations and chromosomal rearrangements. These genomic changes happened immediately after interspecific hybridization, primarily during the early stages

of generation following hybridization. Furthermore, these novel genomic alterations had effects on yield-related traits in this recombinant inbred-line population (Zou *et al.*, 2011). Similarly, in the comparison of a resynthesized *B. napus* and its diploid parents, transposable elements (TE) responses to allopolyploidization were found and were highly site-specific, partly due to the changes in DNA methylation (Sarilar *et al.*, 2013).

One can imagine that when two genomes from different species merge together they conflict each other and undergo genome stabilization processes. The direction, position and degree of genome variation show evidence of varied trends in the new populations. What effects does the genome instability of new-type *B. napus* take on intersubgenomic heterosis? There are two problems when using intersubgenomic heterosis: (i) the continuous variation of genome of new-type *B. napus* in multiple generations delays the time of breeding process; and (ii) the genome instability of new-type *B. napus* can result in the instability of phenotypes. It may thus become very difficult as an ideal genotype for maximizing the heterosis.

There are not many options to overcome these problems. However, two points are worth being investigated: (i) by consciously discarding lines which continue to vary in phenotype over several breeding cycles; and (ii) we may select some lines with high variability to perform the analysis of genome, or gene expression with an aim to hunt for genes for genome instability. Only when we understand the mechanism of genome instability of new-type *B. napus* can we correctly and efficiently select target lines.

Gene Expression Changes

To fully exploit these novel phenotypes it is necessary to understand the regularities of gene expression in new-type *B. napus*. In resynthesized *B. napus*, the homologous chromosome exchanges could lead to the loss and doubling of homologous genes and then to gene expression alterations. Less than 0.3% of genes showed non-additive expression in the S (0:1) lines while only 0.1–0.2% displayed non-additive effects in all S (5:6) lines (Gaeta *et al.*, 2009). In contrast, there were changes of ~4.09 and 6.84% of the sequences in gene expression and DNA methylation, respectively, occurring immediately after interspecific hybridization and polyploidization in synthetic *B. napus* in contrast to the two parents, *B. rapa* and *B. oleracea*. The proportions of gene silencing and changes of DNA methylation in the C genome were far higher than those in the A genome (Xu *et al.*, 2009), suggesting that the change of gene expression could occur. Digital gene expression analysis helped to identify a large number of differentially expressed genes (DEGs) in the first four generations following re-synthesis (F_1–F_4). The 20 most DEGs were involved with DNA binding/transcription factor, cyclin-dependent protein kinase, epoxycarotenoid dioxygenase and glycine-rich protein (Jiang *et al.*, 2013). In addition, extensive transcriptomic changes detected by a high-throughput RNA-Seq method were also found in a *Brassica* allohexaploid compared with its parents (*B. rapa* and *B. carinata*). The 2545 non-additive genes of the *Brassica* hexaploid were involved with response to the stimulus, immune system process, the cellular process and metabolic process (Zhao *et al.*, 2013). In addition, 26–30% of the duplicated genes exhibited alternation of alternative splicing (AS), including loss of the AS event after allopolyploidization as compared to the diploid donors. As changes after polyploidization were more frequent than homologue silencing events, AS changes constituted an essential aspect of the transcriptome shock following allopolyploidization (Zhou *et al.*, 2011). In addition, AS can produce multiple mature mRNAs from a precursor mRNA, which also can regulate gene expression in resynthesized *B. napus*.

The variation in genomes of new-type *B. napus* was also reflected at protein level. Comparing synthetic *B. napus* to its diploid parents *B. rapa* and *B. oleracea*, there was gain or loss of only <1% proteins, but numerous (25–38%) polypeptides showed a quantitative non-additive pattern. These changes were not random and mainly triggered by interspecific hybridization (89%), followed by genome doubling (approximately 3%) and selfing (approximately 9%). Both organ-specific and non-organ-specific regulations existed for non-additive proteins (Albertin *et al.*, 2006). Similarly, comparison of synthetic *B. napus* to its diploid progenitors *B. rapa* and *B. oleracea*, revealed that many polypeptides exhibited non-additivity. The functionally related polypeptides were differentially monitored, but the overall topology of protein networks and metabolic pathways possessed conservative characteristics (Albertin *et al.*, 2007). The differential protein regulation between the diploid *Brassica* progenitors and their synthetic allotetraploid derivatives was mainly governed by post-transcriptional modifications, perhaps regulated by a small RNA pathway (Marmagne *et al.*, 2010). In allopolyploidization, gene expression exhibits a certain variation, particularly the emergence of novel transcripts. They do not follow Mendel's law and cause the difficulty of predicting or manipulating them. Nowadays, these studies are mainly involved with new-type *B. napus per se*, and less with the intersubgenomic hybrids. Revealing the regularity of gene expression of new-type *B. napus* is a basis of dissecting intersubgenomic heterosis. However, similar work is also needed to focus on the comparison between intersubgenomic hybrids and their parents, new-type *B. napus* and natural *B. napus*.

Epigenetic Modifications

New-type *B. napus* also exhibits different epigenetic regulation in contrast to the progenitor species. Nucleolar dominance was reported. Nucleolar dominance refers to an epigenetic phenomenon that leads to the generation of nucleoli around rRNA genes in the offspring of an interspecific hybrid being only inherited from one parent. Nucleolar dominance was not affected with ploidy, maternal effect and rRNA gene dosage (Chen and Pikaard, 1997). DNA methylation is an important regulatory mechanism of gene expression that regulates development and differentiation of organisms. For example, in S_1 isogenic resynthesized *B. napus*, 48% of CpG methylation status showed changes in S_1

synthesized *B. napus* lines in contrast to parental species. The frequency of *de novo* methylation was higher than that of *de novo* demethylation. Very little genetic change occurred in the S_0 resynthesized *B. napus* while the changes of DNA methylation occurred extensively (Lukens *et al.*, 2006). However, there are no more reports regarding the methylation of new-type *B. napus* change in intersubgenomic hybrids in contrast to two parents, but Xiong *et al.* (2013) proposed that the alternation in methylation of 5'-CCGG sites was not simply correlated with heterosis in both the interspecific and intraspecific hybridizations from crosses between *B. rapa* and *B. napus*. The function of other types of epigenetic regulations, such as small RNAs, siRNA and chromatin remodelling, in polyploidization and intersubgenomic heterosis, have not been reported. Interestingly, yield potential can be enhanced through epigenetic selection during breeding. The lines with different physiological and agronomical characteristics could be produced from an isogenic *B. napus* population by respiration intensity test. The genetics of these lines were identical, but their epigeneticals were different. Artificial selection can enhance the yield potential (up to 5%) by selecting lines with specific epigenomic status (Hauben *et al.*, 2009). However, it is not known whether this is true in intersubgenomic heterosis, but at least this study provides one novel alternative to improve intersubgenomic heterosis.

Future Research Directions

Genome-specific markers

Development of genome-specific markers was found helpful to improve the screening efficiency of desirable genotypes in the progeny of interspecific hybridization between different ploidy species. Genome-specific probes derived from direct amplification of mini-satellite DNA by PCR were developed in *B. rapa* (A genome), *B. nigra* (B genome), *Sinapis alba* and several other crucifer species (Somers and Demmon, 2002). For example, pBr17.1.3A, a *B. rapa*-specific probe, had a mini-satellite region consisting of three tandem repeats with each repeat comprised of 2–5 sub-repeats and each sub-repeat contained a highly conserved 29 bp motif. This mini-satellite sequence was also present in the A genome of *B. napus* and *B. juncea*. The development of these probes provides a basis for detecting exotic DNA fragments in the offspring from intergeneric or interspecific hybridizations (Somers and Demmon, 2002). The simple and PCR-based marker system with an aim to detect genome-specific loci is very useful for researchers. *B. oleracea* and *B. napus* genomes are not released, and therefore the comparison between A and C in different species cannot be conducted. Genome-specific markers can be developed only after complete genome sequences of crop *Brassica* species are released.

Homologous recombination

Homologous recombination is an important factor for generating chromosome rearrangements in synthetic *B. napus* and homologous pairing/recombination is closely related to genome stability and change of genomic sequence and gene expression. Efficient recombination in wide hybridization will enforce the introgression of novel alleles from *B. rapa* and *B. oleracea* into *B. napus* and decrease linkage drag (Parkin and Lydiate, 1997). In *B. napus* × *B. carinata* interspecific hybrids, there were 19 A–C, 3 A–B and 10 B–C duplication/deletion events caused by homologous recombination (Mason *et al.*, 2011). The complex exchange between genome structure and allelic composition may affect homologous pairing between A, B and C genomes in *Brassica* interspecific hybrids (Mason *et al.*, 2010). Breeding introgressed *B. napus* with a stable genome requires complete understanding of the causes of genomic instability. This may include the identification of regions of homologous recombination, the detection and function analysis of genes controlling homologous recombination, establishing relationship between homologous recombination and phenotype, and procedures to select for genomic stability. With the continuous inbreeding of new type *B. napus*, the intersubgenomic heterosis gradually declines. Whether it is associated with homologous recombination or not is worth investigation.

Fixed heterosis

Fixed heterosis is derived from the inter-actions between the homologous genes on the different genomes in an allopolyploid, which is the same as the traditional heterosis concept: the interaction between homologous genes in a diploid heterozygous plant. This possibly allowed allopolyploids to gain prominence in the plant kingdom. The comparative QTL mapping in resynthesized *B. napus* and its parental species *B. rapa* and *B. oleracea* have been used to identify 'fixed heterosis' in re-synthesized *B. napus* (Abel *et al.*, 2005). The average fixed heterosis in resynthesized *B. juncea* was high for biomass yield but low for seed yield, and the genetic diversity signifi-cantly contributed to fixed heterosis in terms of biomass yield. This may perhaps equip *de novo* alloploids with the adaptive advan-tage during evolution (Bansal *et al.*, 2012). In addition, allopolyploidization can gen-erate fixed-heterosis loci, and chromosome rearrangements also lead to produce fixed-heterosis loci. The two alleles with an inter-action mode become two non-allelic loci be-cause of chromosomal translocation. Then, the two non-allelic loci still interact, which can be considered as one fixed-heterosis loci because of their high homology. In new type *B. napus* introgressed by *B. rapa*, six fixed-heterosis loci were identified (Fu *et al.*, 2012). Heterosis fixation is the ultimate objective of heterosis utilization. Identification and even cloning fixed-heterosis loci may be an important re-search direction in polyploidy species.

New phenotype

Compared to the parents, synthetic *B. napus* exhibited increased growth, wider ranges for yield per plant as well as yield-related charac-ters. The values of silique length as well as beak length, number of seeds per silique, seed yield per plant and 1000-seed weight in synthetic *B. napus* were greater than those of *B. rapa* and *B. oleracea* (Malek *et al.*, 2012). However, the resynthesized *B. napus* also exhibited unstable morphological traits and produced many off-type plants. These off-type plants were found to have lower pollen fertility, aberrant chromo-some number, and architecture with small chromosome pieces. The frequency of aber-rant meiosis in off-type plants was remarkably higher than in the normal plants. However, both type of plants were multivalent in varied frequencies. There abnormal division in mei-osis may lead to the unstable morphological characters in resynthesized *B. napus* progeny. These factors, normal phenotype and chromo-some structure and least abnormal meiosis are the determinants of stabilizing resynthe-sized *B. napus* progeny (Fujii and Ohmido, 2011). *De novo* phenotypic variation for flowering time can happen rapidly after poly-ploidization. Several sub-lineages of each lin-eage were advanced. The flowering times of self-crossing sixth generation plants from the same one synthetic *B. napus* hybrid from a cross between *B. rapa* and *B. oleracea* ranged from 39 to 75 days and from 43 to 64 days for the two lineages. Further analysis in self-pollinated and test-cross progenies showed that the variation could be heritable. The new pheno-typic variation may bestow natural poly-ploids with greater adaptability and eco-logical niche selection options (Schranz and Osborn, 2000). Likewise, in a resynthesized *B. napus* lineage derived from a cross between diploid *B. rapa* and *B. oleracea*, there were sig-nificant differences within lineages for each of eight life-history traits measured and for performance in different growth conditions. Compared with *B. rapa* and *B. oleracea*, ap-proximately 30% of the phenotypes of each resynthesized *B. napus* line for each trait in each environment were similar to either of two par-ents, 50% were intermediate and 20% differed from parents (Schranz and Osborn, 2004). Reasons for the origin of these novel pheno-types and their role in intersubgenomic heter-osis merit further investigations.

Conclusions

It is well known that *B. napus* has a narrow genetic base, and utilization of intersubgen-omic heterosis is an important field of re-search to broaden the *B. napus* genetic base and to fully exploit alien germplasm. Broadening parental genetic diversity may produce better hybrid combinations, but the parental genetic diversity cannot be used to predict the level of intersubgenomic heterosis (Girke *et al.*, 2012). Nevertheless, abundant genetic diversity in parents covers abundant gene/alleles, which offers us more choices for selecting superior genotypes. Alien germplasm has some 'wide' characteristics, e.g. low seed yield but high resistance to disease. If the whole alien genome is imported into new-type *B. napus*, these disadvantageous genes are also brought into new-type *B. napus*. This complicates the usage of intersubgenomic heterosis. Selecting superior genotypes from large populations, especially a gene pool consisting of alien germplasm, is quite critical to intersubgenomic heterosis. A large population, higher selection pressure and multiple selections under different developmental stages is necessary to breed elite new-type *B. napus*. Actually, intersubgenomic heterosis also can be used in other species. For example, interspecific hybrids between *Gossypium hirsutum* (AADD) × *G. barbadense* (AADD) exhibited higher value for yield, fibre length, fibre fineness and fibre strength in contrast to non-hybrid varieties (Basbag and Gencer, 2007). However, they also encounter similar problems. Complete utilization of intersubgenomic heterosis requires complete understanding of the genetic basis of this concept. In *Brassica* species, intersubgenomic heterosis refers to one specific type of heterosis derived from a cross between natural polyploidy species (e.g. *B. napus*) and another new-type polyploidy species introgressed by alien species (e.g. new-type *B. napus* introgressed by *B. rapa* or *B. carinata*). Intersubgenomic heterosis, as one novel approach of fully utilizing alien germplasm, exhibits stronger heterosis potential than common intraspecific heterosis, which attracts more and more attention.

Acknowledgements

This work was supported financially by National Natural Science Foundation of China (code: 31260335), Research Fund for the Doctoral Program of Higher Education of China (code: 20123603120002), Educational Commission of Jiangxi Province of China (code: GJJ13268) and Jiangxi Science and Technology Support Program (code: 20132BBF60013).

References

Abel, S., Möllers, C. and Becker, H.C. (2005) Development of synthetic *Brassica napus* lines for the analysis of 'fixed heterosis' in allopolyploid plants. *Euphytica* 146, 157–163.

Albertin, W., Balliau, T., Brabant, P., Chevre, A.M., Eber, F., Malosse, C. and Thiellement, H. (2006) Numerous and rapid nonstochastic modifications of gene products in newly synthesized *Brassica napus* allotetraploids. *Genetics* 173,1101–1113.

Albertin, W., Alix, K., Balliau, T., Brabant, P., Davanture, M., Malosse, C., Valot, B. and Thiellement, H. (2007) Differential regulation of gene products in newly synthesized *Brassica napus* allotetraploids is not related to protein function nor subcellular localization. *BMC Genomics* 8, 56.

Bansal, P., Banga, S. and Banga, S.S. (2012) Heterosis as investigated in terms of polyploidy and genetic diversity using designed *Brassica juncea* amphiploid and its progenitor diploid species. *PLoS One* 7, e29607.

Basbag, S. and Gencer, O. (2007) Investigation of some yield and fibre quality characteristics of interspecific hybrid (*Gossypium hirsutum* L. x *G. barbadense* L.) cotton varieties. *Hereditas* 144, 33–42.

Becker, H.C., Engqvist, G.M. and Karlsson, B. (1995) Comparison of rapeseed cultivars and resynthesized lines based on allozyme and RFLP markers. *Theoretical and Applied Genetics* 91, 62–67.

Chen, X., Li, M., Shi, J., Fu, D., Qian, W., Zou, J., Zhang, C. and Meng, J. (2008) Gene expression profiles associated with intersubgenomic heterosis in *Brassica napus*. *Theoretical and Applied Genetics* 117, 1031–1040.

Chen, Z.J. and Pikaard, C.S. (1997) Transcriptional analysis of nucleolar dominance in polyploid plants: biased expression/silencing of progenitor rRNA genes is developmentally regulated in *Brassica*. *Proceedings of National Academy of Sciences USA* 94, 3442–3447.

Cowling, W.A. (2007) Genetic diversity in Australian canola and implications for crop breeding for changing future environments. *Field Crops Research* 104, 103–111.

Cui, C., Ge, X., Gautam, M., Kang, L. and Li, Z. (2012) Cytoplasmic and genomic effects on meiotic pairing in *Brassica* hybrids and allotetraploids from pair crosses of three cultivated diploids. *Genetics* 191, 725–738.

Fu, D.H., Qian, W., Zou, J. and Meng, J.L. (2012) Genetic dissection of intersubgenomic heterosis in *Brassica napus* carrying genomic components of *B. rapa*. *Euphytica* 184, 151.

Fujii, K. and Ohmido, N. (2011) Stable progeny production of the amphidiploid resynthesized Brassica napus cv. Hanakkori, a newly bred vegetable. *Theoretical and Applied Genetics* 123, 1433–1443.

Gaeta, R.T. and Pires, J.C. (2010) Homoeologous recombination in allopolyploids, the polyploid ratchet. *New Phytologist* 186, 18–28.

Gaeta, R.T., Pires, J.C., Iniguez-Luy, F., Leon, E. and Osborn, T.C. (2007) Genomic changes in resynthesized *Brassica napus* and their effect on gene expression and phenotype. *Plant Cell* 19, 3403–3417.

Gaeta, R.T., Yoo, S.Y., Pires, J.C., Doerge, R.W., Chen, Z.J. and Osborn, T.C. (2009) Analysis of gene expression in resynthesized *Brassica napus* allopolyploids using *Arabidopsis* 70mer oligo microarrays. *PLoS One* 4, e4760.

Girke, A., Schierholt, A. and Becker, H.C. (2012) Extending the rapeseed gene pool with resynthesized *Brassica napus* II: Heterosis. *Theoretical and Applied Genetics* 124, 1017–1026.

Hauben, M., Haesendonckx, B., Standaert, E., Van Der Kelen, K., Azmi, A., Akpo, H., Van Breusegem, F., Guisez, Y., Bots, M., Lambert, B., Laga, B. and De Block, M. (2009) Energy use efficiency is characterized by an epigenetic component that can be directed through artificial selection to increase yield. *Proceedings of the National Academy of Sciences USA* 106, 20109–20114.

Jesske, T., Olberg, B., Schierholt, A. and Becker, H.C. (2013) Resynthesized lines from domesticated and wild *Brassica* taxa and their hybrids with *B. napus* L.: genetic diversity and hybrid yield. *Theoretical and Applied Genetics* 126, 1053–1065.

Jiang, J., Shao, Y., Du, K., Ran, L., Fang, X. and Wang, Y. (2013) Use of digital gene expression to discriminate gene expression differences in early generations of resynthesized *Brassica napus* and its diploid progenitors. *BMC Genomics* 14, 72.

Leflon, M., Eber, F., Letanneur, J.C., Chelysheva, L., Coriton, O., Huteau, V., Ryder, C.D., Barker, G., Jenczewski, E. and Chevre, A.M. (2006) Pairing and recombination at meiosis of *Brassica rapa* (AA) x *Brassica napus* (AACC) hybrids. *Theoretical and Applied Genetics* 113, 1467–1480.

Li, M., Qian, W., Meng, J. and Li, Z. (2004) Construction of novel *Brassica napus* genotypes through chromosomal substitution and elimination using interploid species hybridization. *Chromosome Research* 12, 417–426.

Li, M.T., Li, Z.Y., Zhang, C.Y., Qian, W. and Meng, J.L. (2005) Reproduction and cytogenetic characterization of interspecific hybrids derived from crosses between *Brassica carinata* and *B. rapa*. *Theoretical and Applied Genetics* 110, 1284–1289.

Li, M., Chen, X. and Meng, J. (2006) Intersubgenomic heterosis in rapeseed production with a partial new-typed Brassica napus containing subgenome A^r from *B. rapa* and C^c from *Brassica carinata*. *Crop Science* 46, 234–242.

Li, M., Liu, J., Wang, Y., Yu, L. and Meng, J. (2007) Production of partial new-typed *Brassica napus* by introgression of genomic components from *B. rapa* and *B. carinata*. *Journal of Genetics & Genomics* 34, 460–468.

Li, Q., Mei, J., Zhang, Y., Li, J., Ge, X., Li, Z. and Qian, W. (2013) A large-scale introgression of genomic components of *Brassica rapa* into *B. napus* by the bridge of hexaploid derived from hybridization between *B. napus* and *B. oleracea*. *Theoretical and Applied Genetics* 126(8), 2073–2080.

Liu, F., Ryschka, U., Marthe, F., Klocke, E., Schumann, G. and Zhao, H. (2007) Culture and fusion of pollen protoplasts of *Brassica oleracea* var. Italica with haploid mesophyll protoplasts of *B. rapa* ssp. Pekinensis. *Protoplasma* 231, 89–97.

Liu, H.L. (2000) Genetics and breeding in rapeseed. Chinese Agricultural Universitatis, Beijing, pp 144–177.

Liu, R., Qian, W. and Meng, J. (2002) Association of RFLP markers and biomass heterosis in trigenomic hybrids of oilseed rape (*Brassica napus* x *B. campestris*). *Theoretical and Applied Genetics* 105, 1050–1057.

Lukens, L.N., Pires, J.C., Leon, E., Vogelzang, R., Oslach, L. and Osborn, T. (2006) Patterns of sequence loss and cytosine methylation within a population of newly resynthesized *Brassica napus* allopolyploids. *Plant Physiologist* 140, 336–348.

Malek, M.A., Ismail, M.R., Rafii, M.Y. and Rahman, M. (2012) Synthetic *Brassica napus*: development and studies on morphological characters, yield attributes and yield. *The Scientific World Journal*, 416901.

Marmagne, A., Brabant, P., Thiellement, H. and Alix, K. (2010) Analysis of gene expression in resynthesized *Brassica napus* allotetraploids: transcriptional changes do not explain differential protein regulation. *New Phytologist* 186, 216–227.

Mason, A.S., Huteau, V., Eber, F., Coriton, O., Yan, G., Nelson, M.N., Cowling, W.A. and Chevre, A.M. (2010) Genome structure affects the rate of autosyndesis and allosyndesis in AABC, BBAC and CCAB *Brassica* interspecific hybrids. *Chromosome Research* 18, 655–666.

Mason, A.S., Nelson, M.N., Castello, M.C., Yan, G. and Cowling, W.A. (2011) Genotypic effects on the frequency of homoeologous and homologous recombination in *Brassica napus* x *B. carinata* hybrids. *Theoretical and Applied Genetics* 122, 543–553.

Meng, J., Shi, S., Gan, L., Li, Z. and Qu, X. (1998) The production of yellow-seeded *Brassica napus* (AACC) through crossing interspecific hybrids of *B. campestris* (AA) and *B. carinata* (BBCC) with *B. napus. Euphytica* 103, 329–333.

Navabi, Z.K., Parkin, I.A., Pires, J.C., Xiong, Z., Thiagarajah, M.R., Good, A.G. and Rahman, M.H. (2010) Introgression of B-genome chromosomes in a doubled haploid population of *Brassica napus* x *B. carinata. Genome* 53, 619–629.

Nelson, M.N., Mason, A.S., Castello, M.C., Thomson, L., Yan, G. and Cowling, W.A. (2009) Microspore culture preferentially selects unreduced (2n) gametes from an interspecific hybrid of *Brassica napus* L. x *Brassica carinata* Braun. *Theoretical and Applied Genetics* 119, 497–505.

Parkin, I.A. and Lydiate, D.J. (1997) Conserved patterns of chromosome pairing and recombination in *Brassica napus* crosses. *Genome* 40, 496–504.

Qian, W., Chen, X., Fu, D., Zou, J. and Meng, J. (2005) Intersubgenomic heterosis in seed yield potential observed in a new type of *Brassica napus* introgressed with partial *Brassica rapa* genome. *Theoretical and Applied Genetics* 110, 1187–1194.

Rana, D., van den Boogaart, T., O'Neill, C.M., Hynes, L., Bent, E., Macpherson, L., Park, J.Y., Lim, Y.P. and Bancroft, I. (2004) Conservation of the microstructure of genome segments in *Brassica napus* and its diploid relatives. *Plant Journal* 40, 725–733.

Sarilar, V., Palacios, P.M., Rousselet, A., Ridel, C., Falque, M., Eber, F., Chevre, A.M., Joets, J., Brabant, P. and Alix, K. (2013) Allopolyploidy has a moderate impact on restructuring at three contrasting transposable element insertion sites in resynthesized *Brassica napus* allotetraploids. *New Phytologist* 198, 593–604.

Schranz, M.E. and Osborn, T.C. (2000) Novel flowering time variation in the resynthesized polyploid *Brassica napus. Journal of Heredity* 91, 242–246.

Schranz, M.E. and Osborn, T.C. (2004) *De novo* variation in life-history traits and responses to growth conditions of resynthesized polyploid *Brassica napus* (Brassicaceae). *American Journal of Botany* 91, 174–183.

Seyis, F., Friedt, W. and Luhs, W. (2006) Yield of *Brassica napus* L. hybrids developed using resynthesized rapeseed material sown at different locations. *Field Crops Research* 96, 176–180.

Somers, D.J. and Demmon, G. (2002) Identification of repetitive, genome-specific probes in crucifer oilseed species. *Genome* 45, 485–492.

Song, K., Lu, P., Tang, K. and Osborn, T.C. (1995) Rapid genome change in synthetic polyploids of *Brassica* and its implications for polyploid evolution. *Proceedings of the National Academy of Sciences USA* 92, 7719–7723.

Xiao, Y., Chen, L., Zou, J., Tian, E., Xia, W. and Meng, J. (2010) Development of a population for substantial new type *Brassica napus* diversified at both A/C genomes. *Theoretical and Applied Genetics* 121, 1141–1150.

Xiong, W., Li, X., Fu, D., Mei, J., Li, Q., Lu, G., Qian, L., Fu, Y., Disi, J.O., Li, J. and Qian, W. (2013) DNA methylation alterations at 5'-CCGG sites in the interspecific and intraspecific hybridizations derived from *Brassica rapa* and *B. napus. PLoS One* 8, e65946.

Xiong, Z., Gaeta, R.T. and Pires, J.C. (2011) Homoeologous shuffling and chromosome compensation maintain genome balance in resynthesized allopolyploid *Brassica napus. Proceedings of the National Academy of Sciences USA* 108, 7908–7913.

Xu, Y., Zhong, L., Wu, X., Fang, X. and Wang, J. (2009) Rapid alterations of gene expression and cytosine methylation in newly synthesized *Brassica napus* allopolyploids. *Planta* 229, 471–483.

Xu, Y., Xu, H., Wu, X., Fang, X. and Wang, J. (2012) Genetic changes following hybridization and genome doubling in synthetic *Brassica napus. Biochemistry & Genetics* 50, 616–624.

Zhao, Q., Zou, J., Meng, J., Mei, S. and Wang, J. (2013) Tracing the transcriptomic changes in synthetic trigenomic allohexaploids of *Brassica* using an RNA-Seq approach. *PLoS One* 8, e68883.

Zhou, R., Moshgabadi, N. and Adams, K.L. (2011) Extensive changes to alternative splicing patterns following allopolyploidy in natural and resynthesized polyploids. *Proceedings of the National Academy of Sciences USA* 108, 16122–16127.

Zou, J., Zhu, J., Huang, S., Tian, E., Xiao, Y., Fu, D., Tu, J., Fu, T. and Meng, J. (2010) Broadening the avenue of intersubgenomic heterosis in oilseed *Brassica. Theoretical & Applied Genetics* 120, 283–290.

Zou, J., Fu, D., Gong, H., Qian, W., Xia, W., Pires, J.C., Li, R., Long, Y., Mason, A.S., Yang, T.J., Lim, Y.P., Park, B.S. and Meng, J. (2011) *De novo* genetic variation associated with retrotransposon activation, genomic rearrangements and trait variation in a recombinant inbred line population of *Brassica napus* derived from interspecific hybridization with *Brassica rapa. Plant Journal* 68, 212–224.

4 Induced Mutagenesis and Allele Mining

Sanjay J. Jambhulkar*

*Nuclear Agriculture & Biotechnology Division,
Bhabha Atomic Research Centre, Mumbai, India*

Introduction

Wide variation for morphological traits and adaptability has been observed in *Brassica* species. A range of genetic, biochemical and metabolic variation is required to be generated to exploit beneficial alleles for effective crop breeding. Mutation breeding is an important approach to improve a small number of specific characters and to enhance the spectrum of variability for traits of significant agronomic value in otherwise high yielding and adapted varieties. It has been optimally utilized to enhance the production potential of many crop plants (Chopra, 2005). Substantive improvement of qualitative and quantitative traits has been achieved in oleiferous brassicas (Robbelen, 1990; Bhatia *et al.*, 1999; Jambhulkar, 2007) through induced mutagenesis (Jambhulkar and Shitre, 2007). An overview of the induced variability for morphological, biochemical and yield attributes, their use to develop high-yielding varieties, the molecular mechanism of selected mutations and advances in mutation breeding techniques have been attempted in this chapter.

Induced Mutations for Plant Traits

Morphological traits can be used as phenotypic markers to identify genotypes and maintain genetic purity. Extensive variability for morphological characters has been reported in *Brassica* germplasm. Induced mutagenesis brings desirable changes in a specific character without affecting the rest of the genome, and the mutant could be used in molecular analysis and for the development of high-yielding varieties. Induced mutations for the following characters have been successfully isolated and utilized for genetic improvement of brassica crops.

Plant height

The introduction of dwarfing genes into cereal crops was crucial for the success of the green revolution (Khush, 2001). Semi-dwarf plants possessed short, strong stalks and had a high harvest index and resisted lodging following higher applications of nitrogenous fertilizers. Dwarf and semi-dwarf mutations with high harvest index may also be helpful

*Corresponding author, e-mail: sjj@barc.gov.in

to overcome the losses due to lodging of existing tall varieties in rapeseed-mustard crops. Dwarf mutants compared to their parents have been isolated using chemical and physical mutagens in *Brassica rapa* (syn. *Brassica campestris*) (Tyagi *et al.*, 1983; Chauhan and Kumar, 1986; Rai and Singh, 1993; Javed *et al.*, 2003), *Brassica napus* (Zanewich *et al.*, 1991; Shah *et al.*, 1999) and *Brassica juncea* (Das and Rahman, 1988). Khatri *et al.* (2005) treated *B. juncea* L. cv. S-9 with gamma rays and EMS and isolated three short statured mutants. Mei *et al.* (2006) identified a dwarf mutant 99CDAM with a reduced plant height of about 85 cm in *B. napus*. This mutation was early to flower and had a large number of branches, as well as superior productivity and quality traits. Rood *et al.* (1989) isolated a single-gene (*rosette* (*ros/ros*)) mutant from a rapid cycling line of *B. rapa*. In the recent studies in the Bhabha Atomic Research Centre, Mumbai, gamma rays induced dwarf, early and yellow seed-coat mutations with better yield and component traits isolated from cv. Varuna.

Maturity

Flowering time is an important component for crop adaptation to adverse environmental conditions such as drought and high temperature. Thus, early flowering may have a positive effect on the seed yield of crop plants as well as in various cropping patterns. Early-flowering mutants were obtained in *B. napus* (Shah *et al.*, 1999) and *B. juncea* (Nayar and George, 1969). Rai (1958) isolated an early flowering mutant in *B. juncea* using X-rays. Khatri *et al.* (2005) treated *B. juncea* L. cv. S-9 with gamma rays and ethyl methane sulfonate (EMS) and isolated early mutations. Barve *et al.* (2009) developed 43–45 early mutations in late-maturing parent 'Heera' using EMS. Thurling and Depittayanan (1992) treated *B. napus* parent TB8 with EMS and isolated 20 days and 59 days earlier mutants than the parent.

Flower morphology

Rapeseed-mustard belongs to *Cruciferae* family and generally possesses four yellow-coloured petals placed in the opposite direction. An X-ray induced white flower mutation has been isolated in *B. juncea* (Rai and Jacob, 1956). Yu *et al.* (2004) isolated a novel yellow-white flower mutation in male sterile progenies derived from commercial *B. napus* hybrid 'CO 22'. Bhat *et al.* (2001) isolated male sterile mutants in *B. juncea* using chemical mutagens such as EMS, ethyl nitroso urea (ENU) and ethidium bromide (EBr). Similar mutations were also isolated by Chauhan and Singh (1998) using gamma rays and EMS. An apetalous mutant has desirable physiological and disease avoidance characters over a normal petalled plant. Mutants for the apetalous flower were discovered in *B. rapa* (Singh, 1961a, b; Cours and Williams, 1977; Buzza, 1983) and *B. carinata* (Rana, 1985).

Silique characters

Reduction of angle of silique to the raceme branch, called an appressed pod, would help to cease the growth rate of aphids. X-ray-induced appressed pod mutants were isolated in *B. juncea* var. Rai5 (Rai, 1958) and RL9 (Nayar and George, 1969). Lavania (1979) obtained four silique mutants from *B. juncea* var. T-25 using EMS. A three-valve mutant was isolated in *B. rapa* var. Yellow Sarson using gamma rays (Kamala and Rao, 1984). Bhat *et al.* (2001) treated the seeds of var. Pusa Jaikisan with various chemical mutagens such as EMS, ENU, EBr and isolated tri- and tetralocular siliquae and non-shattering mutations. Singh and Sareen (2004) reported bunching and appressed pod mutants in a combination treatment of gamma rays and EMS to var. RH30.

Yellow seed coat

Thin seed coat, high oil content, high protein and lower fibre content in yellow seeded rapeseed-mustard have advantages over brown/black seed (Stringam *et al.*, 1974; Woods, 1980; Xiao, 1982; Shirzadegan and Robbelen, 1985) and has improved nutritive value of the meal after oil extraction (Simbaya *et al.*, 1995; Slominski *et al.*, 1999). Yellow-seeded genotypes are available in *B. rapa*, *B. juncea*, *B. carinata* and *B. napus* (Rahman, 2001). Induced mutation to isolate the yellow seed coat was initiated at BARC, Mumbai, India. Two yellow seed-coat mutants (*YSM1* and *YSM2*) were isolated from blackish brown seed variety Rai5 using S^{35} radioisotope (Jambhulkar, unpublished). The new yellow-seeded mutant was isolated from the same variety Rai5 by using ^{32}P radioisotope (Nayar, 1968) and named as 'Trombay Mustard 1' (TM1). Verma and Rai (1980) isolated 19 yellow seed-coat mutants from three parent varieties using gamma rays.

Root morphology

Tailoring root morphology could result in the development of drought-tolerant genotypes. A high root biomass mutant was isolated upon treatment of 'Rose Red' variety of turnip (*B. rapa* L.) by gamma rays and EMS and gamma ray induced mutations in parent variety 'Varuna' showed preliminary tolerance to drought, possessing greater root length (Basak and Prasad, 2004).

Insect pests

The full seed yield potential of brassica crops cannot be realized due to susceptible germplasm to important pests such as aphids. Mutation breeding could pave the way to resolve this important problem. Srinivasachar and Malik (1972) isolated a non-waxy aphid-resistant mutant in var. Pusa Sweti and PTW Globe of *B. rapa* using gamma rays, ethylene imine (EI) and hydrazine.

Seed yield and yield attributes

Seed yield is the outcome of several inter-related component traits or yield components. Numbers of primary and secondary branches, pods per plant, pods on main shoot, seeds per pod and seed weight are major yield components in rapeseed-mustard. Mutants with

more branches and siliquae per plant have been isolated (Naz and Islam, 1979; Chauhan and Kumar, 1986; Shah *et al.*, 1990; Javed *et al.*, 2003). Work on induced mutations has helped to isolate mutants for desirable economic traits such as plant height, number of pods per plant, number of seeds per siliqua, seed weight, seed yield and oil content (Chauhan and Kumar, 1986; Rehman *et al.*, 1987; Mahla *et al.*, 1990, 1991; Robbelen, 1990; Shah *et al.*, 1990, 1998, 1999; Rehman, 1996; Javed *et al.*, 2003). Verma and Rai (1980) isolated T6342 mutant, which had 12.8% higher 1000-seed weight than the parental variety of *B. juncea*. Khatri *et al.* (2005) treated *B. juncea* L. cv. S-9 with gamma rays and EMS and isolated bold seed size and higher yielding mutants than the parent. Barve *et al.* (2009) developed high yielding early mutation EH1 in *B. juncea* using EMS. Javed *et al.* (2003) isolated high-yielding mutants in *B. rapa* using gamma rays. Malek *et al.* (2012) developed high-yielding mutants with improved yield attributes in *B. napus*. Hao-Jie *et al.* (2005) documented variability in an M_2 generation for seed yield and its components in gamma ray-induced *B. napus* genotypes. Raza *et al.* (2009) observed that seed yields of two mutants, namely ROO-100/6 and ROO-125/14 of the genotype 'Rainbow', were higher by 34% and 32%, respectively. Westar mutant W97-75/11 had 30% higher seed yield than its parent plant.

Oil content

Increasing oil content in a high-yielding variety is the correct approach to increase oil yield. Khatri *et al.* (2005) treated *B. juncea* L. cv. S-9 with gamma rays and EMS and isolated mutants with 3% more oil content. The gamma ray-induced mutants had 1–5% higher oil content than those of their parent var. Pant Rai 5 (Verma and Rai, 1980).

Fatty acid composition and oil quality

In vegetable oils, oleic acid (C18:1), linoleic acid (C18:2) and α-linolenic acid (C18:3) along with other fatty acids such as palmitic and stearic acids form the major constituents of the fatty acids. Rapeseed-mustard oil is an exception to other vegetable oils as it contains a high proportion (~50%) of erucic acid (C22:1). Tailoring oil crops for improved fatty acid composition has been successfully achieved through mutation breeding (Robbelen, 1990). Brassica crops have a wide range of fatty acid composition as a single mutation can result in desirable oil composition. 'Liho', the first zero erucic acid mutant, was reported by Stefansson *et al.* (1961) in *B. napus*, which opened the era of mutant-assisted quality improvement in oilseed crops. Downey (1964) suggested that erucic acid-free oil was expected in *B. rapa* as it is one of the parents of *B. napus*. He reported zero erucic acid natural mutant's variability in *B. rapa*. Two zero erucic acid natural mutant's variability were found in a Chinese accession of *B. juncea* (Kirk and Oram, 1981) and termed *zem1* and *zem2*. Olsson (1984) also reported natural mutant's variability for low erucic acid in *B. juncea*. A high level of erucic acid also has industrial applications; however, a mutation for high erucic acid has not yet been reported.

Linolenic acid is easily oxidized, and oil cannot be stored for a long period. Robbelen and Rakow (1970) reported the first reduced linolenic acid mutant in *B. napus* in Germany followed by Rakow (1973) and Robbelen and Nitsch (1975). High oleic acid in the oil is considered as nutritionally desirable for human health. Auld *et al.* (1992) and Rucker and Robbelen (1997) isolated high oleic acid mutants in *B. napus*. This initial success of mutant isolation has laid the foundation stone for improvement of oil quality in *Brassica* crops. Mutation for various fatty acids has been listed in Table 4.1.

Double Haploids and Induced Mutations

Mutagenesis can be employed effectively in the haploid system for inducing genetic variation. The benefits of the haploid system includes instant fixation of mutated genotypes, *in vitro* selection and increased selection efficiency. Spikes, buds, anthers, microspores and haploid calli, embryos or protoplasts can

be used for mutagenic treatment. A dwarf mutant NDF-1 was induced from a doubled haploid (DH) line '3529' of *B. napus* L. after treating seeds with chemical inducers and fast neutrons (Wang *et al.*, 2004). Prem *et al.* (2012) isolated reduced height, appressed pod, altered fatty acid composition, higher protein proportion and lower glucosinolate content mutants from *B. juncea* parents after treating the microspore with EMS. Liu *et al.* (2005) isolated a resistant mutant to *Sclerotinia sclerotiorum* through the doubled-haploid (DH) mutation technique using EMS in *B. napus* lines. Induction of mutation using a chemical mutagen such as EMS (Beversdorf and Kott, 1987; Barro *et al.*, 2001), ENU (Swanson *et al.*, 1988, 1989), NaN$_3$ (Polsoni *et al.*, 1988), MNU (Jedrzejaszek *et al.*, 1997) and a physical mutagen such as gamma rays (Beversdorf and Kott, 1987; Swanson *et al.*, 1988; MacDonald *et al.*, 1991), X-rays (MacDonald *et al.*, 1991), and ultraviolet (UV) rays (MacDonald *et al.*, 1991; Jedrzejaszek, *et al.*, 1997) were undertaken in various *Brassica* species. Microspore mutagenesis has been most successful in rapeseed for isolating desirable stable mutants.

Improved oil and meal quality can be achieved through microspore mutagenesis (Kott *et al.*, 1996; Kott, 1998). Auld *et al.* (1992) and Wong and Swanson (1991) reported increased level of oleic acid and reduction of linolenic acid through DH mutations. Similarly, Turner and Facciotti (1990) and Huang (1992) isolated mutants for high oleic acid and decreased level of saturated fatty acids from mutagenized microspore culture. Barro *et al.* (2001) reported high and low levels of erucic acid microspore-derived mutations. More recently, Beaith *et al.* (2005) treated microspore with UV light and isolated mutants for reduced palmitic and stearic acid content in *B. napus*.

Isolation of plant for resistance/tolerance to herbicide is comparatively easy, as single point mutations are known to interfere with the uptake, translocation or assimilation of the herbicide. *In vitro* mutagenesis and selection have helped to isolate mutants for herbicide resistance (Beversdorf and Kott, 1987; Polsoni *et al.*, 1988; Swanson *et al.*, 1989; Ahmad *et al.*, 1991; McDonald *et al.*, 1991;

Table 4.1. Mutations for fatty acids in *Brassica* species (Robbelen, 1990; Bhatia *et al.*, 1999; Jambhulkar, 2007).

Fatty acids	*Brassica* species	Parent with fatty acid %	Mutagen	Fatty acid in mutant (%)	References
Palmitic acid	*B. napus*	Wotan (4.7)	EMS	9.0	Rucker and Robbelen, 1997
	B. napus	Wotan (4.5)	EMS	9.2	Schnurbusch et al., 2000
Oleic acid	*B. napus*	Cascade (64.8)	EMS	80.2	Auld et al., 1992
		Wotan (60.3)	EMS	71.0–80.3	Rucker and Robbelen, 1997
		PN3756/93(66)	EMS	76	Spasibionek, 2006
		Wotan (61.6)	EMS	44.2	Schnurbusch et al., 2000
		Rainbow		39.1–66.3	Raza et al., 2009
	B. carinata	C-101 (9.4)	EMS	23.3 and 15.3	Velasco et al., 1997
		PC5 (11.2)	EMS	28.2	Sheikh et al., 2009
		BC-71-A2 (9.8)	EMS	19.0 and 8.8	Barro et al., 2001
Linoleic acid	*B. napus*	Oro (21.5)	EMS	22.40–37.92	Robbelen and Nitsch, 1975
		PN3756/93(17.8)	EMS	8.5 and 26	Spasibionek, 2006
	B. rapa	R-500 (11.9)	EMS	2.1	Auld et al., 1992
	B. carinata	C-101 (18.3)	EMS	8.3 and 9.1	Velasco et al., 1997
		Wotan (19.8)	EMS	26.9	Schnurbusch et al., 2000
		Rainbow		15.3–41.6	Raza et al., 2009
Linolenic acid	*B. napus*	Cascade (17.6)	EMS	3–5	Auld et al., 1992
		Wotan (9.9)	EMS	3.1 and 6.2	Rucker and Robbelen, 1997
		Oro	EMS	3.24–8.42	Robbelen and Nitsch, 1975
		M57 (5.6)	EMS	3.2	Robbelen and Nitsch, 1975
		Rainbow			Rakow, 1973
	B. rapa	PN3756/93(8.5)	EMS	7.5 and 2.6	Spasibionek, 2006
	B. juncea	Wotan (10.1)	EMS	14.8	Schnurbusch et al., 2000
		Rainbow		18.1–28.9	Raza et al., 2009
Erucic acid	*B. rapa*	R-500 (8.6)	EMS	3.0	Robbelen and Nitsch, 1975
	B. juncea	Zem1 (15.0)	EMS/gamma rays	9.0	Auld et al., 1992
	B. carinata	C-101 (12.9)	EMS	6.3, 6.6 and 9.1	Velasco et al., 1997
	B. carinata	C-101 (44.0)	EMS	5–10	Velasco et al., 1995
		N2-6230	EMS	54.9	Velasco et al., 1998
		PC5 (45.9)	EMS	27.5	Sheikh et al., 2009
		BC-71-A2 (42.7)	EMS	20.6 and 53.2	Barro et al., 2001
	B. napus		Natural mutant/variability	0	Stefansson et al., 1961
	B. rapa		Natural mutant/variability	0	Downey, 1964
	B. juncea		Natural mutant/variability	0	Kirk and Oram, 1981

Kott, 1995, 1998; Palmer *et al.*, 1996), disease resistance (Sacriston, 1982; McDonald and Ingram, 1986; Newsholme *et al.*, 1989; Ahmad *et al.*, 1991; Bansal *et al.*, 1998; Liu *et al.*, 2005) and long pod and short plant (Shi *et al.*, 1995) in *B. napus*.

Mutations in Cross Breeding

The yellow seed-coat mutants and their derivatives were used extensively in the cross-breeding programme across India, and a large number of high-yielding bold- and yellow-seeded genotypes has been developed (Nayar, 1976, 1979; Abraham and Bhatia, 1986). Using mutants in cross-breeding, Barve *et al.* (2009) isolated several selections with low glucosinolate, high erucic acid, high oil content and canola-quality recombinants. In recent years, we have also isolated a yellow seed-coat mutant from the most popular variety 'Varuna' using gamma rays. These mutants have been extensively used for cross-breeding to develop yellow, bold and high-yielding varieties (Jambhulkar *et al.*, 2005; Jambhulkar, 2009, 2012).

High-yielding Mutant Varieties

Utilization of induced mutations for crop improvement has resulted in the development of more than 3242 high-yielding varieties in different crop plants. Of these, 163 varieties belong to oilseed crops. The maximum number of varieties has been released in soybean (58), followed by groundnut (44), sesame (16), linseed (15), castor (4) and sunflower (1). A total of 31 high-yielding varieties in rapeseed-mustard, comprising 12 in *B. juncea*, 14 in *B. napus*, 2 in *B. rapa* and 3 in white mustard, have been released for cultivation using the mutation breeding approach. Fifteen varieties have been developed using gamma rays and four by X-rays. Remaining varieties were developed using chemical mutagens and mutants used in hybridization. In *B. juncea*, Labana (1976) obtained a gamma-ray-induced mutant RLM-198 with 25% higher seed yield, moderately resistant to aphids and leaf miners, higher oil

content and 5–6 days earlier maturity. Labana (1981) reported an induced mutant, RLM 214, with 22% higher seed yield, high oil content and shattering resistant through gamma irradiation of cv.RL-18. Development of high-yielding mutant varieties has also been reported (Rehman, 1996; Shah *et al.*, 1999). Backcrossing of M-11, a low linolenic acid mutant, with 'Regent' resulted in the development of high-yielding low-linolenic acid vars 'Stellar' (Scarth *et al.*, 1988) and 'Apollow' (Scarth *et al.*, 1995). The details of the released mutant varieties are presented in Table 4.2.

Biochemical and Molecular Basis of Mutations

Phenotypic expression of any mutation is due to alteration, such as deletions, duplications and substitutions of base pairs, at DNA (deoxyribonucleic acid) level to bring the changes in biochemical products. Molecular mechanism of mutation can be revealed through characterization at protein and DNA levels. The molecular basis of two selected characters, fatty acids and dwarfs, has been mentioned below.

Fatty acids

Low erucic acid (LEA) is characterized by the absence of very long chain fatty acids (VLCFA) in the seed oil, which has been correlated with lack of acyl-CoA elongation activity (Roscoe *et al.*, 2001). To validate the speculation, Katavic *et al.* (2002) used site-directed mutagenesis and observed the substitution of phenylalanine 282 residue with a serine residue in the *FAE1* polypeptide from LEA *B. napus* cv. Westar. This mutated gene was expressed in yeast and led to synthesis of very long chain monounsaturated fatty acids (VLCMFAs). This suggested restoration of elongase activity in the LEA *FAE1* enzyme by the substitution of a single amino acid, serine. Thus, phenylalanine prevented the biosynthesis of eicosenoic and erucic acids in low erucic acid *B. napus*. Gene *fad2* codes an enzyme, endoplasmic delta 12 oleate desaturase, responsible for desaturation of oleic acid (C18:1) to form linoleic acid (C18:2). Single nucleotide substitution of leucine with proline in

Table 4.2. Released varieties in rapeseed-mustard using mutation breeding (http://www-mvgs.iaea.org/AboutMutantVarieties.aspx; Kumar *et al.*, 2000; Jambhulkar, 2007).

Variety	Country and year of release	Mutagen/mutant in cross	Improved characters
Brassica juncea (Indian mustard)			
Shambal (BAU-/248)	Bangladesh, 1984	EMS 0.64%	Early, seed yield
Agrani	Bangladesh, 1991	Gamma rays 700Gy	Early, blight resistance
Safal	Bangladesh, 1991	Gamma rays 700Gy	Seed yield, oil content, blight resistance
RLM 198	India, 1975	Gamma rays	Seed yield, oil content, moderately tolerant to aphid and leaf miners
RLM 514	India, 1980	Gamma rays	Early, bold seed, seed yield, oil content, moderately resistant to aphid and shattering
RLM 619	India, 1983	Gamma rays	Seed yield, oil content, bold seed, moderately tolerant to aphid, downy mildew and white rust
RLM 185	India, 1983	Gamma rays	Seed yield, oil content, tolerant to drought
RL 1359	India, 1987	Gamma rays-induced mutant RLM514 used in cross (RLM514 × Varuna)	Seed yield, bold seed, comparatively dwarf, tolerant to aphids
TM 2	India, 1978	X-rays 750 Gy	Seed yield, appressed pod
TM 4	India, 1987	^{32}P-induced yellow seed coat mutant TM1 used in cross (Varuna × TM1)	Seed yield, oil content, yellow seed, less susceptible to powdery mildew
NIFA-Mustard canola	Pakistan, 2003	Gamma rays 14 Gy	Seed yield
TPM1	India, 2006	^{32}P	Seed yield, early, oil content, tolerant to powdery mildew, reduced erucic acid
Brassica sp. (white mustard)			
Svalof's Primex	Sweden, 1950	X-rays 350 Gy	Seed yield, oil content
Seco	Sweden, 1961	X-ray-induced mutant Primex used in cross (Primex × Rumanian white mustard)	Seed yield, early, stalk stiffness, resistant to shattering
Trico	Sweden, 1967	Selection from Primex	Seed yield, oil content
Brassica napus			
Binasharisha-3	Bangladesh, 1997	Gamma-rays 800 Gy	Seed and oil yields, early
Binasharisha-4	Bangladesh, 1997	Gamma-rays 700 Gy	Seed and oil yield, early
Stellar	Canada, 1987	EMS-induced mutant Mil used in cross (Mil × Regent) × Regent	Seed yield, low linolenic and erucic acids, low glucosinolates
Apollo	Canada, 1993	–	Seed yield, low linolenic acid
Ganyou No.5	China, 1984	Gamma-rays 1400 Gy	Seed yield, short plant height, early, long silique, resistant to white rust and *Sclerotinia*
Huyou No.4	China, 1970	Gamma-rays 600 Gy	Seed yield, cold tolerant

Continued

Table 4.2. Continued.

Variety	Country and year of release	Mutagen/mutant in cross	Improved characters
Xinyou No.1	China, 1979	Gamma-rays 700 Gy	Seed yield, tolerant to cold, drought, saline and alkaline soils
Xiuyou No.1	China, 1979	Gamma-rays 80 Gy, F_1 seeds	Seed yield, early, cold resistant
Hua-Yellow No.1	China, 1990	Gamma-rays	Seed yield, seed viability
Abasin 95	Pakistan, 1996	Gamma-rays 1400 Gy	Seed yield, early
Regina varraps elite A	Sweden, 1953	X-rays 350 Gy	Seed yield, oil content
Regina varraps elite F	Sweden, 1962	X-rays 450 Gy	Seed yield, oil content
Tismenitskii	USSR, 1989	DMS 0.006%	Seed yield, oil content
Ivana	USSR, 1990	MNH 0.0025%	Seed yield, oil content, insect resistance
Brassica rapa			
Haya-natane	Japan, 1961	Colchicine	Seed yield
Hanakkori	Japan, 1999	Colchicine	Seed yield, plant architecture

the *fad2* gene of *B. rapa* precipitated an increased C18:1 content (Tanhuanpaa *et al.*, 1998). Substitution for single nucleotide from C to T in the gene *fad2* responsible for generating stop codon (TAG) and causing premature termination of the peptide chain during translation was identified by Hu *et al.* (2006) in *B. napus*. Consequently, only 185 amino acids could be incorporated into the polypeptide instead of all 384 amino acids representing the full length polypeptide. Thus, the truncated polypeptide is an inactive desaturase for the desaturation of C18:1 to C18:2 and therefore will result in the accumulation of C18:1 in the seeds of the mutant line.

Dwarfs

Dwarfing phenotype in the plants is due to genetic alterations in the biosynthetic pathways of gibberellic acid (GA). Application of exogenous GA revealed the mutations could be either sensitive (Rood *et al.*, 1989) or insensitive (Zanewich *et al.*, 1991; Li *et al.*, 2011). Degradation of DELLA protein in the ubiquitine proteosome pathway is responsible for the promotion of stem growth by GA. Muangprom and Osborn (2004) found that a *dwf2* mutation in *B. rapa* was insensitive to exogenous GA_3 for both plant height and flowering time

and inferred that it is not a mutation in the gibberellin biosynthetic pathway. Muangprom *et al.* (2005) reported substitution of conserved amino acid in the C-terminal domain of DELLA protein in a mutant of *B. rapa* (Brrga1-d). Brrga1-d retained its repressor function in the presence of GA but failed to interact with a protein constituent required for degradation, suggesting that the mutated amino acid caused dwarfism by stopping an interaction needed for degradation. Li *et al.* (2011) observed three mutated bases in the pyrimidine box (P-box) of the BnGID1 promoter, which was associated with the dwarf plant type.

Brassica TILLING

Targeting Induced Local Lesions in Genomes (TILLING) is a reverse genetic strategy for identification of an allelic series of induced mutations in genes of interest, and is a rapid and inexpensive detection of induced point mutations. EcoTILLING (TILLING for germplasm available in nature) is also ideal for examining natural variation. Endonuclease CEL I cut (cut into two pieces or insert gap between two bases of DNA) identifies the multiple mismatches in a DNA heteroduplex of unknown sequence with that of a known sequence and thus the positions of polymorphic sites identifying nucleotide changes and small insertions/deletions. Therefore, TILLING and EcoTILLING are better strategies from functional genomic study to practical crop breeding. This technique was first used in the *Arabidopsis* TILLING Project (ATP) during 2001. Over 1000 mutations in more than 100 genes have been detected, sequenced and delivered in the ATP project (Till *et al.*, 2003).

Wang *et al.* (2008) obtained the first results on TILLING in rapeseed. The screening of 1344 M_2 plants led to the identification of 19 mutants of the *BnFAE1* gene family, which were phenotypically verified by M_3 analysis. Among them, three were functionally established for reduced seed erucic acid content. *B. rapa* with a genome size of 625 Mbp was used as a first EMS TILLING source in the diploid brassicas. The mutation frequency observed in this population was ~1 per 60 kb, making it a most densely mutated diploid (Stephenson *et al.*, 2010). Harloff *et al.* (2012) developed two mutant populations of spring genotype YN01-429 and the winter'type cv. Express 617 of oilseed rape (*B. napus* L.) using EMS as a mutagen. Nearly 5361 and 3488 M_2 plants, respectively, were screened for mutations in the sinapin biosynthesis pathway. Screening helped to identify 229 and 341 mutations within the BnaX.SGT sequences (135 missense and 13 non-sense mutations) and the BnaX.REF1 sequences (162 mis-sense, 3 non-sense, 8 splice site mutations), respectively. The frequencies of mis-sense and non-sense mutations paralleled the frequencies of the target codons. Mutation frequencies in the 'Express 617' population ranged from 1/12 to 1/22 kb and YN01-429 population from 1/27 to 1/60 kb.

Gene/allele mining

Extensive morphological, genetic and genomic diversity has been reported in *Brassica* species. Genomic information has become a handy tool for targeting specific components in future crop improvement programmes. Genomic information is critical for exploiting knowledge gained through gene expression, biochemistry, metabolism and physiology. Finer details between trait and gene are essential for effective crop-based studies and to differentiate locus-specific copies. Molecular marker-based genetic maps are available for all *Brassica* species. Considerable efforts have been made towards developing high-density maps based on DNA sequence-tagged genetic markers (SNPs, SSRs, CAPS, etc.). Navigation amongst different genetic maps, attached with common sequence-tagged markers, helps in the development of genomics tools. An increasing amount of information has become available on sequence-tagged markers that provide links between *Brassica* genetic maps and the *Arabidopsis* genes and loci. QTLs have been identified for a range of morphological, physiological and crop traits in the different brassica crops. Resolution of QTLs depends upon access to a large number of recombinant individuals (segregating populations, substitution or near-isogenic lines), as well as to high-density genetic and physical maps. New high-throughput technologies deliver a wealth of genomic data. The increased complexity and abundance of genomic and physiological data has made it difficult for researchers to comprehend the complete information. Despite the progress made in the genetic understanding of plant traits, limitations remain towards characterizing the behaviour of the underlying genes, genomic regulatory networks and associated metabolism. Harnessing genetic diversity through an information-led approach for crop improvement is still a major challenge. To further improve accessibility, numerous computational tools and databases are now available (Edwards and Batley, 2004). However, diverse data structures and formats limit seamless querying across data types. To overcome some of these limitations, Erwin *et al.* (2007) have developed the BASC bioinformatics system for flow through integration of gene and genome DNA sequences, gene expression, molecular genetic marker, phenotypic trait and population data.

Whole genome sequences are not available for *B. rapa* (Wang *et al.*, 2011), *B. oleracea* (Liu *et al.*, 2014) and *B. napus* (Chalhoub *et al.*, 2014). Based on the needs of the *Brassica* researchers and the bulk genomic data, BRAD (Brassica Database) has been constructed (Cheng *et al.*, 2011). It was developed as a repository for genome-scale genetic and genomic data and related resources of brassica crops. BRAD was designed to act as an initial access point for other related web pages and specialized datasets.

Conclusions

Mutation breeding has been successfully employed to enhance the production and productivity of oleiferous *Brassica* species through genetic modification of quantitative and qualitative traits using physical and chemical mutagens. Mutations for dwarf, early maturity, flower morphology, siliqua characters, yellow seed-coat, root morphology and insect resistance have been isolated in cultivated *Brassica* species. Direct mutations for seed yield through yield components have also been isolated. Mutations for reduced erucic acid with increased oleic and linoleic acids and decreased linolenic acid have been developed. The haploid system can successfully be used for induction of genetic variation through mutagenesis. A wide spectrum of variability for yield-related traits and fatty acid composition has been isolated through mutagenesis. High-yielding mutations can be released directly as a variety whereas mutations for specific character can be introgressed into a high-yielding background. A total of 31 high-yielding varieties, comprising 12 in *B. juncea*, 14 in *B. napus*, 2 in *B. rapa* and 3 in white mustard, have been released for cultivation. Fifteen varieties have been developed using gamma rays and four by X-rays. The remaining varieties were developed using chemical mutagen and mutants used in hybridization. Mutations for dwarf and oleic acid gene are due to base substitution. The Brassica TILLING project has opened up the avenue to the isolation of a large number of alleles related to a specific gene and thus opened up a new arena for brassica crop improvement. Genomic information is critical for exploiting knowledge gained regarding gene expression, biochemistry, metabolism and physiology. For effective crop-based studies, it is essential to navigate across trait and gene(s), and to distinguish different locus-specific copies. Both the BASC and the BRAD bioinformatics system will be helpful for harnessing genetic diversity through an information-led approach for the brassica crop improvement programme.

References

Abraham, V. and Bhatia, C.R. (1986) Development of strains with yellow seed coat in Indian mustard [*Brassica juncea* (L.) Czern. & Coss.]. *Plant Breeding* 97, 86–88.

Ahmad, I., Day, J.P., McDonald, M.V. and Ingram, D.S. (1991) Haploid culture and UV mutagenesis in rapid-cycling *Brassica napus* for the generation of resistance to chlorsulfuron and *Alternaria brassicicola*. *Annals of Botany* 67, 521–525.

Auld, D.L., Heikkinen, M.K., Erickson, D.A., Sernyk, J.L. and Romero, J.E. (1992) Rapeseed mutants with reduced level of polyunsaturated fatty acids and increased levels of oleic acid. *Crop Science* 32, 657–662.

Bansal, V.K., Thiagarajah, M.R., Stringam, G.R. and Hardin, R.T. (1998) Haploid plantlet screening in the development of blackleg resistant DH lines of *Brassica napus*. *Plant Breeding* 117, 103–106.

Barro, F., Fernandez-Escobar, J., De La Vega, M. and Martin, A. (2001) Double haploid lines of *Brassica carinata* with modified erucic acid content through mutagenesis by EMS treatment of isolated microspores. *Plant Breeding* 120, 262–264.

Barve, Y.Y., Gupta, R.K., Bhadauria, S.S., Thakre, R.P. and Pawar, S.E. (2009) Induced mutations for development of *B. juncea* canola quality varieties suitable for Indian agro-climatic conditions. In: Shu, Q.Y. (ed.) *Induced Plant Mutations in the Genomics Era*. Food and Agriculture Organization of the United Nations, Rome, pp. 373–375.

Basak, S. and Prasad, C. (2004) Screening for root mutant in turnip (*Brassica rapa* L.). *Cruciferae Newsletter* 25, 9–10.

Beaith, M.E., Fletcher, R.S. and Kott, L.S. (2005) Reduction of saturated fats by mutagenesis and heat selection in *Brassica napus* L. *Euphytica* 144, 1–9.

Beversdorf, W.D. and Kott, L.S. (1987) An *in vitro* mutagenesis/selection system for *Brassica napus*. *Iowa State Journal of Research* 64, 435–443.

Bhat, S.R., Haque, A. and Chopra, V.L. (2001) Induction of mutations for cytoplasmic male sterility and some rare phenotypes in Indian mustard (*Brassica juncea* L.). *Indian Journal of Genetics* 61, 335–340.

Bhatia, C.R., Nichterlein, K. and Maluszynski, M. (1999) Oilseed cultivars development from induced mutations and mutation altering fatty acid composition. *Mutation Breeding Review* 11, 1–36.

Buzza, G.C. (1983) The inheritance of an apetalous characters in canola (*B. napus* L.). *Cruciferae Newsletter* 8, 11–12.

Chalhoub, B., Denoeud, F., Shengyi, L., Parkin, I. *et al.* (2014) Early allopolyploid evolution in the post-Neolithic *Brassica napus* oilseed genome. *Science* 345, 950–953.

Chauhan, S.V. and Singh, N.K. (1998) Gamma ray and EMS induced male sterility mutants in *Brassica juncea*. *Journal of Cytology and Genetics* 33, 29–34.

Chauhan, Y.S. and Kumar, K. (1986) Gamma ray induced chocolate B seeded mutant in *Brassica campestris*. *Current Science* 55, 410.

Cheng, F., Liu, S., Wu, J., Fang, L., Sun, S., Liu, B, Li, P., Hua, W. and Wang, X. (2011) BRAD, the genetics and genomics database for *Brassica* plants. *BMC Plant Biology* 11, 136–141.

Chopra, V.L. (2005) Mutagenesis: investigating the process and processing the outcome for crop improvement. *Current Science* 89, 353–359.

Cours, B.J. and Williams, P.H. (1977) Genetic studies in *Brassica campestris* L. *Plant Breeding Abstracts* 51, 1533.

Das, M.L. and Rahman, A. (1988) Induced mutagenesis for the development of high yielding varieties in mustard. *Journal of Nuclear Agricultural Biology* 17, 1–4.

Downey, R.K. (1964) Selection of *Brassica campestris* L. containing no erucic acid in its seed oil. *Canadian Journal of Plant Science* 44, 295.

Edwards, D. and Batley, J. (2004) Plant bioinformatics: from genome to phenome. *Trends in Biotechnology* 22, 232–237.

Erwin, T.A., Jewell, E.G., Love, C.G., Lim, G.A.C., Li, X., Chapman, R., Batley, J., Stajich, J.E., Mongin, E., Stupka, E., Ross, B., Spangenberg, G. and Edwards, D. (2007) BASC: an integrated bioinformatics system for *Brassica* research. *Nucleic Acids Research* 35, D870–D873.

Hao-Jie, L.I., Xiao-Bin, P.U., Jin-Fang, Z., Qi-Xing, Z. and Liang-Cai, J. (2005) Primary study on radiation induced mutation in *Brassica napus*. *Southwest China Journal of Agricultural Sciences* 18, 542–546.

Harloff, H.J., Lemcke, S., Mittasch, J., Frolov, A., Guo, J.W., Dreyer, F., Leckband, G. and Jung, C. (2012) A mutation screening platform for rapeseed (*Brassica napus* L.) and the detection of sinapine biosynthesis mutants. *Theoretical and Applied Genetics* 124, 957–969.

Hu, X., Sullivan-Gilbert, M., Gupta, M. and Thompson, S.A. (2006) Mapping of loci controlling oleic and linolenic acid contents and development of *fad2* and *fad3* allele-specific marker in canola. *Theoretical and Applied Genetics* 113, 497–501.

Huang, B. (1992) Genetic manipulation of microspores and microspore derived embryos. *In Vitro Cell Division and Biology* 28, 53–58.

Jambhulkar, S.J. (2007) Mutagenesis: generation and evaluation of induced mutations. In: Gupta, P.C. (ed.) *Advances in Botanical Research (Rapeseed Breeding)*, Vol. 45. Elsevier, New York, pp. 418–428.

Jambhulkar, S.J. (2009) Mutation induction, evaluation and utilization for development of high yielding varieties in Indian mustard and sunflower: an overview of BARC work. In: D'Souza, S.F. and Sharma, A. (eds) *Proceedings of International Conference on Peaceful uses of Nuclear Energy*, New Delhi, pp. 49–54.

Jambhulkar, S.J. (2012) Induced mutagenesis for genetic improvement of Indian mustard. In: *1st National Brassica Conference on Production Barriers and Technological options in Oilseed Brassica*. March 2–3, 2012, CCSHAU Hisar, Haryana, India, pp. 22–23.

Jambhulkar, S.J. and Shitre, A.S. (2007) Genetic improvement of rapeseed mustard through induced mutations. *IANCAS Bulletin VI*, 327–329.

Jambhulkar, S.J., Shitre, A.S. and Pathak, B. (2005) Heterosis in Indian mustard using induced mutations. In: *Conference on resource development and marketing issues in rapeseed mustard*, 28–29 March 2005, Jaipur, India, pp. 27–28.

Javed, M.A., Siddiqui, M.A., Khan, M.K.R., Khatri, A., Khan, I.A., Dahar, N.A., Khanzada, M.H. and Khan, R. (2003) Development of high yielding mutants of *Brassica campestris* var. Toria selection through gamma rays irradiation. *Asian Journal of Plant Science* 2, 192–195.

Jedrzejaszek, K., Kruczkowska, H., Pawlowska, H. and Skucinska, B. (1997) Stimulating effect of mutagens on *in vitro* plant regeneration. *Mutagen Breeding Newsletter* 43, 10–11.

Kamala, T. and Rao, R.N. (1984) Gamma ray induced 3-valved mutant in Yellow Sarson. *Journal of Nuclear Agriculture and Biology* 13, 28.

Katavic, V., Mietkiewska, E., Borton, D.L., Giblin, E.M., Reed, D.W. and Taylor, D.C. (2002) Restoring enzyme activity in nonfunctional low erucic acid *Brassica napus* fatty acid elongase 1 by single amino acid substitution. *European Journal of Biochemistry* 269, 2652–2631.

Khatri, A., Khan, I.A., Siddiqui, M.A., Raza, S. and Nizamani, G.S. (2005) Evaluation of high yielding mutants of *Brassica juncea* cv. S-9 developed through gamma rays and EMS. *Pakistan Journal of Botany* 37, 279–284.

Khush, G.S. (2001) Green revolution: the way forward. *National Review of Genetics* 2, 815–822.

Kirk, J.T.O. and Oram, R.N. (1981) Isolation of erucic acid free lines of *Brassica juncea*: Indian mustard now a potential oilseed crop in Australia. *Journal of Australian Institutional Agricultural Science* 47, 51–52.

Kott, L. (1995) Production of mutants using the rapeseed double haploid system. In: Gichner, T. (ed.) *Induced Mutations and Molecular Techniques for Crop Improvement*. International Atomic Energy Agency, Vienna, pp. 505–515.

Kott, L. (1998) Application of double haploid technology in breeding of oilseed *Brassica napus*. *Agricultural Biotechnology News and Information* 10, 69N–74N.

Kott, L., Wong, R., Swanson, E. and Chen, J. (1996) Mutation and selection for improved oil and meal quality in *Brassica napus* utilizing microspore culture. In: Jain, S.M, Sopory, S.K. and Veilleux, R.E. (eds) In-vitro *Haploid Production in Higher Plants, Vol. 2*. Springer, New York, pp. 151–167.

Kumar, P.R., Chauhan, J.S., Singh, A.K. and Yadav, S.K. (2000) *Rapeseed-Mustard Varieties of India*. National Research Centre on Rapeseed-Mustard, Bharatpur, India, 115 pp.

Labana, K.S. (1976) Release of mutant variety of raya (*Brassica juncea*). *Mutation Breeding Newsletter* 7, 11.

Labana, K.S. (1981) Release of mutant variety in raya (*Brassica juncea*). *Mutation Breeding Newsletter* 17, 13.

Lavania, U.C. (1979) Interphase sensitivity and specificity of siliqua mutations induced by EMS in Indian mustard. *Current Science* 48, 101.

Li, H., Wang, Y., Li, X., Gao, Y., Wang, Z., Zhao, Y. and Wang, M. (2011) A GA-insensitive dwarf mutant of *Brassica napus* L. correlated with mutation in pyrimidine box in the promoter of GID1. *Molecular Biology Reporter* 38, 191–197.

Liu, S.,Y., Liu, Yang, X.,Tong, C., D., Edwards, Parkin, I. A.P., M., Zhao, Ma, J., J., Yu, Huang, S., Wang, X., J., Wang, Lu, K., Z., Fang, Bancroft, I., Yang, T.J., Hu, Q., Wang, X., Yue, Z., Li, H., Yang, L., Wu, J., Zhou, Q., Wang, W., King, G.J., Pires, J.C., Lu, C., Wu, Z., Sampath, P., Wang, Z., Guo, H., Pan, S., Yang, L., Min, J., Zhang, D., Jin, D., W., Li, Belcram, H., Tu, J., Guan, M., Qi, C., Du, D., Li, J., Jiang, L., Batley, J., Sharpe, A.G., Park, B.S., Ruperao, P., Cheng, F., Waminal, N.E., Huang, Y., Dong, C., Wang, L., Li, J., Hu, Z., Zhuang, M., Huang, Y., Huang, J., Shi, J., Mei D., Liu, J., Lee, T.H., Wang, J., Jin, H., Li, Z., Li, X., Zhang, J., Xiao, L., Zhou, Y., Liu, Z., Liu, X., Qin, R., Tang, X., Liu, W., Wang, Y., Zhang, Y., Lee, J., Kim, H.H., Denoeud, F., Xu, X., Liang, X., Hua, W., Wang, X., Wang, J., Chalhoub, B. and Paterson, A.H. (2014) The *Brassica oleracea* genome reveals the asymmetrical evolution of polyploid genomes. *Nature Communications* 5, 3930. DOI: 10.1038/ncomms4930.

Liu, S., Wang, H., Zhang, J., Fitt, B.D.L., Xu, Z., Evans, N., Liu, Y., Yang, W. and Guo, X. (2005) *In vitro* mutation and selection of double haploid *Brassica napus* lines with improved resistance to *Sclerotinia sclerotiorum*. *Plant Cell Reporter* 24, 133–144.

MacDonald, M.V., Ahmad, I., Menten, J.O.M. and Ingram, D.S. (1991) Haploid culture and *in-vitro* mutagenesis (UV light, X-rays and gamma rays) of rapid cycling *Brassica napus* for improved resistance to diseases. In: *Plant Mutations Breeding for Crop Improvement*, Vol. 2. International Atomic Energy Agency, Vienna, pp. 129–138.

Mahla, S.V.S., Mor, B.R. and Yadava, J.S. (1990) Effect of mutagen on yield and its component characters in mustard. *Haryana Agricultural University Journal of Research* 20, 259–264.

Mahla, S.V.S., Mor, B.R. and Yadava, J.S. (1991) Mutagen induced polygenic variability in some mustard (*Brassica juncea* L.) varieties and their hybrids. *Journal of Oilseed Research* 8, 173–177.

Malek, M.A., Ismail, M.R., Monshi, F.I., Mondal, M.M.A. and Alam, M.N. (2012) Selection of promising rapeseed mutants through multi-location trials. *Bangladesh Journal of Botany* 41, 111–114.

McDonald, M.V. and Ingram, D.S. (1986) Towards the selection *in vitro* for resistance to *Alternaria brassicicola* (Schw) Welts., in *Brassica napus* ssp. Oleifera (Metzg.) Sinsk., winter oilseed rape. *New Phytologist* 104, 621–629.

McDonald, M.V., Ahmad, I., Menten, J.O.M. and Ingram, D.S. (1991) Haploid culture and *in-vitro* mutagenesis (UV light, X-rays and gamma rays) of rapid cycling *Brassica napus* for improved resistance to diseases. In: *Plant Mutations Breeding for Crop Improvement*, Vol. 2. International Atomic Energy Agency, Vienna, pp. 129–138.

Mei, D.S., Wang, H.Z., Li, Y.C., Hu, Q., Li, Y.D. and Xu, Y.S. (2006) The discovery and genetic analysis of dwarf mutation 99CDAM in *Brassica napus* L. *Yi Chuan* 28, 851–857.

Muangprom, A. and Osborn, T.C. (2004) Characterisation of dwarf gene in *Brassica rapa*, including the identification of a candidate gene. *Theoretical and Applied Genetics* 108, 1378–1384.

Muangprom, A., Thomas, S.G., Sun, T.P. and Osborn, T.C. (2005) A novel dwarfing mutation in a green revolution gene from *Brassica rapa*. *Plant Physiology* 137, 931–938.

Nayar, G.G. (1968) Seed colour mutation in *Brassica juncea* Hook. F. and Thomas induced by radioactive phosphorus ^{32}P. *Science and Culture* 34, 421–422.

Nayar, G.G. (1976) Studies on radiation induced mutants and their recombinant types in mustard. *Oilseed Journal* (January–June), 46–47.

Nayar, G.G. (1979) Breeding strategy for the improvement of mustard (*Brassica juncea*). In: *Proceedings of the Symposium on the role of induced mutation in crop improvement*. Osmania University, Hyderabad, 10–13 September, 258–270.

Nayar, G.G. and George, K.P. (1969) X-ray induced early flowering, appressed pod mutant in *Brassica juncea*. In: *Proceedings of the Symposium on Radiation and Radiomimetic Substances in Mutation Breeding*. Bombay, 26–29 September, pp. 409–413.

Naz, R.M.A. and Islam, R.Z. (1979) Effect of irradiation of *Brassica trilocularis* (Yellow Sarson). *Pakistan Journal of Agricultural Research* 17, 87–93.

Newsholme, D.M., MacDonald, M.V., and Ingram, D.S. (1989) Studies of selection *in-vitro* for novel resistance to phytotoxic products of *Leptospheria maculans* (Desm.) Ces. & De Not in secondary embryogenic lines of *Brassica napus* ssp. Oleifera (Metzg.) Sinsk., winter oilseed rape. *New Phytologist* 113, 117–126.

Olsson, G. (1984) Selection for low erucic acid in *Brassica juncea*. *Sveriges Utsadesf. Tidskr.* 94, 187.

Palmer, C.E., Keller, W.A. and Arnison, P.G. (1996) Utilization of *Brassica* haploids. In: Jain. S.M., Sopory, S.K. and Veilleux, R.E. (eds) In vitro *Haploid Production in Higher Plants*. Kluwer Academic Publisher, Dordrecht, the Netherlands, pp. 173–192.

Polsoni, L., Kott, L.S. and Beversdorf, W.D. (1988) Large scale microspore culture techniques for mutation/selection studies in *Brassica napus*. *Canadian Journal of Botany* 66, 1681–1685.

Prem, D., Gupta, K. and Agnihotri, A. (2012) Harnessing mutant donor plants for microspore culture in Indian mustard [*Brassica juncea* (L.) Czern & Coss]. *Euphytica* 184, 207–222.

Rahman, M.H. (2001) Production of yellow-seeded *Brassica napus* through interspecific crosses. *Plant Breeding* 120, 463–472.

Rai, B. and Singh, D. (1993) A note on potential sources of dwarfing genes in Indian rapeseed (*Brassica campestris* L.). *Indian Journal of Genetics* 53, 153–156.

Rai, U.K. (1958) X-ray induced appressed pod mutant in *Brassica juncea*. *Science and Culture* 24, 46–47.

Rai, U.K. and Jacob, K.T. (1956) Induced mutation studies in sesamum and mustard. *Science and Culture* 22, 344–346.

Rakow, G. (1973) Selektion auf Linol- und Linolensauregehalt in Rapssamen nach mutagener Behandlung. *Zeitschrift fur Philosophische Forschung* 69, 205–209.

Rana, M.A. (1985) Developmental morphology of flower and inheritance of an apetalous mutant in *Brassica carinata*. Dissertation, University of California Davis, Abstr. Int. B. (Sci. and Engi.), 45, 3425B.

Raza, G., Siddique, A., Khan, I.A., Ashraf, M.Y. and Khatri, A. (2009) Determination of essential fatty acid composition among mutant lines of canola (*Brassica napus*), through high pressure liquid chromatography. *Journal of Integrative Plant Biology* 51, 1080–1085.

Rehman, A. (1996) New mutant cultivar. *Mutation Breeding NewsLetter* 42, 27.

Rehman, A., Das, M.L., Hawlidar, M.A.R. and Mansure, M.A. (1987) Promising mutants in *Brassica campestris*. *Mutation Breeding NewsLetter* 29, 14–15.

Robbelen, G. (1990) Mutation breeding for quality improvement a case study for oilseed crops. *Mutation Breeding Review* 6, 1–43.

Robbelen, G. and Nitsch, A. (1975) Genetical and physiological investigations on mutants for polyeonic fatty acid in rapeseed, *Brassica napus* L. I. Selection and description of new mutants. *Zeitschrift fur Philosophische Forschung* 75, 93–105.

Robbelen, G. and Rakow, G. (1970) Selection for fatty acids in rapeseed. In: *Proceedings of the International Rapeseed Conference*, St. Adele, Canada pp. 476–490.

Rood, S.B., Pearce, D., Williams, P.H. and Pharis, R.P. (1989) A gibberellin-deficient *Brassica* mutant-rosette1. *Plant Physiology* 89, 482–487.

Roscoe, T.J., Lessire, R., Puyaubert, J., Renard, M. and Delseny, M. (2001) Mutations in the fatty acid elongation 1 gene are associated with a loss of beta-ketoacyl-CoA synthase activity in low erucic acid rapeseed. *FEBS Letters* 492, 107–111.

Rucker, B. and Robbelen, G. (1997) Mutants of *Brassica napus* with altered seed lipid fatty acid composition. In: *Proceedings of the 12th International Symposium Plant Lipids*. Kluwer Academic Publisher, Dordrecht, the Netherlands, pp. 316–318.

Sacriston, M.D. (1982) Resistance response to *Phoma lingam* of plants regenerated from selected cell and embryogenic cultures of haploid *Brassica napus*. *Theoretical and Applied Genetics* 61, 193–200.

Scarth, R., McVetty, R.B.E., Rimmer, S.R. and Stefansson, B.R. (1988) Stellar low linolenic-high linoleic acid summer rape. *Canadian Journal of Plant Science* 68, 509–511.

Scarth, R., Rimmer, S.R. and McVetty, P.B.E. (1995) Apollo low linolenic summer rape. *Canadian Journal of Plant Science* 75, 203–204.

Schnurbusch, T., Möllers, C. and Becker, H.C. (2000) A mutant of *Brassica napus* with increased palmitic acid content. *Plant Breeding* 119, 141–144.

Shah, S.A., Ali, I. and Rahman, K. (1990) Induction and selection of superior genetic variables of oilseed rape, *Brassica napus* L. *The Nucleus* 7, 37–40.

Shah, S.A., Ali, I. and Rahman, K. (1998) Use of induced genetic variability for the improvement of oilseed rape (*Brassica napus* L.) In: *Proceedings of the 2nd International Symposium New Genetical Approaches to Crop Improvement*. Nuclear Institute of Agriculture, Tandojam, Pakistan, pp. 229–237.

Shah, S.A., Ali, I., Iqbal, M.M., Khattak, S.U. and Rahman, K. (1999) Evolution of high yielding and early flowering variety of rapeseed (*Brassica napus* L.) through *in-vivo* mutagenesis. In: *Proceedings of the 3rd International Symposium New Genetical Approaches to Crop Improvement*. Nuclear Institute of Agriculture, Tandojam, Pakistan, pp. 47–53.

Sheikh, F.A., Lone, B., Najeeb, S., Shikari, A.B., Parray, G.A., Rather, A.G., Khudwani, R.R. and Khudwani, R.S. (2009) Induced mutagenesis for seed quality traits in Ethiopian mustard (*Brassica carinata* A. braun). *ARPN Journal of Agricultural and Biological Science* 4, 42–46.

Shi, S., Wu, J. and Liu, H. (1995) *In-vitro* mutation for long pod and short plant. *Nucleic Agriculture* 9, 252–253.

Shirzadegan, M. and Robbelen, G. (1985) Influence of seed colour and hull proportions of seeds in *Brassica napus* L. *Fette. Seifen, Anstrichmittel* 87, 235–237.

Simbaya, J., Slominski, B.A., Rakow, G., Campbell, L.D., Downey, R.K. and Bell, J.M. (1995) Quality characteristics of yellow seeded *Brassica* seed meals: protein, carbohydrates and dietary fibre components. *Journal of Agriculture Food Chemistry* 43, 2062–2066.

Singh, D. (1961a) Heredity changes in number of petals in Brown Sarson. *Indian Oilseed Journal* 5, 190–193.

Singh, D. (1961b) An apetalous mutant in Toria, Brown Sarson (*B. campestris* var. Brown Sarson) and its inheritance. *Current Science* 30, 62–63.

Singh, K.P. and Sareen, P.K. (2004) Induced podding mutants of Indian mustard [*Brassica juncea* (L.) Czern & Coss.]. *Cruciferae Newsletter* 25, 17.

Slominski, B.A., Simbaya, J., Campbell, L.D., Rakow, G. and Guenter, W. (1999) Nutritive value for broilers of meals derived from newly developed varieties of yellow-seeded canola. *Animal Feeding Science and Technology* 78, 249–262.

Spasibionek, S. (2006) New mutants of winter rapeseed (*Brassica napus* L.) with changed fatty acid composition. *Plant Breeding* 125, 259–267.

Srinivasachar, D. and Malik, R.S. (1972) An induced aphid resistant non-waxy mutant in turnip, *Brassica rapa*. *Current Science* 41, 820–821.

Stefansson, B.R., Hougen, F.W. and Downey, R.K. (1961) Note on the isolation of rape plants with seed oil free from erucic acid. *Canadian Journal of Plant Science* 41, 218–219.

Stephenson, P., Baker, D., Girin, T., Perez, A., Amoah, S., King, G.J. and Ostergaard, L. (2010) A rich TILLING resource for studying gene function in *Brassica rapa*. *BMC Plant Biology* 10, 62–71.

Stringam, G.R., McGregor, D.I. and Pawlowski, S.H. (1974) Chemical and morphological characteristics associated with seed coat colour in rapeseed. In: *Proceedings of the 4th International Rapeseed Congress*, Giessen, West Germany, pp. 99–108.

Swanson, E.S., Coumans, M.P., Brown, G.L., Patel, J.D. and Beversdorf, W.D. (1988) The characterization of herbicide tolerant plants in *Brassica napus* L. after *in-vitro* selection of microspores and protoplasts. *Plant Cell Reporter* 7, 83–87.

Swanson, E.S., Herrgesell, M.J., Arnoldo, M., Sippell, D. and Wong, R.S.C. (1989) Microspore mutagenesis and selection: canola plants with field tolerance to the imidazolinones. *Theoretical and Applied Genetics* 78, 525–530.

Tanhuanpaa, P., Vikki, J. and Vihinen, M. (1998) Mapping and cloning of *FAD2* gene to develop allele-specific PCR for oleic acid in spring turnip rape (*Brassica rapa* ssp. *oleiferae*). *Molecular Breeding* 4, 543–550.

Thurling, N. and Depittayanan, V. (1992) EMS Induction of early flowering mutants in spring rape (*Brassica napus*). *Plant Breeding* 108, 177–184.

Till, B.J., Colbert, T., Tompa, R., Enns, L.C., Codomo, C.A., Johnson, J.E., Reynolds, S.H., Henikoff, J.G., Greene, E.A., Steine, M.N., Comai, L. and Henikoff, S. (2003) High-throughput TILLING for functional genomics. In: Grotewold, E. (ed.) *Methods in Molecular Biology*, 236, 205–220.

Turner, J. and Facciotti, D. (1990) High oleic acid *Brassica napus* from mutagenized microspores. In: McFerson, J.R., Kresovich, S. and Dwyer, S.G. (eds) *Proceedings of the 6th Crucifer Genetics Workshop*. Geneva, New York, pp. 24.

Tyagi, D.V.S., Rai, B. and Verma, R.B. (1983) A note on the bunching dwarf mutant in toria. *Indian Journal of Genetics* 43, 374–377.

Velasco, L., Fernandez-Martinez, J.M. and De Haro, A. (1995) Isolation of induced mutants in Ethiopean mustard (*Brassica carinata* Braun). *Plant Breeding* 116, 396–397.

Velasco, L., Fernandez-Martinez, J.M. and De Haro, A. (1997) Selection for reduced linolenic acid content in Ethiopian mustard (*Brassica carinata* Braun) with low levels of erucic acid. *Plant Breeding* 114, 454–456.

Velasco, L., Fernandez-Martinez, J.M. and De Haro, A. (1998) Increasing erucic acid content in Ethiopian mustard through mutation breeding. *Plant Breeding* 117, 85–87.

Verma, V.D. and Rai, B. (1980) Mutation in seed-coat colour in Indian mustard. *Indian Journal of Agricultural Sciences* 50, 545–548.

Wang, M.L., Zhao, Y., Chen, F. and Yin, X.C. (2004) Inheritance and potentials of a mutated dwarfing gene ndf1 in *Brassica napus*. *Plant Breeding* 123, 449–453.

Wang, N., Wang, Y., Tian, F., King, G.J., Zhang, C., Long, Y., Shi, L. and Meng, J. (2008) A functional genomics resource for *Brassica napus*: development of an EMS mutagenised population and discovery of FAE1 point mutations by TILLING. *New Phytologist* 180, 751–765.

Wang, X., Wang, H., Wang, J., Sun, R., Wu, J., Liu, S., Bai, Y., Mun, J.H., Bancroft, I. and Cheng, F. (2011) The genome of the mesopolyploid crop species *Brassica rapa*. *Nature Genetics* 43, 1035–1039.

Wong, R.S.C. and Swanson, E. (1991) Genetic modification of canola oil: high oleic acid canola. In: Haberstroh, C. and Morris, C.E. (eds) *Fat and Cholesterol Reduced Food*. Gulf, Houston, Texas, pp. 154–164.

Woods, D.L. (1980) The association of yellow seed coat with other characters in mustard *Brassica juncea*. *Cruciferae Newletter* 5, 23–24.

Xiao, D. (1982) Analysis of the correlation between seed coat colour and oil contents of *Brassica napus* L. *Acta Agronomica Sinica* 8, 24–27.

Yu, C.Y., Hu, S.W., Zhang, C.H. and Yu, Y.J. (2004) The discovery of novel flower color mutation in male sterile rapeseed (*Brassica napus* L.). *Yi Chuan* 26, 330–332.

Zanewich, K.P., Rood, S.B., Southworth, C.E. and Williams, P.H. (1991) Dwarf mutant of *Brassica*: responses to applied gibberellins and gibberellin content. *Plant Growth Regulation* 10, 121–127.

5 Seed Quality Modifications in Oilseed Brassicas

Abha Agnihotri*

Centre for Agricultural Biotechnology, Amity Institute of Microbial Technology, Amity University, Uttar Pradesh, Noida, India

Introduction

Brassica is a large genus belonging to the *Brassicaceae* family. It contain 37 species (Kumar and Tsunoda, 1980), which include important crop plants such as oilseed rape (canola), mustard, cabbage, turnip rape, cauliflower and broccoli. Nutritionally, brassicas are rich in vitamins A and C and are the source of various bioactive agents. They are an important source of edible oils; their seeds are extensively used as condiments and spices. Among oil-bearing *Brassica* species, commonly known as rapeseed-mustard, Indian mustard (*Brassica juncea*) occupies the maximum hectarage followed by *Brassica rapa* and *Brassica napus* (Anonymous, 2005). Rapeseed-mustard provides one of the healthiest edible oils, being commonly consumed in India; however, the biochemical composition of presently cultivated Indian rapeseed-mustard varieties does not match the internationally accepted standards. Therefore, the enhancement of oil quality is aimed at making Indian mustard at par with international standards and competitive in Indian and international markets. This chapter presents the trends in oil quality improvement and bio-analytical techniques towards value-added end products in rapeseed-mustard.

Quality Considerations

The main constituents of rapeseed-mustard oil are the triacylglycerol fatty acids (92–98%) and the rest is the lipid compounds: the unsaponifiable fraction of hydrocarbons, sterols, tocopherols, tarpins, glycolipids and phospholipids (Sutariya *et al.*, 2011). While breeding for quality traits, total oil, protein and fibre contents are routinely used as monitoring parameters. The oil content in the seeds of rapeseed-mustard ranges from 30 to 48%. Taylor *et al.* (2009a) have reported increased seed-oil content (ranging from 2.5 to 7% of dried mass on an absolute basis) by overexpression of *DGAT1* gene in transgenic canola. As well as oil content, attention has also been paid to minor seed constituents such as tocopherols (Marwede *et al.*, 2004) and sinapate esters (Hüsken *et al.*, 2005; Zum Felde *et al.*, 2006). Among the minor constituents, phytosterols form an important group and are known for their cholesterol-reducing property. With concentrations ranging from

---

*Corresponding author, e-mail: agnihotri.abha@gmail.com

0.5 to 1% of the crude oil, rapeseed is one of the richest natural sources of phytosterols (Piironen *et al.*, 2000). While glucosinolates in the oil-free meal are important from an animal feed perspective and fatty acids composition forms the most important parameter for edible purposes, the incorporation of high phytosterol content in the oil will further enhance the nutraceutical advantages of mustard oil.

In the Indian subcontinent and several Asian countries, the traditional rapeseed-mustard varieties grown are high in erucic acid and glucosinolates, whereas in other parts of the world high erucic acid rapeseed is mostly used in the oleochemical industry and 'canola'-quality varieties are grown for edible purposes. The 'canola' oil, commonly known as double low, is internationally considered as having the ideal fatty acid composition for edible purposes and is preferred for human consumption. Canola has less than 2% erucic acid and less than 6% saturated fat, about 65% oleic, 20% linoleic and 9% linolenic acid in the seed oil and less than 30 μm glucosinolate in the seed meal (Downey, 1990). A diet rich in high erucic acid is shown to be associated with fibrotic changes in the myocardium (Gopalan *et al.*, 1974; Sauer and Kramer, 1983), thus having adverse effects on the heart. The glucosinolates are reported to be associated with enlarged goitre and toxicity in birds and animals and thus is considered harmful for consumption for feed purposes (Bille *et al.*, 1983; Bell, 1984). Therefore, due to the problems associated with long-term animal feeding, low glucosinolate content is desired in seed meal. However, on the positive side, glucosinolates have been shown to have a role against fungal and insect pest infestation as well as possessing anticancer properties (Glen *et al.*, 1990; Mitten, 1992).

Breeding for desired oil and de-oiled meal composition was initially restricted due to the non-availability of desirable donor sources in the Indian gene pool as well as the complicated genetic inheritance. Studies have shown that the erucic acid content is governed by two genes with additive effects, determined by the growing embryo (Dorrell and Downey, 1964; Harvey and Downey, 1964; Siebel and Pauls, 1989). Therefore, each seed from the same plant contains a different genetic background resulting

in seed to seed variation in the levels of erucic acid. The glucosinolates, on the other hand, are controlled by maternal genotype. The high glucosinolate content is dominant over low glucosinolate types and is governed by at least three partially recessive genes (Kondra and Stefansson, 1970; Uzunova *et al.*, 1995). Since both erucic acid and glucosinolates are inherited independently, a large number of segregating plant populations need to be screened for the selection of the plants having desired quality parameters. Hence, availability of efficient and accurate screening methods becomes a pre-requisite for quality improvement programmes.

Fatty Acids

The oils or fats are an integral part of our diet; they provide the most concentrated form of energy, act as a vehicle for the fat-soluble vitamins and also play an important role in metabolic functions. They are composed of triglycerides that contain one glycerol molecule linked by covalent bonds to three fatty-acid molecules. The physical and chemical properties of oil are directly dependent upon the composition of its fatty acids that make up the triglycerides. Based upon the occurrence of double bonds between the carbon molecules that make up the fatty acids, these can be classified as saturated fatty acids (SFA), mono-unsaturated (MUFA) and polyunsaturated fatty acids (PUFA). The oils containing high amounts of MUFA, moderate amounts of PUFA and low amounts of SFA are considered good for human consumption.

As per WHO recommendation, a nutritionally healthy edible oil should have a well-balanced ratio of MUFA:PUFA:SFA, the ratio of essential fatty acids linoleic:linolenic ranging from 5:1 to 10:1 and presence of natural antioxidants. Rapeseed-mustard oil is considered one of the healthiest vegetable oils as it contains the lowest amount of harmful SFA (viz. palmitic acid) that increases the levels of low density lipoprotein (LDL) in the blood. These SFA play a significant role in cholesterol deposition and are, therefore, undesired for human nutrition (Gurr, 1992). In addition, rapeseed-mustard oil also contains MUFA (oleic acid), which, being thermostable, increases the shelf life and is preferred for cooking and deep frying (Prabhu, 2000). It also reduces cholesterol and is thus beneficial for health (Bonanone *et al.*, 1992). The oil also contains the two essential PUFAs (linoleic and linolenic) that are not synthesized in humans and need to be supplied in the diet (Newton, 1998; Prakash *et al.*, 2000). Table 5.1 gives the composition of rapeseed-mustard oil and a comparison of dietary fats in Fig. 5.1 indicates the superiority of rapeseed-mustard

oil over others. Promising low-erucic genotypes of *B. juncea* and *B. napus* have been bred in India by embryo rescue-aided hybridization or conventional breeding approaches (Chauhan *et al.*, 2000; Kumar *et al.*, 2009). In recent years, emphasis is being laid on increasing the proportion of very long chain fatty acids (VLCFA) as bioactive agents. A major derivative of erucic acid is erucamide, which is used for industrial purposes as surface coating, plastic films and slip promoting agent. Arachidonic acid, eicosapentaenoic acid, docosahexaenoic acid and nervonic acids are some of the VLCFA required for promotion of human and animal health (Wu *et al.*, 2005; Taylor *et al.*, 2009b).

Table 5.1. Characteristics of rapeseed-mustard oil (Kumar and Agnihotri, 2004).

Component	Brassica napus	Brassica rapa
Seed coat (%)	16.5	18.7
Seed oil (%)	41.5	40.0
Hull oil (%)	16.0	16.2
Kernel oil (%)	47.1	45.0
Crude fibre (%)	11.8	11.7

Screening Techniques for Fatty Acids Analysis

The screening methods employed for the estimation of fatty acids can broadly be classified

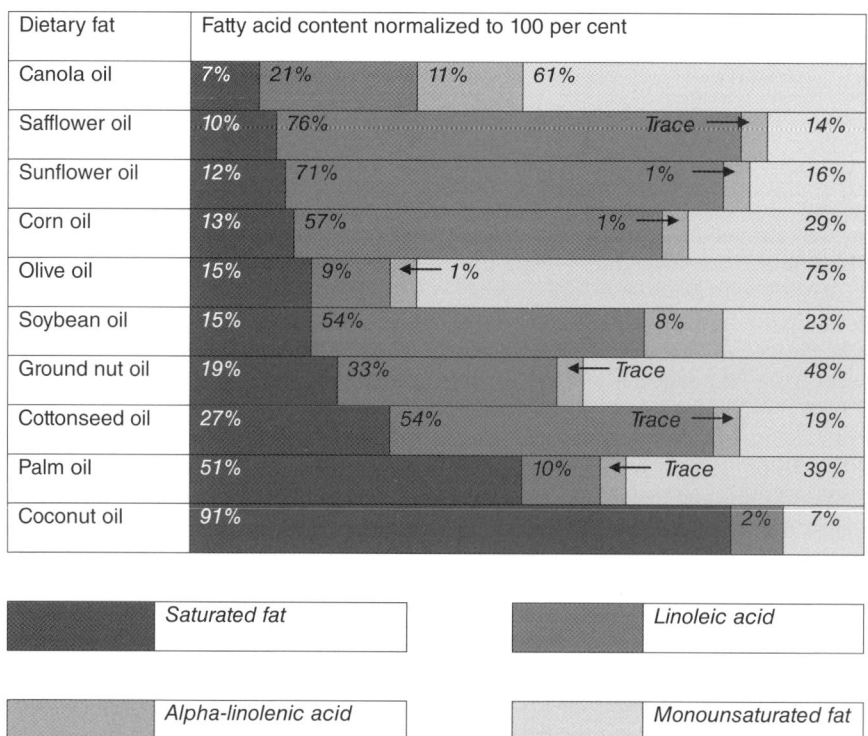

Fig. 5.1. Fatty acids profile of some commonly used edible oils.

into two different types: chromatographic and non-chromatographic methods.

The initial non-chromatographic methods

McGregor (1977) described a method for determination of erucic acid depending on the solubility of oil in absolute ethanol or a mixture of methanol and n-propanol (1.7:2.0; v/v). The time required for a warm alcoholic solution of the oil to turn opaque on cooling was related to its erucic acid content. This method was used for quality control in commercial seed samples, but due to low sensitivity it was not found suitable for quality breeding. Near infrared reflectance spectroscopy (NIRS) offers a rapid and non-destructive method that can be used for simultaneous determination of several quality parameters in intact seeds (Biston et al., 1988). This has been used by Reinhardt and Robbelen (1991) for determination of oleic acid content and for fatty acids by Velasco et al. (1997). NIRS can be used for screening large populations, followed by chromatographic methods for accurate fatty acids composition.

The chromatographic methods

The early methods for the separation of free acids by fractional crystallization in ethanol (Stiepel, 1926; Kaufmann and Fiedler, 1938; McGregor, 1980) and thin layer chromatography (TLC) described by Stahl (1969) were time consuming and lacked sensitivity. Gas chromatography (GC) methods were introduced for separation and quantitative determination of methyl esters of individual fatty acids (Craig and Murty, 1958, 1959; Conacher and Chadha, 1974; McGregor, 1974, 1977). These methods have been improved over the years for sample preparation and automation (Appelqvist, 1968). Downey and Harvey (1963), in a major breakthrough for selection and breeding for low erucic acid, described a method for fatty acid determination through half-seed technique. In this method, after the fatty acid analysis of the outer cotyledon of the

germinating seed, the identified low erucic acid plant was grown from the corresponding inner cotyledon stored in vitro along with the radical. This was further facilitated by exuding the oil with the tip of a warm soldering iron (McGregor, 1974; Stringam and McGregor, 1980).

To date, GC offers the most valuable and efficient method of analysis of fatty acid methyl esters, providing the simultaneous quantitative determination of individual fatty acids. GC sample preparation requires oil extraction and esterification of fatty acids. Downey and Craig (1964) used methanolic hydrochloric acid for fatty acids esterification. Other workers have utilized sodium methoxide for the purpose (Hougen and Bodo, 1973; Stringam and McGregor, 1980). These procedures required the extraction of oil prior to esterification. A one-step trans-esterification using acetylchloride in methanol-benzene (4:1, v/v) was described by Lepage and Roy (1986). Banerjee et al. (1992) used microwave heating to reduce the reaction time. Kaushik and Agnihotri (1997) have further improved the method of Lepage and Roy (1986) for fatty acid analysis, by the use of microwave heating in place of conventional heating, thus substantially reducing the esterification time from 1 h to 2 min.

Different liquid phases can be employed for GC. The non-polar liquid phases having low volatility (viz. SE-30 and UCN-98) are preferred for the separation of long-chain fatty acids. Polar liquid phases, such as diethylene glycol succinate (DEGS) and butane diol succinate (BDS), have higher volatility and thus require short column length. These can be used for the separation for both long- and short-chain fatty acids. For both types of analysis, chromatography is facilitated by use of low volatile columns such as GP3%SP-2310/2%SP-2300 (McGregor, 1980). Kaushik and Agnihotri (1997) could achieve rapid separation on a stainless steel GC column packed with a mixture of 2% SP-2300 and 3% SP-2310 on chromosorb 'W', within 7 min on a 1.8 m column and in 15 min by a 3 m column. This method was found suitable for routine analysis of fatty acids for the quality improvement programme.

Glucosinolate(s)

More than 90 different kinds of glucosinolates have been identified and the known glucosinolates can be classified into three main classes: (i) aliphatic/alkenyl glucosinolates derived from L-methionine; (ii) aromatic glucosinolates derived from L-phenylalanine/L-tyrosine; and (iii) the indolyl glucosinolates derived from L-tryptophan (Sorenson, 1991). They are found in all parts of the plant and up to 15 different kinds of glucosinolates have been found in the same plant (Zukalova and Vasak, 2002). Most of the glucosinolates are located in the embryo, comprising 80–90% of the seed dry weight (Robbelen and Thies, 1980). Bennett *et al.* (1996) have demonstrated that one particular glucosinolate predominates in a particular plant species (Table 5.2). Kaushik *et al.* (2007) have characterized more than 1200 brassica varieties based on their glucosinolate profile.

Glucosinolates, in the presence of moisture, are hydrolysed by the endogenous enzyme myrosinase (thioglucosidase, EC. 3.2.3.1).

Depending upon the structure of glucosides and degradation conditions, the hydrolysis products (Fig. 5.2), in addition to inorganic sulfate and glucose, include thiocyanates, isothiocyanates and nitriles in varying proportions (McGregor and Downey, 1975; Lange, 1987). These degradation products lend the characteristic odour and flavour to brassica plants. Various approaches have been utilized to breed for low glucosinolate varieties including interspecific hybridization via embryo culture to aid conventional breeding (Kumar *et al.*, 2009; Agnihotri *et al.*, 2011), molecular breeding (Ramchiary *et al.*, 2007) and genetic transformation (Elhiti *et al.*, 2012).

Screening Techniques for Glucosinolate(s) Analysis

The analytical methods employed for the isolation, separation, identification and quantification of glucosinolates can be broadly classified into two categories for total or individual glucosinolates.

Table 5.2. Characteristic glucosinolates in *Brassica* species (Kumar and Agnihotri, 2004).

Glucosinolate	Common name	*Brassica* species
Allyl	Sinigrin	*B. juncea, B. nigra, B. carinata, B. oleracea*
3-Butenyl	Gluconapin	*B. rapa, B. juncea*
2-Hydroxy-3-butenyl	Progoitrin	*B. napus, B. oleracea*
4-Pentenyl	Glucobrassica-napin	Minor glucosinolate in all *Brassica* species
2-Hydroxy-4-pentenyl	Napoleiferin	Minor glucosinolate in all *Brassica* species
3-Indolyl-methyl	Glucobrassicin	Minor glucosinolate in all *Brassica* species
4-Hydroxy-3-indolylmethyl	4-Hydroxyglu-cobrassicin	Minor glucosinolate in all *Brassica* species

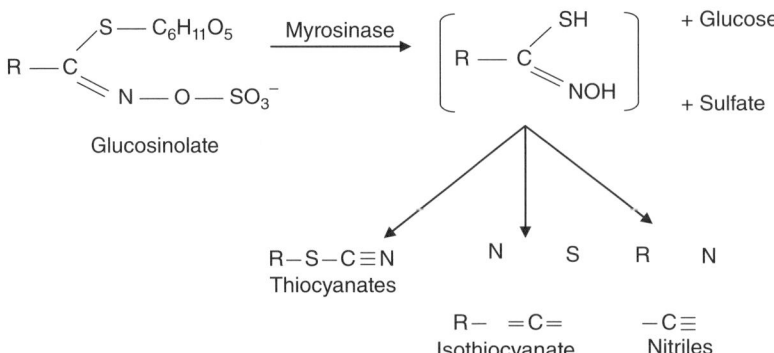

Fig. 5.2. Enzymatic hydrolysis products of glucosinolate.

Methods for total glucosinolate estimation

The initial methods are based on the measurement of glucosinolate hydrolysis products. These include spectrophotometric/colorimetric analysis of reaction products formed between thymol-sulfuric acid and intact/desulfoglucosinolates, or determination of glucose liberated by myrosinase catalysed hydrolysis. The total glucosinolate can also be directly determined in seed samples by NIRS (Biston et al., 1988) and X-ray fluorescence (Schnug and Haneklaus, 1988) or X-ray reflectance (Tholen et al., 1993). However, more accurate methods are required to differentiate individual glucosinolates and varieties with very low glucosinolate content (Bjerg et al., 1987).

The initial spectrophotometric methods based on measurement of degradation product are the argentimetric method involving steam distillation and titration of the volatile isothiocyanates (Wetter, 1955) combined with UV spectroscopy of the oxazolidinethiones (Wetter, 1957), gas chromatography of the volatile isothiocyanates combined with UV spectroscopy of the oxazolidine-thiones (Youngs and Wetter, 1967) and UV spectroscopy of thio-urea derivatives of the isothiocyanates (Appelqvist and Josefsson, 1967; Daxenbichler et al., 1970; Wetter and Youngs, 1976). Out of these, the thio-urea-UV method of Wetter and Youngs showed maximum reproducibility of results (McGregor, 1980). Subsequently, Brzezinski and Mendelewski (1984) reported the total glucosinolate content measurement using thymol reagent. It showed the technical difficulty for use of concentrated sulfuric acid and high degrees of variation in replicate analyses. The modified thymol method reported by Truscott and Shen (1987) eliminated the underestimation of glucosinolate content in the high glucosinolate-containing rapeseed. In this method, sinigrin (a glucosinolate) reacts with sulfuric acid and is first hydrolysed to thioglucose that is further dehydrated to thiofurfural derivative. Furfural then reacts with thymol to form a coloured compound whose intensity is related to total glucosinolate concentration.

Lein (1970) described two methods based on measurement of myrosinase-catalysed hydrolysis products in raw extracts of B. napus seeds without defatting the seeds prior to analysis. The glucose released was quantitatively measured by the hexokinase + ATP/ glucose-6-phosphotydel dehydrogenase + NADP enzyme-system, or the glucose-UV-test or by the glucose-oxidase/peroxidase system (Glukotest). The 'Glukotest' was found more efficient by analysing 200 plants per day while only 50 plants per day could be analysed by the 'glucose-UV-test'; however, the former was found to be comparatively less sensitive and precise. A single cotyledon or a single seed of B. napus could be measured with ±1% accuracy by the glucose-UV-test. Bjorkman (1972) described the colorimetric determination of glucosinolates in purified rapeseed-extracts released through glucose oxidase, peroxidase and the chromogen O-dianiside. However, Van Etten et al. (1974) noticed colour inhibition while analysing the crude extracts; they colorimetrically measured the glucose released by enzymatic oxidation of glucosinolates after removal of interfering substances with charcoal.

Among the methods used for quick screening of large plant populations, the semi-quantitative glucose Test-tape technique developed by Comer (1956) has gained much popularity. Lein (1970) applied the glucose test paper for determination of approximate glucosinolate level in rapeseed. Van Etten et al. (1974) separated the colour development-interfering substances from glucose through capillary action. Subsequently, the modified test-tape method, suitable for commercial screening, was reported by McGregor and Downey (1975). The glucose-specific test paper containing glucose oxidase, peroxidase and the chromogen O-tolidine, was used as described by Comer (1956) and the glucose released from glucosinolates hydrolysis by the endogenous myrosinase in aqueous seed extracts was measured after removal of interfering substances with charcoal. Hassan et al. (1988) reported an enzyme-linked immune absorbent assay (ELISA) for determination of alkenyl glucosinolates, which was not specific for quantitative determination of sinigrin. Doorn et al. (1998) further improved the ELISA method for estimation of sinigrin and progoitrin glucosinolates by specific antibody assays in Brussels sprouts. When glucosinolates are incubated with tetrachloro-palladate (II) (Pd Cl$_4^{2-}$) in water a colour intensification from light to dark brown is observed and measured in brassica seeds (Kumar et al., 2004).

Methods for individual glucosinolate(s) estimation

The chromatographic techniques can overcome the limitation of the above-mentioned methods for accurate quantification of individual glucosinolates. The gas liquid chromatography (GLC) technique is used extensively to analyse the hydrolysis products of glucosinolates. Underhill and Kirkland (1971) analysed the trimethyl-silyl (TMS) derivatives of glucosinolates by GLC in rapeseed defatted meal; however, the direct desulfation of glucosinolates in this method resulted in poor sensitivity. Persson (1974) and Thies (1974, 1976, 1977, 1978, 1979) further improved the method by ion-exchange chromatography to remove impurities. Thies (1980) introduced on-column desulfation to improve the derivatization, but determination of indole glucosinolate was still not up to the mark. Heaney and Fenwick (1980, 1982) overcame this limitation by use of high temperature for derivatization and successfully separated all of the major glucosinolates in rapeseed, including the two indoles, glucobrassicin (3-indolylmethyl glucosinolate) and neoglucobrassicin (1-methoxy-3-indolylmethyl glucosinolate). However, the 4-hydroxy-glucobrassicin (4-hydroxy-3-indolylmethyl glucosinolate), the major indole glucosinolate in rapeseed (Truscott et al., 1982, 1983), showed a two-fold difference in its quantification than that calculated by the TMS method (Sosulski and Dabrowski, 1984). This could be due to its sensitivity to high temperature, oxygen and heavy metals (Thies, 1985). Brzezinski et al. (1986) on the comparison noticed higher values of glucosinolate by the thymol method than with GLC. Slominski and Campbell (1987) showed that a substantial loss of indoles occurred during the heat inactivation of myrosinase enzyme contributing to the underestimation of indole glucosinolates by GLC. To avoid decomposition of indole glucosinolates they recommend 15 min dry heat followed by 3 min wet heat treatment for myrosinase deactivation. Although there are some limitations in quantification of 4-OH-glucobrassicin (Thies, 1985) and all important glucosinolates do not convert to volatile derivatives (Christensen et al., 1982), GLC was found adequate for the glucosinolate analysis as compared to the earlier described methods.

Truscott et al. first described the high performance liquid chromatography (HPLC) method in 1983 and the technique has been used since by many scientists (Sang and Truscott, 1984; Spinks et al., 1984; McGregor, 1985; Palmer et al., 1987; Kaushik and Agnihotri, 1999) for quantification of individual glucosinolates. Heaney et al. (1986) treated the glucosinolates with hot methanol followed by incubation with sulfatase enzyme and analysed desulfoglucosinolates by HPLC. Buchner and Thies (1987) further optimized the indolyl glucosinolates analysis. Lichter et al. (1988) designed an automated HPLC system that helped in accurate and efficient analysis in a short time. Iqbal et al. (1995) further modified the desulfated glucosinolate method of Kraling et al. (1990) by adjusting the volumes of extraction, injection and internal standard's dilution resulting in better resolution of individual glucosinolates.

Most of the major known glucosinolates can be resolved and quantified in a single chromatogram by the reversed phase HPLC of intact glucosinolates or desulfoglucosinolates. The desulfoglucosinolate separation by HPLC can be applied to either seed or vegetative samples; however, some glucosinolates may escape detection (Bjerg and Sorenson, 1986). These limitations can be resolved by intact glucosinolates analysis by HPLC (Moller et al., 1985). However, higher column temperature and expensive ion-pairing reagents are required for intact glucosinolates separation (Lichter et al., 1988). These disadvantages can be overcome by use of an aqueous mobile phase with specific salt solution in ion-suppression chromatography. Bjorkqvist and Hase (1988) have used aqueous ammonium acetate–acetonitrile mixture as the mobile phase. The defatting of seeds, followed by proteins precipitation with 0.03 M barium acetate and lead acetate was employed for the analysis. The method has been further improved by Kaushik and Agnihotri (1999). They have reported separation of glucosinolates after dry heat inactivation of myrosinase, followed by a single step extraction in boiling water and ammonium sulfate as a specific salt in the mobile phase, using Novapack RP C18. The availability of pure reference compound is a pre-requisite for calibration, which is generally performed with a commercially available sinigrin or benzoyl glucosinolate. Rapeseed meals from different species with specific known contents of glucosinolates can be utilized for calculation of relative response factors. The individual glucosinolates are then quantified by the peak area integration of their specific response factors as determined by Buchner (Thies, 1976; Lichter *et al.*, 1988; Iqbal *et al.*, 1995).

Other Minor Constituents of Importance

The phytosterols

Phytosterols are natural steroid alcohols found in plants, amphiphilic in nature and are vital constituents of all membranes. Phytosterols are products of the isoprenoid biosynthetic pathway occurring exclusively in the cytoplasm and consisting of more than 25 enzyme-catalysed reactions (Benveniste, 2002). Phytosterols are well known to lower total serum and LDL cholesterol levels (Eskin *et al.*, 1996) by reducing its absorption in the gut by up to 50% and thus reducing the risk of cardiovascular disorders.

Two decades ago, increased levels of plasma cholesterol have been recognized as one of the main risk factors of cardiovascular diseases; the leading cause of mortality in Western countries (Castelli, 1984). One of the main mechanisms for cholesterol reduction is prevention of cholesterol absorption by its replacement with phytosterols in the intestinal-micellar phase (Nissinen *et al.*, 2002; Trautwein *et al.*, 2003). Following consumption, phytosterols reduce the absorption of dietary and endogenous cholesterol by about 50% (Law, 2000). Trautwein *et al.* (2003) concluded that phytosterol-enriched foods increase the effectiveness of a heart-healthy diet as it lowers

LDL cholesterol and can be an important strategy in reducing the risk of coronary heart diseases. They have also been reported to have anticancerous properties (Woyengo *et al.*, 2009). The phytosterol content in the food items has been reported to be either maintained or increased up to 10%, while the cholesterol levels were decreased to about 50–80% upon deep frying of foods with rapeseed oil. Today, milk and dairy products fortified with natural phytosterols are available in many countries. In most cases, vegetable oils are being used as a source for phytosterol extraction; they are obtained as a by-product during vegetable oil refining.

In oilseed brassicas, phytosterols are the minor components (0.5 to 1%) that constitute the un-saponifiable fraction of the oil along with sterols and have antioxidant properties (Sil *et al.*, 1990). Sitosterol is the most abundant phytosterol (50–53%) in rapeseed, followed by campesterol (19%), brassicasterol (13–15%), avenasterol (2–4%) and stigmasterol (traces) (Appelqvist *et al.*, 1981). These are structurally similar to each other and have close molecular weights (Table 5.3). Recent research shows that multiple genes with an additive effect on total accumulation of individual phytosterols in the seed embryo control the phytosterols in rapeseed.

Screening techniques for phytosterol analysis

The Leibermann-Burchard method has been widely used for colorimetric analysis of total phytosterol content by UV-visible spectrophotometer using LB reagent (Kim and Goldberg, 1969). The GC (Samija *et al.*, 2007) and GC- MS (Liu *et al.*, 2007; Scialabba *et al.*, 2009) or MS with SIM-selective ion monitoring (Mohamed Ahmidaa *et al.*, 2006) have been used by scientists for estimation of individual phytosterols in rapeseed-mustard (Table 5.4). Scientists have also reported the use of HPLC-MS (Zhang *et al.*, 2009), HPLC with evaporative light-scattering detection (Warner and Mounts, 1990) for appropriate resolution of phytosterols, except for one study of analysis by HPLC and UV-visible detector alone (Table 5.5). Careri *et al.* (2001) have used liquid chromatography–UV determination

Table 5.3. Molecular weights and structure of major phytosterols found in *B. juncea.*

Phytosterol	Molecular weight	Chemical formula	Structure
β-sitosterol	414.70	$C_{29}H_{50}O$	
Campesterol	400.68	$C_{28}H_{48}O$	
Brassicasterol	398.66	$C_{28}H_{46}O$	
Avenasterol	412.69	$C_{29}H_{48}O$	
Stigmasterol	412.69	$C_{29}H_{48}O$	

Table 5.4. Analysis of phytosterols using gas chromatography.

Detector used	Carrier gas	Column used	Standards used	References
FID (Flame Ionization Detector)	Hydrogen	50 m fused silica capillary column	200 µl of cholesterol as internal standard	Samija et al. (2007)
FID (for quantification) UV spectrum and GC-MS (for structure elucidation)	–	HP 5 capillary GC column	Squalane as internal standard	Liu et al. (2007)
MS (mass spectrometer) with SIM	Helium	60 m capillary column	Cholestane as internal standard and cholesterol, beta-sitosterol, campesterol, lathosterol	Mohamed Ahmidaa et al. (2006)
Mass spectrometer	Helium	Capillary column	Cholesterol	Scialabba et al. (2009)

Table 5.5. Analysis of phytosterols by HPLC.

Detector used	Mobile phase	Column used	Standards used	References
Mass spectrometer (APCI Probe)	Phase A: water acetic acid 0.01%; Phase B : acetonitrile	C18, 5 µm column	Cholesterol, β-sitosterol, stigmasterol, sitostanol, uvaol, fucosterol	Zhang et al. (2009)
Evaporating light scattering detection (ELSD)	Methanol:water (98:2)	C18, 5 µm column	β-sitosterol, campesterol, stigmasterol, brassicasterol	Warner and Mounts (1990)
UV detector + ELSD	Hexane/2-propanol/ acetic acid	5 µm column	–	Singh et al. (2003)
UV-Vis detector + MS	Acetonitrile:water (86:14)	C8 column	Sitosterol, stigmasterol	Careri et al. (2001)
Semipermeative HPLC + GC-MS	Acetonitrile:2-propanol: water (2:1:1)	C8-A column	Cholesterol	Careri et al. (2001)

and liquid chromatography–atmospheric pressure chemical ionization for phytosterols analysis in soybean oil and Singh et al. (2003) and Zhang et al. (2006) achieved separation of Δ5- and Δ7-phytosterols by adsorption chromatography and semi-preparative reversed phase high performance liquid chromatography. Mortuza (2006) has reported a wide variation in the four types of phytosterols and their lipid concentration in micrograms per gram in 20 genotypes of rapeseed mustard in Bangladesh analysed through HPLC; sitosterol (3183 to 4463), campesterol (1306 to 3471), brassicasterol (582 to 1368), avenasterol (138 to 1135) and cycloartenol (119 to 1170).

In a recent study, Kilam et al. (2015) analysed Indian B. juncea varieties using HPLC and UPLC (ultra-performance liquid chromatography; courtesy IGIB, New Delhi) in combination with MS to resolve the individual phytosterols whose molecular weights are very close to each other. For estimation of phytosterols in twenty B. juncea varieties, saponification and isolation of unsaponifiable fraction was performed according to the method described by the Official Journal of the European Community (European Union Commission, 1991, in Zhang et al., 2009). The total phytosterol content was analysed through a colorimetric method (Kim and Goldberg,

1969; Sabir *et al.*, 2003); the phytosterol content showed high variability ranging from 170 to 319 µg/g seed, the lowest in *B. varuna* var. TM2 and highest in var Pusa Bahar (DIRA 367).

Screening techniques for oil, protein and fibre contents analysis

The protein quality is determined by the availability of essential amino acids for body protein synthesis (Chadd *et al.*, 2002). Rapeseed-mustard seed meal contains about 40% proteins with a well-balanced aminogram (Miller *et al.*, 1962). It is rich in the essential amino acids lysine and methionine that are not found in cereal grains. The carbohydrates found in the seed meal are sugars (sucrose, stachyose, raffinose, glucose and fructose), polysaccharides and glucosinolates. The seed meal is also a good source of minerals such as calcium, manganese, magnesium, potassium, phosphorous, sulfur and zinc. It is a rich source of vitamin E and also contains other important vitamins/precursors such as niacin, pantothenic acid, riboflavin and thiamine. The oil content ranges from 39 to 42% and a simple switchover to yellow-seeded cultivars can bring about a 2% increase in oil content (Banga, 1996). A negative correlation has been shown to exist between seed oil and protein or carbohydrate content (Grami *et al.*, 1977; Mitra and Bhatia, 1979). The moisture content is also important for maintaining seed quality upon storage (Bandel *et al.*, 1991), but it has mainly been used for documentation of seed quality characteristics (Anonymous, 2005). Among the conventional methods used for the determination of oil content, both Soxhlet and cold percolation methods are destructive methods thus not very suitable for screening breeding populations. So far the nuclear magnetic resonance (NMR; Madsen, 1976) spectrophotometry is the most widely used, efficient and accurate, non-destructive method for determination of oil content. The NIRS (Bengtsson, 1985) technique is gaining popularity as it is more economical as compared to NMR and requires a small amount of seed samples. For protein content, the micro-Kjeldahl digestion followed by an automated colorimetric analysis has been the most commonly used method (Stringam *et al.*, 1974). The protein content can also be determined by the NIRS method (Tkachuk, 1981; Velasco and

Mollers, 2002). Bengtsson (1985) reported the NIRS method to be equally accurate as that of the Kjeldahl method, thus found it suitable for large-scale screening (Williams, 1975). Kumar *et al.* (2003) have used the NIRS screening for both oil and protein content. It is less time-consuming and the sample preparation involves only drying and cleaning of seeds. However, calibration requires quite a large number of seed samples representing the working range for both oil and protein contents. With the availability of NIRS and automated elemental nitrogen analysers, greater selection pressure can be employed for high protein content selection without reduction in oil content (Downey and Rimmer, 1993; Leckband *et al.*, 2003). Recently Prem *et al.* (2012) successfully used NIRS for simultaneous determination of oil, protein and erucic acid in *B. juncea*. The hydrolysis of oilseed meal improves its value as the hydrolysates have antimicrobial, antioxidant, blood pressure lowering and immune-modulatory properties. Xuyan *et al.* (2011) have used continuous microwave-assisted enzymatic hydrolysis of proteins with the aim to obtain efficient protein hydrolysates. The crude fibre content is mainly determined by the micro-acid-base procedure (AOAC, 1975). However, quick visual screening for yellow seed-coat colour is often utilized as a selection tool for reduced fibre content in plant breeding populations (Stringam *et al.*, 1974). A new NIRS calibration has been established as a non-destructive high-throughput method for selecting rapeseed varieties with reduced levels of antinutritive fibre fractions, particularly lignin (Wittkop *et al.*, 2011).

Screening techniques for sinapate esters and tocopherols analysis

The phenolic compounds are the secondary metabolites that play an important role in the plant defence system. Rapeseed contains higher phenolic compounds as compared to other oilseeds (Nowak *et al.*, 1992). The accumulation of sinapate esters in seeds, i.e. sinapoylcholine (sinapine) and sinapoylmalate that plays a role in UV-B tolerance in leaves, is considered antinutritional and hampers the use of oilseed protein for animal feed and human consumption. Thus work has been in progress towards metabolic engineering for low sinapate ester lines in *B. napus* (Hüsken *et al.*, 2005; Milkowski and Strack, 2010). Tocopherols (vitamin E) are present mostly in green leafy parts of brassicas and play an important role as antioxidants. Both sinapine and tocopherols are routinely analysed by spectrophotometric methods (El-Beltagi and Mohamed, 2010). Annunziata *et al.* (2012) have used an efficient fluorimetric HPLC method for quantification of tocopherols. Guzman *et al.* (2012) have described a novel UPLC method for simultaneous determination of carotenoids, chlorophylls and tocopherols in brassica vegetables with the aim of enhancing phytonutrient profile. Recently, Hussain *et al.* (2013) compared three methods of sample preparation and concluded that trimethylsilylation is most suitable for tocopherol extraction followed by detection by GC (with Flame Ionization Detector; FID).

Conclusions

The perspectives of breeding for seed quality in *Brassica* oilseed crops have been reviewed by Rakow and Raney (2003). The comparative advantages of available protocols for biochemical evaluations are summarized in Table 5.6 (Prem, 2006; Agnihotri *et al.*, 2007). Various review meetings/symposia have been held to discuss and decide on the most suitable method for analysis of biochemical constituents in oilseed brassicas, such as the Symposium on Analytical Chemistry of Rapeseed and its Products (Winnipeg, Manitoba, organized by Canola Council of Canada; McGregor, 1980) and the Interlaboratory Harmonization of Analytical Methods (under the Indo-Swedish Collaborative Research Program on Rapeseed-Mustard Improvement and Oil and Protein Utilization, India, 1984).

It has been concluded that a two-step screening, i.e. a quick non-destructive screening of breeding populations followed by more precise and accurate quantification, is most suited for the quality breeding programmes (McGregor *et al.*, 1983). Initial steps in quality breeding require the screening of a large number of samples, many of which contain only a few seeds from a single plant. Thus non-destructive

efficient analytical methods are most suited for screening large breeding populations. At present, NIRS offers such an alternative for simultaneous evaluation of most of the biochemical traits in many agricultural products (Reinhardt and Robbelen, 1991; Shenk and Westerhaus, 1993; Daun and Williams, 1995; Velasco *et al.*, 1997; Mika *et al.*, 2003). However, species-specific and multiple species calibrations are required to overcome the variations in NIR absorbance of the whole seed and specific baseline shifts in NIR spectrums (Shenk and Westerhaus, 1993). The accuracy of analysis by NIRS, as in the case of other secondary methods, is dependent on the precision of the primary wet chemistry method used to develop calibrations. In the case of brassicas, the calibration set should include both yellow and brown coloured seed-coat seeds, harvested from multiple agroclimatic zones/years to nullify the seasonal/environmental influences and to achieve a reliable spectroscopic evaluation (Van Deynze and Pauls, 1994; Dardanne, 1996; Prem *et al.*, 2012). Following NIR, NMR or solvent extraction can be used for oil content estimation, however, both require a large seed sample and the former is preferred as the seeds can be used further. GC has remained the most preferred method for FA profile estimations. While Test-Tape and ELISA offers a good alternative for quick screening of total glucosinolates in breeding populations, to date, HPLC remains the method of choice for individual glucosinolates quantification.

Among the several concerns associated with the practical applications of value added oilseeds, segregation of traditional, nutritionally improved or genetically modified transgenic varieties are of considerable importance. Growth and maintenance of such crop produce separately is a major hurdle, especially in developing nations where the regulatory procedures are still in their nascent stage. Diversification of edible oilseeds to industrial feed stocks will not be feasible in nations having an edible oilseeds shortfall for human consumption. Another area of concern is the development of commodity-based niche markets for dissemination of nutritionally enriched oils at a premium price. Effective marketing strategies with the participation of private partners, which also benefits oilseed farmers, will need policy intervention.

Table 5.6. Techniques for estimation of biochemical components in oilseed brassicas (Prem, 2006; Agnihotri *et al.*, 2007).

Parameter/trait and method or basic principal	Advantage/drawback	Reference
Seed oil content		
Solvent extraction	Accurate but requires large seed sample (2–5 g) and is destructive	Anonymous (2000)
Nuclear magnetic resonance (NMR)	No sample preparation required, rapid, accurate and non-destructive but requires expensive equipment	Tiwari *et al.* (1974)
Near infrared reflectance spectroscopy (NIRS)	No sample preparation required, rapid, accurate and non-destructive but requires expensive equipment and is species-specific	Greenwood *et al.* (1999); Velasco *et al.* (1999a); Mika *et al.* (2003)
Seed FA profile		
Chromatographic methods		
Erucic acid co-precipitation with SFAs as lead or magnesium salts followed by separation of free FA by fractional crystallization in ethanol	Time consuming and lack sensitivity	Stiepel (1926); Kaufmann and Fiedler (1938)
Thin layer chromatography (TLC)	Low sensitivity	Stahl (1969)
Paper chromatography for separation of FA using 95% acetic acid as mobile phase	Quick screening method for selection but suitable only for initial screening	Thies (1971)
Gas chromatography – quantitative estimation of methyl esters of individual FA	Accurate estimation of FA profile, however cumbersome sample preparation since it requires extraction of oil prior to esterification	Craig and Murty (1958, 1959); Appelqvist (1968); Conacher and Chada (1974); McGregor (1974, 1977); Stringam and McGregor (1980)
GC estimation with modified sample preparation		
Methanolic hydrochloric acid used for esterification	Oil extraction required prior to esterification	Downey and Craig (1964)
Sodium methoxide used for esterification	Oil extraction required prior to esterification	Hougen and Bodo (1973); Stringam and McGregor (1980)
One step transesterification using acetylchloride in methanol-benzene (4:1, v/v)	Reduced reaction time thus rapid and accurate estimation	Lepage and Roy (1986); Kaushik and Agnihotri (1997)
Non-chromatographic methods		
Determination of erucic acid based on the solubility of oil in absolute ethanol or mixture of methanol and n-propanol (1.7:2, v/v) – time required for warm alcoholic solution to turn opaque on cooling related to erucic acid content	Not suitable for breeding purposes due to low sensitivity	McGregor (1977)

Continued

Table 5.6. Continued.

Parameter/trait and method or basic principal	Advantage/drawback	Reference
NIRS	No sample preparation required, rapid and accurate but species-specific and requires large seed sample with high representative variability for standardization. The equipment used is expensive	Reinhardt and Robblen (1991); Velasco *et al.* (1995, 1999a, b, 1997); Pallot *et al.* (1999)
Meal protein content		
Kjeldahl nitrogen estimation-based protein content evaluation	Low sensitivity, time – labour consuming, involves hazardous reagents and is destructive	AOAC (1995)
Combustion Nitrogen Analysis (CNA) or DUMAS nitrogen estimation-based protein evaluation	Accurate and rapid but destructive and involves expensive instruments	Simonne *et al.* (1997)
NIRS	No sample preparation required, rapid, accurate and non-destructive but requires expensive equipment and is species-specific	Velasco and Mollers (2002); Kumar *et al.* (2003)
Meal glucosinolate content		
Spectrophotometric methods based on measurement of glucosinolate degradation products or glucosinolate–reagent colour complex – methods for estimation of total glucosinolate		
Steam distillation and titration of volatile isothiocyanates combined with UV spectroscopy of oxazolidinethiones	Low sensitivity and low reproducibility of results	Wetter (1955, 1957)
Gas chromatography of volatile isothiocyanates combined with UV spectroscopy of oxazolidinethiones	Low sensitivity for low glucosinolate content since reactions are best if glucosinolate breakdown products are in large volume	Youngs and Wetter (1967)
UV spectroscopy of thio-urea derivatives of the isothiocyanates	Low sensitivity but good reproducibility of results	Appelqvist and Josefsson (1967); Daxenbichler *et al.* (1970); Wetter and Youngs (1976)
Glucosinolate–palladate coloured complex-based spectroscopic determination	Efficient, fast and reliable method; however, it involves expensive reagents	Thies (1983); Kolovrat (1988)
Thymol method – based on determination of glucosinolate content using thymol reagent	Low repeatability of results and hazardous reagents involved	Brzezinski and Mendelewski (1984); Truscott and Shen (1987); DeClercq and Daun (1989); Tholen *et al.* (1989)
Glucose–UV test and Glukotest: myrosinase catalysed glucosinolate breakdown glucose estimation	Accurate estimation but enzyme system based glucose estimation requires stringent maintenance experiment conditions for consistence in results	Lein (1970); Bjorkman (1972)
Test-tape method	Rapid and efficient method suitable for commercial screening	Comer (1956); Lein (1970); Van Etten *et al.* (1974); McGregor and Downey (1975)

Continued

Table 5.6. Continued.

Parameter/trait and method or basic principal	Advantage/drawback	Reference
ELISA-based colorimetric estimation using glucosinolate – specific antibody complex or glucosinolate – sodium tetrachloropalladate complex	Rapid and efficient method for quantitative estimation	Hassan et al. (1988); Kumar et al. (2004)
Spectroscopy-based glucosinolate estimation		
NIRS	Rapid, accurate and non-destructive but requires large sample representing the high variability in glucosinolate content, is species specific and requires expensive equipment	Daun and Williams (1995); Mika et al. (2003)
X-ray fluorescence or reflectance	Non-destructive but requires expensive equipment and the risk of handling hazardous radiations	Schnug and Haneklaus (1988); Tholen et al. (1993)
Chromatographic methods for determination of individual glucosinolates		
GLC technique for analysis of hydrolysis products of glucosinolates using myrosinase enzyme or as their trimethyl-silyl-desulfo derivatives	Accurate estimation of individual glucosinolate	Underhill and Kirkland (1971); Persson (1974); Thies (1980); Brzezinski et al. (1986); Slominski and Campbell (1987)
High performance liquid chromatography (HPLC) based on separation of desulfoglucosinolates	Accurate but requires time consuming enzymatic desulfatation step due to which some glucosinolates may escape detection	Sang and Truscott (1984); Spinks et al. (1984); McGregor (1985); Bjerg and Sorenson (1986); Palmer et al. (1987)
HPLC-based estimation of intact glucosinolates		
Reverse phase HPLC of intact glucosinolates	Efficient and accurate estimation of individual glucosinolates	Kaushik and Agnihotri (1999)

References

Agnihotri, A., Prem, D. and Gupta, K. (2007) The chronicles of oil and meal quality improvement in oilseed rape. In: Gupta, S.K. (ed.) *Advances in Botanical Research: Oilseed Rape Breeding.* Elsevier Publishers, Academic Press, New York, pp. 45, 49–97.

Agnihotri, A., Kumar, A. and Singh, N.B. (2011) The oil and meal quality improvements in rapeseed-mustard: status in India. In: *Proceedings 13th International Rapeseed Congress: Brassica 2011*, Prague, Czech Republic, 5–9 June.

Annunziata, M.G., Attico, A., Woodrow, P., Alessandra liva, M., Fuggi, A. and Carillo, P. (2012) An improved fluorimetric HPLC method for quantifying tocopherols in *Brassica rapa* L. subsp. *sylvestris* after harvest. *Journal of Food Composition and Analysis* 27(2), 145–150.

Anonymous (2000) *AOCS official method Am 2-93.* Sampling and analysis of vegetable oil source materials – Determination of oil content in oilseeds American Oil Chemist Society International, CD-ROM. Available at: http://www.aoac.org/pubs/omacd_rom_revised.htm (accessed July 2005).

Anonymous (2005) *Annual Progress Report, All India Coordinated Research Project on Rapeseed- mustard, ICAR.* National Research Centre on Rapeseed-Mustard, Bharatpur, Rajasthan, India.

AOAC (1975) *Official Methods of Analysis of the Association of Analytical Chemists*, 12th edn. AOAC, Washington, DC, pp. 136–137.

AOAC (1995) *Official Methods of Analysis of the Association of Analytical Chemists*, 16th edn. AOAC, Washington, DC, p. 1141.

Appelqvist, L.A. (1968) Rapid methods of lipid extraction and fatty acid methyl ester preparation for seed and leaf tissue with special remarks on preventing the accumulation of lipid contaminants. *Arkiv Kemi* 28, 551–570.

Appelqvist, L.A. and Josefsson, E. (1967) Method for quantitative determination of isothiocyanates and ox-azolidinethiones in digests of seed meals of rapeseed and turnip rape. *Journal of Science and Food Agriculture* 18, 510–519.

Appelqvist, L.A.D., Kornfeldt, A.K. and Wennerholm, J.E. (1981) Sterols and steryl esters in some *Brassica* and *Sinapis* seeds. *Photochemistry* 20, 207–210.

Banerjee, P., Dawson, G. and Dasgupta, A. (1992) Enrichment of saturated fatty acid containing phospholipids in sheep brain serotonin receptor preparations: use of microwave irradiation for rapid transesterification of phospholipids. *Biochemistry and Biophysics* 1110, 65–74.

Bandel, V.A., Mulford, F.R., Ritter, R.L., Kantzes, G.J. and Hellman, J.L. (1991) Fact Sheet 635: Canola production guidelines University of Maryland and USDA. Available at: http://www.agnr.umd.edu/MCE/Publications/PDFs/FS635.pdf (accessed July 2005).

Banga, S.K. (1996) Breeding for oil and meal quality. In: Chopra, V.L. and Prakash, S. (eds) *Oilseed and Vegetable Brassicas: Indian Perspective*. Oxford and IBH publishing Company, New Delhi, pp. 234–249.

Bell, J.M. (1984) Nutrients and toxicants in rapeseed meal: a review. *Journal of Animal Science* 58, 996–1010.

Bengtsson, L. (1985) Some experiences of using different analytical methods in screening for oil and protein content in rapeseed. *Fette, Seifen, Anstrichm* 87, 262.

Bennett, R.N., Kiddle, G., Hick, A.J., Dawson, G.W. and Wallsgrove, R.M. (1996) Distribution and activity of microsomal NADPH-dependent monooxygenases and amino acid decarboxylases in cruciferous and non-cruciferous plants and their relationship to foliar glucosinolate content. *Plant, Cell and Environment* 19, 801–812.

Benveniste, P. (2002) Sterol metabolism. In: Meyerowitz, E.M. and Somerville, C. (eds) The *Arabidopsis* book. American Society of Plant Biologists, pp. 1–31. Available at: http://www.aspb.org/publications/Arabidopsis (accessed 12 June 2007).

Bille, N., Eggum, B.O., Jacobsen, I., Olseno, O. and Sorensen, N. (1983) Antinutritional and toxic effects in rats of individual glucosinolates (+) myrosinases added to a standard diet I. Effects on protein utilization and organ weight. *Tierphysiol Tierer nahar Futter-mittelkd* 49, 195–210.

Biston, R., Dardenne, P., Cwikowski, M., Marlier, M., Severin, M. and Wathelet, J.P. (1988) Fast analysis of rapeseed glucosinolates by near infrared reflectance spectroscopy. *Journal of the American Oil Chemists' Society* 65, 1599–1600.

Bjerg, B. and Sorenson, H. (1986) Quantitative analysis of glucosinolates in oilseed rape based on HPLC of desulfoglucosinolates and HPLC of intact glucosinolates. *Proceedings of CEC Workshop glucosinolates in rapeseeds*, Gembloux 1–3 October, Springer, Netherlands.

Bjerg, B., Larsen, L.M. and Sorensen, H. (1987) Reliability of analytical methods for quantitative determination of individual glucosinolates and total glucosinolate content in double low oilseed rape. In: *Proceedings 7th International Rapeseed Congress, Poznan, Poland*, 11–14 May 1987, pp. 1330–1341.

Bjorkman, R. (1972) Preparative isolation and ^{35}S-labelling of glucosinolates from rapeseed (*Brassica napus* L.). *Acta Chimica Scandinavica* 26, 1111–1116.

Bjorkqvist, B. and Hase, A. (1988) Separation and determination of intact glucosinolates in rapeseed by high performance liquid chromatography. *Journal of Chromatography* 435, 501–507.

Bonanone, A., Pagnan, A., Biffani, S., Opportuno, A., Sogato, H., Dorella, M., Maiorino, M. and Potts, D.A. (1992) Effects of dietary monounsaturated and polyunsaturated fatty acids on susceptibility of plasma LDL to oxidative modification. *Arteriosclerosis and Thrombosis* 12, 529–533.

Brzezinski, W. and Mendelewski, P. (1984) Determination of total glucosinolate content in rapeseed meal with thymol reagent. *Zeitschrift für Pflanzenzüchtung* 93, 177–183.

Brzezinski, W., Mendelewski, P. and Musse, B.G. (1986) Comparative study on determination of glucosinolates in rapeseed. *Cruiceferae Newsletter* 11, 128–129.

Buchner, R. and Thies, W. (1987) HPLC analysis of glucosinolates in OO-rapeseed. In: *Proceedings 7th International Rapeseed Congress, Poznan, Poland*, 11–14 May 1987, pp. 1322–1329.

Careri, M., Elviri, L. and Mangia, A. (2001) Liquid chromatography–UV determination and liquid chromatography–atmospheric pressure chemical ionization mass spectrometric characterization of sitosterol and stigmasterol in soybean oil. *Journal of Chromatography* 935, 249–257.

Castelli, W.P. (1984) Epidemiology of coronary heart disease: The Framingham Study. *American Journal of Medicine* 76, 4–12.

Chadd, S.A., Davies, W.P. and Koivisto, J.M. (2002) Practical production of protein for food animals. In: Proceedings Protein sources for the animal feed industry–expert consultation and workshop 29 April–3 May 2002, Bangkok. Available at: http://www.fao.org/documents/show_cdr.asp?url_file=/docrep/007/y5019e/y5019e00.htm (accessed July 2005).

Chauhan, J.S., Tyagi, M.K., Kumar, P.R., Tyagi, P., Singh, M. and Kumar, S. (2000) Breeding for oil and seed meal quality in rapeseed mustard in India – a review. *Agricultural Reviews* 23(2), 71–90.

Christensen, B.W., Kjar, A., Madsen, J.O., Olsen, C.E., Olsen, O. and Sorensen, H. (1982) Mass-spectrometric characteristics of some per-trimethylsilylated desulfoglucosinolates. *Tetrahedron* 38, 353–357.

Comer, J.P. (1956) Semi-quantitative specific test paper for glucose in urine. *Annals of Chemistry* 28, 1748–1750.

Conacher, H.B.S. and Chadha, R.K. (1974) Determination of docosenoic acids in fats and oils by gas-liquid chromatography. *Journal of the American Oil Chemists' Society* 57, 1161–1164.

Craig, B.M. and Murty, N.L. (1958) The separation of saturated and unsaturated fatty acid esters by gas-liquid chromatography. *Canadian Journal of Chemistry* 36, 1297–1301.

Craig, B.M. and Murty, N.L. (1959) Quantitative fatty acids analysis of vegetable oils by gas chromatography. *Journal of the American Oil Chemists' Society* 36, 549–552.

Dardanne, P. (1996) Stability of NIR spectroscopy equations. *NIR News* 7, 8–9.

Daun, J.K. and Williams, P.C. (1995) Use of NIR spectroscopy to determine quality factors in harvest surveys of canola. In: *Proceedings 9th International Rapeseed Congress Cambridge*, UK, 4–7 July 1995, pp. 864–866.

Daxenbichler, M.E., Spencer, G.F., Kleiman, R., Van Etten, C.H. and Wolff, I.A. (1970) Gas liquid chromatographic determination of products from the progoitrins in crambe and rapeseed meal. *Annals of Biochemistry* 38, 373–382.

DeClercq, D.R. and Daun, J.K. (1989) Determination of the total glucosinolate content in Canola by reaction with thymol and sulfuric acid. *Journal of American Oil Chemist Society* 66(6), 788–791.

Doorn, H.E., Van, Holst, G.J., Van, Kruk, G.C., Vander, Raaijmakers-Ruijs, N.C.M.E. and Postma, E. (1998) Quantitative determination of the glucosinolates sinigrin and progoitrin by specific antibody EliSA assays in Brussels sprouts. *Journal of Agriculture Food Chemistry* 46, 793–800.

Dorell, D.C. and Downey, R.K. (1964) The inheritance of erucic acid content in rapeseed (*Brassica campestris*). *Canadian Journal of Plant Science* 44, 499–504.

Downey, R.K. (1990) *Brassica* oilseed breeding – achievements and opportunities. *Plant Breeding Abstracts* 60(10), 1165–1170.

Downey, R.K. and Craig, B.M. (1964) Genetic control of fatty acid biosynthesis in rapeseed (*Brassica napus L.*). *Journal of the American Oil Chemists' Society* 41, 475–478.

Downey, R.K. and Harvey, B.L. (1963) Methods of breeding for oil quality in rape. *Canadian Journal of Plant Science* 43, 271–275.

Downey, R.K. and Rimmer, S.R. (1993) Agronomic improvements in oilseed Brassicas. *Advances in Agronomy* 50, 1–66.

El-Beltagi, H.S. and Mohamed, A.A. (2010) Variations in fatty acid composition, glucosinolate profile and some phytochemical contents in selected oil seed rape (*Brassica napus* L.) cultivars. *Grasas Y Aceites* 61, 143–150.

Elhiti, M., Yang, C., Chan, A., Durnin, D.C., Belmonte, M.F., Ayele, B.T., Tahir, M. and Stasolla, C. (2012) Altered seed oils and glucosinolate levels in transgenic plants over-expressing the *Brassica napus* SHOOT MERISTEM LESS gene. *Experimental Botany* 1–15.

Eskin, N.A., McDonald, B.E., Przybylski, R., Malcolmson, L.J., Scarth, R., Mag, T., Ward, K. and Adolf, D. (1996) Canola oil. In: Hui, Y.H. (ed.) *Baily's Industrial Oil and Fat Products*. John Wiley & Sons, New York, pp. 1–95.

Glen, D.M., Jones, H. and Fieldsend, J.K. (1990) Damage to oilseed rape seedlings by the field slug *Deroceras reticulatum* in relation to glucosinolate concentration. *Annals of Applied Biology* 117, 197–207.

Gopalan, C.D., Krishnamurthi, D., Shenolikar, I.S. and Krishnamachari, K.A.V.R. (1974) Myocardial changes in monkey fed on mustard oil. *Nutrition and Metabolism* 6, 352–365.

Grami, B., Baker, R.J. and Stefansson, B.R. (1977) Genetics of protein and oil content in summer rape. Heritability, number of effective factors and correlations. *Canadian Journal of Plant Science* 57, 937–943.

Greenwood, C.F., Allen, J.A., Leong, A.S., Pallot, T.N., Golder, T.M. and Golebiowski, T. (1999) An investigation of the stability of NIRS calibration for the analysis of oil content in whole seed canola In: *Proceedings 10th International Rapeseed Congress,* 26–29 September, Canberra, Australia.

Gurr, M. (1992) Dietary lipid and coronary heart disease. *Journal of Lipid Research* 31, 195–243.

Guzman, I., Gad, G.Y. and Brown, A.F. (2012) Simultaneous extraction and quantitation of carotenoids, chlorophylls and tocopherols in Brassica vegetables. *Journal of Agricultural and Food Chemistry* 60(29), 7238–7244.

Harvey, B.L. and Downey, R.K. (1964) The inheritance of erucic acid content in rapeseed (*Brassica napus*). *Canadian Journal of Plant Science* 44, 104–111.

Hassan, F., Rothnie, N.E., Yeung, S.P. and Palmer, M.V. (1988) Enzyme linked immunosorbent assays for alkenyl glucosinolates. *Journal of Agriculture and Food Chemistry* 36, 398–403.

Heaney, R.K. and Fenwick, G.R. (1980) The quantitative analysis of indole glucosinolates by gas chromatography – the importance of the derivatisation conditions. *Journal of Science and Food Agriculture* 31, 593–599.

Heaney, R.K. and Fenwick, G.R. (1982) The analysis of glucosinolates in *Brassica* species using gas chromatography. Direct determination of the thiocyanate ion precursors, glucobrassicin and neoglucobrassicin. *Journal of Science and Food Agriculture* 33, 68–70.

Heaney, R.K., Spinks, E.A., Hanley, A.B. and Fenwick, G.R. (1986) *Analysis of glucosinolates in rapeseed. Technical Bulletin*. AFRC Food Research Institute, Norwich.

Hougen, F.W. and Bodo, V. (1973) Extraction and methanalysis of oil from whole or crushed rapeseed for fatty acid analysis. *Journal of the American Oil Chemists' Society* 50, 230–234.

Hüsken, A., Baumert, A., Strack, D., Becker, H.C., Möllers, C. and Milkowski, C. (2005) Reduction of sinapate ester content in transgenic oilseed rape (*Brassica napus* L.) by dsRNAi-based suppression of BnSGT1 gene expression. *Molecular Breeding* 16, 127–138.

Hussain, N., Jabeen, Z., Li, Y.L., Chen, M.X., Li, Z.L., Guo, W.L. and Jiang, L.X. (2013) Detection of tocopherol in oilseed rape (*Brassica napus* L.) using gas chromatography with flame ionization detector. *Journal of Integrative Agriculture* 12(5), 803–814.

Iqbal, M.C.M., Robbelen, G. and Mollers, C. (1995) Biosynthesis of glucosinolates by microspore derived embryoids and plantlets *in vitro* of *B. napus* L. *Plant Science* 112, 107–115.

Kaufmann, H.P. and Fiedler, H. (1938) Untersuehungen fiber die Brauehbarkeit des Maekey-Schmutzes in den Laugen Fette u. Seif. 46, 292–299 [cited in McGregor, D.I. (1980) Analytical chemistry of rapeseed and its products. In: *Proceedings of a symposium*, 5–6 March, Winnipeg, Canada, pp. 59–66].

Kaushik, N. and Agnihotri, A. (1997) Evaluation of improved method for determination of rapeseed-mustard FAMES by GC. *Chromatographia* 44, 97–99.

Kaushik, N. and Agnihotri, A. (1999) High performance liquid chromatographic method for separation and quantification of intact glucosinolates. *Chromatographia* 49(5/6), 281–284.

Kaushik, N., Gautam, S. and Agnihotri, A. (2007) Variability of glucosinolates in *Brassica* germplasm collections. In: Fu, T.-D. and Guan, C.-Y. (eds) *Proceedings of the 12th International Rapeseed Congress, Wuhan, China; Sustainable Development in Cruciferous Oilseed Crops Production*. Science Press USA Inc., 1, 371–373.

Kilam, D., Aneja, J.K., Goyal, P. and Agnihotri, A. (2015) Piriformospora indica mediated enhancement in phytosterol content in *Brassica juncea*. In: *Proceedings National Seminar on Strategic Interventions to Enhance Oilseeds Production in India*, 19–21 February, DRMR (ICAR), Bharatpur, pp. 281–283.

Kim, E. and Goldberg, M. (1969) Serum cholesterol assay using a stable Liebermann-Burchard reagent. *The American Association of Clinical Chemists* 15(12), 1171–1179.

Kolovrat, O. (1988) Use of a modified palladium test in winter swede rape breeding. *Rostlinna-Vyroba* 34(6), 667–672.

Kondra, Z.P. and Stefansson, B.R. (1970) Inheritance of the major glucosinolates of rapeseed (*Brassica napus*) meal. *Canadian Journal of Plant Science* 50, 643–647.

Kraling, K., Robbelen, G., Thies, W., Herrmann, H. and Ahmadi, M.R. (1990) Variation of seed glucosinolates in lines of *B. napus*. *Plant Breeding* 105, 33–39.

Kumar, A., Sharma, P., Thomas, L., Agnihotri, A. and Banga, S.S. (2009) Canola cultivation in India: scenario and future strategy. In: *Proceedings of the 16th Australian Research Assembly on Brassicas*, Ballarat, Australia, 14–16 September, pp. 5–9.

Kumar, P.R. and Tsunoda, S. (1980) Variation in oil content and fatty acid composition among seeds from the Crucifereae. In: Tsunoda, S., Hinata, K. and Gomez-Campo, C. (eds) *Brassica Crops and Wild Allies*. Japan Scientific Societies Press, Tokyo, pp. 235–252.

Kumar, S. and Agnihotri, A. (2004) Value addition in Rapeseed Mustard: Status and Prospects. In: Kumar, A. and Singh, N.B. (eds) *Rapeseed-Mustard Research in India*. National Research Centre on Rapeseed-Mustard, Indian Council of Agricultural Research, Sewar, Bharatpur, India, pp. 212–231.

Kumar, S., Singh, A.K., Kumar, M., Yadav, S.K., Chauhan, J.S. and Kumar, P.R. (2003) Standardization of near-infrared spectroscopy (NIRS) for determination of seed oil and protein content in rapeseed-mustard. *Journal of Food Science and Technology* 40, 306–309.

Kumar, S., Yadav, S.K., Chauhan, J.S., Singh, A.K., Khan, N.A. and Kumar, P.R. (2004) Total glucosinolate estimation by complex formation between glucosinolate and tetrachloropalladate (II) using ELISA reader. *Journal of Food Science and Technology* 41, 63–65.

Lange, R. (1987) Glucosinolates and their breakdown products in seeds of standard and high quality rape species. In: *Proceedings 7th International Rapeseed Congress, Poznan, Poland, 11-14 May* 1987, pp. 1509–1513.

Law, M. (2000) Plant sterol and stanol margarines and health. *British Medical Journal* 320, 861–864.

Leckband, G., Rades, H., Frauen, M. and Friedt, W. (2003) *NAPUS* 2000 – a research programme for the improvement of the whole rapeseed. In: *Proceedings of the 11th International Rapeseed Congress*, 6–10 July, The Royal Veterinary and Agricultural University, Copenhagen, Denmark, pp. 209–211.

Lein, K.A. (1970) Quantitative Bestimmungsmethoden for samenglucosinolate in *Brassica* Arten and Ihre Anwendung in der Zuchtung Von Glucosinolataramen Raps. *Z. Pflanzenzuchtg* 63, 137–154.

Lepage, G. and Roy, C.C. (1986) Direct transesterification of all classes of lipids in a one-step reaction. *Journal of Lipid Research* 27, 114–120.

Lichter, R., Groot, E. de, Fiebig, D., Schweiger, R. and Gland, A. (1988) Glucosinolates determined by HPLC in the seeds of microspore-derived homozygous lines of rapeseed (*Brassica napus* L.). *Plant Breeding* 100, 209–221.

Liu, X., Chen, M., He, S. and Wu, C. (2007) Study on extraction, isolation and bioactivities of phytosterol from rapeseed. *Proceedings of the 12th International Rapeseed Congress*, 26–30 March, Wuhan, China, pp. 150–151.

Madsen, E. (1976) Nuclear magnetic resonance spectrometry as a quick method of determination of oil content in rapeseed. *Journal of the American Oil Chemists' Society* 53, 467–469.

Marwede, V., Schierholt, A., Möllers, C. and Becker, H.C. (2004) Genotype × environment interactions and heritability of tocopherols content in canola. *Crop Science* 44, 728–731.

McGregor, D.I. (1974) A rapid and sensitive spot test for linolenic acid levels in rapeseed. *Canadian Journal of Plant Science* 54, 211–213.

McGregor, D.I. (1977) A rapid and simple method of screening rapeseed and mustard seed for erucic acid content. *Canadian Journal of Plant Science* 57, 133–142.

McGregor, D.I. (1980) In: Daun, J.K., McGregor, D.I. and McGregor, E.E. (eds) *Analytical Chemistry of Rapeseed and its Products*. A symposium, 5–6 May 1980, Winnipeg, Canada, pp. 59–66.

McGregor, D.I. (1985) Determination of glucosinolate in *Brassica* seed. *Cruciferae Newsletter* 10, 132–136.

McGregor, D.I. and Downey, R.K. (1975) A rapid and simple assay for identifying low glucosinolate in rapeseed. *Canadian Journal of Plant Science* 55, 191–196.

McGregor, D.I., Mullin, W.J. and Fenwick, G.R. (1983) Review of analysis of glucosinolates: Analytical methodology for determining glucosinolate composition and content. *Journal of the Association of Official Analytical Chemists* 66, 825–849.

Mika, V., Tillman, P., Koprna, R., Nerusil, P. and Kucera, V. (2003) Fast prediction of quality parameters in whole seeds of oilseed rape (*Brassica napus* L.). *Plant and Soil Environment* 49(4), 141–145.

Milkowski, C. and Strack, D. (2010) Sinapate esters in brassicaceous plants: biochemistry, molecular biology, evolution and metabolic engineering. *Planta* 232(1), 19–35.

Miller, R.W., Van Etten, C.H., Mc Grew, C.E., Wolf, I.A. and Jones, Q. (1962) Amino acid composition of seed meals from 41 species of Cruciferae. *Journal of Agriculture and Food Chemistry* 10, 426–430.

Mitra, R. and Bhatia, C.R. (1979) Bioenergetic considerations in the improvement of oil content and quality in oilseed crop. *Theoretical and Applied Genetics* 54, 41–47.

Mitten, R. (1992) Leaf glucosinolate profiles and their relationship to pests and disease resistance in oilseed rape. *Euphytica* 63, 71–83.

Mohamed Ahmidaa, H.S., Bertucci, P., Franzo, L., Massouda, R., Cortese, C., Lala, A. and Federici, G. (2006) Simultaneous determination of plasmatic phytosterols and cholesterol precursors using gas chromatography mass spectrometry (GCMS) with selective ion monitoring (SIM). *Journal of Chromatography B* 842, 43.

Moller, P., Olsen, O., Ploger, A., Rasmussen, K.W. and Sorensen, H. (1985) Quantitative analysis of individual glucosinolates in double low oilseed rape by HPLC of intact glucosinolates. In: Sorensen, H. (ed.) *Advances in the Production and Utilization of Cruciferous Crops with Special Emphasis to Oilseed Rape*. Copenhagen, pp. 111–126.

Mortuza, M.J. (2006) Tocopherol and sterol contents of some rapeseed mustard cultivars developed in Bangladesh. *Pakistan Journal of Biological Sciences* 9(9), 1812–1816.

Newton, I.S. (1998) Long chain polyunsaturated fatty acids – the new frontier for nutrition. *Lipid Technology* 10, 77–81.

Nissinen, M., Gylling, H., Vuoristo, M. and Miettinen, T.A. (2002) Micellar distribution of cholesterol and phytosterols after duodenal plant stanol ester infusion. *American Journal of Physiology – Gastrointestinal and Liver Physiology* 282, 1009–1015.

Nowak, H., Kujava, R., Zadernowski, R., Rocznaik, B. and Kowzlowska, H. (1992) Antioxidative and bactericidal properties of phenolic compounds in rapeseeds. *Fat Science Technology* 94, 149–152.

Pallot, T.N., Leong, A.S., Allen, J.A., Golder, T.M., Greenwood, C.F. and Golebiowski, T. (1999) Precision of fatty acid analyses using near infrared spectroscopy of whole seed Brassicas. In: Wratten, N. and Salisbury, P.A. (eds) *New Horizons for an old crop. Proceedings of the 10th International Rapeseed Congress*, 26–29 September 1999, Canberra, Australia.

Palmer, M.V., Yeung, S.P. and Sang, J.P. (1987) Glucosinolate content of seedlings, tissue cultures and regenerant plants of *B. juncea* (Indian mustard). *Journal of Agriculture and Food Chemistry* 35, 262–265.

Persson, S. (1974) A method for determination of glucosinolates in rapeseed as TMS-derivatives. In: *Proceedings of 4th International Rapeseed Conference*, 4–8 June, Giessen, West Germany, pp. 381–386.

Piironen, V., Lindsay, D.G., Miettinen, T.A., Toivo, J. and Lampi, A.M. (2000) Review sterols: Biosynthetis, biological function and their importance to human nutrition. *Journal of the Science of Food and Agriculture* 80, 939–966.

Prabhu, H.R. (2000) Lipid peroxidation in culinary oils subjected to thermal stress. *Indian Journal of Clinical Biochemistry* 15, 1–5.

Prakash, S., Kumar, P.R., Sethi, M., Singh, C. and Tandon, R.K. (2000) Mustard oil – the ultimate edible oil. *The Botanica* 50, 94–101.

Prem, D. (2006) Induction of genetic variability for agro-morphological and biochemical traits in Indian mustard [*Brassica juncea* (L.) Czern & Coss.] through chemical mutagenesis in conjugation with doubled haploid technology. Dissertation PhD thesis. TERI University, New Delhi.

Prem, D., Gupta, K., Sarkar, G. and Agnihotri, A. (2012). Determination of oil, protein and moisture content in whole seeds of three oleiferous *Brassica* species using near-infrared reflectance spectroscopy. *Journal of Oilseed Brassicas* 3(2), 88–98.

Rakow, G. and Raney, J.P. (2003) Present status and future perspectives of breeding for seed quality in *Brassica* oilseed crops. In: *Proceedings 11th International Rapeseed Congress*, 6–10 July, The Royal Veterinary and Agricultural University, Copenhagen, Denmark, pp. 181–185.

Ramchiary, N., Padmaja, K.L., Sharma, S., Gupta, V., Sodhi, Y.S., Mukhopadhyay, A., Arumugam, N., Pental, D. and Pradhan, A.K. (2007) Mapping of yield influencing QTL in *Brassica juncea*: Implications for breeding of major oilseed crop of dryland areas. *Theoretical and Applied Genetics* 115, 807–817.

Reinhardt, T.C. and Robbelen, G. (1991) Quantitative analysis of fatty acids in intact rapeseed by Near Infrared Reflectance Spectroscopy. In: *Proceedings of the 8th International Rapeseed Congress*, Saskatoon, Canada, pp. 1380–1384.

Robbelen, G. and Thies, W. (1980) Variation in rapeseed glucosinolates and breeding for improved meal quality. In: Tsunoda, S., Hinata, K. and Gomez-Campo, C. (eds) *Brassica Crops and Wild Allies*. Japan Science Society Press, Tokyo, pp. 285–299.

Sabir, S.M., Hayat, I. and Gardezi, S.D. (2003) Estimation of sterols in edible fats and oils. *Pakistan Journal of Nutrition* 2(3), 178–181.

Samija, A., Heiko, C., Becker and Christian, M. (2007) Genetic variation and genotype × environment interactions for phytosterol content in rapeseed (*Brassica napus* L.). *Proceedings of the 12th International Rapeseed Congress*, 26–30 March, Wuhan, China, pp. 340–342.

Sang, J.P. and Truscott, R.J.W. (1984) Liquid chromatographic determination of glucosinolates in rapeseed as desulfoglucosinolates. *Journal of Association of Official Analytical Chemist* 67, 829–833.

Sauer, F.D. and Kramer, J.K.G. (1983) The problems associated with the feeding of high erucic acid rapeseed oils and some fish oils to experimental animals. In: Kramer, J.K.G., Sauer, F.D. and Figden, W.J. (eds) *High and Low Erucic Acid Rapeseed Oils*. Academic Press, Toronto, Canada, pp. 254–292.

Schnug, E. and Haneklaus, S. (1988) Theoretical principles for the indirect determination of the total glucosinolate content in rapeseed and meal quantifying the sulphur concentration via X-ray fluorescence (X-RF method). *Journal of Science and Food Agriculture* 45, 243–254.

Scialabba, A., Salvini, L., Faqi, A.S. and Bellani, L.M. (2009) Tocopherols, fatty acids and phytosterols content in seeds of nine wild taxa of Sicilian *Brassica* (*Cruciferae*). *Plant Biosystemics* 144(3), 626–633.

Shenk, J. and Westerhaus, M.O. (1993) *Analysis of Agriculture and food products by Near-infrared Reflectance Spectroscopy*. Infrasoft International, Port Matilda, Pennsylvania, pp. 41–51.

Siebel, J. and Pauls, K.P. (1989) Inheritance pattern of erucic acid content in populations of *Brassica napus* microspore-derived spontaneous diploids. *Theoretical and Applied Genetics* 77, 489–494.

Sil, S., Chkrabarti, J. and Gandhi, R.S. (1990) Sterol profile of the seed oils of a few members of the compositae family. *Journal of Food Science Technology* 27, 234–235.

Simonne, A.H., Simonne, E.H., Eitenmiller, R.R., Mills, H.A. and Cresman, C.P. (1997) Could the Dumas method replace the Kjeldahl digestion for nitrogen and crude protein determinations in foods? *Journal of Science Food and Agriculture* 73, 39–45.

Singh, V., Moreau, R.A. and Hicks, K.B. (2003) Yield and phytosterol composition of oil extracted from grain sorghum and its wet-milled fractions. *Cereal Chemistry* 80(2), 126–129.

Slominski, B.A. and Campbell, L.D. (1987) Gas chromatographic determination of Indole glucosinolates – A re-examination. *Journal of Science and Food Agriculture* 40, 131–143.

Sorenson, H. (1991) Glucosinolates: structure-properties-function. In: Shahidi, F. (ed.) *Canola and Rapeseed*. Van Nostrand Rheinhold, New York, pp. 149–172.

Sosulski, F.W. and Dabrowski, K.J. (1984) Determination of glucosinolates in canola meal and protein products by desulfatation and capillary gas-liquid chromatography. *Journal of Agriculture and Food Chemistry* 32, 1172–1175.

Spinks, E.A., Sones, K. and Fenwick, G.R. (1984) The quantitative analysis of glucosinolates in cruciferous vegetables, oilseeds and forage crops using high performance liquid chromatography. *Fette, Seifen, Anstrichmittel* 86, 228–231.

Stahl, E. (1969) *Thin Layer Chromatography - A laboratory handbook*. Springer, New York.

Stiepel, M. (1926) Geschichte der chinesischen Philosophie, *Bol.1* Das klassische Zeitalzer bis zur Han Dynastie [cited in Mc Groger, D.I. (1980) Analytical chemistry of rapeseed and its products. In: *Proceedings of a Symposium*, 5–6 March, Winnipeg, Canada, pp. 59–66].

Stringam, G.R. and McGregor, D.I. (1980) Inheritance and fatty acid composition of a yellow-embryo mutant in turnip rape (*Brassica campestris* L.). *Canadian Journal of Plant Science* 60, 97–102.

Stringam, G.R., Mc Gregor, D.I. and Pawlowski, S.H. (1974) Chemical and morphological characteristics associated with seed colour in rapeseed In: *Proceedings of the 4th International Science Congress*, Giessen, West Germany, pp. 99–108.

Sutariya, D.A., Patel, K.M., Bhadauria, H.S., Vaghela, P.O., Prajapati, D.V. and Parmar, S.K. (2011) Genetic diversity for quality traits in Indian mustard (*Brassica juncea* L.). *Journal of Oilseed Brassica* 2(1), 44–47.

Taylor, D.C., Zhang, Y., Kumar, A., Francis, E., Giblin, E.M., Barton, D.L., Ferrie, J.R., Laroche, A., Shah, S., Zhu, W., Snyder, C.L., Hall, L., Rakow, G., Harwood, J.L. and Weselake, R.J. (2009a) Molecular modification of triacylglycerol accumulation by over-expression of DGAT1 to produce canola with increased seed oil content under field conditions. *Botany* 87, 533–543.

Taylor, D.C., Guo, Y., Katavic, V., Mietkiewska, F.T. and Bettger, W. (2009b) New seed oils for improved human and animal health and as industrial feed stocks; genetic manipulation of the Brassicaceae to produce oils enriched in nervonic acid. In: Krishnan, A.B. (ed.) *Modification of Seed Composition to Promote Health and Nutrition*. ASA-CSSA-SSSA Publishing, Medison, Wisconsin, pp. 219–233.

Thies, W. (1971) Rapid and simple analysis of the fatty acid composition in individual rape cotyledons. I. *Methods of Gas and Paper Chromatography Z Pflanzenzuchtung* 65, 181–202.

Thies, W. (1974) New methods for the analysis of rapeseed constituents. *Proceedings of the 4th International Rapeseed Conference*, Giessen, West Germany, 4–8 June, pp. 275–282.

Thies, W. (1976) Quantitative gas liquid chromatography on a microliter scale. *Fette Seifen Anstrichmittel* 78, 231–234.

Thies, W. (1977) Analysis of glucosinolates in seeds of rapeseed (*B. napus* L.): concentration of glucosinolates by ion exchange. *Zeitschrift für Pflanzenzüchtung* 79, 331–335.

Thies, W. (1978) Quantitative analysis of glucosinolates after their enzymatic desulfation on ion exchange columns. *Proceedings of the 5th International Rapeseed Conference*, Malmo, Sweden, 1, 136–139.

Thies, W. (1979) Detection and utilization of a glucosinolate sulphohydrolase in the edible snail *Helix pomatia*. *Naturwissenschaften* 66, 364–365.

Thies, W. (1980) Analysis of glucosinolates via 'on-column' desulfation. In: Daun, J., McGregor, D.I. and McGregor, E.E. (eds) *Analytical Chemistry of Rapeseed and its Products*. The Canola Council of Canada, Winnipeg.

Thies, W. (1983) Complex formation between glucosinolates and tetra chloropalladate (II) and its utilization in plant breeding *Fette Seifen Anstrichm* 84, 338–342.

Thies, W. (1985) Determination of the glucosinolate content in commercial rapeseed loads with a pocket reflectometer. *Fette, Seifen, Anstrichmittel* 87, 347–350.

Tholen, J.T., Shiefeng, S., Truscott, R.J.W. and Roger, J.W. (1989) The thymol method for glucosinolate determination. *Journal of the Science of Food and Agriculture* 49, 157–165.

Tholen, J.T., Buzza, G., McGregor, D.I. and Truscott, R.J.W. (1993) Measuring of the glucosinolate content in rapeseed using the trubluglu meter. *Plant Breeding* 110, 137–143.

Tiwari, P.N., Gambhir, P.N. and Rajan, T.S. (1974) Rapid and non-destructive determination of seed oil by pulsed NMR technique. *Journal of American Oil Chemist Society* 51, 104–109.

Tkachuk, R. (1981) Oil and protein analysis of whole rapeseed kernels by near infrared reflectance spectoscopy. *Journal of the American Oil Chemists' Society* 58, 819.

Trautwein, E.A., Duchateau, G., Lin, Y., Melcnikov, S.M., Molhuizen, H. and Ntanios, F.Y. (2003) Proposed mechanisms of cholesterol-lowering action of plant sterols. *European Journal of Lipid Science Technology* 105, 171–185.

Truscott, R.J.W. and Shen, S. (1987) Thymol method for glucosinolate estimation. *Cruciferae Newsletter* 9, 116–118.

Truscott, R.J.W., Burke, D.G. and Minchinton, I.R. (1982) The characterization of a novel hydroxyindole glucosinolate. *Biochemistry and Biophysics Research Communication* 107, 1258–1264.

Truscott, R.J.W., Minchinton, J. and Sang, J. (1983) The isolation and purification of indole glucosinolates from *Brassica* species. *Journal of Science and Food Agriculture* 34, 247–254.

Underhill, E.W. and Kirkland, D.F. (1971) Gas chromatography of trimethlysilyl derivatives of glucosinolates. *Journal of Chromatography* 57, 47–54.

Uzunova, M., Ecke, W., Weibleder, K. and Robbelen, G. (1995) Mapping the genome of rapeseed (*Brassica napus* L.). Construction of an RFLP linkage map and localization of QTLs for seed glucosinolate content. *Theoretical Appllied Genetics* 90, 194–204.

Van Deynze, A.E. and Pauls, K.P. (1994) Seed colour assessment in *Brassica napus* using a near-infrared reflectance spectrometer adapted for visible light measurement. *Euphytica* 76, 45–51.

Van Etten, C.H., McGrew, C.E. and Daxenbichler, M.E. (1974) Glucosinolate determination in cruciferous seeds and meals by measurement of enzymatically released glucose. *Journal of Agriculture and Food Chemistry* 22, 483–487.

Velasco, L. and Mollers, C. (2002) Non-destructive assessment of protein content in single seed of rapeseed (*Brassica napus* L.) by near-infrared reflectance spectroscopy. *Euphytica* 123, 89–93.

Velasco, L., Fernández-Martínez, J.M. and De Haro, A. (1995) The applicability of NIRS for estimating multiple seed quality components in Ethiopian mustard. In: *Proceedings 9th International Rapeseed Congress*, Cambridge, UK, pp. 867–869.

Velasco, L., Fernandez Martinez, J.M. and De Haro, A. (1997) Determination of the fatty acid composition of the oil in intact-seed mustard by near-infrared reflectance spectroscopy. *Journal of the American Oil Chemists' Society* 74, 1595–1602.

Velasco, L., Goffman Fernando, D. and Becker Heiko, C. (1999a) Development of calibration equations to predict oil content and fatty acid composition in *Brassica*ceae germplasm by near-infrared reflectance spectroscopy. *Journal of American Oil Chemist Society* 76, 25–30.

Velasco, L., Pérez-Vich, B. and Fernández-Martínez, J.M. (1999b) Estimation of seed weight, oil content and fatty acid composition in intact single seeds of rapeseed (*Brassica napus* L.) by near-infrared reflectance spectroscopy. *Euphytica* 106, 79–85.

Warner, K. and Mounts, T.L. (1990) Analysis of tocopherols and phytosterols in vegetable oils by HPLC with evaporative light-scattering detection. *Journal of the American Oil Chemists' Society* 67(11), 827–831.

Wetter, L.R. (1955) The determination of mustard oils in rapeseed meal. *Canadian Journal of Biochemistry and Physiology* 33, 980–984.

Wetter, L.R. (1957) The estimation of substituted thiooxazolidines in rapeseed meals. *Canadian Journal of Biochemistry and Physiology* 35, 293–297.

Wetter, L.R. and Youngs, C.G. (1976) A thiourea-UV assay for total glucosinolate content in rapeseed meal. *Journal of the American Oil Chemists' Society* 53, 162–164.

Williams, P.C. (1975) Application of near-infrared reflectance spectroscopy to analysis of cereal grains and oilseeds. *Cereal Chemistry* 52, 561–576.

Wittkop, B., Friedt, W. and Snowdon, R. (2011) Improvement of oilseed rape meal using near-infrared reflection spectroscopy based selection methods. In: *Proceedings 13th International Rapeseed Congress*, 5–9 June, Prague, Czech Republic.

Woyengo, T.A., Ramprasath, V.R. and Jones, P.J. (2009) Anticancer effects of phytosterols. *European Journal of Clinical Nutrition* 63(7), 813–820.

Wu, G., Truksa, M., Datla, N., Vrinten, P., Bauer, J., Zank, T., Cirpus, P., Heinz, E. and Qiu, X. (2005) Stepswise engineering to produce high yields of very long-chained polyunsaturated fatty acids in plants. *Nature Biotechnology* 23, 1013–1017.

Xuyan, D., Bi, X., Fang, W., Mulan, J., Fenghong, H., Guangming, L., Yuandi, Z. and Hong, C. (2011) Continuous microwave-assisted enzymatic hydrolysis system for oilseed protein: design and performance evaluation. In: *Proceedings of the 13th International Rapeseed Congress*, 5–9 June, Prague, Czech Republic.

Youngs, C.G. and Wetter, L.R. (1967) Microdetermination of the major individual isothiocyanates and oxazolidinethiones in rapeseed. *Journal of the American Oil Chemists' Society* 44, 551–554.

Zhang, L.F., Liu, C.F. and Meng, Y.B. (2009) Isolation and determination of sterols in rapeseed oil by HPLC-MS. *Analytical Letters* 42, 1650–1661.

Zhang, X., Cambrai, A., Miesch, M., Roussi, S., Raul, F., Aoude-Werner, D. and Marchioni, E. (2006) Separation of ¢5- and ¢7-Phytosterols by adsorption chromatography and semipreparative reversed phase high performance liquid chromatography for quantitative analysis of phytosterols in food. *Journal of Agriculture and Food Chemistry* 54, 1196–1202.

Zukalova, H. and Vasak, J. (2002) The role and effect of glucosinolates of *Brassica* species – a review. *Rostlinna Vyroba* 48(4), 175–180.

Zum Felde, T., Becker, H.C. and Möllers, C. (2006) Genotype × environment interactions, heritability and trait correlations of sinapate ester content in winter rapeseed (*Brassica napus* L.). *Crop Science* 46, 2195–2199.

6 Genomics of Brassica Oilseeds

Venkatesh Bollina,[1] Yogendra Khedikar,[1] Wayne E. Clarke[1,2] and Isobel A.P. Parkin[1]*

[1]*Agriculture and Agri-Food Canada, Saskatoon;* [2]*Department of Plant Sciences, University of Saskatchewan, Saskatoon, Saskatchewan, Canada*

Introduction

Brassica species are important for oilseed production worldwide and represent a significant agricultural commodity for a number of countries (http://www.fao.org). All brassica crops belong to tribe *Brassiceae* of the family *Brassicaceae*. These are commonly known as mustards due to their natural production of high levels of the secondary metabolites glucosinolates, which contribute to the distinct pungent taste of the seed. In the 1970s breeding efforts to lower the levels of the perceived antinutritionals, glucosinolates and the long-chain saturated fatty acid, erucic acid, from *Brassica napus* (rapeseed) seed led to the development of the most widely grown and economically important brassica crop type, canola (Canadian oil low acid). Brassica oilseeds are currently grown predominantly for human consumption either as canola or mustard types, although recent interest in developing sustainable sources of oil for the production of biofuel and bioproducts has led to increased interest in some of the lesser-studied species, such as *Brassica carinata* (Taylor *et al.*, 2010).

Brassica research has benefited greatly from the close relationship between the brassica crops and the crucifer model *Arabidopsis thaliana*, which was the first plant species to have a fully sequenced and annotated genome (The Arabidopsis Genome Initiative, 2000). The availability of the *A. thaliana* sequence allowed the rapid identification of candidate genes for traits of interest and through comparative mapping provided novel insights into the larger genomes of the brassica crops (Wang, X. *et al.*, 2011; Town *et al.*, 2006; Navabi *et al.*, 2013). However, the genome of *A. thaliana* is still an imperfect model for the crops and the current promise of access to whole genome sequences for each agronomically important *Brassica* species will revolutionize research and development of brassica crops in the coming decade. This chapter will define our current knowledge and tools for brassica research and describe the potential impact of the genome sequence data on brassica crop improvement.

The Foundation of U's Triangle

Foundational work in the 1940s by a Korean scientist, Nagaharu U, defined the accepted yet unique relationship between the six most

*Corresponding author, e-mail: isobel.parkin@agr.gc.ca

commonly grown brassica crop species, referred to as U's triangle (Fig. 6.1). Based on cytogenetic analyses and interspecific crosses, U (1935) determined that the three *Brassica* diploid species (*B. rapa*, *B. oleracea* and *B. nigra*) had paired to form all possible interspecific hybrids to generate three allotetraploid species (*B. napus*, *B. carinata* and *B. juncea*). Brassica researchers have adopted the nomenclature of U such that the genomes of the diploids, named A, B and C for the *B. rapa*, *B. nigra* and *B. oleracea* genomes, respectively, are often used interchangeably with the species name. It is possible to generate newly resynthesized *Brassica* allotetraploids in a laboratory setting yet they are prone to fertility issues resulting from improper homologous pairing between the related diploid genomes (Szadkowski *et al.*, 2010). Allotetraploid variants that could control chromosome pairing would have had a selective advantage and the increased gene complement and overall size resulting from the polyploid genome would have provided

opportunities for colonizing niche environments or responding to periods of environmental stress (Comai, 2005). Similar to many successful crop species, the most widely grown brassica oilseed crops are derived from the polyploid species (*B. napus* and *B. juncea*) presumably selected by farmers due to their increased productivity in comparison to their diploid relatives. Today, *Brassica* species are extensively studied not only for their economic importance but also for the insights that they offer into the genome evolution of polyploid species.

Molecular Markers

Molecular markers have become essential tools for uncovering aspects of plant evolution and their application to crop improvement through marker assisted breeding (MAS) (Xu and Crouch, 2008). In research and development of *Brassica* species markers are

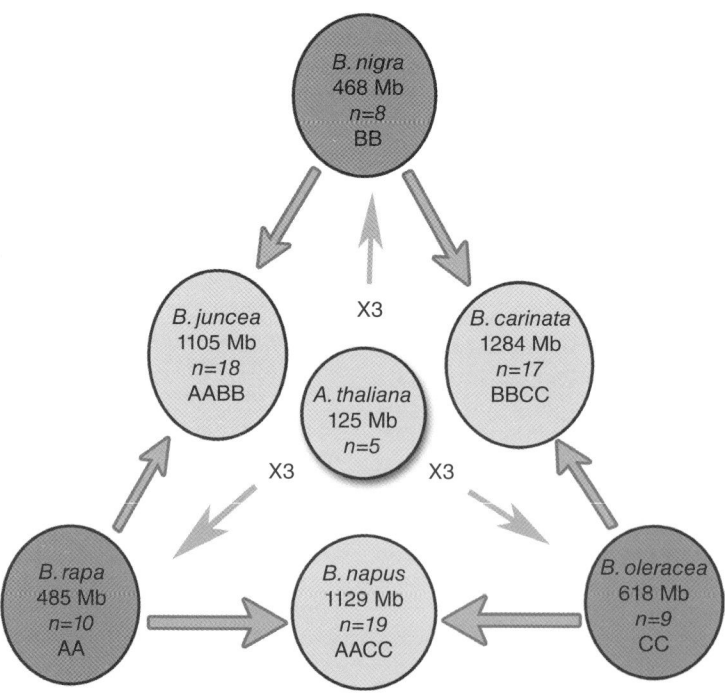

Fig. 6.1. U's triangle of *Brassica* species with their estimated genome sizes. The inferred relationship with the model species *A. thaliana* is indicated, but it should be noted that each of the species shown share a common ancestor with an increased chromosome number.

no less important and dramatic changes in marker technology over the last 20 years have led to the development of multiple marker types. The first genetic maps for *Brassica* species were generated using restriction fragment length polymorphisms (RFLP) (Ferreira *et al.*, 1994). The maps confirmed the work of U (1935), provided insights into the evolution of allotetraploid genomes and uncovered evidence of multiple ancient genome duplication events that predated the formation of the allotetraploids. Although robust and highly reproducible, the use of RFLP markers has been supplanted by alternatives that are less labour intensive and have options for multiplexing. A recent review described the history of marker development in brassica research (Gali and Sharpe, 2012); in the following section, two marker types that are now the most widely adopted in brassica research are introduced. These largely benefitted from increased access to genome sequence data.

Simple sequence repeats markers

Simple sequence repeats (SSRs) or microsatellites were second-generation markers designed to take advantage of developments in DNA amplification through PCR and widespread features of genome architecture. SSRs are short tandem repeats of nucleotides from 1 to 6 bp in length, which are ubiquitous in transcribed and non-transcribed regions (Morgante *et al.*, 2002). SSRs are extensively used in plant genetic studies since they are co-dominant, highly polymorphic, multi-allelic, reproducible and transferable across species. The sequence flanking SSRs can be highly conserved, which allows specific primers to be designed to amplify homologous SSR loci in not only the same species but also across related species (Selkoe and Toonen, 2006).

Earlier efforts to isolate SSRs from *Brassica* species were limited by the approach, which required probe hybridization of size-fractionated genomic libraries to identify clones potentially containing SSR sequences. All clones were sequenced yet this generally yielded low numbers of microsatellites, with between 6 and 15% success rate (Kresovich *et al.*, 1995; Szewc-McFadden *et al.*, 1996). Although improvements

were made to the approach, it remained time consuming and relatively expensive per marker developed. These methods have been largely circumvented by increasing access to genome sequence information, which can be mined for SSRs.

A number of studies have used available sequence data to assess the range and distribution of SSRs in genome sequence data. Burgess *et al.* (2006) exploited shotgun sequences of the *B. oleracea* genome for *in silico* data mining of SSRs, identifying 46,949 SSRs ranging from 10 to 228 bp in repeat length. Ling *et al.* (2007) used end sequences from 110,336 large insert BAC clones generated as part of the *B. rapa* genome sequencing project to identify 12,647 SSRs with repeat lengths ranging from 11 to 153 bp. In both instances the majority, approximately 70%, of the identified repeats were di- or tri-nucleotide repeats with a prevalence of AG and AAG repeats. With the currently available *B. rapa* genome sequence (Wang, X. *et al.*, 2011) it is possible to compare previous efforts, which analysed relatively limited snapshots of the genome, to the actual distribution of repeats found in a comprehensive coverage of the genome. The most marked difference is in the density of SSRs derived from the A/T-rich regions of the genome, which become the most common repeats found. Generation of clone libraries enriched in A/T sequences was notoriously difficult and bias in sequencing or culling of sequences containing repetitive elements may have limited the identification of such SSR types in previous work.

With the ever increasing amounts of publicly available sequence data, mining for SSR markers is relatively simple; however, the utility of the SSR loci in downstream analyses is limited by a number of factors. Each of the SSR types appears to show variation in the levels of polymorphism detected, generally di- and tri-nucleotide repeats tend to be more polymorphic (Sharma *et al.*, 1995). Similarly, intergenic SSR loci appear to show more allelic variation than their genic counterparts (Morgante *et al.*, 2002), presumably due to pressures incurred to maintain the coding regions. In addition, notwithstanding the benefits of cross-species amplification, the conserved nature of the primers used to amplify the SSR loci and the duplicated architecture of plant genomes often results in the amplification of multiple genomic loci, leading to complex and irresolvable marker patterns. In an attempt to assess the available markers for their value for genome analyses in *Brassica* species, we compiled a list over 9000 published SSR primer pairs (Lagercrantz *et al.*, 1993; Kresovich *et al.*, 1995; Szewc-McFadden *et al.*, 1996; Uzanova and Ecke, 1999; Smith and King, 2000; Lowe *et al.*, 2002, 2004; Suwabe *et al.*, 2002; Burgess *et al.*, 2006; Batley *et al.*, 2007; Hopkins *et al.*, 2007; Ling *et al.*, 2007; Iniguez-Luy *et al.*, 2008; Cheng *et al.*, 2009; Kim *et al.*, 2009, 2012; Yu *et al.*, 2009; Parida *et al.*, 2010; Xu *et al.*, 2010; Ge *et al.*, 2011; Li *et al.*, 2011; Ramchiary *et al.*, 2011; Wang, F. *et al.*, 2012) that have been mapped in a range of populations or tested in diversity collections. These markers were aligned to the available genome sequences for the diploid *Brassica* A and C genomes, 2775 and 2069 markers were aligned to the A and C genome, respectively, and their distribution is shown in Plate 1. Although the various methods have identified SSR markers across all regions of the genome, there are notable segments with a paucity of markers that appear to coincide with regions of low gene density.

Single nucleotide polymorphisms

Single nucleotide polymorphisms (SNPs) are generally single base-pair differences between the DNA sequences of any two individuals in a population. Their prevalence throughout the genome makes them an ideal marker for genetic studies. SNPs can be categorized as transversion (A/T, C/G, C/A and T/G) or transition (C/T or G/A) and insertion/deletion (Indel). Biallelic SNPs are most frequent, but tri-allelic and tetra-allelic have also been reported. Development of SNP markers in polyploid species is challenging due to the amplification of multiple fragments caused by orthologous and paralogous gene copies. Recently, large-scale identification of SNPs in brassica oilseeds was undertaken by several groups and has been further utilized in bi-parental or linkage disequilibrium (LD) or association mapping (Table 6.1).

Initial efforts to identify SNP loci in *Brassica* species required the re-sequencing of amplicons derived from genic regions of lines of interest (Westermeier *et al.*, 2009; Durstewitz *et al.*, 2010). The identification of SNP loci is particularly problematic in the *Brassica* allotetraploids where the strong sequence identity between the two constituent genomes leads to co-amplification of homologous loci. However, these methods successfully identified and mapped SNP loci in *B. napus*, suggesting a density of one SNP every 42–216 bp in genic regions. A number of the identified loci were tested using a SNP genotyping array (Illumina GoldenGate) (Durstewitz *et al.*, 2010). Such high-throughput methods rely on discrimination through differential hybridization, which also can be confounded by the presence of homologous genomes. More recently a high density SNP array (Illumina Infinium) was developed for *B. napus*, which allowed over 7000 SNP loci to be queried simultaneously (Delourme *et al.*, 2013). Array-based SNP genotyping provides a robust common platform for genetic analysis, which will provide opportunities for accurate integration of *Brassica* genetic maps, which has historically been a problem for brassica researchers. Although SNP arrays benefit from standardized software tools for genotype calls, which ensures cross-lab compatibility, they are currently limited to use in only the *Brassica* A and C genomes and their utility also will be dependent on available polymorphism of the designed assays.

With the arrival of cost-efficient next generation sequencing (NGS) technologies, it has become feasible to adopt these tools not only to discover SNPs but also simultaneously map such loci. Since the beginning of marker development for molecular plant breeding, it has been a daunting task to develop sufficient markers to saturate a genetic map effectively. The advent of NGS has revolutionized this step, making it possible to determine tens and thousands of molecular markers, especially SNPs, across the genome, with or without access to a reference genome. Despite improvements in sequencing technologies, the generally large and complex nature of any plant genome makes it a challenge to deduce the useful nucleotide diversity,

Table 6.1. Details of currently available SNP data for *Brassica* species.

Species	Source	Panel	No. of SNP	Indels	Presence/absence variations (PAV)	No. of SNP mapped/used in analysis	Frequency	Methods	Platform	Reference
B. napus	DNA	6 (Express, Artus, Aviso, Smart, Talent, Falcon)	87	6	–	87	1 SNP/247 bp; 1 Indel/3,583 bp	*In silico*	ABI Prism 377	Westermeier *et al.*, 2009
B. napus	RNA	2 (Tapidor, Ningyou 7)	23,330–41,593	–	–	7,980–18,130	1 SNP/2,130 bp (read depth of 8)	RNA-seq	Illumina sequencing	Trick *et al.*, 2009
B. napus	ESTs	16 (Express, Lirajet, Milena, Samo, ACElect, HiQ, Yickedee, Dakini andor, 3 resynthesized lines and pools of winter types, spring types and diploids)	604	–	–	507	1 SNP/42 bp	Amplicon sequencing	Illumina golden gate assay	Durstewitz *et al.*, 2010
B. napus	RNA	TNDH lines	23,037	–	–	23,037	–	RNA-seq	Illumina sequencing	Bancroft *et al.*, 2011
B. napus	RNA	84 accessions	101,644	–	–	62,980	–	RNA-seq	Illumina sequencing	Harper *et al.*, 2012
B. napus	DNA	8 (Express 617, PSA 12, Devon Champion, Tira, SWU Chinese 9, PI 271452)	20,835 (113,221 RAD clusters)	125	–	20,835	1 SNP/446 bp; 1 Indel/67,027 bp	RAD	Illumina sequencing	Bus *et al.*, 2012
B. napus	DNA	2 (ZY821 and No2127)	15,586	–	20,856	8,780 SNPs and 12,423 PAV	5–25 kb RAD tags	ddRADseq	Illumina sequencing	Chen, X. *et al.*, 2013

Continued

Table 6.1. Continued.

Species	Source	Panel	No. of SNP	Indels	Presence/absence variations (PAV)	No. of SNP mapped/used in analysis	Frequency	Methods	Platform	Reference
B. rapa	RNA	4 (YSPB-24, Tetralocular, Candle, Chiifu)	20,310–346,189	–	–	594	–	RNA-seq	Illumina sequencing	Paritosh *et al.*, 2013
B. napus	DNA	5 (Aviso, Montego, Aburamasari, Tapidor, Ningyou7)	7,322	–	–	5,764	–	454 sequencing libraries	Illumina Infinium and Golden Gate technology	Delourme *et al.*, 2013
B. napus	DNA	10 (DH12075, PSA12, Ningyou7, Tapidor, Express, V8, Rainbow, YN-429, CGNA1, CGNA2)	589,367	–	–	4,333	1 SNP/421 bp	Sequence capture array	NimbleGen sequence capture, 454 and Illumina Sequencing, Illumina Infinium assay	Clarke *et al.*, 2013

thus a number of approaches have been developed to reduce the complexity of the genome being sampled thus increasing the efficiency of identifying putative SNPs (Trick *et al.*, 2009; Bancroft *et al.*, 2011; Harper *et al.*, 2012; Higgins *et al.*, 2012; Clarke *et al.*, 2013; Delourme *et al.*, 2013; Paritosh *et al.*, 2013).

By focusing only on expressed genes, in a single tissue type more than 23,000 putative SNPs were discovered between a winter and semi-winter *B. napus* cultivar through NGS re-sequencing of their leaf transcriptome (Trick *et al.*, 2009). However, emphasizing the strong homology between the constituent genomes and limitations of short sequence reads for discriminating closely related genomes, the majority of the identified SNPs (>80%) were defined as hemi-SNPs that are due to polymorphism between homologous sequences. Clarke *et al.* (2013) took a different approach, using a sequence capture array to target 47 genomic regions (51.2 Mb), previously found to be associated with agronomic traits of interest, from the A and C genomes of *B. napus*. Ten genetically diverse genotypes (six winter and four spring types) underwent sequence capture and NGS re-sequencing. Sequence reads were reference mapped at stringent parameters (98% identity) to draft A and C genome reference sequences, this allowed for separation of homologous sequences, reducing the number of hemi-SNPs identified. More than 589,000 putative SNPs were discovered, with 46% mapping to the A genome and 54% mapping to the C genome.

Alternatively, there are now well-established approaches that reduce the complexity of the whole genome, viz. GBS (genotyping by sequencing; Elshire *et al.*, 2011) and RAD (restriction site associated DNA marker; Baird *et al.*, 2008) (Fig. 6.2). The two approaches rely on capturing regions of the genome through endonuclease restriction and in the case of RAD, random shearing. NGS re-sequencing of the captured fragments allows identification of SNP variation adjacent to the cut sites. There have been many instances where RAD technology was applied successfully to crop species that range dramatically in genome size, including maize (Nipper *et al.*, 2010), aubergine (Barchi *et al.*, 2011), barley

(Chutimanitsakun *et al.*, 2011), ryegrass (Pfender *et al.*, 2011; Hegarty *et al.*, 2013), sorghum (Nelson *et al.*, 2011), artichoke (Scaglione *et al.*, 2012), grape (Wang, N. *et al.*, 2012) and dwarf birch (Wang *et al.*, 2013). Application of RAD methodology to identify SNPs in *Brassica* is still in its infancy when compared to other crop species. To date only one publication has reported on genome-wide SNP identification using RAD technology in *Brassica*. Bus *et al.* (2012) sequenced a diverse set of eight *B. napus* breeding lines, identifying more than 20,000 SNPs in both coding and non-coding regions across the genome. To address the challenges associated with the presence of homologous sequences in polyploid crops like *B. napus*, Chen, X. *et al.* (2013) used a modified RAD approach, ddRADseq (double digestion) to identify and genotype SNPs. Sequencing of 91 double haploids (DH) identified 8780 SNPs and thousands of presence/absence variations. The authors identified miss-assembled and miss-ordered scaffolds in the current *B. rapa* genome and also placed 44 unassigned scaffolds comprising an additional 8.5 Mb onto the *B. rapa* genome. Table 6.1 provides a summary of the SNP development that has been presented for *Brassica* species to date.

Although similar to RAD, protocols have been developed for GBS that allow development of sequencing libraries for genotypes of interest in a streamlined fashion, with a reduction in handling steps and inbuilt multi-sample indexing (Poland *et al.*, 2012) (Fig. 6.2). These improvements are likely to increase the application of this technology to further crops and although it has yet to be applied widely in brassicas its efficacy in polyploids has been proven in the allohexaploid wheat (Poland *et al.*, 2012). More recently, both RAD and GBS were used to anchor the genome sequence of *B. oleracea* to the genetic linkage map, effectively generating the nine pseudomolecules of the C genome (Parkin *et al.*, 2014). The density of these markers allows megabase regions of the genome to be accurately anchored and oriented, only limited by the distribution of recombination events across the genome. Although these approaches are possible in the absence of a reference genome (Bus *et al.*, 2012; Poland *et al.*, 2012; Chen, X. *et al.*, 2013),

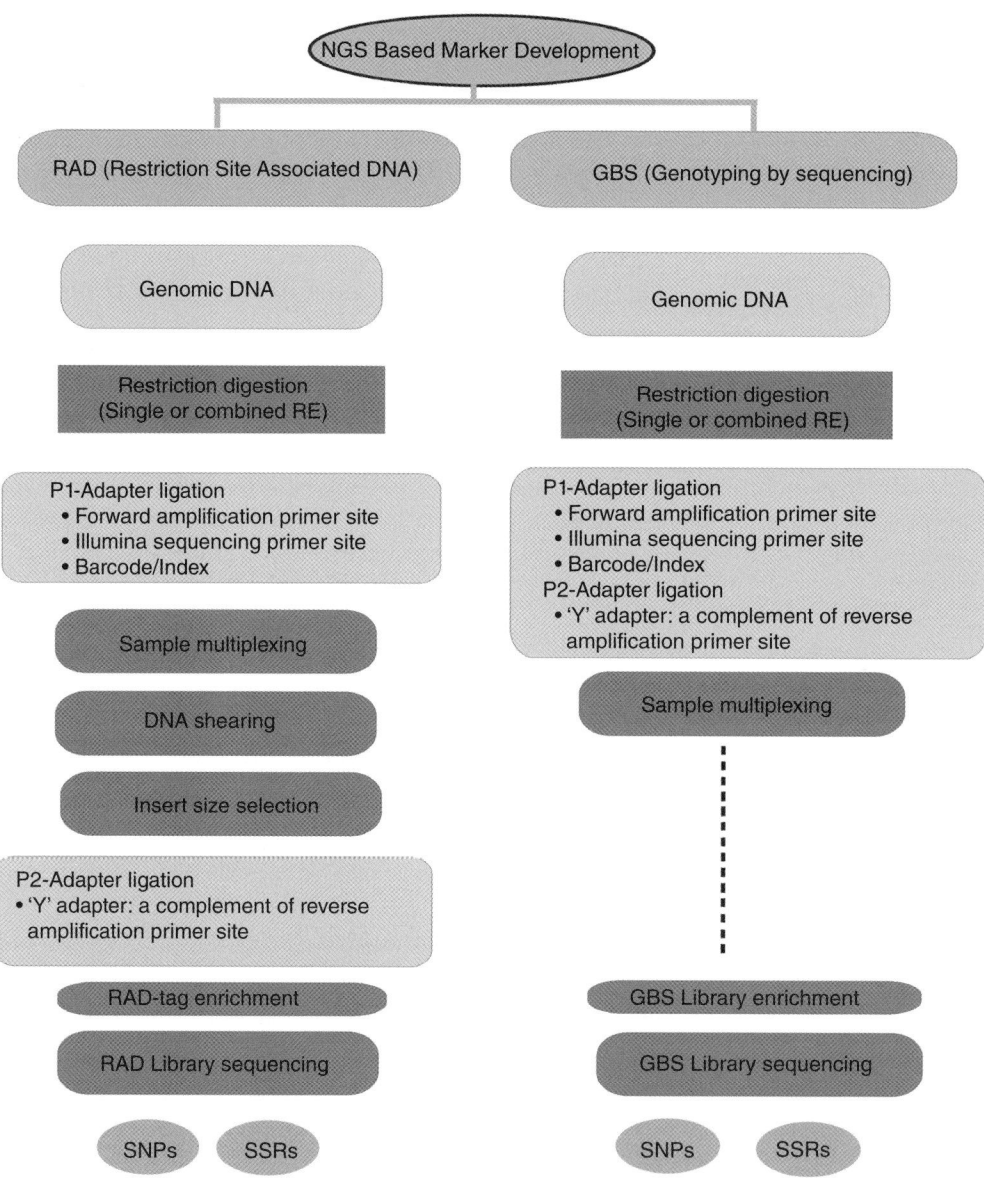

Fig. 6.2. Next generation sequencing (NGS)-based approaches for rapid genome-wide high-density marker development.

they become much more powerful with access to such. With the release of the A and C genome of the *Brassica* diploids, *B. rapa* and *B. oleracea* (Wang, X. *et al.*, 2011; Parkin *et al.*, 2014), full realization of these approaches in half of U's species is possible. Over the next 2 years, it is anticipated that the genomes of all U's *Brassica* species will be available,

which will facilitate all avenues of molecular marker development.

Application of Molecular Markers

Molecular markers have become essential tools for brassica researchers. In their first inception, they provided tools for tagging genomic regions controlling traits of interest, gave a snapshot of the level of diversity available within the tribe and as they evolved they provided insights into the relationship between the different genomes of U's triangle and beyond to the model species *A. thaliana*. These latter analyses have determined some of the key evolutionary steps determining the structure of the crop genomes and in so doing have allowed the inference of conserved genes controlling key phenotypes across all *Brassica* species.

Quantitative trait loci mapping

Many of the important traits relevant to the agronomic improvement of crops are quantitative in nature, controlled by multiple genomic elements of varying strength or quantitative trait loci (QTL). Molecular markers combined with well-structured mapping populations have proved invaluable in identifying such regions of the genome, which allows their further manipulation in marker-assisted breeding strategies. In addition, with the soon to be available genome sequences for each of the brassica oilseeds, it will be possible to study the regions underlying important QTL, potentially elucidating the genes and associated pathways controlling key traits. There have been numerous papers published detailing QTL controlling a range of traits for *Brassica* species including disease resistance, heterosis, yield and quality traits. Since *Brassica* species are major sources of vegetable oil globally, much work has focused on identifying QTLs pertaining to oil content and oil profiles. In an attempt to summarize efforts from published work and to target regions of the genome for marker discovery, Clarke *et al.* (2013) defined regions of the *B. napus* genome that have been consistently uncovered in trait mapping studies. We have expanded this analysis to show the regions of homology between the A and C genomes and where possible the underlying genome collinearity with *A. thaliana*, as shown in Plate 2. It is notable that in particular quality-trait QTLs mapped in independent experiments are often located to similar regions of the genome. In addition, QTLs for the same trait are often identified in

homologous regions of the A and C genome in *B. napus*, suggesting that they are controlled by duplicate loci. Such mapping efforts have been further refined with the application of dense marker sets to assist in identifying candidate genes that could be controlling these traits. With the development of the new sequence-based marker technologies and the alignment of these markers to the *Brassica* genome sequences, the accurate identification of conserved QTL positions across populations will be facilitated, allowing the resolution of robust QTLs as targets for marker assisted breeding.

Capturing genetic diversity

Capturing favourable genetic variation is the key for any successful plant breeding programme. The genus *Brassica* consists of 49–54 species in the *Brassiceae* tribe belonging to the crucifer family (Bailey *et al.*, 2006; Beilstein *et al.*, 2006). As a whole, the family displays extreme genetic diversity for morphology, phytochemistry and response to environmental stresses; this variation has allowed a number of the crucifer species to be exploited as sources of not only edible oil, but also condiments, vegetables, fodder and health-promoting phytochemicals. Understanding the genetic diversity of *Brassica* species is important to broaden the available genetic base, which in particular for the more widely grown crops has been narrowed through constant selection for a small number of essential traits. For example, it is well documented that the development of the double low canola lines has generated genetic bottlenecks associated with the maintenance of the low erucic acid and low glucosinolate phenotype (Hasan *et al.*, 2008).

It is often useful to classify an experimental sample into different populations that have become separated within a species. These sub-groups can result from the impact of breeding bottlenecks, geographical separation, or differential maturity rates. In understanding levels of diversity and for downstream applications, in particular association mapping, it is important to define such relationships, since any locus with uneven allele distribution between populations will associate with traits that differentiate the populations, irrespective of genomic location. Bayesian clustering algorithms played a prominent role in the last decade in providing such information. Pritchard *et al.* (2000) developed a population genetics tool, 'POPULATION STRUCTURE' to designate each individual into a certain population using allele frequencies determined from sets of unlinked genetic markers. Since then, due to the simplicity and versatility of the software, 'STRUCTURE' analysis has been widely applied to the study of plant breeding material.

Like others, although limited thus far, the brassica research community has seized the benefit of 'STRUCTURE' analysis to understand their genetic material for breeding and genetic studies. Mainly due to the economic significance of the crop and the earlier application of markers, the majority of the studies thus far have targeted *B. napus* to delineate the available genetic diversity, in order to improve integration of these resources in breeding programmes. In the first of these studies Hasan *et al.* (2008) analysed 94 *B. napus* gene bank accessions originating from all continents except Africa using SSR markers, defining three subpopulations which represented distinct differences in growth habit, namely, spring oilseed, winter oilseed and fodder or vegetable rape types. Since then, a number of similar studies have been completed focusing on different sets of breeding lines and landraces (Bus *et al.*, 2012; Li *et al.*, 2012; Wang, X. *et al.*, 2012; Gyawali *et al.*, 2013). For example, Gyawali *et al.* (2013) focused on including Asian germplasm which has had more limited application in the larger breeding programmes and may provide novel sources of resistance for common brassica diseases, in particular Sclerotinia stem rot (Zhao *et al.*, 2009). In each case, generally two populations were distinguished among oilseed types based on the requirement for vernalization, with the majority of the winter types from Europe and semi-winter types from China and northern Asia clustered in one and spring-type cultivars from Canada and Australia found in subpopulation two. Most of the structure-based analysis in *B. napus* has been based on SSR marker analysis; however, a perhaps more robust NGS approach

using more than 100,000 SNPs identified the same pattern with two main clusters; cluster one consisted of all but one winter oilseed types and the second contained spring, swede, fodder and Chinese accessions (Harper *et al.*, 2012). There have been limited attempts to study the relationship between population structure and traits of interest. However, studies of flowering time in *B. napus* grouped 95 accessions into three major clusters which explained over 50% of the observed variation, underlying the effect of flowering time on the population structure of *B. napus* (Wang, N. *et al.*, 2011). The impact of current breeding strategies in *B. napus* crop development has become apparent through such analyses indicating a relatively small gene pool contributes to the two main populations. Although there are no known wild populations of *B. napus*, feral populations that have colonized less favourable areas, generally along transport routes such as roadsides, railway banks, waste lands and riverbanks, are quite common. Interestingly, diversity analyses of such populations in comparison with commercial *B. napus* varieties from which they have presumably evolved showed a clear differentiation between groups (Pascher *et al.*, 2010). These observations could reflect the age of the feral populations, suggesting that the modern cultivars have made minimal impact into such areas and that the today's herbicide-resistant varieties cannot compete without the advantage of continual selection through spraying of non-resistant neighbours.

Brassica juncea, commonly called Indian mustard, is mainly cultivated as an oilseed but also as a condiment. Analysis of genetic diversity among *B. juncea* accessions has been less intensive when compared to *B. napus*. Using ALFP markers Srivastava *et al.* (2001) studied the levels of genetic diversity among 21 natural and nine re-synthesized lines of *B. juncea* from India, Canada, Australia and Russia. Cluster analysis performed through the more traditional pairwise similarity method of hierarchical clustering, UPGMA (unweighted pair group method with arithmetic mean), revealed two distinctive clusters, with all Indian accessions, including the re-synthesized lines developed in India, in one cluster and all remaining accessions in the second, suggesting two possible origins of diversity for *B. juncea*. Hierarchical cluster analysis again using AFLP markers among a wider collection of 77 elite canola-quality *B. juncea* breeding lines and 15 other genotypes again grouped Indian genotypes separately from the remaining genotypes (Burton *et al.*, 2004). Although two generalized gene pools have been defined for *B. juncea* the centre of origin of the various *B. juncea* genotypes remains unclear. S. Chen *et al.* (2013) studied 119 *B. juncea* varieties collected from India, China, Australia and Europe using 99 SSR markers. In both hierarchical cluster analysis and STRUCTURE analysis, Indian and Chinese accessions were both found in either of two clusters whereas European and Australian accessions were restricted to one cluster. These analyses suggested that both India and China are secondary centres of origin for *B. juncea*.

A number of studies have analysed collections of *B. rapa* accessions that have largely separated the lines based on gross morphological differences (Zhao *et al.*, 2007; Yu *et al.*, 2010; Del Carpio *et al.*, 2011). Although *B. rapa* could be a source of additional genetic diversity for both *B. napus* and *B. juncea* oilseed breeding, the majority of the diversity studies of *B. rapa* have, unfortunately, utilized AFLP markers, which prohibits integration or comparison with similar analyses in *B. napus* or *B. juncea*. With the release of *Brassica* genome sequences and as GBS-based analyses become more widely adopted, it is anticipated that it will be possible to capture common allelic variation across all species of U's triangle, which should provide novel insights into the impacts of breeding and germplasm transfer across continents among the brassica crop species.

Comparative mapping

The earliest comparative analyses of *Brassica* genomes could be dated back to U's seminal publication that defined the relationship between the six *Brassica* species. This analysis determined that each of the three extant *Brassica* diploid species had paired in all possible combinations to generate the three amphidiploid species (Fig. 6.1). However, although determining the gross sympathies between the genomes without the application of markers it was not possible to resolve whether fusion of the diploid genomes had resulted in major chromosomal rearrangements. Since chromosomal pairing had been observed in haploids derived from the allotetraploids (Prakash and Tsunoda, 1983), it was commonly inferred that diploidization of the nucleus following polyploidy might necessitate chromosomal restructuring. Comparison of genetic linkage maps developed using common marker sets for both naturalized and resynthesized *B. napus* determined that no such rearrangements had occurred (Parkin *et al.*, 1995) and similarly the genome of *B. juncea* could not be distinguished from the extant diploid progenitors (Axelsson *et al.*, 2000).

A major breakthrough in the study of *Brassica* genome evolution came with the availability of the *A. thaliana* genome sequence. Although distantly related, it was apparent through *in silico* analyses of sequences from *A. thaliana* and *Brassica* species that there was strong conservation of genic identity at the nucleotide level (~85%) (Parkin *et al.*, 2005). By sequencing a core collection of *Brassica* markers and mapping their positions in the *A. thaliana* genome, it was possible to build a map of the *Brassica* A and C genome defined by blocks that were conserved across the two lineages (Parkin *et al.*, 2005) (Plate 2). Although there have been minor refinements to the block structure determined in Parkin *et al.* (2005), this publication led to the general acceptance of a core set of evolutionarily conserved ancestral blocks (A–X) for the *Brassicaceae* that can be duplicated and rearranged to form each of the genomes within the family (Schranz *et al.*, 2006). Analysis of the organization of the underlying ancestral block structure in the *Brassica* genomes also provided evidence for a whole genome triplication (WGT) event that had first been suggested from limited mapping data in the *B. nigra* genome (Lagercrantz and Lydiate, 1996). This WGT is now widely accepted with evidence supporting the event having been found in all three *Brassica* diploid genomes (Town *et al.*, 2006; Wang, X. *et al.*, 2011; Navabi *et al.*, 2013).

Despite the disparity in divergence times for the three genomes, with the B genome having separated from the A/C genome earlier than they diverged from each other, the WGT is believed to be a shared event common to the *Brassica* lineage (Navabi *et al.*, 2013).

Identification of candidate gene(s) for important agronomic traits

The comparative mapping studies made it possible to associate genomic regions of interest of *Brassica* species with sequenced regions of *A. thaliana* (Parkin *et al.*, 2005). In turn, the strong genome conservation between *A. thaliana* and *Brassica* species suggested that it would be possible to identify potential candidate genes for traits of interest (Parkin, 2011).

There are two major growth types among brassica crops that are differentiated by a requirement for vernalization: winter and spring types. Winter (or biennial) types require vernalization, a period of low, but generally non-freezing temperatures, in order to initiate flowering. Understanding the genetic basis of flowering time control is essential in order to grow crops in a specific environment. Extensive research has led to the identification of multiple genes important for the control of flowering time in *A. thaliana* (Koornneef *et al.*, 2004; Putterill *et al.*, 2004). One such gene is *FRIGIDA*, which controls the expression of a major repressor of flowering in *A. thaliana*. Inferring conservation of function across the two species, Wang, N. *et al.* (2011) cloned four homologues of *FRIGIDA* from *B. napus* and studied the expression and available sequence variation for each copy. There appeared to be evidence of tissue-specific expression for the different copies but perhaps more compelling was the fact that one of the copies co-localized with a major flowering time QTL on *B. napus* linkage group A3 suggesting *FRIGIDA* is similarly a major determinant of flowering time in *B. napus* (Wang, N. *et al.*, 2011).

Seed yield is an important target for improvement in brassica crops; however, it is a typical quantitative trait controlled by multiple loci generally of small effect (Udall *et al.*, 2006;

Chen *et al.*, 2007; Kramer *et al.*, 2009). Although seed weight as a component of yield has been shown to have relatively high heritability, due to the challenges involved in map-based cloning of QTLs in polyploid species, there have been no candidate genes as yet identified for seed weight or indeed other seed yield-related traits using traditional approaches. Cai *et al.* (2012) used collinearity between *A. thaliana* and *B. napus* to infer the positions of 43 *A. thaliana* genes which have been shown to impact either seed or fruit size in *A. thaliana*, tomato, maize or rice. In total, candidate genes were implicated for eight seed weight QTLs, although it was noted that, for two major seed weight QTLs in *B. napus* no candidate could be identified (Cai *et al.*, 2012). Such analyses of QTL regions will be further resolved with access to the full genome sequence for *B. napus*.

Levels of seed oil content and relative composition of the different fatty acids have a tremendous impact on the marketing of brassica oils, for human and animal consumption. High levels of erucic acid were a problem in early varieties of *B. napus*, which contained up to 50% erucic acid, since this was considered to be a health concern for human consumption. Targeted breeding approaches led to the development of modern brassica oil with low erucic acid (canola). Erucic acid is produced by elongation of oleic acid by fatty acid elongase1 (FAE1), an enzyme that has been characterized at the molecular level in both *A. thaliana* and *B. napus* (James *et al.*, 1995; Han *et al.*, 2001). By eco-TILLING for allelic variation in homologues of *FAE1* in a wide range of germplasm, Wang *et al.* (2008) demonstrated that specific allelic variants of two orthologues of *FAE1*, namely a single C to T SNP and a two base-deletion (AA), were consistently isolated in low erucic acid lines, indicating the influence of selection for particular traits within breeding programmes, since all low erucic *B. napus* lines studied to date appear to have inherited the same variants of *FAE1*.

Additional factors related to brassica seed composition could warrant improvement. The content and composition of tocopherol (vitamin E) is important for both oil stability and health claims related to the marketing of canola. Fritsche *et al.* (2012) utilized the established pathway for tocopherol biosynthesis in *A. thaliana* to carry out a targeted association mapping study of tocopherol content in 229 *B. napus* lines. Re-sequencing of 13 candidate genes and association of identified nucleotide variation with tocopherol content identified two major candidate genes, *BnaX.VTE3.a* and *BnaA.PDS1.c*, which were in close proximity to QTLs mapped to chromosomes A7 and A10, respectively.

Glucosinolates are secondary plant metabolites synthesized by *Brassica* species as part of the plant defence against attack by pests and pathogens. Importantly, in higher concentrations these compounds also act as antinutritional factors for animal consumption, which can limit the use of the protein rich meal as a feed (Walker and Booth, 2001). Accumulation of seed glucosinolates is a polygenic trait in *Brassica* species and a number of QTLs have been identified controlling this trait in both *B. napus* and *B. juncea* (Sodhi *et al.*, 2002; Howell *et al.*, 2003; Sharpe and Lydiate, 2003; Basunanda *et al.*, 2007; Ramchiary *et al.*, 2007; Bisht *et al.*, 2009; Harper *et al.*, 2012). Although the glucosinolate biosynthesis pathway had been relatively well elucidated in *A. thaliana* for a number of years (Hull *et al.*, 2000; Bak *et al.*, 2001; Hansen *et al.*, 2001), there had been no association of candidate genes with QTLs for glucosinolate accumulation until relatively recently. An initial comparative mapping analysis of potential candidate genes in *B. juncea* identified orthologues of two candidate genes (*GSL-ELONG* and *MYB28*) that co-located with QTLs for glucosinolate accumulation (Bisht *et al.*, 2009). An associative transcriptomics study of a range of *B. napus* lines, in addition to associating SNP variation with known QTL loci for glucosinolate accumulation on A9 and C2, concomitantly identified regions of low transcript abundance that correlated with low glucosinolate content. Interestingly, comparative analysis of the resolved regions found that an orthologue of *A. thaliana* gene *HAG1* (or *MYB28*) was deleted from accessions of *B. napus* with low glucosinolate content. The *HAG1* gene of *A. thaliana* is a transcription factor that regulates the biosynthesis of seed aliphatic glucosinolates and makes a logical candidate as a key gene controlling the glucosinolate phenotype of multiple *Brassica* species

(Hirai *et al.*, 2007). In one of the first *in planta* tests of the candidate gene approach, Augustine *et al.* (2013) used RNAi-based transcription suppression of a *HAG1* orthologue (*BjMYB28*) in *B. juncea* to replicate the low seed glucosinolate content phenotype.

Blackleg disease caused by the *Leptosphaeria maculans* fungus is a major disease in brassica crops and the leading source of yield losses globally in canola (Fitt *et al.*, 2006; Kutcher *et al.*, 2010). Host crop resistance for blackleg is considered the most cost effective and environmentally sustainable strategy to manage the disease. This approach has been supported by considerable research worldwide, evaluating resistant germplasm, identifying qualitative and quantitative resistance loci and resistance gene cloning and candidate gene identification (Rimmer, 2006). Perhaps due to the novel evolutionary paths of disease resistance genes, which tend to be in relatively unstable regions of the genome that undergo significant gene expansion through tandem duplication and often do not follow predicted patterns of gene synteny, candidate gene identification for blackleg resistance loci in *B. napus* is still in its infancy. Only one study has been published where the gene underlying a qualitative trait locus for blackleg resistance in *B. napus* was identified. A map-based cloning approach in *B. napus* was still required, yet the authors exploited the collinearity between the *A. thaliana* and *B. rapa* genomes to saturate the target region with markers (Larkan *et al.*, 2013). An alternative but as yet unproven approach to identifying candidate genes for blackleg resistance was suggested by Tollenaere *et al.* (2012). NGS data of resistant and susceptible *B. napus* lines were used to study nucleotide variation in 18 candidate *A. thaliana* genes underlying the resistance locus *Rlm4* (Raman *et al.*, 2012). Two potential candidates were identified, which either varied in read depth or SNP density between the resistance and susceptible parents; however, functional confirmation is still required. These two approaches, high-resolution mapping and targeted association mapping have been made possible through access to the *B. rapa* genome sequence (Wang, X. *et al.*, 2011). Once the genomes of each of the species of U's triangle become available, such analyses will become routine and more robust since the true extent of maintenance of disease resistance clusters across *Brassicaceae* species will be resolved.

Another devastating disease of *B. napus* is Sclerotinia rot caused by *Sclerotinia sclerotiorum*. This disease is controlled solely by QTL with no classic *R* genes identified, thus generating lines that are resistant to the disease has been far more challenging. Wu *et al.* (2013) identified relatively large effect QTLs, LRA9 and SRC6, which controlled 8.54–15.86% and 29.01–32.61% of the observed variation for resistance, respectively. A previous transcriptome analysis of susceptible and resistant *B. napus* lines had identified a number of genes that were highly responsive to *Sclerotinia* infection and listed their closest *A. thaliana* homologue (Zhao *et al.*, 2009). By aligning QTL regions with *A. thaliana* and searching the collinear regions for those genes that were differentially expressed in response to infection, the candidate gene *BnaC.IGMT5.a*, which encodes an indole glucosinolate methyltransferase (IGMT), was co-localized with the SRC6 locus. It was hypothesized that IGMT's role in monolignol biosynthesis is contributing to the plant's ability to limit spread of the fungus.

Club root caused by *Plasmodiophora brassicae* has also been the focus of a number of studies attempting to identify candidate resistance genes in *Brassica* species. Comparative mapping between *B. rapa* and *A. thaliana* demonstrated that the genomic regions underlying two major QTLs, Crr1 and Crr2 in *B. rapa*, shared a region of common synteny with *A. thaliana* chromosome 4 region, moreover this region contained a major disease resistance cluster in *A. thaliana* (Suwabe *et al.*, 2006). Though many QTLs have been identified for club root resistance the molecular mechanism of resistance is unknown, yet recent studies have suggested that some resistance components may be controlled by genes that closely resemble those involved in classic gene-for-gene (resistance gene–avirulence gene) interactions. Map-based gene cloning and characterization revealed that one resistant locus in *B. rapa*, Crr1a, encodes a Toll-Interleukin-1 receptor/nucleotide-binding site (NBSs)/leucine-rich repeat (LRR) (TIR-NBS-LRR) protein (Hatakeyama *et al.*, 2013). Similarly, the candidate gene for *CRa*, an additional resistance locus in *B. rapa*, was also found to encode a TIR-NBS-LRR protein (Ueno *et al.*, 2012).

Genome Sequencing of U's Triangle Species

In 2011, the first genome sequence of a *Brassica* species was published, that of *B. rapa*; the genome assembly represented 283 Mb, only 53% of the estimated 529 Mb genome (Wang, X. *et al.*, 2011). Although this sequencing project began as a traditional BAC (bacterial artificial chromosome) by BAC approach, where physical maps are generated for each chromosome and each BAC on a chromosome is sequenced using high quality first generation Sanger sequencing, the rapid incursion of NGS prevailed and the final genome was largely created from high coverage of short read second generation sequences. The traditional approach, although more expensive, at the time it required the commitment of multiple groups from around the world, would have generated a somewhat more complete genome sequence. The *B. rapa* sequence was one of the first crop genomes sequenced and no doubt there would have been significant delays in generating the genome if NGS had not been employed. Today, due to efficiencies of cost and time, most plant genomes are sequenced using NGS strategies. Improvements in sequencing chemistry that have led to longer and more accurate reads and the ability to carry out high density genotyping to anchor genomes to genetic maps has significantly enhanced the quality of the resultant assemblies. Such improvements led to the development of a new version of the *B. rapa* genome, with an additional 100 Mb of sequence and a further 7000 genes annotated, which is anticipated to be released in 2015 (Xiaowu Wang, personal communication).

The *B. rapa* genome represented a milestone for brassica research, since researchers and breeders were no longer forced to use the model *A. thaliana* as a surrogate for gene identification. In addition, access to the genome provided greater insights into the influence of the WGT event on the maintenance of gene copies, which has direct implications in trait manipulation. Subsequent to polyploidy events there is a period of adjustment termed diploidization that ensures normal meiotic pairing occurs, reducing the impact of the chromosome doubling event on the fertility of further generations (Comai, 2005). How the return to the diploid state is achieved is still an area of much study; however, evidence suggests that extensive gene loss or fractionation of the duplicated genomes contributes to this process. Although as indicated above the gene number is slightly underestimated, *B. rapa* was annotated with just over 41,000 protein coding regions, significantly less than what would be anticipated from the triplication of a genome with a similar composition to *A. thaliana*. The sequencing of the *B. rapa* genome revealed the three ancestral genomes which make up the diploid genome following the WGT event. The three genomes were shown to have differential fractionation rates; one termed the least fractionated genome maintained the highest number of genes (75% of the expected gene copies based on synteny) while the two additional genomes had much higher levels of gene loss (Tang *et al.*, 2012). It has been suggested that the least fractionated genome maintains genome dominance over the additional copies, which is reflected in the levels of gene expression observed for the duplicate gene copies (Cheng *et al.*, 2012). Differential maintenance of functional genes and divergent expression of these copies could contribute to the morphological and phytochemical variation observed in *Brassica* species.

Genome sequencing has now been completed for both the diploid *B. oleracea* (C genome) and the amphidiploid *B. napus* (AC genome) (Chalhoub *et al.*, 2014; Parkin *et al.*, 2014), which are starting to provide insights into the evolution of young amphidiploid genomes that have been selected in the last 10,000 years (Allender and King, 2010). *Brassica oleracea*, with ~59,000 annotated protein coding regions, was found to have a higher gene content than *B. rapa*; this difference is partly a reflection of improvements in genome sequencing technologies and variations in annotation methods, yet underlying this are apparent species-specific differences that reflect species divergence. It was known that the *B. napus* genome had

undergone limited major chromosomal rearrangements since the fusion of the constituent genomes (Parkin *et al.*, 1995); this has been corroborated with the sequencing of the *B. napus* genome (Chalhoub *et al.*, 2014). Moreover, it appears that this maintenance of the diploid genomes is also reflected at the gene level, with relatively little gene loss having been incurred. Interestingly, gene loss in *B. napus* has been implicated in the evolution of the canola phenotype, since the control of glucosinolate accumulation was found to result from deletion of key regulatory genes (Harper *et al.*, 2012). The evolution of the low glucosinolate phenotype has now been shown to be associated with a larger chromosomal homoeologous recombination event, which took place between the A and C genomes of *B. napus* (Chalhoub *et al*, 2014). It appears that the plasticity of the two nuclear genomes in allowing such events to take place provides a source of novel variation upon which pressures of adaptation can apply, either natural selection or through human intervention in the case of the low glucosinolate phenotype.

The genomes of the remaining members of U's triangle are in development and are scheduled for completion and release in 2015 (Parkin and Sharpe, personal communication). Initial work in the B genome has indicated that some of the observed phenotypes such as increased resistance to blackleg and other common brassica pests and pathogens may result from maintenance of specific gene families that have been lost from the A and C genomes, presumably due to the lack of an adaptive function (Navabi *et al.*, 2013). Thus the B genome could provide access to a plethora of useful genes for breeding purposes. It will be interesting to see if the more intensive breeding efforts in *B. napus* have influenced the resultant genome architecture compared to that of *B. juncea* and, in particular, the lesser studied *B. carinata*.

Future Developments in Brassica Genomics

The promise of genome sequences for all members of U's triangle will create new opportunities for both basic research and crop improvement. As a fundamental resource the genomes will provide access to the gene complement (potentially with elements unique to each species) will provide a platform for next-generation genotyping technologies and will allow a comprehensive study of the impact of polyploidy in relatively new species. In order to exploit the value of this information for breeding, further developments are envisioned through the adoption of strategies for accelerating detection of variation, for reducing breeding cycles and for the dissection of complex agronomic traits.

TILLING in Brassica oilseeds

TILLING (Targeting Induced Local Lesion In Genomes) is a reverse genetic approach to discover a series of allelic variants for a gene of interest. TILLING has been successfully employed in model plants as well as cultivated crops, for example, *A. thaliana* (McCallum *et al.*, 2000), maize (Till *et al.*, 2004), rice (Till *et al.*, 2007), soybean (Cooper *et al.*, 2008) and in polyploids such as wheat (Slade *et al.*, 2004). TILLING populations are generated using the mutagens ethyl methane sulfonate (EMS) or N-ethyl-N-nitrosourea (ENU) that causes point mutations/lesions randomly in the genomes, with results ranging from no change in gene function due to synonymous mutations to complete loss of function (Slade and Knauf, 2005). TILLING is thought to be well suited to the improvement of polyploids since the complex genomes with their evolved functional redundancy conferred by paralogous sets of genes are more tolerant of induced mutations (Slade and Knauf, 2005). On the other hand, manipulating any one trait may involve eliminating or limiting the expression of multiple genes, requiring time-consuming pyramiding of mutated gene copies. One of the first attempts to apply TILLING in *B. napus* was carried out by Wang *et al.* (2008), wherein two EMS TILLING populations were generated and screened for mutations in the *FAE1* (*fatty acid elongase 1*) gene, which is ascribed to play a key role in erucic acid biosynthesis. The study yielded 19 mutations by screening 1344 M_2 plants, of which three showed altered functions that resulted in changes in the accumulation of erucic acid. A similar population has been developed for *B. rapa* (Stephenson *et al.*, 2010) that showed a useful mutation rate for analysing gene function. TILLING of the diploid genome may prove more amenable to the identification of mutations, since developing locus-specific primers for genes of interest is complicated by the presence of the second homologous genome in *B. napus*. However, a recent development has shown that NGS technologies can be exploited to identify mutations in the allopolyploid genome more efficiently (Gilchrist *et al.*, 2013); the authors used NGS to calculate the mutation rate for 26 gene targets in almost 2000 mutagenized lines of *B. napus*.

Genomic selection

Over the years, a large number of plant studies have focused on identifying loci affecting traits of interest, but there has been limited practical application for this work since the majority of the traits are complex in nature and in any one study only a fraction of the genetic variation was explained. The concept of genomic selection (GS) has been developed in animal breeding to increase the efficiency of what are generally expensive breeding programmes (Smith, 1967; Soller and Beckmann, 1983). GS requires genomic estimated breeding values (GEBV) to be calculated, which is the sum of the effects of all loci, haplotypes or markers across the entire genome, essentially capturing all possible QTL in one analysis (Meuwissen *et al.*, 2001) (Fig. 6.3). GEBV is calculated from training populations (large populations of individuals with associated phenotypic and genotypic data), and these estimated breeding values are used in the subsequent selection of genotypes for their further advancement in breeding cycles (Meuwissen *et al.*, 2001; Heffner *et al.*, 2009). In contrast to marker-assisted selection (MAS) that generally targets one or a few loci, GS incorporates all available marker estimates for a population, avoiding effects of marker bias and incorporating variation due to small effects QTLs.

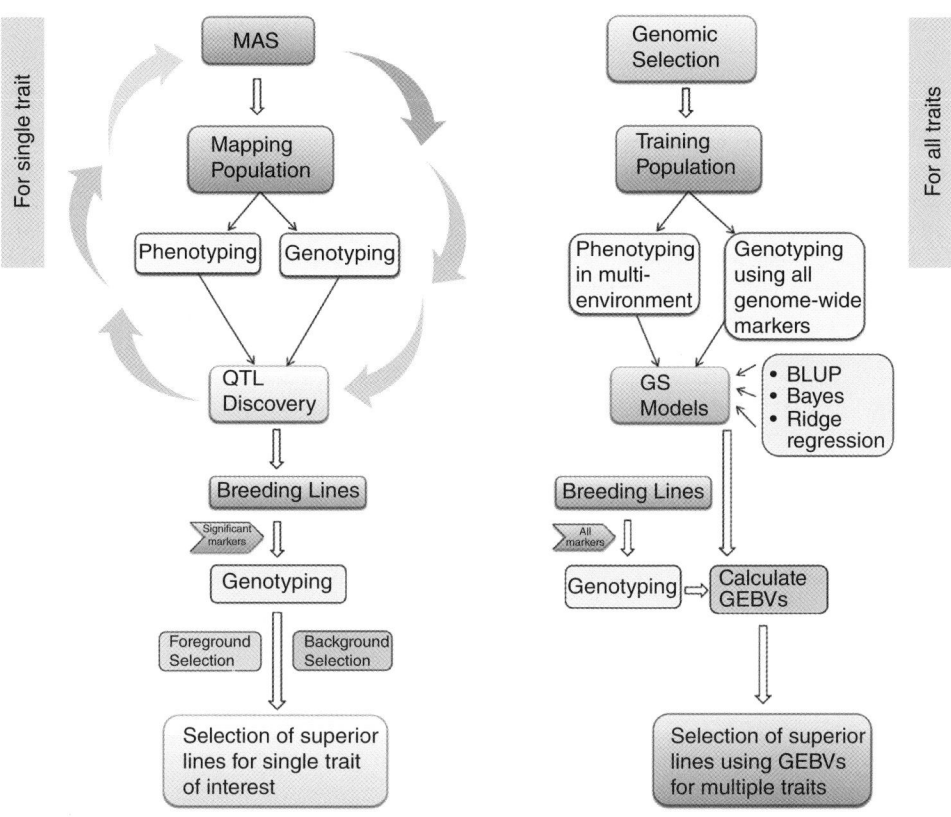

Fig. 6.3. Comparison between marker assisted selection (MAS) and genomic selection (GS). Best linear unbiased prediction (BLUP).

GS can also reduce the role of phenotyping in the selection of elite lines, which provides both time and cost benefits.

GS is in its infancy in plant breeding. Nevertheless, prediction accuracy is being tested in a few crop species. For instance, average prediction accuracy for the selection of winter wheat lines, using information from 374 advanced breeding lines and based on 13 agronomic traits, resulted in a 28% greater average prediction accuracy compared to MAS and was 95% as accurate as phenotypic selection (Heffner *et al.*, 2011). Similarly, Zhao *et al.* (2012) reported high prediction accuracy for grain moisture (0.90) and grain yield (0.58) in European maize, which corresponds to the precision of phenotyping in un-replicated field trials in three to four locations. Progress in the discovery of genome-wide marker sets for *Brassica* species through exploitation of the genome sequences will help to establish

these novel genomic approaches, allowing selection of elite genotypes with the ideal combination of favourable alleles in a timely and efficient fashion.

Next-generation populations

Advances in sequencing technology and the increasing availability of *Brassica* genome sequences has already and will continue to yield large numbers of markers for *Brassica* species (Trick *et al.*, 2009; Bus *et al.*, 2012; Chen, X. *et al.*, 2013; Paritosh *et al.*, 2013). The availability of these data offers new opportunities for the study of important agronomic traits; however, the available germplasm resources can limit our understanding of the genetic control of such traits. It has been suggested that QTL effects have been overestimated in QTL mapping studies where small numbers of progeny have been used, although statistical tests have been employed to attempt to reduce the so-called Beavis effect; this has probably contributed to the limited application of MAS studies in crop improvement (Xu, 2003). In order to dissect the genetic basis of complex quantitative traits further, a number of novel population development strategies have been proposed, which could be applied to *Brassica* species. For example, nested association mapping (NAM) which generates a core mapping population that combines the high statistical power per allele provided through linkage analysis with the high mapping resolution offered through association mapping. Such a population was generated for maize using 25 highly diverse inbred lines crossed with the maize reference line B73 and has been used successfully to dissect a number of complex traits (McMullen *et al.*, 2009). A similar approach, which may also prove effective for *Brassica* species, has been employed in hexaploid wheat: multiparent advanced generation inter-cross (MAGIC) lines were developed and tested as a resource for mapping of complex traits (Huang *et al.*, 2012).

Conclusions

In the 1990s, the first marker studies were completed for *Brassica* species and provided insights into the complexities of the underlying genome structure clarifying the multilocus nature of trait inheritance for most agronomically important traits. The sequencing of the first plant genome, that of *A. thaliana*, provided distinct advantages to *Brassica* researchers due to its close relationship and startling synteny with the *Brassica* genomes. However, our understanding of *Brassica* genome organization and our ability to functionally manipulate key traits utilizing the latest advances in genomics technologies will be significantly advanced with access to genome sequences for each of the *Brassica* species.

References

Allender, C.J. and King, G.J. (2010) Origin of the amphidiploid species *Brassica napus* L. Investigated by chloroplast and nuclear molecular markers. *BMC Plant Biology* 10, 54.

Augustine, R., Mukhopadhyay, A. and Bisht, N.C. (2013) Targeted silencing of BjMYB28 transcription factor gene directs development of low glucosinolate lines in oilseed *Brassica juncea*. *Plant Biotechnology Journal* doi: 10.1111/pbi.12078

Axelsson, T., Bowman, C.M., Sharpe, A.G., Lydiate, D.J. and Lagercrantz, U. (2000) Amphidiploid *Brassica juncea* contains conserved progenitor genomes. *Genome* 43, 679–688.

Bailey, C.D., Koch, M.A., Mayer, M., Mummenhoff, K., O'Kane, S.L., Warwick, S.I., Windham, M.D. and Al-Shehbaz, I.A. (2006) Toward a global phylogeny of the *Brassicaceae*. *Molecular Biology and Evolution* 23, 2142–2160.

Baird, N.A., Etter, P.D., Atwood, T.S., Currey, M.C., Shiver, A.L., Lewis, Z.A., Selker, E.U., Cresko, W.A. and Johnson, E.A. (2008) Rapid SNP discovery and genetic mapping using sequenced RAD markers. *PLoS One* 3, e3376.

Bak, S., Tax, F.E., Feldmann, K.A., Galbraith, D.W. and Feyereisen, R. (2001) CYP83B1, a cytochrome P450 at the metabolic branch point in auxin and indole glucosinolate biosynthesis in *Arabidopsis*. *The Plant Cell* 13, 101–111.

Bancroft, I., Morgan, C., Fraser, F., Higgins, J., Wells, R., Clissold, L., Baker, D., Long, Y., Meng, J. and Wang, X. (2011) Dissecting the genome of the polyploid crop oilseed rape by transcriptome sequencing. *Nature Biotechnology* 29, 762–766.

Barchi, L., Lanteri, S., Portis, E., Acquadro, A., Vale, G., Toppino, L. and Rotino, G. (2011) Identification of SNP and SSR markers in eggplant using RAD tag sequencing. *BMC Genomics* 12, 304.

Basunanda, P., Spiller, T.H., Hasan, M., Gehringer, A., Schondelmaier, J., Lühs, W., Friedt, W. and Snowdon, R.J. (2007) Marker-assisted increase of genetic diversity in a double-low seed quality winter oilseed rape genetic background. *Plant Breeding* 126, 581–557.

Batley, J., Hopkins, C.J., Cogan, N.O.I., Hand, M., Jewell, E., Kaur, J., Kaur, S., Li, X.I., Ling, A.E. and Love, C. (2007) Identification and characterization of simple sequence repeat markers from *Brassica napus* expressed sequences. *Molecular Ecology Notes* 7, 886–889.

Beilstein, M.A., Al-Shehbaz, I.A. and Kellogg, E.A. (2006) *Brassicaceae* phylogeny and trichome evolution. *American Journal of Botany* 93, 607–619.

Bisht, N.C., Gupta, V., Ramchiary, N., Sodhi, Y.S., Mukhopadhyay, A., Arumugam, N., Pental, D. and Pradhan, A.K. (2009) Fine mapping of loci involved with glucosinolate biosynthesis in oilseed mustard (*Brassica juncea*) using genomic information from allied species. *Theoretical Applied Genetics* 118, 413–421.

Burgess, B., Mountford, H., Hopkins, C.J., Love, C., Ling, A.E., Spangenberg, G.C., Edwards, D. and Batley, J. (2006) Identification and characterization of simple sequence repeat (SSR) markers derived *in silico* from *Brassica oleracea* genome shotgun sequences. *Molecular Ecology Notes* 6, 1191–1194.

Burton, W.A., Ripley, V.L., Potts, D.A. and Salisbury, P.A. (2004) Assessment of genetic diversity in selected breeding lines and cultivars of canola quality *Brassica juncea* and their implications for canola breeding. *Euphytica* 136, 181–192.

Bus, A., Hecht, J., Huettel, B., Reinhardt, R. and Stich, B. (2012) High-throughput polymorphism detection and genotyping in *Brassica napus* using next-generation RAD sequencing. *BMC Genomics* 13, 281.

Cai, G., Yang, Q., Yang, Q., Zhao, Z., Chen, H., Wu, J., Fan, C. and Zhou, Y. (2012) Identification of candidate genes of QTLs for seed weight in *Brassica napus* through comparative mapping among *Arabidopsis* and *Brassica* species. *BMC Genetics* 13, 105.

Chalhoub, B., Denoeud, F., Liu, S., Parkin, I.A.P., Tang, H., Wang, X., Chiquet, J., Belcram, H., Tong, C., Samans, B., Corréa, M., Da Silva, C., *et al.* (2014) Early allopolyploid evolution in the post-Neolithic *Brassica napus* oilseed genome. *Science* 22, 950–953.

Chen, S., Wan, Z., Nelson, M.N., Chauhan, J.S., Redden, R., Burton, W.A., Lin, P., Salisbury, P.A., Fu, T. and Cowling, W.A. (2013) Evidence from genome-wide simple sequence repeat markers for a polyphyletic origin and secondary centers of genetic diversity of *Brassica juncea* in China and India. *Journal of Heredity* 104, 416–427.

Chen, W., Zhang, Y., Liu, X., Chen, B., Tu, J. and Fu, T. (2007) Detection of QTL for six yield-related traits in oilseed rape (*Brassica napus*) using DH and immortalized F2 populations. *Theoretical Applied Genetics* 115, 849–858.

Chen, X., Li, X., Zhang, B., Xu, J., Wu, Z., Wang, B., Li, H., Younas, M., Huang, L. and Luo, Y. (2013) Detection and genotyping of restriction fragment associated polymorphisms in polyploid crops with a pseudo-reference sequence: a case study in allotetraploid *Brassica napus*. *BMC Genomics* 14, 346.

Cheng, F., Wu, J., Fang, L., Sun, S., Liu, B., Lin, K., Bonnema, G. and Wang, X. (2012) Biased gene fractionation and dominant gene expression among the subgenomes of *Brassica rapa*. *PLoS One* 7, e36442.

Cheng, X.M., Xu, J., Xia, S., Gu, J., Yang, Y., Fu, J., Qian, X., Zhang, S., Wu, J. and Liu, K. (2009) Development and genetic mapping of microsatellite markers from genome survey sequences in *Brassica napus*. *Theoretical Applied Genetics* 118, 1121–1131.

Chutimanitsakun, Y., Nipper, R., Cuesta-Marcos, A., Cistué, L., Corey, A., Filichkina, T., Johnson, E. and Hayes, P. (2011) Construction and application for QTL analysis of a restriction site associated DNA (RAD) linkage map in barley. *BMC Genomics* 12, 4.

Clarke, W.E., Parkin, I.A., Gajardo, H.A., Gerhardt, D.J., Higgins, E., Sidebottom, C., Sharpe, A.G., Snowdon, R.J., Federico, M.L. and Iniguez-Luy, F. (2013) Genomic DNA enrichment using sequence capture micro-arrays: a novel approach to discover sequence nucleotide polymorphisms (SNP) in *Brassica napus* L. *PLoS One* 8(12), 1371.

Comai, L. (2005) The advantages and disadvantages of being polyploid. *Nature Reviews Genetics* 6, 836–846.

Cooper, J., Till, B., Laport, R., Darlow, M., Kleffner, J., Jamai, A., El-Mellouki, T., Liu, S., Ritchie, R. and Nielsen, N. (2008) TILLING to detect induced mutations in soybean. *BMC Plant Biology* 8, 9.

Del Carpio, D.P., Basnet, R.K., De Vos, R.C.H., Maliepaard, C., Visser, R. and Bonnema, G. (2011) The patterns of population differentiation in a *Brassica rapa* core collection. *Theoretical Applied Genetics* 122, 1105–1118.

Delourme, R., Falentin, C., Fomeju, B.F., Boillot, M., Lassalle, G., André, I., Duarte, J., Gauthier, V., Lucante, N. and Marty, A. (2013) High-density SNP-based genetic map development and linkage disequilibrium assessment in *Brassica napus* L. *BMC Genomics* 14, 120.

Durstewitz, G., Polley, A., Plieske, J., Luerssen, H., Graner, E.M., Wieseke, R. and Ganal, M.W. (2010) SNP discovery by amplicon sequencing and multiplex SNP genotyping in the allopolyploid species *Brassica napus*. *Genome* 53, 948–956.

Elshire, R.J., Glaubitz, J.C., Sun, Q., Poland, J.A., Kawamoto, K., Buckler, E.S. and Mitchell, S.E. (2011) A robust, simple genotyping-by-sequencing (GBS) approach for high diversity species. *PLoS One* 6, e19379.

Ferreira, M.E., Williams, P.H. and Osborn, T.C. (1994) RFLP mapping of *Brassica napus* using doubled haploid lines. *Theoretical Applied Genetics* 89, 615–621.

Fitt, B.D.L., Brun, H., Barbetti, M.J. and Rimmer, S.R. (2006) World-wide importance of Phoma stem canker (*Leptosphaeria maculans* and *L. biglobosa*) on oilseed rape (*Brassica napus*). *European Journal of Plant Pathology* 114, 3–15.

Fritsche, S., Wang, X., Li, J., Stich, B., Kopisch-Obuch, F.J., Endrigkeit, J., Leckband, G., Dreyer, F., Friedt, W. and Meng, J. (2012) A candidate gene-based association study of tocopherol content and composition in rapeseed (*Brassica napus*). *Frontiers in Plant Science* 3, 129.

Gali, K.K. and Sharpe, A.G. (2012) Molecular linkage maps: strategies, resources and achievements. In: Edwards, D., Batley, J., Parkin, I. and Kole, C. (eds) *Genetics, Genomics and Breeding of Oilseed Brassicas*. Science Publishers, New Hampshire, pp. 85–129.

Ge, Y., Ramchiary, N., Wang, T., Liang, C., Wang, N., Wang, Z., Choi, S.R., Lim, Y.P. and Piao, Z.Y. (2011) Development and linkage mapping of unigene-derived microsatellite markers in *Brassica rapa* L. *Breeding Science* 61, 160–167.

Gilchrist, E.J., Sidebottom, C., Koh, C.H., MacInnes, T., Sharpe, A.G. and Haughn, G.W. (2013) A mutant *Brassica napus* (Canola) population for the identification of new genetic diversity via TILLING and next generation sequencing. *PLoS One* 8(12), e84303.

Gyawali, S., Hegedus, D.D., Isobel, A.P., Poon, J., Higgins, E., Horner, K., Bekkaoui, D., Coutu, C. and Buchwaldt, L. (2013) Genetic diversity and population structure in a world collection of *Brassica napus* accessions with emphasis on South Korea, Japan and Pakistan. *Crop Science* 53, 1537–1545.

Han, J., Lühs, W., Sonntag, K., Zähringer, U., Borchardt, D.S., Wolter, F.P., Heinz, E. and Frentzen, M. (2001) Functional characterization of β-ketoacyl-CoA synthase genes from *Brassica napus* L. *Plant Molecular Biology* 46, 229–239.

Hansen, C.H., Du, L., Naur, P., Olsen, C.E., Axelsen, K.B., Hick, A.J., Pickett, J.A. and Halkier, B.A. (2001) CYP83B1 is the oxime-metabolizing enzyme in the glucosinolate pathway in *Arabidopsis*. *Journal of Biological Chemistry* 276, 24790–24796.

Harper, A.L., Trick, M., Higgins, J., Fraser, F., Clissold, L., Wells, R., Hattori, C., Werner, P. and Bancroft, I. (2012) Associative transcriptomics of traits in the polyploid crop species *Brassica napus*. *Nature Biotechnology* 30, 798–802.

Hasan, M., Friedt, W., Pons-Kühnemann, J., Freitag, N.M., Link, K. and Snowdon, R.J. (2008) Association of gene-linked SSR markers to seed glucosinolate content in oilseed rape (*Brassica napus* ssp. *napus*). *Theoretical Applied Genetics* 116, 1035–1049.

Hatakeyama, K., Suwabe, K., Tomita, R.N., Kato, T., Nunome, T., Fukuoka, H. and Matsumoto, S. (2013) Identification and characterization of Crr1a, a gene for resistance to Club root disease (*Plasmodiophora brassicae* Woronin) in *Brassica rapa* L. *PLoS One* 8, e54745.

Heffner, E.L., Sorrells, M.E. and Jannink, J.L. (2009) Genomic selection for crop improvement. *Crop Science* 49, 1–12.

Heffner, E.L., Jannink, J.L. and Sorrells, M.E. (2011) Genomic selection accuracy using multifamily prediction models in a wheat breeding program. *The Plant Genome* 4, 65–75.

Hegarty, M., Yadav, R., Lee, M., Armstead, I., Sanderson, R., Scollan, N., Powell, W. and Skøt, L. (2013) Genotyping by RAD sequencing enables mapping of fatty acid composition traits in perennial ryegrass (*Lolium perenne* (L.)). *Plant Biotechnolgy Journal* 11, 572–581.

Higgins, J., Magusin, A., Trick, M., Fraser, F. and Bancroft, I. (2012) Use of mRNA-seq to discriminate contributions to the transcriptome from the constituent genomes of the polyploid crop species *Brassica napus*. *BMC Genomics* 13, 247.

Hirai, M.Y., Sugiyama, K., Sawada, Y., Tohge, T., Obayashi, T., Suzuki, A., Araki, R., Sakurai, N., Suzuki, H. and Aoki, K. (2007) Omics-based identification of *Arabidopsis* Myb transcription factors regulating aliphatic glucosinolate biosynthesis. *Proceedings of the National Academy of Sciences* 104, 6478–6483.

Hopkins, C.J., Cogan, N.O.I., Hand, M., Jewell, E., Kaur, J., Li, X.I., Lim, G.A.C., Ling, A.E., Love, C. and Mountford, H. (2007) Sixteen new simple sequence repeat markers from *Brassica juncea* expressed sequences and their cross-species amplification. *Molecular Ecology Notes* 7, 697–700.

Howell, P.M., Sharpe, A.G. and Lydiate, D.J. (2003) Homoeologous loci control the accumulation of seed glucosinolates in oilseed rape (*Brassica napus*). *Genome* 46, 454–460.

Huang, B.E., George, A.W., Forrest, K.L., Kilian, A., Hayden, M.J., Morell, M.K. and Cavanagh, C.R. (2012) A multiparent advanced generation inter-cross population for genetic analysis in wheat. *Plant Biotechnology Journal* 10, 826–839.

Hull, A.K., Vij, R. and Celenza, J.L. (2000) *Arabidopsis* cytochrome P450s that catalyze the first step of tryptophan-dependent indole-3-acetic acid biosynthesis. *Proceedings of the National Academy of Science* 97, 2379–2384.

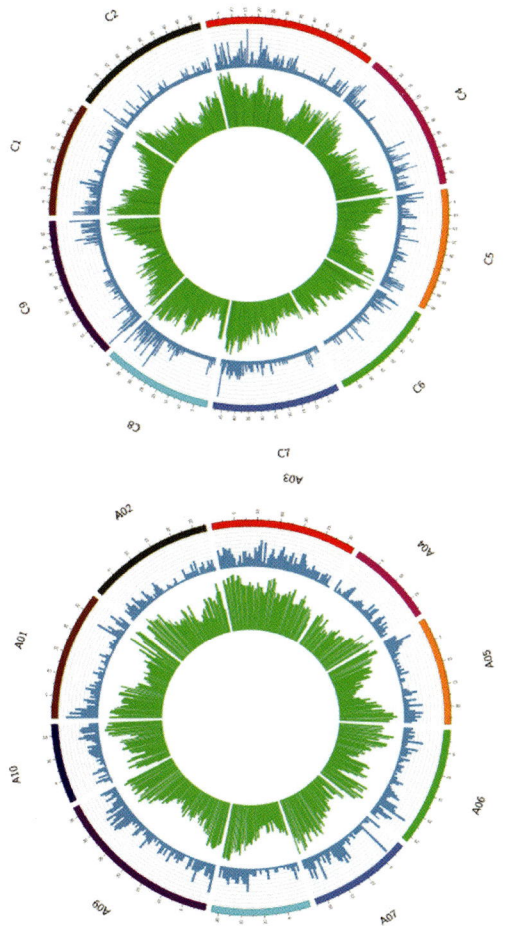

Plate 1. Distribution of mapped publicly available *B. napus* SSR loci across the diploid A and C genomes. Each of chromosomes is shown at the edges of the two circles, the physically positioned SSR loci are indicated in blue and the gene density is shown in green.

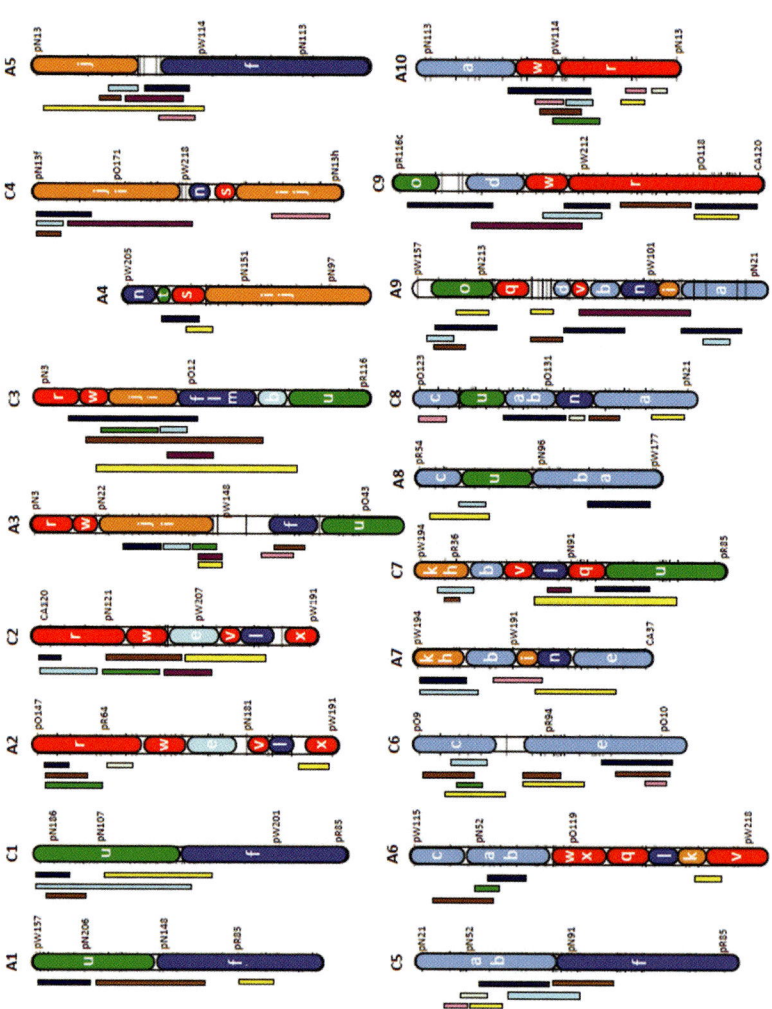

Plate 2. Genomic regions underlying traits of agronomical and nutritional interest in *B. napus* (redrawn from Clarke et al., 2013). The *B. napus* genome is ordered according to regions of homoeology between the A and C genome. QTL loci are indicated as coloured rectangles to the left of each group: yield traits (dark blue), yield component traits (light blue), plant height (brown), flowering time (green), seed quality (yellow), seedling vigour (dark purple), disease incidence (pink) and lodging (white). Where possible the blocks (a-x) which represent the ancestral conserved crucifer blocks have been identified and are coloured according to their collinearity with *A. thaliana* chromosomes: 1 (light blue); 2 (orange); 3 (dark blue); 4 (green); 5 (red).

Iniguez-Luy, F.L., Voort, A.V. and Osborn, T.C. (2008) Development of a set of public SSR markers derived from genomic sequence of a rapid cycling *Brassica oleracea* L. genotype. *Theoretical Applied Genetics* 117, 977–985.

James, D.W., Lim, E., Keller, J., Plooy, I., Ralston, E. and Dooner, H.K. (1995) Directed tagging of the *Arabidopsis* FATTY ACID ELONGATION1 (FAE1) gene with the maize transposon activator. *The Plant Cell* 7, 309–319.

Kim, H., Choi, S.R., Bae, J., Hong, C.P., Lee, S.Y., Hossain, M.J., Van Nguyen, D., Jin, M., Park, B.S. and Bang, J.W. (2009) Sequenced BAC anchored reference genetic map that reconciles the ten individual chromosomes of *Brassica rapa*. *BMC Genomics* 10, 432.

Kim, K.S., Chang, Y.S., Lee, S.I., Lee, Y.S., Son, J.H., Ha, M.W., Kim, N.S. and Park, K.C. (2012) Development of EST-SSRs of *Brassica napus*. *Korean Journal of Breeding Science* 44, 490–502.

Koornneef, M., Alonso-Blanco, C. and Vreugdenhil, D. (2004) Naturally occurring genetic variation in *Arabidopsis thaliana*. *Annual Review of Plant Biology* 55, 141–172.

Kramer, C.C., Polewicz, H. and Osborn, T.C. (2009) Evaluation of QTL alleles from exotic sources for hybrid seed yield in the original and different genetic backgrounds of spring-type *Brassica napus* L. *Molecular Breeding* 24, 419–431.

Kresovich, S., Szewc-McFadden, A.K., Bliek, S.M. and McFerson, J.R. (1995) Abundance and characterization of simple-sequence repeats (SSRs) isolated from a size-fractionated genomic library of *Brassica napus* L. (rapeseed). *Theoretical Applied Genetics* 91, 206–211.

Kutcher, H.R., Yu, F. and Brun, H. (2010) Improving blackleg disease management of *Brassica napus* from knowledge of genetic interactions with *Leptosphaeria maculans*. *Canadian Journal of Plant Pathology* 32, 29–34.

Lagercrantz, U. and Lydiate, D.J. (1996) Comparative genome mapping in *Brassica*. *Genetics* 144, 1903–1910.

Lagercrantz, U., Ellegren, H. and Andersson, L. (1993) The abundance of various polymorphic microsatellite motifs differs between plants and vertebrates. *Nucleic Acids Research* 21, 1111–1115.

Larkan, N.J., Lydiate, D.J., Parkin, I.A.P., Nelson, M.N., Epp, D.J., Cowling, W.A., Rimmer, S.R. and Borhan, M.H. (2013) The *Brassica napus* blackleg resistance gene LepR3 encodes a receptor-like protein triggered by the *Leptosphaeria maculans* effector AVRLM1. *New Phytologist* 197, 595–605.

Li, H., Chen, X., Yang, Y., Xu, J., Gu, J., Fu, J., Qian, X., Zhang, S., Wu, J. and Liu, K. (2011) Development and genetic mapping of microsatellite markers from whole genome shotgun sequences in *Brassica oleracea*. *Molecular Breeding* 28, 585–596.

Li, W., Jiang, W., Zhao, H.X., Vyvadilova, M., Stamm, M. and Hu, S.W. (2012) Genetic diversity of rapeseed accessions from different geographic locations revealed by expressed sequence tag-simple sequence repeat and random amplified polymorphic DNA markers. *Crop Science* 52, 201–210.

Ling, A.E., Kaur, J., Burgess, B., Hand, M., Hopkins, C.J., Li, X.I., Love, C.G., Vardy, M., Walkiewicz, M. and Spangenberg, G. (2007) Characterization of simple sequence repeat markers derived *in silico* from *Brassica rapa* bacterial artificial chromosome sequences and their application in *Brassica napus*. *Molecular Ecology Notes* 7, 273–277.

Lowe, A.J., Jones, A.E., Raybould, A.F., Trick, M., Moule, C.L. and Edwards, K.J. (2002) Transferability and genome specificity of a new set of microsatellite primers among *Brassica* species of the U triangle. *Molecular Ecology Notes* 2, 7–11.

Lowe, A.J., Moule, C., Trick, M. and Edwards, K.J. (2004) Efficient large-scale development of microsatellites for marker and mapping applications in *Brassica* species. *Theoretical Applied Genetics* 108, 1103–1112.

McCallum, C.M., Comai, L., Greene, E.A. and Henikoff, S. (2000) Targeting induced local lesions in genomes (TILLING) for plant functional genomics. *Plant Physiology* 123, 439–442.

McMullen, M.D., Kresovich, S., Villeda, H.S., Bradbury, P., Li, H., Sun, Q., Flint-Garcia, S., Thornsberry, J., Acharya, C. and Bottoms, C. (2009) Genetic properties of the maize nested association mapping population. *Science* 325, 737–740.

Meuwissen, T.H.E., Hayes, B.J. and Goddard, M.E. (2001) Prediction of total genetic value using genome-wide dense marker maps. *Genetics* 157, 1819–1829.

Morgante, M., Hanafey, M. and Powell, W. (2002) Microsatellites are preferentially associated with nonrepetitive DNA in plant genomes. *Nature Genetics* 30, 194–200.

Navabi, Z.K., Huebert, T., Sharpe, A.G., Bancroft, I. and Parkin, I.A.P. (2013) Conserved microstructure of the *Brassica* B Genome of *Brassica nigra* in relation to homologous regions of *Arabidopsis thaliana*, *B. rapa* and *B. oleracea*. *BMC Genomics* 14, 250.

Nelson, J., Wang, S., Wu, Y., Li, X., Antony, G., White, F. and Yu, J. (2011) Single-nucleotide polymorphism discovery by high-throughput sequencing in sorghum. *BMC Genomics* 12, 352.

Nipper, R.W., Atwood, T.S., Boone, J.Q., Gribbin, J.M. and Johnson, E.A. (2010) SNP discovery in *Zea mays* using sequenced restriction-site associated DNA markers. *Acta Horticulturae* 859, 129–133.

Parida, S.K., Yadava, D.K. and Mohapatra, T. (2010) Microsatellites in *Brassica* unigenes: relative abundance, marker design and use in comparative physical mapping and genome analysis. *Genome* 53, 55–67.

Paritosh, K., Yadava, S.K., Gupta, V., Panjabi-Massand, P., Sodhi, Y.S., Pradhan, A.K. and Pental, D. (2013) RNA-seq based SNPs in some agronomically important oleiferous lines of *Brassica rapa* and their use for genome-wide linkage mapping and specific-region fine mapping. *BMC Genomics* 14, 463.

Parkin, I.A.P. (2011) Chasing ghosts: Comparative mapping in the Brassicaceae. In: Bancroft, I. and Schmidt, R. (eds) *Genetics and Genomics of the Brassicaceae.* Springer, New York, pp. 153–170.

Parkin, I.A.P., Sharpe, A.G., Keith, D.J. and Lydiate, D.J. (1995) Identification of the A and C genomes of amphidiploid *Brassica napus* (oilseed rape). *Genome* 38, 1122–1131.

Parkin, I.A.P., Gulden, S.M., Sharpe, A.G., Lukens, L., Trick, M., Osborn, T.C. and Lydiate, D.J. (2005) Segmental structure of the *Brassica napus* genome based on comparative analysis with *Arabidopsis thaliana.* *Genetics* 171, 765–781.

Parkin, I.A.P., Koh, C., Tang, H., Robinson, S.J., Kagale, S., Clarke, W.E., Town, C.D., Nixon, J., Krishnakumar, V., Bidwell, S.L., Denoeud, F., Belcram, H., *et al.* (2014) Transcriptome and methylome profiling reveals relics of genome dominance in the mesopolyploid *Brassica oleracea.* *Genome Biology* 15, R77.

Pascher, K., Macalka, S., Rau, D., Gollmann, G., Reiner, H., Glössl, J. and Grabherr, G. (2010) Molecular differentiation of commercial varieties and feral populations of oilseed rape (*Brassica napus* L.). *BMC Evolution Biology* 10, 63.

Pfender, W.F., Saha, M.C., Johnson, E.A. and Slabaugh, M.B. (2011) Mapping with RAD (restriction-site associated DNA) markers to rapidly identify QTL for stem rust resistance in *Lolium perenne.* *Theoretical Applied Genetics* 122, 1467–1480.

Poland, J.A., Brown, P.J., Sorrells, M.E. and Jannink, J.L. (2012) Development of high-density genetic maps for barley and wheat using a novel two-enzyme genotyping-by-sequencing approach. *PLoS One* 7, e32253.

Prakash, S. and Tsunoda, S. (1983) Cytogenetics of *Brassica.* In: Swaminathan, M.S., Gupta, P.K. and Sinha, U. (eds) Cytogenetics of Crop Plants. MacMillan India Ltd, pp. 481–514.

Pritchard, J.K., Stephens, M. and Donnelly, P. (2000) Inference of population structure using multilocus genotype data. *Genetics* 155, 945–959.

Putterill, J., Laurie, R. and Macknight, R. (2004) It's time to flower: the genetic control of flowering time. *Bioessays* 26, 363–373.

Raman, R., Taylor, B., Marcroft, S., Stiller, J., Eckermann, P., Coombes, N., Rehman, A., Lindbeck, K., Luckett, D. and Wratten, N. (2012) Molecular mapping of qualitative and quantitative loci for resistance to *Leptosphaeria maculans* causing blackleg disease in canola (*Brassica napus* L.). *Theoretical Applied Genetics* 125, 405–418.

Ramchiary, N., Bisht, N.C., Gupta, V., Mukhopadhyay, A., Arumugam, N., Sodhi, Y.S., Pental, D. and Pradhan, A.K. (2007) QTL analysis reveals context-dependent loci for seed glucosinolate trait in the oilseed *Brassica juncea*: importance of recurrent selection backcross scheme for the identification of 'true' QTL. *Theoretical Applied Genetics* 116, 77–85.

Ramchiary, N., Li, X., Hong, C.P., Dhandapani, V., Choi, S.R., Yu, G., Piao, Z.Y. and Lim, Y.P. (2011) Genic microsatellite markers in *Brassica rapa*: development, characterization, mapping and their utility in other cultivated and wild *Brassica* relatives. *DNA Research* 18, 305–320.

Rimmer, S.R. (2006) Resistance genes to *Leptosphaeria maculans* in *Brassica napus.* *Canadian Journal of Plant Pathology* 28, S288–S297.

Scaglione, D., Acquadro, A., Portis, E., Tirone, M., Knapp, S.J. and Lanteri, S. (2012) RAD tag sequencing as a source of SNP markers in *Cynara cardunculus* L. *BMC Genomics* 13, 3.

Schranz, M.E., Lysak, M.A. and Mitchell-Olds, T. (2006) The ABC's of comparative genomics in the Brassicaceae: building blocks of crucifer genomes. *Trends in Plant Science* 11, 535–542.

Selkoe, K.A. and Toonen, R.J. (2006) Microsatellites for ecologists: a practical guide to using and evaluating microsatellite markers. *Ecological Letters* 9, 615–629.

Sharma, P.C., Winter, P., Bünger, T., Hüttel, B., Weigand, F., Weising, K. and Kahl, G. (1995) Abundance and polymorphism of di-, tri-and tetra-nucleotide tandem repeats in chickpea (*Cicer arietinum* L.). *Theoretical Applied Genetics* 90, 90–96.

Sharpe, A.G. and Lydiate, D.J. (2003) Mapping the mosaic of ancestral genotypes in a cultivar of oilseed rape (*Brassica napus*) selected via pedigree breeding. *Genome* 46, 461–468.

Slade, A.J. and Knauf, V.C. (2005) TILLING moves beyond functional genomics into crop improvement. *Transgenic Research* 14, 109–115.

Slade, A.J., Fuerstenberg, S.I., Loeffler, D., Steine, M.N. and Facciotti, D. (2004) A reverse genetic, nontransgenic approach to wheat crop improvement by TILLING. *Nature Biotechnology* 23, 75–81.

Smith, C. (1967) Improvement of metric traits through specific genetic loci. *Animal Production* 9, 349–358.

Smith, L.B. and King, G.J. (2000) The distribution of BoCAL-a alleles in *Brassica oleracea* is consistent with a genetic model for curd development and domestication of the cauliflower. *Molecular Breeding* 6, 603–613.

Sodhi, Y.S., Mukhopadhyay, A., Arumugam, N., Verma, J.K., Gupta, V., Pental, D. and Pradhan, A.K. (2002) Genetic analysis of total glucosinolate in crosses involving a high glucosinolate Indian variety and a low glucosinolate line of *Brassica juncea*. *Plant Breeding* 121, 508–511.

Soller, M. and Beckmann, J.S. (1983) Genetic polymorphism in variety identification and genetic improvement. *Theoretical Applied Genetics* 67, 25–33.

Srivastava, A., Gupta, V., Pental, D. and Pradhan, A.K. (2001) AFLP-based genetic diversity assessment amongst agronomically important natural and some newly synthesized lines of *Brassica juncea*. *Theoretical Applied Genetics* 102, 193–199.

Stephenson, P., Baker, D., Girin, T., Perez, A., Amoah, S., King, G.J. and Østergaard, L. (2010) A rich TILLING resource for studying gene function in *Brassica rapa*. *BMC Plant Biology* 10, 62.

Suwabe, K., Iketani, H., Nunome, T., Kage, T. and Hirai, M. (2002) Isolation and characterization of microsatellites in *Brassica rapa* L. *Theoretical Applied Genetics* 104, 1092–1098.

Suwabe, K., Tsukazaki, H., Iketani, H., Hatakeyama, K., Kondo, M., Fujimura, M., Nunome, T., Fukuoka, H., Hirai, M. and Matsumoto, S. (2006) Simple sequence repeat-based comparative genomics between *Brassica rapa* and *Arabidopsis thaliana*: the genetic origin of clubroot resistance. *Genetics* 173, 309–319.

Szadkowski, E., Eber, F., Huteau, V., Lodé, M., Huneau, C., Belcram, H., Coriton, O., Manzanares-Dauleux, M.J., Delourme, R., King, G.J., Chalhoub, B., Jenczewski, E. and Chèvre, A.M. (2010) The first meiosis of resynthesized *Brassica napus*, a genome blender. *New Phytologist* 186, 102–112.

Szewc-McFadden, A.K., Kresovich, S., Bliek, S.M., Mitchell, S.E. and McFerson, J.R. (1996) Identification of polymorphic, conserved simple sequence repeats (SSRs) in cultivated *Brassica* species. *Theoretical Applied Genetics* 93, 534–538.

Tang, H., Woodhouse, M.R., Cheng, F., Schnable, J.C., Pedersen, B.S., Conant, G., Wang, X., Freeling, M. and Pires, J.C. (2012) Altered patterns of fractionation and exon deletions in *Brassica rapa* support a two-step model of paleohexaploidy. *Genetics* 190, 1563–1574.

Taylor, D.C., Falk, K.C., Palmer, C.D., Hammerlindl, J., Babic, V., Mietkiewska, E., Jadhav, A., Marillia, E.F., Francis, T. and Hoffman, T. (2010) *Brassica carinata* – a new molecular farming platform for delivering bio-industrial oil feed stocks: case studies of genetic modifications to improve very long-chain fatty acid and oil content in seeds. *Biofuels, Bioproducts and Biorefining* 4, 538–561.

The Arabidopsis Genome Initiative (2000) Analysis of the genome sequence of the flowering plant *Arabidopsis thaliana*. *Nature* 408, 796–815.

Till, B.J., Reynolds, S.H., Weil, C., Springer, N., Burtner, C., Young, K., Bowers, E., Codomo, C.A., Enns, L.C. and Odden, A.R. (2004) Discovery of induced point mutations in maize genes by TILLING. *BMC Plant Biology* 4, 12.

Till, B.J., Cooper, J., Tai, T.H., Colowit, P., Greene, E.A., Henikoff, S. and Comai, L. (2007) Discovery of chemically induced mutations in rice by TILLING. *BMC Plant Biology* 7, 19.

Tollenaere, R., Hayward, A., Dalton-Morgan, J., Campbell, E., Lee, J.R.M., Lorenc, M.T., Manoli, S., Stiller, J., Raman, R. and Raman, H. (2012) Identification and characterization of candidate *Rlm4* blackleg resistance genes in *Brassica napus* using next-generation sequencing. *Plant Biotechnology Journal* 10, 709–715.

Town, C.D., Cheung, F., Maiti, R., Crabtree, J., Haas, B.J., Wortman, J.R., Hine, E.E., Althoff, R., Arbogast, T.S. and Tallon, L.J. (2006) Comparative genomics of *Brassica oleracea* and *Arabidopsis thaliana* reveal gene loss, fragmentation and dispersal after polyploidy. *The Plant Cell* 18, 1348–1359.

Trick, M., Long, Y., Meng, J. and Bancroft, I. (2009) Single nucleotide polymorphism (SNP) discovery in the polyploid *Brassica napus* using Solexa transcriptome sequencing. *Plant Biotechnology Journal* 7, 334–346.

U, N. (1935) Genome analysis in *Brassica* with special reference to the experimental formation of *B. napus* and peculiar mode of fertilization. *Japan Journal of Botany* 7, 389–452.

Udall, J.A., Quijada, P.A., Lambert, B. and Osborn, T.C. (2006) Quantitative trait analysis of seed yield and other complex traits in hybrid spring rapeseed (*Brassica napus* L.): 2. Identification of alleles from unadapted germplasm. *Theoretical Applied Genetics* 113, 597–609.

Ueno, H., Matsumoto, E., Aruga, D., Kitagawa, S., Matsumura, H. and Hayashida, N. (2012) Molecular characterization of the CRa gene conferring clubroot resistance in *Brassica rapa*. *Plant Molecular Biology* 80, 621–629.

Uzunova, M.I. and Ecke, W. (1999) Abundance, polymorphism and genetic mapping of microsatellites in oilseed rape (*Brassica napus* L.). *Plant Breeding* 118, 323–326.

Walker, K.C. and Booth, E.J. (2001) Agricultural aspects of rape and other *Brassica* products. *European Journal of Lipid Science Technology* 103, 441–446.

Wang, N., Wang, Y., Tian, F., King, G.J., Zhang, C., Long, Y., Shi, L. and Meng, J. (2008) A functional genomics resource for *Brassica napus*: development of an EMS mutagenized population and discovery of FAE1 point mutations by TILLING. *New Phytologist* 180, 751–765.

Wang, F., Wang, X., Chen, X., Xiao, Y., Li, H., Zhang, S., Xu, J., Fu, J., Huang, L. and Liu, C. (2012) Abundance, marker development and genetic mapping of microsatellites from unigenes in *Brassica napus*. *Molecular Breeding* 30, 731–744.

Wang, N., Qian, W., Suppanz, I., Wei, L., Mao, B., Long, Y., Meng, J., Müller, A.E. and Jung, C. (2011) Flowering time variation in oilseed rape (*Brassica napus* L.) is associated with allelic variation in the FRIGIDA homologue BnaA.FRI.a. *Journal of Experimental Botany* 62, 5641–5658.

Wang, N., Fang, L., Xin, H., Wang, L. and Li, S. (2012) Construction of a high-density genetic map for grape using next generation restriction-site associated DNA sequencing. *BMC Plant Biology*, 12.

Wang, N., Thomson, M., Bodles, W.J.A., Crawford, R.M.M., Hunt, H.V., Featherstone, A.W., Pellicer, J. and Buggs, R.J.A. (2013) Genome sequence of dwarf birch (*Betula nana*) and cross-species RAD markers. *Molecular Ecology* 22, 3098–3111.

Wang, X., Wang, H., Wang, J., Sun, R., Wu, J., Liu, S., Bai, Y., Mun, J.H., Bancroft, I., Cheng, F., Huang, S., Li, X., *et al.* (2011) The genome of the mesopolyploid crop species *Brassica rapa*. *Nature Genetics* 43, 1035–1039.

Wang, X., Zhang, C., Li, L., Fritsche, S., Endrigkeit, J., Zhang, W., Long, Y., Jung, C. and Meng, J. (2012) Unraveling the genetic basis of seed tocopherol content and composition in rapeseed (*Brassica napus* L.). *PLoS One* 7:e50038

Westermeier, P., Wenzel, G. and Mohler, V. (2009) Development and evaluation of single-nucleotide polymorphism markers in allotetraploid rapeseed (*Brassica napus* L.). *Theoretical Applied Genetics* 119, 1301–1311.

Wu, J., Cai, G., Tu, J., Li, L., Liu, S., Luo, X., Zhou, L., Fan, C. and Zhou, Y. (2013) Identification of qtls for resistance to Sclerotinia stem rot and bnac. igmt5.a as a candidate gene of the major resistant qtl src6 in *Brassica napus*. *PLoS One* 8, e67740.

Xu, J., Qian, X., Wang, X., Li, R., Cheng, X., Yang, Y., Fu, J., Zhang, S., King, G. and Wu, J. (2010) Construction of an integrated genetic linkage map for the A genome of *Brassica napus* using SSR markers derived from sequenced BACs in *B. rapa*. *BMC Genomics* 11, 594.

Xu, S. (2003) Theoretical basis of the Beavis effect. *Genetics* 165, 2259–2268.

Xu, Y. and Crouch, J.H. (2008) Marker-assisted selection in plant breeding: from publications to practice. *Crop Science* 48, 391–407.

Yu, S., Zhang, F., Yu, R., Zou, Y., Qi, J., Zhao, X., Yu, Y., Zhang, D. and Li, L. (2009) Genetic mapping and localization of a major QTL for seedling resistance to downy mildew in Chinese cabbage (*Brassica rapa* ssp. *Pekinensis*). *Molecular Breeding* 23, 573–590.

Yu, S., Zhang, F., Wang, X., Zhao, X., Zhang, D., Yu, Y. and Xu, J. (2010) Genetic diversity and marker-trait associations in a collection of Pak-choi (*Brassica rapa* L. ssp. *Chinensis* Makino) accessions. *Genes & Genomics* 32, 419–428.

Zhao, J., Paulo, M.J., Jamar, D., Lou, P., van Eeuwijk, F., Bonnema, G., Vreugdenhil, D. and Koornneef, M. (2007) Association mapping of leaf traits, flowering time and phytate content in *Brassica rapa*. *Genome* 50, 963–973.

Zhao, J., Buchwaldt, L., Rimmer, S.R., Sharpe, A., McGregor, L., Bekkaoui, D. and Hegedus, D. (2009) Patterns of differential gene expression in *Brassica napus* cultivars infected with *Sclerotinia sclerotiorum*. *Molecular Plant Pathology* 10, 635–649.

Zhao, Y., Gowda, M., Liu, W., Würschum, T., Maurer, H.P., Longin, F.H., Ranc, N. and Reif, J.C. (2012) Accuracy of genomic selection in European maize elite breeding populations. *Theoretical Applied Genetics* 124, 769–776.

7 Diseases

C. Chattopadhyay[1]* and S.J. Kolte[2]
[1]*National Centre on Integrated Pest Management, Pusa Campus, New Delhi;*
[2]*Ex-Professor (Plant Pathology), Kothrud, Pune, India*

Introduction

Rapeseed-mustard crops are confronted by numerous diseases, insects, drought, high temperature, salinity and frost, etc. Fungal diseases are a major hurdle towards achieving higher production. The intensive cultivation of rapeseed-mustard crops with more inputs has further compounded the problem and now the occurrence of diseases has become more frequent and widespread. Severe outbreak of diseases deteriorates the quantity as well as quality of seed and oil content drastically in different oilseed brassica crops. Expression of full inherent genetic potential of a genotype is governed by inputs that go in to the production system. This can be very well illustrated with examples that involve disease management of rapeseed-mustard. The yield reduction in oilseed brassica crops due to biotic stresses is about 19.9%, out of which diseases cause severe yield reduction at various plant growth stages. Various plant pathogens have been found to distress the crop. Of these, 18 are commercially damaging in different parts of the world. It is essential to know the causal agents, their behaviour and means to attack the vulnerable stage of the pathogen to avoid yield losses. A retrospective of researches on major disease problems in rapeseed-mustard crops, particularly in Indian conditions, and their management is discussed here along with brief perspectives based on emerging trends for the future. Breeding for disease resistance is an important method of protecting crops from damages due to biotic factors (Chattopadhyay and Séguin-Swartz, 2005). Inherited resistance is more valuable as it is economical and environmentally safe.

Damping Off and Seedling Blight

Several fungi species can cause seed rot and seedling blight around the world. Among them, *Rhizopus stolonifer* is reported to be an important cause (Petrie, 1973a). Post-emergence mortality is not frequent, with *Pythium aphanidermatum* (Mahmud, 1950), *P. butleri* (Aulakh, 1971), *Rhizoctonia solani* (Srivastava, 1968), *Sclerotium rolfsii* (Upadhyay and Pavgi, 1967), *Macrophomina phaseolina* (Srivastava and Dhawan, 1979) and *Fusarium* spp. being the pathogens involved in India, causing 6–15% incidence (Khan and Kolte, 2002).

*Corresponding author, e-mail: chirantan_cha@hotmail.com

They mostly survive on crop debris and soil as different resting structures to infect the following crop.

Symptoms

A necrotic lesion 1–2 cm long can occur initially at the base of the stem, with sometimes girdling taking place near soil level. The taproot may be discoloured, and sometimes wire-stem symptoms become visible. Salmon-coloured spore masses of *Fusarium* are often observed on affected tissues. Sometimes the symptoms are restricted to roots consisting of light-brown lesions on the taproot and at the bases of larger lateral roots. Girdling of the main root may take place, which may lead to loss of the entire root system. Damping-off and seedling blight are mostly encountered due to use of infested seed.

Management

Drainage from the crop field should be ensured at the time of sowing the crop in order to avoid water stagnation. Clean cultivation and removal of crop debris before sowing is important to manage the problem. Significant combined effect of *Brassica napus* green manuring as well as of *Trichoderma* seed treatment against different soil pathogenic fungi (*Pythium* and *Rhizoctonia*) could be useful (Galletti *et al.*, 2006). Seed treatment with thiophanate methyl 70 WP at 2g/kg ensured better plant stand with protection against *S. rolfsii, R. solani* and *F. oxysporum* (Khan and Kolte, 2002). Seed treatment with metalaxyl 35 SD 6 g/kg + carbendazim 1 g active ingredient/kg or with any other suitable seed protectant fungicide may be helpful in increasing the stand of the crop.

Alternaria Blight

Rapeseed-mustard crops are ravaged by Alternaria blight, or black spot, which is caused mostly by *Alternaria brassicea* (Berk.) Sac. The pathogen infects all the above-ground parts of the plant. It has been reported all over the world and is an important constraint for oilseed brassica cultivation in India. *Alternaria brassicicola* and *A. raphani* are also rarely encountered on the crop. Losses due to this disease usually range between 5 and 15% at harvest. However, losses may reach up to 47% at high disease severity (Kolte *et al.*, 1987) accompanied by reduction in seed quality, viz. seed size, viability, etc. Alternaria blight severity on oilseed brassicas fluctuates with seasons and regions depending upon favourable weather conditions for pathogen development.

Symptoms

These are characterized by light brown to black spots on leaves, stem and siliquae (Fig. 7.1a, b). The pathogen can affect seed germination, oil quality and quantity (Meena *et al.*, 2010a). Symptoms on seedlings included dark stem lesions immediately after germination that can cause damping-off or stunted growth. In general, disease appears at 40–45 days after seeding and the most critical stage has been reported after 75 and 45 days of plant growth (Meena *et al.*, 2004). Lesions formed by *A. brassicae* incidence are normally grey in colour while *A. brassicicola* produced black sooty velvety spots. *A. raphani* form spots with distinctive yellow halos around them. The disease symptoms can vary with host species and the environment.

Disease symptoms first show on the lower leaves as black points, which later expand into prominent, round, concentric spots of different sizes. Symptoms become visible on the middle and upper leaves with small spots. Defoliation of lower leaves takes place with the progression of the disease; later, round black noticeable spots appear on siliquae and stem. These spots may coalesce, causing siliquae blackening or weakening of the stem. Seed decay may be visible below the black spot on siliquae of *B. rapa* var. Yellow Sarson and *B. rapa* var. Brown Sarson. Brownish black spots with a grey middle can appear on the mustard siliqua. Symptoms mostly arise on the senescing leaves, since they are closer to the soil and are easily infected due to rain splash or wind-blown rain. Silique-bearing

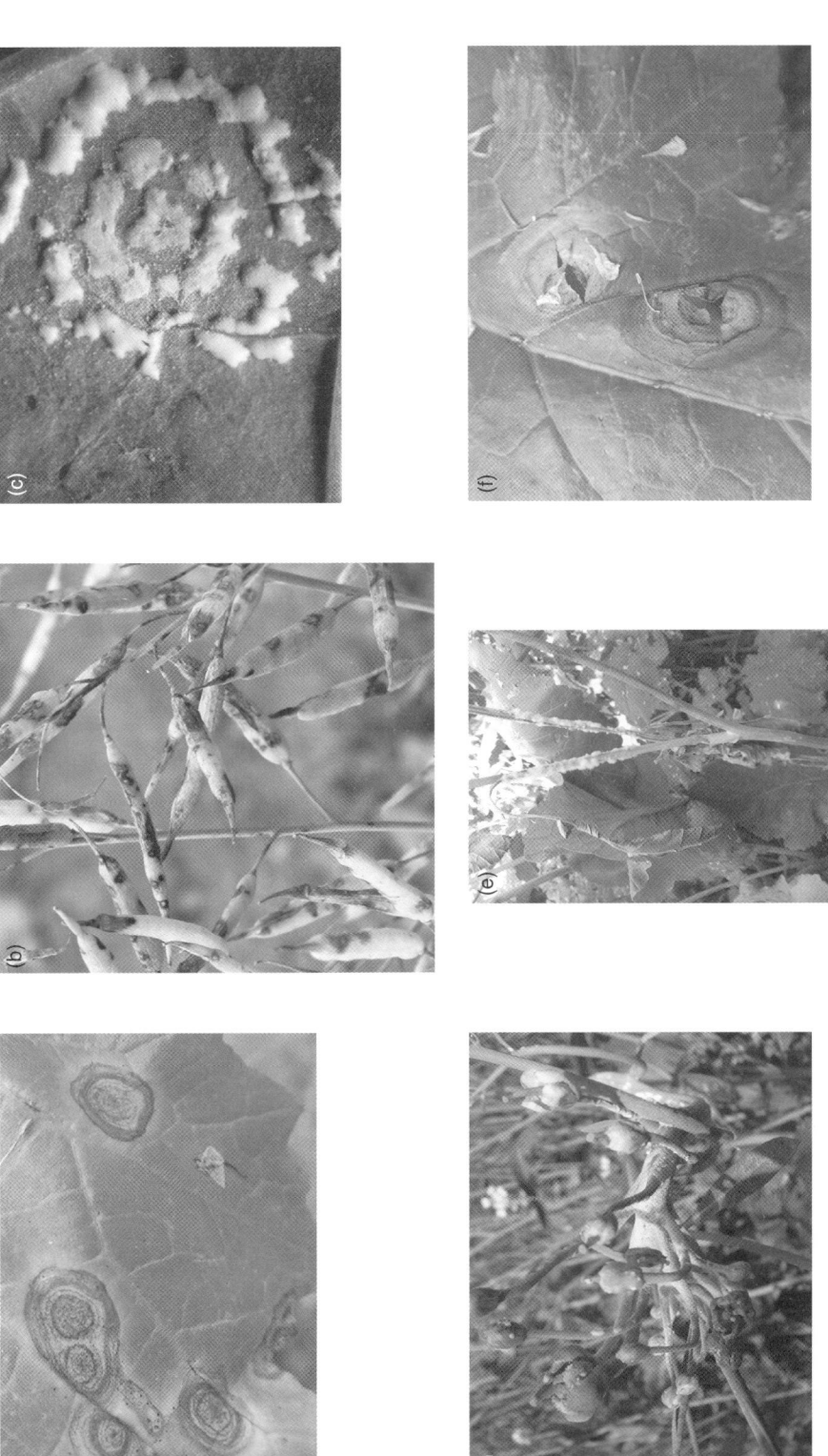

Fig. 7.1. Symptoms of diseases of rapeseed-mustard: (a) Alternaria blight on leaves; (b) Alternaria blight on pods; (c) white rust on leaves; (d) staghead (hypertrophied inflorescence) caused by *Albugo*; (e) Sclerotinia rot-affected stem; (f) Sclerotinia rot-affected leaf; (g) Sclerotinia proliferating on fallen petals; carpogenic germination of *Sclerotinia* apothecia; (h) powdery mildew; (i) downy mildew affected leaves at seedling stage; (j) downy mildew affected pods; (k) staghead (hypertrophied inflorescence) caused by *Hyaloperonospora*; and (l) club root.

Continued

Fig. 7.1. Continued.

branches and siliquae showing dark or blackened spots may cause yield loss due to the premature ripening and shattering of the silique. Alternaria blight infection on leaves and silique reduces the available photosynthetic area. Silique infection adversely affects normal seed development, seed weight, colour of seeds and seed oil content.

Pathogen

Alternaria sp. are recognizable by large and catenate conidia, solitary in occurrence, or forming chains, typically ovoid to obclavate, frequently beaked, light brown to brown, multicellular and muriform (Simmons, 2007). Chlamydospore formation is reported in *A. brassicae* and *A. raphani*, whereas microsclerotia are formed by *A. brassicae*. Although the use of species-group designation may not resolve ultimate species boundaries within *Alternaria*, it permits organization of the pathogen at the subgeneric level. Morphological diversity of *Alternaria* species permits comparison with morphologically similar species. Variation among *A. brassicae* isolates on a morphological, pathogenic and molecular basis has been discussed (Saharan, 1992; Vishwanath *et al.*, 1999; Goyal, 2009; Goyal *et al.*, 2011a). Cellulase enzymes (Nehemiah and Deshpande, 1976) are produced by *A. brassicae*; however, their real role in pathogenesis is not known. *Alternaria longipes* and *A. napiformae* have also been reported on rapeseed-mustard from India (Rao, 1977).

Survival

Unlike in temperate conditions (Humpherson-Jones and Maude, 1982), the pathogen cannot generally survive on infected plant rubble or diseased seeds in tropical and subtropical parts of the world. Airborne spores of *A. brassicae* form the principal source of inoculum in tropical and subtropical regions (Kolte, 1985; Mehta *et al.*, 2002). Further, fungal pathogen surviving on vegetable brassica crops and alternative hosts (*Anagallis arvensis, Convolvulus arvensis*) generally facilitate the carry-over of

the *A. brassicae* across crop seasons (Tripathi and Kaushik, 1984; Verma and Saharan, 1994; Mehta *et al.*, 2002). *Alternaria brassicae, A. brassicicola, A. raphani* and *A. alternata* primarily penetrate the unaffected tissues of several brassicaceous hosts directly by germ tube from germinated spores (Changsri and Weber, 1963; Czyzewska, 1970), although indirect penetration through stomata was also reported from *A. brassicae* (Changsri, 1961; Tsuneda and Skoropad, 1978). Black spot lesions develop in about 48 h after inoculation. Tewari (1986) opined that *A. brassicae* in rapeseed becomes sub-cuticular after direct infiltration. Colonization of epidermal and the mesophyll cells occurs later. The pathogen colonizes the necrotic centre only in rapeseed leaves and is not present in the chlorotic area, suggesting that a diffusible metabolite may result in leaf chlorosis. It is likely that plasma membrane is the initial target of the diffusible metabolites followed by chloroplasts, precipitating leaf chlorosis. Goyal (2009) affirmed that *A. brassicae* conidia may germinate on the upper epidermis of *Brassica juncea* leaf by producing germ tubes which penetrate the host directly without formation of an appressorium. The mycelia then ramify and colonize mesophyll and palisade tissue, leading to cell necrosis by producing toxins or metabolites that lead to the formation of necrotic spots and reduction of available photosynthetic area. Other than phenolic compounds, the infection decreases cell constituents such as lignin, lipids, suberin and protein.

Epidemiology

Ecofriendly and economic control of Alternaria blight requires knowledge regarding appearance of disease and its interaction with weather factors. This may permit better prediction of disease epidemics, facilitating farmers to initiate timely disease control measures. Weather is a critical factor defining Alternaria blight severity. The influence of temperatures, relative humidity (RH) and daylength on incidence of the blight of oilseed brassicas has been widely reported (Saharan and Kadian, 1984; Sinha *et al.*, 1992; Awasthi and Kolte, 1994; Dang *et al.*, 1995). Empirical models have been prepared that depict relationships between different weather factors and Alternaria blight occurrence. Alternaria blight occurrence on leaves (Meena *et al.*, 2002) and pods (Sandhu *et al.*, 1985) is generally greater in later sown crops. Delayed sowing causes sensitive growth stage of plants to coincide with environmental conditions of temperature (maximum temperature: 18–26°C; minimum temperature: 8–12°C) and humidity (mean RH >70%) that favour rapid multiplication and spread of the pathogen. Initiation of Alternaria blight on mustard leaves mostly occurs during 36–139 days after sowing (DAS), the maximum occurrence being at 45 and 75 DAS. Initiation of the disease symptoms on pod occurs between 67 and 142 DAS, the highest being at 99 DAS. Maximum temperatures of 20–30°C, mean temperature >14°C, morning RH >90%, mean RH >70%, daylength >9 h and >10 h of leaf wetness are considered favourable for the disease severity on pods. Region-specific models can help to predict the crop age at which Alternaria blight first appears on the leaves and pods. Similarly, the highest blight severity can be predicted at least 1 week ahead of first appearance of the disease (Chattopadhyay *et al.*, 2005). Darkness favours sporulation of *A. brassicae* (Kadian and Saharan, 1984; Humpherson-Jones and Phelps, 1989).

Management

Alternaria spores can survive on leaves for 8–12 weeks and on stem tissue for about 23 weeks. Hence, fields replanted soon after the first crop tend to encounter large quantities of inoculum, which is expected to affect the crop germination and initial phases of crop growth (Humpherson-Jones, 1992). Rotation with non-brassicaceous crops and suppression of alternate hosts can control the pathogen. Fungicide application should also be initiated. Iprodione (Rovral) spray helped to check silique infection due to *A. brassicae* (Cox *et al.*, 1983). Mancozeb could reduce the disease severity on mustard leaves. Application of 1% (w/v) aqueous bulb extract of *Allium sativum* at 45 and 75 DAS also checked the

disease severity on leaves and pods (Meena *et al.*, 2004, 2008, 2011b). Application of the GR isolate of *Trichoderma viride* was at par with Mancozeb in checking blight severity. Soil application of K as basal dose can check Alternaria blight disease in mustard (Sharma and Kolte, 1994). Timely planting (Meena *et al.*, 2002) of pathogen-free seeds after deep tilling, weeding, optimum plant population, irrigation at flowering and silique development stages may reduce the disease incidence. *Alternaria* growth can be reduced by treating seeds with hot water (Humpherson-Jones and Maude, 1982).

A number of sources of Alternaria resistance are available (AICRP-RM, 1986–2010; Gupta *et al.*, 2001; Chattopadhyay and Séguin-Swartz, 2005). A short stature *B. juncea* cultivar Divya was reported tolerant to Alternaria blight (Kolte *et al.*, 2000). Among the *Brassica* species, *B. juncea* and *B. rapa* show more susceptibility than *B. carinata* and *B. napus* (Skoropad and Tewari, 1977).

Other sources of resistance include *B. juncea* PAB 9511, PAB-9534, JMM-915, EC-399296, EC-399301, EC-399299, PHR-2 and Divya; *B. carinata* HC-1, PBC-9221 (Kiran), NRCDR-515 and DLSC-1; *B. napus* PBN-9501, PBN-9502, PBN-2001 and PBN-2002 (AICRP-RM, 1986-2010; Kolte *et al.*, 2006). Some wild crucifers such as *B. alba* (Dueck and Degenhardt, 1975; Brun *et al.*, 1987; Hansen and Earle, 1997), *Capsella bursa-pastoris*, *Eruca sativa*, *Neslia paniculata* and *Camelina sativa* (Tewari, 1991; Tewari and Conn, 1993; Westman and Dickson, 1998), *Brassica desnottesii*, *Coincya pseuderucastrum*, *Diplotaxis berthautii*, *D. catholica*, *D. cretacea*, *D. erucoides* and *Erucastrum gallicum* (Sharma *et al.*, 2002) show resistance. Resistance appeared to be associated with polyphenol oxidase, peroxidase, catalase in leaves and higher sugar content (Gupta *et al.*, 1990; Singh *et al.*, 1999). Epicuticular wax on leaves may form a physical barrier of hydrophobic coating to reduce deposition of waterborne inoculum, reduce conidia germination and germ-tube formation (Saharan, 1992). *Brassica napus* (Tower, HNS-3), *B. carinata* (HC-2) and *B. alba* possess more wax on the plant/leaf surface as compared to *B. rapa* (BSH-1, YSPB-24) or *B. juncea* (RH-30) (Conn *et al.*, 1984; Tewari, 1986).

Studies have highlighted the role of additive genes or polygenes in governing resistance (Saharan and Kadian, 1983; Krishnia *et al.*, 2000) with resistance being partially dominant (Zhang *et al.*, 1997). Components of resistance seemed significantly correlated to slow blighting (Kumar and Kolte, 2001). Disease onset appeared to be under control of dominance (h), while progression seemed to be under additive × dominance control (Kant and Gulati, 2002; Meena *et al.*, 2011a). Wild crucifers have been shown to elicit phytoalexins following challenge inoculation (Conn *et al.*, 1988). Activities of camalexin ($C_{11}H_8N_2S$) and 6-methoxy-camalexin ($C_{12}H_{10}N_2SO$) were toxic to *A. brassicae* (Dzurilla *et al.*, 1998). Phytotoxin dextruxin B elicits phytoalexin response in *B. alba* (Pedras and Smith, 1997), *B. rapa* and *B. napus* (Pedras and Khallaf, 2012). A layered and multicomponent mechanism of resistance to *A. brassicae* has been reported. Quantitative and qualitative elicitation of phytoalexins, hypersensitive reaction and Ca-sequestration possibly determined the fate of host–pathogen interaction (Tewari, 1991). Systemic acquired resistance (SAR) could be induced by inoculation with avirulent *A. brassicae* isolate (Vishwanath *et al.*, 1999). Due to a complex inheritance of resistance, breeding for resistance may involve pyramiding of minor genes. Introgression of alien genes, reciprocal recurrent selection or di-allele selective mating (Sigareva and Earle, 1999; Sigareva *et al.*, 1999; Krishnia *et al.*, 2000), gametic selection (Shivanna and Sawhney, 1993) and transgenic expression of *Trichoderma harzianum* endo-chitinase gene (Mora and Earle, 2001) have been found useful. *Brassica juncea* transgenics with a cDNA encoding hevein (chitin-binding lectin from *Hevea brasiliensis*) showed longer incubation and latent period for fungal germination, reduced necrotic lesions, lower disease intensity and delayed senescence (Kanrar *et al.*, 2002). Cramer and Lawrence (2004) have identified a gene, *P3F2*, that only expressed during infection in *Arabidopsis*.

White Rust

Plants of 241 species belonging to 63 genera of the Cruciferae family have been reported to be infected by *A. candida* (Biga, 1955). The disease has been reported from Brazil, many European and South Asian countries, Canada and Australia as reviewed by Saharan *et al.* (2014). White rust caused by *Albugo candida* (Pers. ex Fr.) Kuntz. causes up to 47% yield loss in India (Kolte, 1985). Every percentage increase of disease severity and staghead formation results in yield losses of about 82 kg/ha and 22 kg/ha, respectively (Meena *et al.*, 2002).

Albugo candida (Pers.) Roussel, earlier considered as an exclusive white rust pathogen of the *Brassicaceae*, infecting 63 genera and 241 species (Choi *et al.*, 2009), is now considered genetically diverse (Voglmayr and Riethmüller, 2006; Choi *et al.*, 2008) with the possibility that many of the observed lineages might constitute distinct species (Choi *et al.*, 2011). Following the recent lectotypification of *A. candida* (Choi *et al.*, 2007), two specialized *Albugo* species parasitic to *Brassicaceae* have been reported within *Albugo* (Choi *et al.*, 2007, 2008). It was also shown that *A. candida* has a broad host range, encompassing over a dozen genera of the *Brassicaceae* and into the *Cleomaceae*. A type of *Albugo chardonii* W. Weston (Vanev *et al.*, 1993) was found to be nested within *A. candida* (Choi *et al.*, 2007). Capers (*Capparis spinosa*) are affected by white blister rust attributed to *Albugo capparidis*, or applying a broad species concept, to *A. candida* (Choi *et al.*, 2009).

Symptoms

The disease on the leaves is characterized by white or creamy yellow pustules up to 2 mm in diameter, which later coalesce to form patches (Fig. 7.1c). The pustules are scattered across the lower surface of the leaves. The part of the upper surface corresponding to the lower surface is tan-yellow, enabling recognition of the affected leaves. After complete development, the pustule ruptures and releases a chalky dust of spores (sporangia). With ageing of white rust pustules, affected leaves become senescent with necrosis around or in the pustules. Such rust pustules are also formed on the surface of well-developed siliquae. Unlike other crucifers (Mundkur, 1959), thickening or hypertrophy of the affected leaves is generally not observed in rapeseed-mustard. Systemic infection through stem or flower, hypertrophy and hyperplasia results in the formation of stagheads (Fig. 7.1d) (Petrie, 1973b). Affected flowers become malformed and petals turn green like sepals; stamens may be transformed to leaf-like club-shaped sterile or carpelloid structures, which persist in the flower. Ovules and pollen grains are usually atrophied causing complete sterility. Association of downy mildew symptoms with that of white rust occurs routinely. Whole plant systemic infection results in a stunted, thickened stem with no branching and bearing white rust pustules on the surface. Thickening of stem may result from a modification of cortex into large thin-walled cells with few intercellular spaces. In floral parts there is also an increase in size and number of cells of parenchymatous tissue with few intercellular spaces, poor differentiation of tissue and organs and increased accumulation of nutrients. Multiplication and spore production by the pathogen results in consumption of the accumulated nutrients leading to collapse and death of cells and drying of affected plants (Vasudeva, 1958).

Pathogen

Albugo candida or *Cystopus candidus* mycelium is aseptate, intercellular with nuclei-free globular haustoria (Coffey, 1975). Masses of mycelia beneath the host epidermis form a palisade of cylindrical-shaped sporangiophores, which are thick-walled at the base and free laterally. The sporangiophores form chains of the spherical hyaline, smooth, 12–18 μm diameter sporangia in a basipetal succession, which germinate to give rise to concave biflagellate zoospores or at times to germ tubes (Walker, 1957). Oogonia and antheridia are formed from the mycelium in intercellular spaces, particularly in systemically affected plants (Webster, 1980). The heterothallic fungus produces resting spores or oospores, which are highly differentiated with a five-layered cell wall (Tewari and Skoropad, 1977) and at maturity are tuberculate, 40–55 μm in diameter. The processes of oospore germination have been well described (DeBary, 1887;

Vanterpool, 1959; Petrie and Verma, 1974; Verma and Petrie, 1975). Physiological specialization of the pathogen has been widely reported (Napper, 1933; Togashi and Shibasaki, 1934; Biga, 1955; Pound and Williams, 1963). Zoospores produced from germinating oospores constitute the primary source of inoculum for infection of rapeseed-mustard (Petrie and Verma, 1974; Verma and Petrie, 1975, 1980; Verma *et al.*, 1975), particularly when mixed with seeds. There is no oospore dormancy as they can germinate just 2 weeks after their isolation from affected tissues. The emerging cotyledon is the most likely primary infection site. The production of zoosporangia masses on cotyledons requires the establishment of a large mycelial base inside the host tissue, and in the *Albugo–Brassica* system such a base apparently develops with a minimum disturbance of the host's synthetic abilities (Harding *et al.*, 1968). The sporangia ruptures out of the epidermis as a white powdery mass, which is readily dispersed by wind to precipitate the secondary infection. Sporangia, if they alight on a suitable host leaf or stem surface, are capable of germinating within a few hours in films of water to form biflagellate zoospores. After initial swimming, a zoospore encysts and forms a germ tube, which penetrates the host epidermis. A further crop of sporangia may be formed within 10 days. The establishment and maintenance of a compatible relationship between *A. candida* and its hosts hinges on the successful formation of the first haustorium. In the susceptible host, such as *B. juncea*, the first haustorium forms within 16–18 h after inoculation (Verma *et al.*, 1975). Haustoria are small and capitate with spherical heads averaging 4 μm in diameter. These are connected to hyphae by slender stalks about 2 μm in length. A haustorium usually originates near the tip of a young hypha, which then continues its growth leaving the haustorium as a side branch. After the formation of the first haustorium in the susceptible host–parasite combination, hyphal growth rate increases rapidly. An encapsulation similar to that observed by Fraymouth (1956) and Berlin and Bowen (1964) in *Raphanus sativus* is seen only infrequently around haustoria in a virulent *Albugo*-susceptible *B. juncea* system. In a susceptible host, the hyphae grow around palisade mesophyll cells as a downward spiral, penetrating individual cells with a variable number of haustoria. Verma *et al.* (1975) have found as many as 14 haustoria in a single cell, in the green island tissue of artificially infected *B. juncea* cotyledons. In the susceptible host, most of the intercellular spaces appear to become occupied by mycelium within 3 days after inoculation. Studies on quantitative changes in amino acid content of white rust-induced hypertrophies of the mustard plant indicated possible breakdown of protein caused by the pathogen to release tryptophan and subsequently increasing IAA content of such tissues (Kamlesh *et al.*, 1970). However, a decrease in IAA (Srivastava *et al.*, 1962), free proline, total proteins and phenolic compounds (Dhingra *et al.*, 1982) in the infected host tissue are also reported. Much work remains to be done on mutual interaction of *Albugo* and associated microorganisms. Some of the *Alternaria* species are known to produce toxins (Degenhardt, 1973), and these could conceivably have an adverse effect on survival of *A. candida*. Host-range and phylogenetic relationships of *A. candida* from cruciferous hosts in Western Australia has been reported recently (Kaur *et al.*, 2011a).

Epidemiology

The optimum temperature for disease development varies between 12 and 18°C. Only 3 h of wetness was required for disease development at 12–22°C. White rust disease (*A. candida*) on leaves and pods (staghead formation) of Indian mustard first appears between 36 and 131 DAS (highest being at 50 and 70 DAS), and 60–123 DAS, respectively. Disease severity is favoured on the leaves by afternoon (minimum) RH >40%, morning (maximum) RH >97% and 16–24°C day maximum temperature. Staghead formation was positively influenced by 20–29°C daily maximum temperature and further aided by minimum daily temperature >12°C and morning (maximum) RH >97%. Maximum white rust severity can be predicted during initial weeks after sowing by stepwise regression models (Chattopadhyay

et al., 2011). Soil application of the herbicide trifluralin tended to increase disease incidence (Berkenkamp, 1980).

Management

Sources of host resistance have been recognized in *B. juncea* (PWR-9541, JMMWR 941-1-2, PAB-9534, PAB 9511, PHR-1, PHR-2, EC-129126, EC- 399299, EC-399301, EC-399300, EC-399296, BIO YSR), *B. rapa* (PT-303, Tobin), *B. carinata* (HC-1, 2, 3, 4, 5, NRCDR-515, PBC-9921, BC-2, DLSC-1), *B. napus* (Tower, GS-7055, HNS-4, GSL-441, PBN-2001, PBN-2002), *E. sativa* (RTM-1471) and *B. alba* (Kolte, 1985; AICRP-RM, 1986–2011; Pal *et al.*, 1999; Mukherjee *et al.*, 2001). Their success has been limited due to the volatile race pattern of the pathogen in India. Yellow-seeded *B. juncea* cvs T4 (Parui and Bandyopadhyay, 1973), YRT-3184 and rapeseed *B. rapa* var. Yellow Sarson, Type 6 (Kolte and Tewari, 1979) have been reported to be resistant. Transfer of white rust resistance from *B. carinata* to *B. juncea* could be partially successful by repeated backcrossing with *B. juncea* and selection for resistance to white rust (Singh *et al.*, 1988). Resistance to white rust in rapeseed-mustard is dominant, governed by one or two genes with either dominant-recessive epistasis or complete dominance at both gene pairs but either gene when dominant epistatic to the other. These genes could be located on the same locus or different loci (Kumar *et al.*, 2002). Resistance to the disease at true leaf infection and susceptibility at the cotyledonary leaf stage of the *B. juncea* genotype EC 399301 was governed by two independent genes. In view of these reports, screening for white rust resistance at the cotyledonary leaf stage need to be carefully considered (Mishra *et al.*, 2009). The resistance of *B. napus* var. Regent was conditioned by independent dominant genes at three loci-designated as AC 7-1, AC 7-2, and AC 7-3 (Fan *et al.*, 1983). A role of two independent loci conferring resistance to *Albugo* has also been reported (Panjabi-Massand *et al.*, 2010). Complete resistance to leaf rust and staghead symptoms caused by *A. candida* has been noted in some Australian germplasm (Li *et al.*, 2009).

However, breakdown of resistance can occur due to mixed infection with *Hyaloperonospora parasitica* (Singh *et al.*, 2002). A tightly linked marker for white rust resistance was developed using amplified fragment length polymorphism (Varshney *et al.*, 2004). A PCR-based cleaved amplified polymorphic sequence (CAPS) marker for closely linked RAPD marker $OPB06_{1000}$ was also developed. Results on 94 recombinant inbred lines (RILs) showed that the CAPS marker for $OPBO6_{1000}$ and AFLP marker E-AAC/M-CAA$_{350}$ flanked the *Ac2*(t) gene at 3.8 centimorgan (cM) and 6.7 cM, respectively. Validation of the CAPS marker in two F_2 populations from crosses, Varuna × BEC-144 and Varuna × BEC-286 established its utility in marker-assisted selection for white rust resistance. The use of both flanking markers in MAS provided greater selection efficiency than traditional approaches (Varshney *et al.*, 2004). Timing of expression of defence-related gene appeared crucial in determining the fate of pathogenesis (Kaur *et al.*, 2011b).

The fungicides captafol, benomyl (Gupta and Sharma, 1978) and chlorothalonil (Verma and Petrie, 1979) have been reported to reduce both foliar and staghead phase infection. Mancozeb (Dueck and Stone, 1979; Verma and Petrie, 1979; Mehta *et al.*, 1996) or combination of metalaxyl 35 ES at 6 ml/kg seed treatment + 0.2 g/l spray of metalaxyl + mancozeb at 50, 65 DAS (Parui and Bandyopadhyay, 1973; Verma and Petrie, 1979; Berkenkamp, 1980; Fan *et al.*, 1983; Meena *et al.*, 2003) have been reported to control the disease. The effectiveness of an aqueous bulb extract of *A. sativum* 1% (w/v), an isolate of *T. viride* as seed treatment and in combination as respective foliar sprays were statistically at par with that of mancozeb, combination of metalaxyl 35 ES 6 ml/kg seed treatment + 0.2 g/l spray of combination of metalaxyl + mancozeb in checking the rust severity on leaves and number of stagheads per plant (Meena *et al.*, 2003). Soil application of K as basal at 40 kg/ha resulted in significantly ($P < 0.05$) lesser white rust on leaves and number of stagheads than control. Four Chinese genotypes (CBJ-001, CBJ-002, CBJ-003 and CBJ-004) and one Australian genotype (JR049) were consistently resistant to an *A. candida* pathotype prevailing in Australia throughout the different plant growth stages (Li *et al.*, 2007a, 2008, 2009).

Sclerotinia Rot

Sclerotinia sclerotiorum (Lib.) de Bary is a pathogen with a very wide host range. It is known to infect about 408 plant species (Boland and Hall, 1994) with no genetically characterized source of resistance. The disease affects broad-leaved crop species and is most common in temperate regions of the world. It was first reported on rapeseed and mustard crops in India (Shaw and Ajrekar, 1915). Since then, it has been reported from Brazil (Neto, 1955), Canada (Dueck, 1977), China (Yang, 1959), Denmark (Buchwald, 1947), Finland (Jamalainen, 1954), France (Hims, 1979a), Germany (Kruger, 1976), India (Butler and Bisby, 1960; Roy and Saikia, 1976), Sweden (Loof and Appleqvist, 1972) and the UK (Rawlinson and Muthyalu, 1979). It is now a serious threat to oilseed rape production in Australia, Europe, India and North America (McCartney *et al.*, 1999; Hind *et al.*, 2003; Koch *et al.*, 2007; Malvarez *et al.*, 2007; Singh *et al.*, 2008). In India, disease has become very damaging with yield losses ranging up to 39.9% in the key mustard-growing areas of the country (Chattopadhyay *et al.*, 2003; Jha and Sharma, 2003). Yield losses vary with the proportion of plants infected and the crop stage at the time of infection. Plants infected at the early flowering stage produce little or no seeds and those infected at the late flowering stage set seed and may suffer little yield reduction.

Symptoms

The disease is known by the names of white blight (Roy and Saikia, 1976), white rot (Rai and Dhawan, 1976a), stem blight (Vasudeva, 1958), stalk break, stem canker, or rape canker. Usually, under natural conditions, the plant stem is seen to be affected more frequently, though all above-ground parts are subject to disease attack. Symptoms on the stem become visible as elongated water-soaked lesions, which later on are covered with a mycelial growth of the fungus (Fig. 7.1e). Infected plants at times appear normal until the fungus grows completely throughout the stem to rot it. After the stem is girdled by lesions, the plant wilts and dies. Foliage may show little sign of attack while at times it may start on the leaves, which wilt and droop downwards, and then moves onto the stem (Fig. 7.1f, g). The infection is often restricted to a smaller area of pith, which results in slow stunting of the plant and premature ripening rather than the sudden collapse of the affected plants. The affected stem tends to shred, and numerous greyish-white to black, spherical sclerotia appear either on the surface or in the pith of the affected stem. When the crop is at seed maturation stage, the plants began to lodge, with siliquae touching the soil level. Such plants, though remaining free from the stem or aerial infection throughout, show rotting of the siliquae with profuse fungal growth, along with sclerotial bodies just above the soil level. Appearance of the disease during the early stage of crop growth causes the death of the whole plant (Vasudeva, 1958). Hims (1979b) reported damping-off of *B. napus* seedlings. Root rot from the disease has also been reported (Berkenkamp and Vaartnou, 1972).

Pathogen

Sclerotinia sclerotiorum (Lib.) de Bary (Syn. *S. libertiana* Fuckel; *Whetzelinia sclerotiorum* (Lib.) Korf and Dumont) forms thin mycelium, 9–18 µm in diameter with lateral branches being narrower than the main hyphae. The vegetative hyphae are multi-nucleate (n=8). Mycelial growth rate on solid agar media is fast and forms moderate to abundant aerial mycelia. The sclerotia are black, round or semi-spherical in shape measuring 3–10 µm. These are formed terminally. The sclerotia can be easily detached from the medium. The fungus does not have a true conidial stage, though the formation of microconidia in culture media has been reported (Willets and Wong, 1980). The mature sclerotium consists of an outer-pigmented rind and a medulla of prosenchymatous tissues partly embedded in a gelatinous matrix (Willets and Wong, 1971). Structural, physiological and biochemical aspects of sclerotia formation and maturation

have been reviewed (Willets and Wong, 1980). The sclerotial germination is mycelogenic (by mycelium) or carpogenic (by formation of apothecia). On germination, the sclerotia form stalked apothecia. One to several apothecia may grow from a single sclerotium. Kosasih and Willets (1975) have described the structure of apothecia of the fungus. The hymenium is made of palisades of asci and paraphyses. The asci measure 119–162.4 µm × 6.4–10.9 µm in size. These are inoperculate, cylindrical, narrow and rounded at the apex with eight ascospores in each ascus. Ascospores are uniform in size (n=8). They measure 10.2–14.0 µm × 6.4–7.7 µm in size. Each ascospore is hyaline, ellipsoid and shows smooth walls. The spores are bi- or tri-guttulate. The paraphyses are about 100 µm long, 1–2 µm in diameter, slightly swollen at their tips, multinucleate, sparsely septate, and occasionally branched at the bases. The effects of some factors, such as age of sclerotium, temperature, light and moisture, on apothecial production and ontogeny of apothecia have been reviewed (Willets and Wong, 1980). There is no or little evidence of physiological specialization (Price and Colhoun, 1975). However, *S. sclerotiorum* isolates vary widely in their pathogenicity.

Discharge of ascospores from the basal apothecia constitutes a primary source of infection. Soil mycelia or mycelia arising from sclerotia may be less important for initial infection due to a poor competitive ability of the fungus (Newton and Sequeria, 1972). The ascospore can germinate in the presence of a thin film of water, in less than 24 h at 5–30°C, optimum being 5–10°C. The ascospore gives rise to the infection hypha, and initial penetration of the tissue takes place directly by mechanical pressure through the cuticle, or the infection hypha may penetrate already wounded or injured tissue. After the entrance of the fungus in the host, mycelia cause enzymatic dissolution of the cell wall in advance, and cells die some distance ahead of the invading hyphae. Pectolytic enzymes are responsible for tissue maceration indirectly damaging the cell membrane, which results in subsequent death of cells (Morrall *et al.*, 1972). Rai and Dhawan (1976b) reported production of polymethyl galacturonase

(PMG) and cell (C_x) enzymes by *S. sclerotiorum* infecting brassica plants. According to them, virulence of different isolates is associated with the activity of PMG and C_x enzymes. The role of protease activity towards infection of *B. juncea* is well known (Dhawan, 1980). Tissue invasion is related to the infection process and is mediated by production of oxalic acid (Rai and Dhawan, 1976a). The oxalic acid-like is formed in a culture filtrate as well as in infected *B. juncea* plants. It is thermostable and translocatable. The treatment of the host plant with culture filtrate results in infection. Potato dextrose agar medium was the best for supporting mycelia growth of the fungus and produced maximum number of sclerotia (Khan, 1976). Differences in the morphology of *S. sclerotiorum* isolates have been observed by Li *et al.* (2003), where isolates producing tan sclerotia were identified. Molecular diversity has been reported (Sexton *et al.*, 2006) among *S. sclerotiorum* isolates of oilseed rape crops from south-east Australia, India (Sharma *et al.*, 2009) and Pakistan (Akram *et al.*, 2008). Reports also described dark-pigmented isolates of *S. sclerotiorum*, such as those from Canada and south-western USA (Lazarovits *et al.*, 2000; Sanogo and Puppala, 2007). Molecular biology approaches can be very useful to classify pathogen variability.

Survival

The pathogen primarily survives in the soil through sclerotia. Such sclerotial bodies become mixed with the soil through affected plant debris after the crop is harvested, or when seeds contaminated with the sclerotial bodies are sown. Samples may contain up to 432 sclerotia/kg seed, and a certain level of sclerotia can be retained in the soil by formation of secondary sclerotia (Williams and Stelfox, 1980). Fungus can also survive either through mycelium or ascospores in dead or live plants (Newton and Sequeria, 1972; Hims, 1979a; Willets and Wong, 1980). Survival of the pathogen is possible through seeds in the form of mycelial infection of the testa (Neergaard, 1958). Hims (1979a) first reported the role of

wild plants as sources of primary inoculum as ascospores for infecting rapeseed crops in the UK. These include: hog weed (*Heracelum sphondylium* L.) and cow parsley (*Anthriscus sylvestris* (L) Holfm.), *Chenopodium* spp. and *Asphondilia* spp.

Epidemiology

The disease is known to be airborne (Williams and Stelfox, 1979) and soil-borne (Henderson, 1962). Since aerial infection (apart from that from soil) depends entirely on continued production and dissemination of ascospores, epidemics are common in areas of continuously cool, moist weather concurrent with the susceptible stage of crop, particularly during flowering. Fields planted with rapeseed for 2 years favour more germination of sclerotia than fields sown over 1 year (Williams and Stelfox, 1980). Pollen, like rapeseed petals (Kapoor, 1983), can also stimulate ascospore germination. Williams and Stelfox (1979) have reported that rapeseed crops did not restrict movement of airborne ascospores in Canada. The ascospores were carried into the air current as high as 147.0 cm above the soil level. They could also trap the spores at a horizontal distance of 150 m from the source and cause disease spread. No correlation between total rainfall and ascospore incidence has been reported (Williams and Stelfox, 1980). Honeybee-carried pollen and pollen in honeycombs are known to carry ascospores; however, the relative importance is very limited (Stelfox *et al.*, 1978). Combination of cool weather and high soil moisture during the critical stage (60–70 days old) of the crop favoured higher incidence on Indian mustard (Sharma *et al.*, 2009). Sharma *et al.* (2010) observed rainfall as an important factor in carpogenic infection of *S. sclerotiorum* in *B. juncea*. Detection of healthy Indian mustard crop and its early differentiation from Sclerotinia rot-affected *B. juncea* plants was possible using remote sensing techniques, which could help in disease forecasting (Dutta *et al.*, 2006). Multiple linear regression models have been described. The equation of the fitted model is percentage Sclerotinia rot incidence

= −11.2351 + 0.9529*BSSH + 4.93924*Eva + 3.83308*pH + 0.60885*RF (mm) − 0.406458*RH 720 + 0.524095*RH1420 + 0.17386*Soil moisture(%) − 0.30461*T_{max} − 0.677744*T_{min} − 2.19556*WS (DRMR, 2010). The ScleroPro system is fully computerized and is based on the weather and field-site-specific data (Koch *et al.*, 2007).

Herbicide Barban, when sprayed on rapeseed crop, increased susceptibility to infection by *S. sclerotiorum*, possibly through altering the physiology of the plant (Berkenkamp and Friesen, 1973). With thio-urea spray, however, intensity of the disease was lower (Dhawan, 1979).

Management

Disease management is tedious, inconsistent and uneconomical due to a very wide host range and long-term survival of resting structures. Burning of infected stubbles helps to kill the sclerotia (Vasudeva, 1958). Sclerotia-free cleaned seeds should be used for sowing. Due to the airborne nature of infection through ascospores and a wide host-range, crop rotation may prove inadequate (Hims, 1979a). However, deep summer ploughing and crop rotation with non-hosts (rice, maize), optimized use of N fertilizers, irrigation and normal plant population, and soil flooding can minimize the sclerotial load in the soil and control soil-borne inoculum (Williams and Stelfox, 1980).

Eliminating broad-leaf weeds like *Chenopodium* spp. is important for checking the disease. Late sowing can control disease in Canada by reducing the overlap between phenological susceptibility and maximum ascospore load (Morrall and Dueck, 1983). Soil application of compost inhibited carpogenic germination of *S. sclerotiorum* and reduced *Sclerotinia* infection (Couper *et al.*, 2001).

Due to a wide host range and absence of tissue specificity, genetic options have been mostly unsuccessful. However, differences in general growth habit and morphological characters of plants can be useful for tolerance of the disease. A high degree of resistance in cv. Isuzu of *B. napus* has also been reported from Japan (Iwata and Igita, 1972).

Responses of some genotypes (e.g. cv. Charlton) were observed relatively constant irrespective of the pathogen isolates, whereas inconsistent responses were observed in some other cultivars (e.g. Zhongyou-ang No. 4, Purler) against the same isolates. Although complete resistance is yet to be identified in canola, partial field resistance may be available in Chinese var. Zhongyou 821 (Li *et al.*, 1999). A cultivar Zhongyou 821 has been reported resistant to Sclerotinia rot (Wang *et al.*, 2003). Other resistant genotypes, 06-6-3792 (China), ZY004 (China) and RT 108 (Australia), showed less mean stem lesion lengths <3.0 cm (Li *et al.*, 2007b, 2008). In addition, *B. juncea* cvs JM06018 and JM 06006 also appeared tolerant with mean stem lesion lengths of 4.8 cm (Li *et al.*, 2008). Garg *et al.* (2010) have recently reported introgression of resistance in *B. juncea* from *Erucastrum cardaminoides*, *Diplotaxis tenuisiliqua* and *E. abyssinicum*. Genotypes showing consistent resistant reaction (e.g. cv. Charlton) across different isolates can be ideal for commercial exploitation (Garg *et al.*, 2010). Population improvement can help to improve resistance in breeding lines from Australia, India and China (Barbetti *et al.*, 2014). Recently, genotype Ringot I of *B. juncea* was reported resistant to the rot (Goyal *et al.*, 2011b). Sequential activations of salicylic acid and the jasmonic acid signalling pathway has been associated with resistance to *S. sclerotiorum* in oilseed rape (Wang *et al.*, 2012).

As no single method may control *S. sclerotiorum*, integration of various measures can be ideal to control the pathogen. Certain chemicals such as quintozene, fentin acetate and calcium cyanamide are known to inhibit the apothecial development of the fungus. The efficacy of calcium cyanamide in controlling the disease by 40–90% has been confirmed under field conditions in Germany (Hara and Yanagita, 1967; Kruger, 1973). Aside from seed contamination, viable sclerotia present a potential quarantine hazard in export of seed. Fumigation of an infested seed-lot with methyl bromide helped to partially eradicate viable sclerotia from infested seed of oilseed brassica (Richardson and Bond, 1978). In order to check the secondary spread of the disease, the possibility of control of the disease through foliar sprays of chemicals has been investigated. Three foliar sprays of benomyl (0.025%) gave the best control of the disease with an increase in seed yield (Roy and Saikia, 1976). A single aerial spray application of benomyl at the early bloom stage in the disease-prone regions of Canada was also suggested (Morrall and Dueck, 1983).

Earlier workers reported management of Sclerotinia rot of mustard by fungicides (Singh *et al.*, 1994). These include a foliar spray of carbendazim at full bloom stage (Sharma *et al.*, 2010) and Boscalid (trade name 'Cantus' in China), a broad-spectrum fungicide belonging to carboximides class, which has been found to control the pathogen by inhibiting the enzyme succinate ubiquinone reductase (complex II), also known as succinate dehydrogenase (SDH), in the mitochondrial electron transport chain. Use of such methyl benzimidazole fungicides in oilseed brassicas to manage Sclerotinia rot has been reported to result in widespread fungicide-resistant strains of *S. sclerotiorum* (Penaud *et al.*, 2003). Chattopadhyay *et al.* (2002) reported seed treatment with *Trichoderma viride* and *Allium sativum* aqueous bulb extract to be effective against the disease (Meena *et al.*, 2006). Integration of the seed treatment with foliar sprays was found promising (Chattopadhyay *et al.*, 2004, 2007).

Powdery Mildew

Incidence of powdery mildew on rapeseed and mustard has been reported from France (Darpoux, 1946), Germany (Cook, 1975), India (Bhander *et al.*, 1963; Sankhla *et al.*, 1967), Japan (Hirata, 1966), Sweden, Turkey, the UK (Cook, 1975) and the USA (Yarwood, 1949). It is generally believed that the disease does not cause much damage to oilseed brassica crops except during severe outbreaks, when all the leaves and siliquae become covered with the powdery growth of the fungus at early phonological stage. Although the exact data on yield losses are not available, loss seems to be proportional to the disease intensity, which varies considerably depending on the stage at which it occurs.

Symptoms

According to Saharan and Kaushik (1981), the symptoms appear in the form of dirty-white, circular, floury patches on both sides of lower leaves of the infected plants (Fig. 7.1h). Under favourable environmental conditions (relatively higher temperature) the floury patches increase in size and coalesce to cover the entire stem and leaves. The severely affected plants show retarded growth and produce less siliquae. The green siliquae also show white patches in the initial stage of infection. Later on such siliquae become completely covered with a white mass of mycelia and conidia. Severely diseased siliquae show reduced size and produce small, shrivelled seeds. Such siliquae form few seeds at the base with twisted sterile tips. Under favourable conditions, cleistothecia may be formed on both sides of affected leaves, stems and siliquae, which become visible in the form of black scattered and/or concentrated bodies (Vasudeva, 1958).

Pathogen

The pathogen is *Erysiphe cruciferarum* Opiz ex. Junell. It forms an ectophytic mycelium. Penetration is through haustoria, which remain confined to the epidermal cells, the remainder of the fungus being extra-matricial. The conidiophores arise from the superficial hyphae on the host surface. The conidiophore is septate, and conidia are borne singly. The conidia are ellipsoid to cylindrical. The ripe conidia fall off quickly and are disseminated by wind. The conidial size shows a range of 8.3–20.8 µm × 20.8–45.8 µm with an average range of 12.58–14.9 µm × 31.0–36.9 µm. Cleistothecia are globose to subglobose with numerous hypha-like brownish septate appendages. They are pinkish-brown first when young and turn brown to dark brown on reaching maturity. Cleistothecia measure from 83.2 to 137.3 µm in diameter (av. 104.4–119.1 µm) on different species and varieties of *Brassica.* The number of asci varies from three to eight per cleistothecium, each ascus producing from two to six ascospores. Asci are subglobose to broadly ovate, not stalked, light brown to yellowish in colour, and measure 25.0–37.4 µm × 41.6–66.6 µm with an average range of 31.7–34.5 µm × 52.3–62.0 µm on different species and varieties. Ascospores are ovoid and measure 19–22 µm × 11–13 µm.

Survival

Brassica rapa, B. nigra, B. juncea, Capsella bursa-pastoris, Coronopus didymil and *Raphanus sativus* have been found susceptible to *E. cruciferarum* (Sharma, 1979; Saharan and Kaushik, 1981). Since the fungus has been reported to form cleistothecia, it is likely to carry over from season to season through cleistothecia or as mycelium in volunteer host plants.

Epidemiology

Relatively dry weather conditions favour the disease development. Initiation of powdery mildew disease in Indian mustard occurs during 50–120 DAS. Disease severity increases with >5 days of ≥9.1 h of daylength, >2 days of morning RH of <90%, afternoon RH 24–50%, minimum temperature >5°C and a maximum temperature of 24–30°C. Maximum temperature and afternoon RH of the week preceding the observation was positively and negatively linked, respectively, to the disease severity (R^2: 0.9) (Desai *et al.*, 2004). It is possible to forecast the disease onset and intensity using weather-based models (Laxmi and Kumar, 2011). Cleistothecial formation appears to be favoured by alternating low and moderate temperature, low nutrition of the host, low relative humidity, dry soil, and ageing of the host (Saharan and Kaushik, 1981).

Management

Choice of suitable planting dates appears to offer a promising method of control of the disease. A limited form of resistance against powdery mildew has been reported in *B. alba, B. alboglabra, B. rapa* var. Brown Sarson, *B. chinensis,*

B. japonica and *E. sativa* (Narain and Siddiqui, 1965). There appears to be a varied level of resistance in rape, although immunity is not apparent. If serious, the disease can be controlled by dusting plants with sulfur or by spraying the plants with wettable sulfur fungicides. Karathane when sprayed three times at 10-day intervals also gives good control of the disease (Singh and Solanki, 1974). Some trends of efficacy of seed treatment and foliar sprays by *T. viride* and aqueous bulb extract of *A. sativum* against the disease have been reported by Meena *et al.* (2003), which may need confirmation.

Downy Mildew

The disease has been reported from Canada, several European countries, Japan and South Asia (Kolte, 1985). Reports of its occurrence either alone (Porter, 1926) or in association with white rust on leaves or inflorescence (Vasudeva, 1958; Bains and Jhooty, 1979; Kolte and Tewari, 1979) are available.

Symptoms

Symptoms appear on all above-ground parts but more so on leaves and inflorescence. Small angular translucent light green lesions first appear on the cotyledon or the first true leaves in a seedling stage (Fig. 7.1i). At times, the disease could be restricted only to initial leaves with later emerging leaves showing no sign of any symptom. Such lesions can enlarge and develop in to greyish white, irregular necrotic patches on the under-surface of the leaves (conidia and conidiophores). In case of a severe attack, the affected leaves dry up and shrivel. The extent to which the necrosis occurs depends upon the type of crop species. Leaf symptoms at the seedling stage are more conspicuous on *B. juncea* compared to *B. rapa*. Downy growth may also appear on siliquae, late in the season (Fig. 7.1j). Thickening of the peduncle/inflorescence due to the disease causes hypertrophy of affected cells, the pith of the stem being more affected than the cortex (Fig. 7.1k)(Vasudeva, 1958). Formation of oospores in the inflorescence takes place as it dries up. The disease may also be found associated with white rust symptoms on leaves and inflorescence. Systemic infection results in thickened stunted growth of the plant bearing profuse sporulation (Bains and Jhooty, 1979; Kolte and Tewari, 1979).

Pathogen

The causal pathogenic fungus *Hyaloperonospora parasitica* is an obligate parasite affecting all crucifers though there exists variation in conidial size and other fungal structures among strains infecting different *Brassicaceae* species. Mycelium is hyaline and coenocytic. It remains inter-cellular in the host and produces large, lobed intra-cellular haustoria, often branched, which nearly fill the entire cell. Erect conidiophores singly or in groups of determinate growth emerge vertically through the epidermis on the under-surface of the leaves through the stomata. Conidiophores are hyaline with a flattened base, stout main axis, twisted at a point crossing the stomata and measure 100–300 μm. At the tip, conidiophores are dichotomously branched six to eight times, the sterigmata are slender, acutely pointed. A single conidium is borne at the tip of each branch, and the same is deciduous. Detachment of conidia is possibly caused by hygroscopic twisting of the conidiophore related to changes in humidity. The conidiophore wall is uniformly thick. Spherical oogonia and tendril-like antheridia are developed on hyphae in hypertrophied tissue to produce oospores. This enables survival of the pathogen for long periods, notwithstanding harsh conditions. On the germination of the oospore or conidia, the germ tube penetrates the host tissue directly or through the stomata.

Epidemiology

The disease is favoured by cool (8–16°C) and moist weather with low light intensity and high (152 mm) rainfall (Porter, 1926; Bains and Jhooty, 1979).

Management

Germplasm accessions such as EC-129126 (*B. juncea*), PBN-9501, PBN-2002 and GSL-1 (*B. napus*) are reported to be resistant to the disease (AICRP-RM, 1986–2012; Gupta *et al.*, 2001; Nashaat *et al.*, 2004). However, there is a danger of breakdown of resistance due to mixed infection of *Hyaloperonospora parasitica* prior to *A. candida* (Kaur *et al.*, 2011c). Selective picking of affected hypertrophied racemes immediately after formation followed by their destruction, and rotation with non-cruciferous crops could also be helpful. Suitable planting times need to be worked out as per location (Bains and Jhooty, 1979). Seed treatment with metalaxyl 35 SD at 6 g/kg and spray of metalaxyl at 0.01% a.i. was found effective in managing the disease (Sharma, 1980).

Club Root

Incidence and severity are higher in regions with severe winters than in regions with spring-type climates. It frequently occurs in soils that are acidic and poorly drained. More damage due to the disease results on vegetable crops such as cabbage (*Brassica oleracea* L.) and turnip (*B. rapa* var. *rapifera*) than on oilseed rape (*B. rapa* var. *oleracea*) and mustard (*B. juncea*). Woronin (1878) and Walker (1952) were among earlier scientists who described the disease in more detail. Disease is reported to occur on oilseed brassicas in Germany, Malaysia, New Zealand, Poland, Sweden, the UK and the USA (Stout *et al.*, 1954). In India, disease has been reported from the hills of Darjeeling (Chattopadhyay and Sengupta, 1952) and Nilgiri (Rajappan *et al.*, 1999) on vegetable brassicas. The disease has been reported from West Bengal and Orissa, respectively, on *B. rapa* var. Yellow Sarson (Laha *et al.*, 1985) and *B. rapa* var. Toria (Das *et al.*, 1987), with yield losses up to 50% (Chattopadhyay, 1991). In southern districts of New Zealand, losses due to club root on rape have been reported to be very high (Lobb, 1951). However, exact information on losses caused by the disease on rapeseed-mustard is not known.

Symptoms

Infected plants continue to show healthy growth during initial stages. Growth, however, becomes stunted as the disease develops, with leaves becoming pale-green or yellowish. The plant then dies within a short time. On pulling the plants, overgrowth (hypertrophy/hyperplasia) of the main and lateral roots becomes visible in the form of small or spindle or spherical-shaped knobs or clubs (Fig. 7.1l). The shape of the club varies with the root type. When many infections occur in close proximity, the root system is transformed into various-shaped malformations. The swollen roots contain large numbers of resting spores. The older, especially the larger, clubbed roots disintegrate before the end of the season.

Pathogen

The pathogen is *Plasmodiophora brassicae* Woronin, which is an obligate biotrophic fungus. Woronin (1878) first described the life cycle of the fungus and its relation to host tissue in detail. Other aspects of its biology have been reviewed by Colhoun (1958) and Karling (1968). There is no evidence that pathotypes of *Plasmodiophora* exist with a single genus or group of related genera within *Brassicaceae*. Hence, the taxonomic concept *formae speciales* has not been applied to *P. brasssicae* as it has for certain other obligate biotrophic pathogens. There is also large phenotypic variation to justify the taxonomic division of the species on the basis of morphology. The fungus has a plasmodial vegetative stage characterized by a naked, amoeboid, multinucleate protoplast without a definite cell wall. The plasmodium is produced only in the cells of the host plant and remains intra-cellular, with two distinct phases. The first, the primary one, usually results from infection by primary zoospores derived from the resting spores, and the secondary phase results from infection by secondary zoospores derived from a zoosporangium.

The resting spore is hyaline, spherical, and measures up to 4 μm in diameter. It germinates by giving rise to single biflagellate primary zoospores (the first motile stage)

having one long and one short flagellum. The zoospore swims by means of its flagella, the long flagellum trailing and short flagellum pointing forward. This zoospore penetrates the host root hairs and develops into a primary plasmodium in the affected cell. The plasmodium produced in this manner later cleaves into multinucleate portions surrounded by separate membranes, with each portion developing into zoosporangia. The zoosporangia come out of the host tissue through pores formed in the host cell wall. About four to eight biflagellate secondary zoospores are formed upon germination of a single zoosporangium. Each secondary zoospore is indistinguishable from the primary zoospore. The exact role of the secondary zoospores is not known, but it is possible that the secondary zoospores pair and unite to produce a zygote to cause fresh infection of the roots, producing a new plasmodium called a secondary or zoosporangial plasmodium, which in turn forms resting spores. Tommercup and Ingram (1971) have studied the behaviour of the pathogen in callus tissue culture of *B. napus* and have presented the interpretation of the life cycle.

Though there are no *formae speciales* in *P. brassicae*, the fungus shows a lot of variation in pathogenicity. Physiologic specialization in *P. brassicae* was first demonstrated by Honig (1931). Attempts to classify *P. brassicae* in races have been made worldwide (Walker, 1942; Ayers, 1957; Lammerink, 1965). Information on the variation of the fungus has been reviewed and a uniform set of differential hosts is available for identification of physiologic races of *P. brassicae* (Buczaki *et al.*, 1975). Such a set of host genotypes is referred to as the European Clubroot Differential (ECD) set. The set consists of 15 different host varieties, five each of *B. rapa*, *B. napus* and *B. oleracea*. Using the ECD set, 34 physiologic races have been identified in Europe.

The resting spores in soil serve as the primary source of inoculum. Infection of the host takes place when uni-nucleate primary biflagellate zoospores are released on germination of the resting spores. The zoospores may collide several times with a root hair before becoming attached. Finally, these are attached at a point opposite to the origin of the flagella through an adhesorium. The zoospores then encyst and penetrate the root hair or epidermal cells. The process of penetration of such cells seems direct, but the role of enzymes or toxins in the pathogenesis is still not known. After the entrance of the pathogen through root hairs, the formation of plasmodium and the subsequent development of zoosporangia take place in the infected tissue as described earlier. The zoospores derived from the zoosporangia may then re-infect the root and initiate the formation of secondary plasmodia. Whether the secondary plasmodia penetrate the cell wall or whether such plasmodia are transformed passively from cell to cell during cell division is not certain. The plasmodium has no specialized feeding structure such as haustoria. It remains immersed in the host cytoplasm surrounded by a thin plasmodial envelope. There is also no evidence for phagocytic inclusion of the host cell organelles with the plasmodium (Williams and Yukawa, 1967). Hypertrophy of the host cells is apparently brought about by increased DNA synthesis and restriction of the cell division process. Tommercup and Ingram (1971) demonstrated that the presence of plasmodia in the host (*B. napus*) cell is associated with increased nuclei, at least in callus culture. Butcher *et al.* (1974) suggested that galling of susceptible *B. rapa* roots is the result of *P. brassicae* infection, enabling the enzyme glucosinolase to act on glucobrassicin, the indole glucosinolate. It is due to the formation of the auxins, 3-indole acetonitrile and/or 3-indole acetic acid, that the characteristic extensive proliferation of tissue takes place. Since crucifers commonly contain indole glucosinolates, it has been suggested that this explains their susceptibility to galling (Buczacki and Ockendon, 1979). It appears that there is a close correlation between the increase in the oxidative process and gall growth. As the galls develop on roots of the rape plant, the activity of glucose-6-phospho-gluconate dehydrogenase, aldolase, triose phosphate isomerase, isocitrate dehydrogenase and malate dehydrogenase increase, to reach a peak at 28–33 days after sowing. There is then accumulation of glucose-6-phosphate, pyruvate, ketoglutarate and malate in the affected cells. At sporulation, the activity of the above enzymes and concentration of the metabolites is decreased. During the patho-genesis, these phenomena parallel the vegetative growth of the fungus. Expression of nitrilases in *B. juncea* in root galls caused by *P. brassicae* has been reported (Liu *et al.*, 2012).

Survival

The fungus survives in the form of resting spores in soil. After the death of the galls, the resting spores are released in the soil; the pathogen thus becomes soil-borne and dispersed in the soil as resting spores through farm implements, footwear, floodwater, etc. There is no evidence that the fungus lives as a saprophyte, yet soils are known to remain infested for 10 years or longer without the presence of a host. The pathogen can also survive on cruciferous weeds such as shepherd's purse (*Capsella bursa-pastoris*). Some of the non-cruciferous hosts are also affected by *P. brassicae*. They are *Agrostis* sp., *Dactylis* sp., *Holcus* sp., *Lolium* sp., *Papaver* sp. and *Rumex* sp. Whether these non-cruciferous plants play any part in maintaining the continuity of the disease in the absence of a cruciferous host is not known. The half-life of spore inoculum for a field with 100% infestation was determined to be 3.6 years (Wallenhammar, 2010). The level of infestation declined below the detectable level after a period of 17.3 years. Multiplication of club root was moderate in fields of moderately resistant cultivars of spring turnip rape (*B. rapa* L.).

Management

In Germany, pot experiments conducted (under field conditions) by Budzier (1956) indicated that club root in yellow mustard can be reduced from 100% to 66% by mixing 50% compost into naturally infested soil. In view of the long viability of resting spores in soil, short-term crop rotation is not feasible. Since *P. brassicae* also infects cruciferous weeds such as shepherd's purse (*Capsella bursa-pastoris*), it may be important to control the weeds in order to check the incidence of the disease. Use of 10 to 30 mg/kg boron and calcium nitrate in soil with pH 6.5 or 7.3 was effective in reducing club root severity (Ruaro *et al.*, 2009).

Growing in the infected fields should be avoided. Improved drainage and application of lime brings about control of the disease. Spores of *P. brassicae* do not germinate or germinate very poorly in alkaline soils. The soil amendment to raise the pH of the soil to 7.2 by treating with lime 1 kg/m² area has been reported to control club root in mustard (Haenseler and Moyer, 1937; AICRP-RM, 2000).

Disease control is difficult because of the longevity of resting spores in the soil. Among different methods, use of resistant varieties seems a most feasible method of disease control. Certain kinds of *Brassica* spp. seem to have a natural resistance to the disease. It appears that the resistance in *B. rapa* lines of a known genotype is associated with hypersensitive cortical cell death following invasion of *P. brassicae* from infected root hairs (Dekhuijen, 1979). Black mustard (*B. nigra* L.) is commonly reported as a resistant host due to volatile mustard oil (Walker *et al.*, 1937). However, any such role of imparting resistance to infection is not proven (Hooker *et al.*, 1945).

There are several physiologic races of *P. brassicae*, which vary in their ability to infect *Brassica* spp., and this complicates the problem of breeding resistant varieties. Resistant varieties bred for club root resistance in one country may be completely susceptible to strains of the pathogens derived from another country. Differential hosts used in ECD set are resistant to some races and susceptible to some others. Development of resistant varieties through interspecific hybridization appears to be logical (Johnston, 1974; Chiang *et al.*, 1978). Resistance to *P. brassicae* race 3 was successfully transferred from the turnip rape (*B. rapa*) var. Wasslander to rape (*B. napus*) var. Nevin by production of the fertile species *B. napocampestris* followed by two generations of backcrossing of Nevin (Johnston, 1974). Some cultures of *B. napus* (GSL-1, WBBN-1, -2, PCRS-80, WW-1507, ISN-700, MNS-3), *B. carinata* (HC-1, -4, -5, 9221, PC-3, PCC-2, PPSC-1, PC-5) and *B. nigra* (ACCBN-479) are reported resistant to the disease (Chattopadhyay *et al.*, 2001). However, the durability of resistance seems unlikely to be profitable due to development of newer pathotypes of *P. brassicae* keeping in view the faster rate of sexual reproduction in the pathogen.

Although certain chemicals such as benomyl, quintozene and other soil fumigants are known to be effective against *P. brassicae*, such disease control methods are not feasible and economical because of the high cost of chemicals and their application.

Fusarium Wilt

Fusarium wilt of mustard is caused by *Fusarium oxysporum* f.sp. *conglutinans* (Wr.) Snyder and Hansen. Rai and Singh (1973) first reported *F. oxysporum* f.sp. *conglutinans* as the cause of the disease in *B. juncea* from India. Later it was reported to occur quite severely on *B. nigra* in India (Kanaujia and Kishore, 1981).

Symptoms

The affected plants show drooping of the leaves, vein clearing and chlorosis, followed by wilting and drying, resulting in the death of the plant. The symptoms progress from the base upward. The expression of the disease symptoms varies with the age of the plants (Rai and Singh, 1973). In the early stage of development, affected plants do not show all the typical symptoms. Plants affected in pre-flowering and early flowering stages show defoliation, and stems of such plants develop longitudinal ridges and furrows externally, which are generally not observed in the later stages. Diseased plants often show stunting, which is more pronounced when the plants are attacked in pre-flowering stages. Unilateral development of the disease is also observed in some of the cases when only one side of the plant shows symptoms of the disease. Roots of the diseased plants show no external abnormality or decay of the tissue until the plants are completely dried. Vascular tissues of stem and root show the presence of the mycelium and/or microconidia of the pathogen. Such tissues show browning of their walls and their plugging with a dark gummy substance, which is one of the characteristic symptoms of vascular wilts. At later stages of the disease, epidermis of roots sloughs off.

Pathogen

On the basis of host-range studies, Rai and Singh (1973) identified the causal fungus *Fusarium oxysporum* f.sp. *conglutinans*. They isolated two types of cultures as group A and B isolates and found that both were pathogenic to *B. rapa* var. Toria, *B. rapa* var. Yellow Sarson, *B. oleracea* var. *botrytis*, *B. oleracea* var. *capitata*, *Eruca sativa*, *Mattiola incana* and *Raphanus sativus*. The fungus is pathogenic on *B. rapa*, *B. nigra*, *E. sativa*, *Sinapis alba* and *Sisymbrium officinale*. Susceptibility of *B. carinata*, *Crambe abyssinica* and *C. hispanica* has been reported from the USA (Armstrong and Armstrong, 1974). It seems there is a complex of *F. oxysporum* causing yellows in oilseed brassicas, among which one that causes yellowing in *B. rapa* is proposed to be caused by a novel *forma specialis*, *F. oxysporum* f. sp. *rapae* (Enya *et al.*, 2008).

Management

Seed treatment with carbendazim at 0.1% a.i. or a suitable biofungicide could be effective in managing the disease.

Bacterial, Viral and Phytoplasmal Diseases

Bacterial stalk rot

Root rot caused by *Erwinia carotovora* pv. *carotovora* (Jones) Bergy is an emerging as a major threat for rapeseed-mustard production system in India. The first report of stalk rot occurrence on *B. juncea* caused by *E. carotovora* (Jones) Holland was made by Bhowmik and Trivedi (1980). Disease also afflicts fodder varieties of *Brassica* species. Due to an extra dose of nitrogen, vigorously growing succulent as well as weak plants in poorly drained soil are affected severely (Meena *et al.*, 2010b).

Symptoms

Symptoms of the disease are characterized by water-soaked lesions at the collar region of plants. It is usually accompanied by a white frothing. The tender branches are also affected as the lesions advance. The leaves show dehydration and wither. The affected stem and branches, especially the pith tissues, become soft, pulpy, and produce dirty white ooze with a foul smell. The infected collar region becomes sunken and turns buff-white to pale brown. Badly affected plants topple down at the basal region.

Bacterial rot

Patel *et al.* (1949) first observed black rot symptoms on *B. juncea* in India under field conditions. The disease is now reported to occur in a severe form in the state of Haryana (Vir *et al.*, 1973; Gandhi and Parashar, 1977). Occurrence of the disease has also been reported in Brazil, Canada (Conners and Savile, 1946), Germany, Sweden and the USA (Bain, 1952). In certain years the disease has been reported to take a heavy toll of the crop in Haryana with records up to 60% incidence in certain varieties of mustard (Vir *et al.*, 1973). The exact information on the yield losses caused by the disease is, however, not known. In fact, monoculture is currently a form of crop management worldwide, which plays a major role in disease progression (Zhu *et al.*, 2000).

Symptoms

Symptoms appear when the plants are 2 months old. In the initial stages, dark streaks of varying length are observed either near the base of the stems or 8 to 10 cm above the ground level. These streaks gradually enlarge and girdle the stem. Finally, the diseased stem becomes very soft and hollow due to severe internal rotting, and this often results in total collapse of the plant. Sometimes cracking of the stem is observed before the toppling of the plant. Occasionally symptoms appear on leaves; lower leaves show the symptoms first, which include midrib cracking and browning of the veins, and, when extensive, brings about withering of the leaves. Profuse exudation of yellowish fluid from affected stems and leaves may also occur. The affected plants, on stripping, show a dark-brown crust full of bacterial ooze. Black rot does not cause any disagreeable odour.

Pathogen

The pathogen is *Xanthomonas campestris* pv. *campestris* (Pammel) Dowson. The bacterium is a short rod with rounded ends, occurring singly, rarely in pairs. In culture on potato dextrose agar, it measures 1.5 µm (1.2–2.1 µm) × 0.7 µm (0.5–1.0 µm). It is motile with one polar flagellum and is gram negative, not acid fast, is aerobic, and capsulated without spore formation. On nutrient dextrose agar, the colony is dark yellow, circular, non-fluidic, convex and opaque. The thermal death point is 50–58°C. The host range includes *B. alba*, *B. rapa* var. Brown Sarson, *B. rapa* var. Yellow Sarson, *B. carinata*, *B. chinensis*, *B. hirta*, *B. napus*, *B. nigra*, *B. oleracea*, *B. rapa*, *B. tournefortii* and *Raphanus sativus*. The pathogen does not infect *E. sativa* and *Camelina sativa* (Gandhi and Parashar, 1977). Details of the mode of penetration and the infection process have not been studied using rapeseed-mustard plants. However, it is believed that the pathogen overwinters in diseased plant refuse or in seed and penetrates the host either through stomata or hydathodes and establishes the infection in a similar manner as in other crucifers (Walker, 1952). Soaking of seeds for about 24 h in a concentrated liquid culture of *X. campestris* resulted in 3.6% infection in the emerging seedlings of *B. nigra*. Isolates of *X. campestris* pv. *campestris* have been characterized in some African countries (Lizybenchidamba and Carlosbezuidenhout, 2012).

Management

Captafol spray (0.2% a.i.) at 20-day intervals is reported to give good control of the disease (Vir *et al.*, 1973). Gandhi and Parashar (1978) tested the efficiency of certain fungicides and antibiotics under field conditions and concluded that aureomycin (chlorotetracycline) 200 µg/ml is most effective in reducing the infection from 85% to about 15% resulting in an increase in yield by 60%. Among fungicides, carboxin is most effective, reducing the infection by 79% with a corresponding increase in yield by 49%. Spray application of copper oxychloride is also reported to give a considerable degree of control of the disease.

Mosaics

Some of the more common crucifer mosaic diseases are caused by viruses included in Turnip Virus I group (Larson *et al.*, 1950). The earliest report of a virus disease on rapeseed was made by Chamberlain (1936) from New Zealand. Mosaic diseases caused by this virus group are described under different names: (i) rape mosaic in China (Ling and Yang, 1940) and Canada (Rao *et al.*, 1977); (ii) mustard mosaic in the USA (Demski, 1973) and Trinidad (Dale, 1948); (iii) Chinese sarson mosaic in India (Azad *et al.*, 1963); (iv) *Brassica nigra* virus in the USA (Sylvester, 1954); and (v) turnip mosaic in China (Shen and Pu, 1965), Germany, Hungary, Russia and the UK (Rawlinson and Muthyalu, 1975). Up to 90% loss in yield has been reported in China (Ling and Yang, 1940; Wei *et al.*, 1960). Shen and Pu (1965) described infection of rape by a necrotic strain of turnip mosaic virus (TuMV) as lethal to the crop.

Symptoms

In *B. juncea* these appear as vein clearing, green vein banding, mottling and severe puckering of the leaves. The affected plants become stunted and fail to produce flowers. Siliquae, if formed, remain poorly filled and show seed shrivelling (Azad and Sehgal, 1959). According to Ling and Yang (1940), the symptoms appear as systemic conspicuous vein clearing, commencing at or near the base of the leaf and gradually spreading over the entire leaf. During the later stages of infection, raised or non-raised dark-green islands of irregular outline appear in the chlorotic area between the veins, giving rise to a mottled appearance. Curvature of the midrib and distortion of the leaf blade on affected leaves may be a prominent symptom. Plants infected early are usually stunted and killed, but those infected late show reduced growth. Similar symptoms have been described by Dale (1948) and Rao *et al.* (1977). A new alternative oxidase gene, designated as *BjAOX1a*, of mustard (*B. juncea*) has been cloned, which might alleviate reactive oxygen species (ROS) and enhance resistance of mustard plants to TuMV, and could serve as a critical component in TuMV resistance (Zhu *et al.*, 2012).

Phyllody

The disease has been reported to occur in India on *B. rapa* var. Toria and *B. rapa* var. Yellow Sarson (Vasudeva and Sahambi, 1955; Sandhu *et al.*, 1969).

Symptoms

The characteristic symptom is the transformation of the floral parts into leafy structures. The corolla becomes green and sepaloid. The stamens turn green and indehiscent. The gynoecium is borne on a distinct gynophore and produces no ovules in the ovary. The affected plants may show varying degrees of severity of the disease, and the affected part of the raceme fails to form siliquae. Some plants may show symptoms only on the terminal portion of the affected branches whereas in others the whole branches show the symptoms. Yield loss may go up to 90%. Losses in yield over a large area would be tremendous if the average percentage of diseased plants was high.

Pathogen

The disease is reported to be caused by the jassid transmissible phytoplasma-like organism, which causes the phyllody disease of sesamum (Klein, 1977). Transmission, detection and identification of a potential vector plant hopper (*Laodelpax striatellus*) for the phyllody disease of toria (*B. rapa* subsp. *dichotoma*) has been reported by Azadvar *et al.* (2011). Molecular characterization and phylogeny of a phytoplasma associated with the phyllody disease of toria has been reported (Azadvar and Baranwal, 2010).

Epidemiology

Under Indian conditions, early planting during late August or at its normal planting time in September has been shown to favour the development of the disease in *B. rapa* var. Toria. This has been investigated through field experiments, though the incidence of the disease might vary from year to year. As high as 24% incidence of the disease was observed, depending on the variety, in plants sown in August (Sandhu *et al.*, 1969).

Conclusions

Pathogens are major yield-limiting factors in rapeseed-mustard. With the identification of critical stages for growth of some foliar diseases, it is now possible to recommend need-based sprays of botanical and chemical fungicides for economic disease management. Breeding for disease resistance has not received enough attention. Genetics of disease resistance and its biochemical basis have been studied only in few cases. Characterization of new sources of resistance to Sclerotinia stem rot is required. There is also a need to transfer resistance, available in wild species. Generation of adequate information on bio-ecology, disease diagnostics, population dynamics, mass multiplication and artificial maintenance techniques, race/biotype pattern of pathogens, functional genomics of host–pathogen interactions and rapid, reliable screening and evaluation techniques against specific diseases are important. Plant immunization through induced resistance is also emerging as an attractive approach. A systems approach in IDM (Integrated Disease Management) needs to be considered for a better crop management. The modern techniques may be very useful to develop novel biocontrol agents and even disease-resistant transgenics, viz. by use of R-gene mediated resistance (Boys *et al.*, 2011). Risks related to the incidence of the disease should be accounted for while programming models for whole-farm planning.

References

AICRP-RM (All India Coordinated Research Project Rapeseed-Mustard) (1986–2012) *Annual Reports*. Directorate of Rapeseed-Mustard Research (ICAR), Bharatpur, Rajasthan, India.

Akram, A., Iqbal, S.M., Ahmed, N., Iqbal, U. and Ghafoor, A. (2008) Morphological variability and mycelia compatibility among the isolates of *Sclerotinia sclerotiorum* associated with stem rot of chickpea. *Pakistan Journal of Botany* 40, 2663–2668.

Alam, A., Tandon and Himanshu, S.K. (2002) *Technologies from ICAR (for industrial liason)*. Indian Council of Agricultural Research, 353 pp.

Armstrong, G.M. and Armstrong, J.K. (1974) Wilt of *Brassica carinata*, *Crambe abyssinica* and *C. hispanica* caused by *Fusarium oxysporum* f. sp. *conglutinans* race 1 or 2. *Plant Disease Reporter* 58, 479–480.

Aulakh, K.S. (1971) Damping-off of toria seedlings due to *Pythium butleri*. *Indian Phytopathology* 24, 611–612.

Awasthi, R.P. and Kolte, S.J. (1994) Epidemiological factors in relation to development and prediction of Alternaria blight of rapeseed and mustard. *Indian Phytopathology* 47, 395–399.

Ayers, G.W. (1957) Races of *Plasmodiophora brassicae*. *Canadian Journal of Botany* 35, 923–932.

Azad, R.N. and Sehgal, O.P. (1959) A mosaic disease of Chinese sarson (*Brassica juncea* L.) var. *rugosa* Roxb. *Indian Phytopathology* 12, 45.

Azad, R.N., Nagaich, B.B. and Sehgal, O.P. (1963) Chinese sarson mosaic: virus-vector relationship. *Indian Phytopathology* 16, 21.

Azadvar, M. and Baranwal, V.K. (2010) Molecular characterization and phylogeny of a phytoplasma associated with phyllody disease of toria (*Brassica rapa* L. subsp. *dichotoma* (Roxb.) in India. *Indian Journal of Virology* 21, 133–139.

Azadvar, M., Baranwal, V.K. and Yadava, D.K. (2011) Transmission and detection of toria [*Brassica rapa* L. subsp. *dichotoma* (Roxb.)] phyllody phytoplasma and identification of a potential vector. *Journal of General Plant Pathology* 77, 194–200.

Bain, D.C. (1952) Reaction of *Brassica* seedlings to black rot. *Phytopathology* 42, 497–500.

Bains, S.S. and Jhooty, J.S. (1979) Mixed infections by *Albugo candida* and *Peronospora parasitica* on *Brassica juncea* inflorescence and their control. *Indian Phytopathology* 32, 268–271.

Barbetti, M.J., Banga, S.K., Fu, T.D., Li, Y.C., Singh, D., Liu, S.Y., Ge, X.T. and Banga, S.S. (2014) Comparative genotype reactions to *Sclerotinia sclerotiorum* within breeding populations of *Brassica napus* and *B. juncea* from India and China. *Euphytica* 197(1), 47–59.

Berkenkamp, B. (1980) Effect of fungicides on staghead of rape. *Canadian Journal of Plant Science* 60, 1039–1040.

Berkenkamp, B. and Friesen, H.A. (1973) Effect of barban on stem rot of rape. *Canadian Journal of Plant Science* 53, 917.

Berkenkamp, B. and Vaartnou, H. (1972) Fungi associated with rape root rot in Alberta. *Canadian Journal of Plant Science* 52, 973–976.

Berlin, J.D. and Bowen, C.C. (1964) The host-parasite interface of *Albugo candida* on *Raphanus sativus*. *American Journal of Botany* 51, 445–452.

Bhander, D.S., Thakur, R.N. and Husain, A. (1963) A new disease of rapeseed and mustard. *Plant Disease Reporter* 47, 1039.

Bhowmik, T.P. and Trivedi, B.M. (1980) A new bacterial stalk rot of *Brassica*. *Current Science* 49, 674–675.

Biga, M.L.B. (1955) Riesaminazione della speciedel *genere Albugo* in base alia morfologia dei conidi. *Sydowia* 9, 339–358.

Boland, G.J. and Hall, R. (1994) Index of plant hosts of *Sclerotinia sclerotiorum*. *Canadian Journal of Plant Pathology* 16, 94–108.

Boys, E.F., Roques, S.E., West, J.S., Werner, C.P., King, G.J., Dyer, P.S. and Fitt, B.D.L. (2011) Effects of R gene-mediated resistance in *Brassica napus* (oilseed rape) on asexual and sexual sporulation of *Pyrenopeziza brassicae* (light leaf spot). *Plant Pathology doi:* 10.1111/j.1365-3059.2011.02529.x

Brun, H., Plessis, J. and Renard, M. (1987) Resistance of some crucifers to *Alternaria brassicae* (Berk.) Sacc. In: *Proceedings of the 7th International Rapeseed Congress,* Poznan, Poland, pp. 1222–1227.

Buchwald, N.F. (1947) Sclerotiniaceae of Denmark, a floristic systematic survey of the sclerotial Cl found in Denmark. Part I. Ciboria, Rutstroemia, *Myriosclerotinia* and *Sclerotinia*. *Friesia* 3, 235.

Buczacki, S.T. and Ockendon, J.G. (1979) Role of glucosinolate incidence and clubroot susceptibility of three cruciferous weeds. *Transactions of the British Mycological Society* 72, 156–157.

Buczacki, S.T., Toxopeaus, H., Mattusch, P., Johnston, T.D., Dixon, G.R. and Hobolth, L.A. (1975) Study of physiologic specialization in *Plasmodiophora brassicae:* proposals for attempted rationalization through international approach. *Transactions of the British Mycological Society* 65, 295–303.

Budzier, H.H. (1956) *Fluoreszenzmikroskopische Untersuchungen an Sporen des Kolhernieerregers Plasmodiophora brassicæ* Wor. *Nachr'bl. d'tsch. Pfll'sch'dienst.* Berlin, n.f. 10, 33–55.

Butcher, D.M., El-Tigani, S. and Ingram, D. (1974) The role of indole glucosinolates in the clubroot disease of the *Cruciferae. Physiological Plant Pathology* 4, 127–141.

Butler, E.J. and Bisby, G.R. (1960) *The Fungi of India* (revised by R.S. Vasudeva). Indian Co Agricultural Research, New Delhi, 552 pp.

Chamberlain, E.E. (1936) Turnip mosaic, a virus disease of crucifers. *New Zealand Journal of Agriculture* 53, 321–330.

Changsri, W. (1961) Studies of *Alternaria* spp. pathogenic on Cruciferae. *Disease Abstracts* 21, 1698.

Changsri, W. and Weber, G.F. (1963) Three *Alternaria* species pathogenic on Cruciferae. *Phytopathology* 53, 643–648.

Chattopadhyay, A.K. (1991) Studies on the control of clubroot disease of rapeseed-mustard in West Bengal. *Indian Phytopathology* 44, 397–398.

Chattopadhyay, A.K., Moitra, A.K. and Bhunia, C.K. (2001) Evaluation of *Brassica* species for resistance to *Plasmodiophora brassicae* causing club root of rapeseed mustard. *Indian Phytopathology* 54, 131–132.

Chattopadhyay, C. and Séguin-Swartz, G. (2005) Breeding for disease resistance in oilseed crops in India. *Annual Review Plant Pathology* 3, 101–142.

Chattopadhyay, C., Meena, P.D. and Kumar, S. (2002) Management of Sclerotinia rot of Indian mustard using ecofriendly strategies. *Journal of Mycology Plant Pathology* 32, 194–200.

Chattopadhyay, C., Meena, P.D., Kalpana Sastry, R. and Meena, R.L. (2003) Relationship among pathological and agronomic attributes for soilborne diseases of three oilseed crops. *Indian Journal of Plant Protection* 31, 127–128.

Chattopadhyay, C., Meena, P.D. and Meena, R.L. (2004) Integrated management of Sclerotinia rot of Indian mustard. *Indian Journal of Plant Protection* 32, 88–92.

Chattopadhyay, C., Agrawal, R., Kumar, A., Bhar, L.M., Meena, P.D., Meena, R.L., Khan, S.A., Chattopadhyay, A.K., Awasthi, R.P., Singh, S.N., Chakravarthy, N.V.K., Kumar, A., Singh, R.B. and Bhunia, C.K. (2005) Epidemiology and forecasting of Alternaria blight of oilseed *Brassica* in India – a case study. *Zeitschrift für Pflanzenkrankheiten und Pflanzenschutz* 112, 351–365.

Chattopadhyay, C., Vijay Kumar, R. and Meena, P.D. (2007) Biomanagement of Sclerotinia rot of *Brassica juncea* in India – a case study. *Phytomorphology* 57, 71–83.

Chattopadhyay, C., Agrawal, R., Kumar, A., Meena, R.L., Faujdar, K., Chakravarthy, N.V.K., Kumar, A., Goyal, P., Meena, P.D. and Shekhar, C. (2011) Epidemiology and development of forecasting models for White rust of *Brassica juncea* in India. *Archives of Phytopathology and Plant Protection* 44, 751–763.

Chattopadhyay, S.B. and Sengupta, S.K. (1952) Addition to fungi of Bengal. *Bulletin of Botanical Society of Bengal* 6, 57–61.

Chiang, B.Y., Grant, W.F. and Chiang, M.S. (1978) Transfer of resistance to race 2, of *Plasmodiophora brassicae* from *Brassica napus* to cabbage (*B. oleracea var. capitata*). II. Meiosis in the interspecific hybrids between *B. napus* and 2x and 4x cabbage. *Euphytica* 27, 81–93.

Choi, Y.J., Shin, H.D., Hong, S.B. and Thines, M. (2007) Morphological and molecular discrimination among *Albugo candida* materials infecting *Capsella bursa-pastoris* world-wide. *Fungal Diversity* 27, 11–34.

Choi, Y.J., Shin, H.D., Ploch, S. and Thines, M. (2008) Evidence for uncharted biodiversity in the *Albugo candida* complex, with the description of a new species. *Mycological Research* 112, 1327–1334.

Choi, Y.J., Shin, H.D. and Thines, M. (2009) The host range of *Albugo candida* extends from Brassicaceae through Cleomaceae to Capparaceae. *Mycological Progress* 8, 329–335.

Choi, Y.J., Shin, H., Ploch, S. and Thines, M. (2011) Three new phylogenetic lineages are the closest relatives of the widespread species *Albugo candida. Fungal Biology* 115, 598–607.

Coffey, M.D. (1975) Ultra structural features of the haustorial apparatus of the white blister fungus, *Albugo candida. Canadian Journal of Botany* 53, 1285–1299.

Colhoun, J. (1958) Clubroot disease of crucifers caused by *Plasmodiophora brassicae. Phytopathological Papers* (Coommonwealth Mycol. Inst.) 3, 1–108.

Conn, K.L., Tewari, J.P. and Hadziyev, D. (1984) The role of epicuticular wax in canola in resistance to *Alternaria brassicae. Phytopathology* 74, 851.

Conn, K.L., Tewari, J.P. and Dahiya, J.S. (1988) Resistance to *Alternaria brassicae* and phytoalexin-elicitation in rapeseed and other crucifers. *Plant Science* 55, 21–25.

Conners, I.L. and Savile, D.B.O. (1946) *Twenty-fifth Annual Report of the Canadian Plant Disease Survey* 1945.

Cook, R.J. (1975) *Diseases of oilseed rape in Europe: Report of a study tour undertaken in Germany and France.* Ministry of Agriculture, Fisheries and Food, Agriculture Development Service (mimeographed).

Couper, G., Litterick, A. and Leifert, C. (2001) Control of *Sclerotinia* within carrot crops in NE Scotland: the effects of irrigation and compost application on sclerotia germination. Available at: http://www.abdn. ac.uk/organic/pdfs/sclerotia.pdf (accessed 15 January 2014).

Cox, T., Souche, J.L. and Grapel, H. (1983) The control of *Sclerotinia, Alternaria* and *Botrytis* on oilseed rape with spray treatments of flowable formulation of iprodione. *Abstracts of the 6th International Rapeseed Congress,* Paris, France, 17–19 May 1983, pp. 928–933.

Cramer, R.A. and Lawrence, C.B. (2004) Identification of *Alternaria brassicicola* genes expressed in planta during pathogenesis of *Arabidopsis thaliana. Fungal Genetics and Biology* 41, 115–128.

Czyzewska, S. (1970) Effect of temperature on the growth and sporulation of *Alternaria* species isolated from *Crambe abyssinica. Acta Mycologia* 6, 261–276.

Dale, W.T. (1948) Observations on a virus disease of certain crucifers in Trinidad. *Annals of Applied Biology* 35, 598–604.

Dang, J.K., Kaushik, C.D. and Sangwan, M.S. (1995) Quantitative relationship between Alternaria leaf blight of rapeseed-mustard and weather variables. *Indian Journal of Mycology and Plant Pathology* 25, 184–188.

Darpoux, H. (1946) A contribution to study of the diseases of oleaginous plants in France. *Annales des Epiphyties* 11, 71–103.

Das, S.N., Mishra, S.K. and Swain, P.K. (1987) Reaction of some toria varieties to *Plasmodiophora brassicae. Indian Phytopathology* 40, 120.

DeBary, A. (1887) *Comparative Morphology and Biology of the Fungi, Mycetezoa and Bacteria* (English translation). Clarendon Press, Oxford, 136 pp.

Degenhardt, K.J. (1973) *Alternaria brassicae* and *A. raphani*: sporulation in culture and their effects on yield and quality of rapeseed. MSc Thesis, Univiversity of Alberta, Edmonton.

Dekhuijen, H.M. (1979) Electron microscopic studies in the root hairs and cortex of a susceptible and a resistant variety of *Brassica campestris* infected with *Plasmodiophora brassicae. Netherland Journal of Plant Pathology* 85, 1–18.

Demski, J.W. (1973) Identity and prevalence of virus diseases of turnip and mustard in Georgia. *Plant Disease Reporter* 57, 974.

Desai, A.G., Chattopadhyay, C., Agrawal, R., Kumar, A., Meena, R.L., Meena, P.D., Sharma, K.C., Rao, M.S., Prasad, Y.G. and Ramakrishna, Y.S. (2004) *Brassica juncea* powdery mildew epidemiology and weather based forecasting models for India – a case study. *Zeitschrift für Pflanzenkrankheiten und Pflanzenschutz* 111, 429–438.

Dhawan, S. (1979) Effect of sulphur on the pathogenesis of *Sclerotinia sclerotiorum. Geobios* 6, 196.

Dhawan, S. (1980) Protease activity in *B. juncea* plants infected with *Sclerotinia sclerotiorum. Current Science* 49, 291.

Dhingra, R.K., Chauhan, N. and Chauhan, S.V.S. (1982) Biochemical changes in the floral parts of *Brassica campestris* infected by *Albugo candida. Indian Phytopathology* 35, 177–179.

DRMR (2010) *Annual Report.* Directorate of Rapeseed-Mustard Research, Bharatpur (Raj), India.

Dueck, J. (1977) *Sclerotinia* in rapeseed. *Canadian Agriculture* 22, 7–9.

Dueck, J. and Degenhardt, K. (1975) Effect of leaf age and inoculum concentration on reaction of oilseed *Brassica* spp. to *Alternaria brassicae. Proceedings of American Phytopathological Society* 2, 59.

Dueck, J. and Stone, J.R. (1979) Evaluation of fungicides for control of *Albugo candida* in turnip rape. *Canadian Journal of Plant Science* 59, 423–427.

Dutta, S., Bhattacharya, B.K., Rajak, D.R., Chattopadhyay, C., Patel, N.K. and Parihar, J.S. (2006) Disease detection in mustard crop using EO-1 hyperion satellite data. *Journal of the Indian Society of Remote Sensing (Photonirvachak)* 34, 325–329.

Dzurilla, M., Kutschy, P., Tewari, J.P., Ruzinsky, M., Senvicky, S. and Kovacik, V. (1998) Synthesis and antifungal activity of new indolypthiazinone derivatives. *Collection Czechoslovakia Chemists Community* 63, 94–102.

Enya, J., Togawa, M., Takeuchi, T., Yoshida, S., Tsushima, S., Arie, T. and Sakai, T. (2008) Biological and phylogenetic characterization of *Fusarium oxysporum* Complex, which causes yellows on *Brassica* spp. and proposal of *F. oxysporum* f. sp. *rapae*, a novel forma specialis pathogenic on *B. rapa* in Japan. *Phytopathology* 98, 475–483.

Fan, Z., Rimmer, S.R. and Stefansson, B.R. (1983) Inheritance of resistance to *Albugo candida* in rape (*Brassica napus* L.). *Canadian Journal of Genetics and Cytology* 25, 420–424.

Fraymouth, J. (1956) Haustoria of Peronosporales. *Transactions of British Mycological Society* 39, 79–107.

Galletti, S., Burzi, P.L., Sala, E., Marinello, S. and Cerato, C. (2006) Combining Brassicaceae green manure with *Trichoderma* seed treatment against damping-off in sugarbeet. *Bulletin-OILB/SROP* 29(2), 71–75.

Gandhi, S.K. and Parashar, R.D. (1977) Bacterial rot of raya (*Brassica juncea*). *Indian Phytopathology* 30, 24–27.

Gandhi, S.K. and Parashar, R.D. (1978) Evaluation of some fungicides and antibiotics against *Xanthomonas campestris* causing bacterial rot of raya. *Indian Phytopathology* 31, 210.

Garg, H., Atri, C., Sandhu, P.S., Kaur, B., Renton, M., Banga, S.K., Singh, H., Singh, C., Barbetti, M.J. and Banga, S.S. (2010) High level of resistance to *Sclerotinia sclerotiorum* in introgression lines derived from hybridization between wild crucifers and the crop *Brassica* species *B. napus* and *B. juncea*. *Field Crops Research* 117, 51–58.

Goyal, P. (2009) Host-pathogen interaction in *Brassica juncea–Alternaria brassicae* and variability in the pathogen. PhD thesis, University of Rajasthan, Jaipur, India

Goyal, P., Chahar, M., Mathur, A.P., Kumar, A. and Chattopadhyay, C. (2011a) Morphological and cultural variation in different oilseed Brassica isolates of *Alternaria brassicae* from different geographical regions of India. *The Indian Journal of Agricultural Science* 81, 1052–1059.

Goyal, P., Chahar, M., Barbetti, M., Liu, S.Y. and Chattopadhyay, C. (2011b) Resistance to Sclerotinia rot caused by *Sclerotinia sclerotiorum* in *Brassica juncea* and *B. napus* germplasm. *Indian Journal of Plant Protection* 39, 60–64

Gupta, I.J. and Sharma, B.J. (1978) Chemical control of white rust of mustard. *Pesticides* 12, 45–46.

Gupta, K., Saharan, G.S. and Singh, D. (2001) Sources of resistance in Indian mustard against white rust and Alternaria blight. *Cruciferae Newsletter* 23, 59–60.

Gupta, S.K., Gupta, P.P., Yadava, T.P. and Kaushik, C.D. (1990) Metabolic changes in mustard due to Alternaria leaf blight. *Indian Phytopathology* 43, 64–69.

Haenseler, C.M. and Moyer, T.R. (1937) Effect of calcium cyanamide on the soil microflora with special reference to certain plant parasites. *Soil Science* 43, 133–152.

Hansen, L.H. and Earle, E.D. (1997) Somatic hybrids between *Brassica oleracea* (L.) and *Sinapis alba* (L.) with resistance to *Alternaria brassicae* (Berk.) Sacc. *Theoretical Applied Genetics* 94, 1078–1085.

Hara, K. and Yanagita, Y. (1967) Using calcium cyanamide to control rape Sclerotinia rot (in Japanese). *Proceedings of the Kyushu Association for Plant Protection* 13, 13.

Harding, H., Williams, P.H. and McNabola, S.S. (1968) Chlorophyll changes, photosynthesis and ultra structure of chloroplasts in *Albugo candida* induced 'green islands' on detached *Brassica juncea* cotyledons. *Canadian Journal of Botany* 46, 1229–1234.

Henderson, R.M. (1962) Some aspects of the life cycle of the plant pathogen *Sclerotinia sclerotiorum* in Western Australia. *Journal of Royal Society of Western Australia* 45, 133–135.

Hims, M.J. (1979a) Wild plants as a source of *Sclerotinia sclerotiorum* infecting oilseed rape. *Plant* 28, 197–198.

Hims, M.J. (1979b) Damping-off of *Brassica napus* (mustard and cress) by *Sclerotinia sclerotiorum*. *Plant* 28, 201–202.

Hind, T.L., Ash, G.J. and Murray, G.M. (2003) Prevalence of Sclerotinia stem rot of canola in New South Wales. *Australian Journal of Experimental Agriculture* 43, 163–168.

Hirata, K. (1966) Host range and geographical distribution of powdery mildews. Niigata University, Japan.

Honig, F. (1931) Der Kohlkropferreger (*Plasmmodiophora brassicae* Wor.). *Gartenbauwenissenchaft* 5, 116–225.

Hooker, W.J., Walker, J.C. and Link, K.P. (1945) Effects of two mustard oils on *Plasmodiophora brassicae* and their relation to resistance to clubroot. *Journal of Agricultural Research* 70, 63–78.

Humpherson-Jones, F.M. (1992) The development of weather-related disease forecasts for vegetable crops in the UK: problems and prospects. *European Plant Protection Organisation Bulletin* 21, 425–429.

Humpherson-Jones, F.M. and Maude, R.B. (1982) Studies on the epidemiology of *Alternaria brassicicola* in *Brassica oleracea* seed production crops. *Annals Applied Biology* 100, 61–71.

Humpherson-Jones, F.M. and Phelps, K. (1989) Climatic factors influencing spore production in *Alternaria brassicae* and *Alternaria brassicicola*. *Annals Applied Biology* 114, 449–458.

Iwata, I. and Igita, K. (1972) On the growth characteristics of direct sowing rape on upland field. *Bulletin of Kyushu Agricultural Experimentation Station* 16, 207.

Jamalainen, E.A. (1954) Overwintering of cultivated plants under snow. *FAO Plant Protection Bulletin* 2, 102–105.

Jha, S.K. and Sharma, A.K. (2003) Problem identification in rapeseed-mustard. *Sarson News* 7, 7.

Johnston, T.D. (1974) Transfer of disease resistance from *Brassica campestris* L. to rape (*B. napus*). *Euphytica* 23, 681–683.

Kadian, A.K. and Saharan, G.S. (1984) Studies on spore germination and infection of *Alternaria brassicae* of rapeseed and mustard. *Journal of Oilseeds Research* 1, 183–188.

Kamlesh, K., Varghese, T.M. and Suryanarayana, D. (1970) Quantitative changes in amino-acid contents of hypertrophied organs in mustard due to *Albugo candida*. *Current Science* 39, 240–241.

Kanaujia, R.S. and Kishore, R. (1981) A new wilt disease of *Brassica nigra* caused by *Fusarium* f. sp. *conglutinans*. *Indian Phytopathology* 34, 84.

Kanrar, S., Venkateswari, J.C., Kirti, P.B. and Chopra, V.L. (2002) Transgenic expression of hevein, the rubber tree lectin, in Indian mustard confers protection against *Alternaria brassicae*. *Plant Science* 162, 441–448.

Kant, L. and Gulati, S.C. (2002) Inheritance of components of horizontal resistance to *Alternaria brassicae* (Berk.) Sacc. in Indian mustard, *Brassica juncea* (L.) Czern and Coss. *Journal of Oilseeds Research* 19, 17–21.

Kapoor, K.S. (1983) Some aspects of the host–parasite relations between *Sclerotinia sclerotiorum* (Lib) de Bary and rapeseed. *Abstracts of the 6th International Rapeseed Congress,* Paris, France, 17–19 May 1983, pp. 188.

Karling, J.S. (1968) *The Plasmodiophorales*, 2nd edn. Hafner, New York.

Kaur, P., Sivasithamparam, K. and Barbetti, M.J. (2011a) Host range and phylogenetic relationships of *Albugo candida* from cruciferous hosts in Western Australia, with special reference to *Brassica juncea*. *Plant Disease* 95, 712–718.

Kaur, P., Jost, R., Sivasithamparam, K., Li, H. and Barbetti, M.J. (2011b) Proteome analysis of the *Albugo candida-Brassica juncea* pathosystem reveals that the timing of the expression of defence-related genes is a crucial determinant of pathogenesis. *Journal of Experimental Botany* 62, 1285–1298.

Kaur, P., Sivasithamparam, K., Li, H. and Barbetti, M.J. (2011c) Pre-inoculation with *Hyaloperonospora parasitica* reduces incubation period and increases the severity of disease caused by *Albugo candida* in a *Brassica juncea* variety resistant to downy mildew. *Journal of General Plant Pathology* 77, 101–106.

Khan, M.A. (1976) Studies on stalk rot of cauliflower caused by *Sclerotinia sclerotiorum*. MSc thesis, H.P. University, College of Agriculure, Solan.

Khan, R.U. and Kolte, S.J. (2002) Some seedling diseases of rapeseed-mustard and their control. *Indian Phytopathology* 55, 102–103.

Klein, M. (1977) Sesamum phyllody in Israel. *Journal of Phytopathology* 88, 165–171.

Koch, S., Dunker, S., Kleinhenz, B., Röhrig, M. and Tiedemann, A. (2007) A crop loss-related forecasting model for Sclerotinia stem rot in winter oilseed rape. *Phytopathology* 97, 1186–1194.

Kolte, S.J. (1985) *Diseases of Annual Edible Oilseed Crops*, Vol. II, *Rapeseed-Mustard and Sesame Diseases*. CRC Press, Boca Raton, Florida.

Kolte, S.J. and Tewari, A.N. (1979) Note on the susceptibility of certain oleiferous *Brassicae* to downy mildew and white blister diseases. *Indian Journal of Mycology and Plant Pathology* 10, 191–192.

Kolte, S.J., Awasthi, R.P. and Vishwanath (1987) Assessment of yield losses due to Alternaria blight in rapeseed and mustard. *Indian Phytopathology* 40, 209–211.

Kolte, S.J., Awasthi, R.P. and Vishwanath (2000) Divya mustard: a useful source to create Alternaria black spot tolerant dwarf varieties of oilseed brassicas. *Plant Variety Seeds* 13, 107–111.

Kolte, S.J., Nashaat, N.I., Kumar, A., Awasthi, R.P. and Chauhan, J.S. (2006) Indo-UK collaboration on oilseeds: Towards improving the genetic base of rapeseed-mustard in India. *Indian Journal of Plant Genetic Resources* 19, 346–351.

Kosasih, B.D. and Willets, H.J. (1975) Ontogenetic and histochemical studies of the apothecium of *Sclerotinia sclerotiorum*. *Annals of Botany* 39, 185–191.

Krishnia, S.K., Saharan, G.S. and Singh, D. (2000) Genetic variation for multiple disease resistance in the families of the interspecific cross of *Brassica juncea* x *Brassica carinata*. *Cruciferae Newsletter* 22, 51–53.

Kruger, W. (1973) Control measures for *Sclerotinia sclerotiorum* in rape. *Journal of Phytopathology* 77, 125–137.

Kruger, W. (1976) Important root and stalk diseases of rape in Germany. *Gesunde Pflanzen* 28, 39–41.

Kumar, B. and Kolte, S.J. (2001) Progression of Alternaria blight of mustard in relation to components of resistance. *Indian Phytopathology* 54, 329–331.

Kumar, S., Saharan, G.S. and Singh, D. (2002) Inheritance of resistance in inter and intraspecific crosses of *Brassica juncea* and *Brassica carinata* to *Albugo candida* and *Erysiphe cruciferarum*. *Journal of Mycology and Plant Pathology* 32, 59–63.

Laha, J.N., Naskar, I. and Sharma, B.D. (1985) A new record of clubroot disease of mustard. *Current Science* 54, 1247.

Lammerink, J. (1965) A survey of pathogenic races of club root in the South Island of New Zealand. *New Zealand Journal of Agricultural Research* 8, 867.

Larson, R.H., Matthews, R.E.F. and Walker, J.C. (1950) Relationships between certain viruses and the genus *Brassica. Phytopathology* 40, 955–962.

Laxmi, R.R. and Kumar, A. (2011) Forecasting of powdery mildew in mustard (*Brassica juncea*) crop using artificial neural network approach. *Indian Journal of Agricultural Sciences* 81, 855–860.

Lazarovits, G., Starratt, A.N. and Huang, H.C. (2000) The effect of tricyclazole and culture medium on the production of the melanin precursor 1, 8-dihydroxynaphthalene by *Sclerotinia sclerotiorum* isolate SS7. *Pesticide Biochemistry and Physiology* 67, 54–62.

Li, C.X., Sivasithamparam, K., Walton, G., Salisbury, P., Burton, W., Banga, S.S., Banga, S., Chattopadhyay, C., Kumar, A., Singh, R., Singh, D., Agnohotri, A., Liu, S.Y., Li, Y.C., Fu, T.D., Wang, Y.F. and Barbetti, M.J. (2007a) Expression and relationships of resistance to white rust (*Albugo candida*) at cotyledonary, seedling and flowering stages in *Brassica juncea* germplasm from Australia, China and India. *Australian Journal of Agricultural Research* 58, 259–264.

Li, C.X., Li, H., Siddique, A.B., Sivasithamparam, K., Salisbury, P., Banga, S.S., Banga, S., Chattopadhyay, C., Kumar, A., Singh, R., Singh, D., Agnihotri, A., Liu, S.Y., Li, Y.C., Tu, J., Fu, T.D., Wang, Y. and Barbetti, M.J. (2007b) The importance of the type and time of inoculation and assessment in the determination of resistance in *Brassica napus* and *B. juncea* to *Sclerotinia sclerotiorum. Australian Journal of Agricultural Research* 58, 1198–1203.

Li, C.X., Sivasithamparam, K., Walton, G., Fels, P. and Barbetti, M.J. (2008) Both incidence and severity of white rust disease reflect host resistance in *Brassica juncea* germplasm from Australia, China and India. *Field Crops Research* 106, 1–8.

Li, C.X., Sivasithamparam, K. and Barbetti, M.J. (2009) Complete resistance to leaf and staghead disease in Australian *Brassica juncea* germplasm exposed to infection by *Albugo candida* (white rust). *Australasian Plant Pathology* 38, 63–66.

Li, G.Q., Huang, H.C., Laroche, A. and Acharaya, S.N. (2003) Occurrence and characterization of hypovirulence in the tan sclerotial isolates of S10 of *Sclerotinia sclerotiorum. Mycological Research* 107, 1350–1360.

Li, Y.C., Chen, J., Bennett, R., Kiddle, G., Wallsgrove, R., Huang, Y.J. and He, Y.H. (1999) Breeding, inheritance and biochemical studies on *Brassica napus* cv. Zhougyou 821: tolerance to *Sclerotinia sclerotiorum* (stem rot). In: Wratten, N. and Salisbury, P.A. (eds) *Proceedings of the 10th International Rapeseed Congress*, Canberra, Australia, pp. 61.

Ling, L. and Yang, J.Y. (1940) A mosaic disease of rape and other crucifers in China. *Phytopathology* 30, 338–342.

Liu, Y., Yin, Y., Wang, Z. and Luo, Y. (2012) Expression of nitrilases in *Brassica juncea* var. *tumida tsen* in root galls caused by *Plasmodiophora brassicae. Journal of Integrative Agriculture* 11, 100–108.

Lizybenchidamba and Carlosbezuidenhout, C. (2012) Characterisation of *Xanthomonas campestris* pv. *campestris* isolates from South Africa using genomic DNA fingerprinting and pathogenicity tests. *European Journal of Plant Pathology* 133, 811–818.

Lobb, W.R. (1951) Resistant type of rape for areas with club root. *New Zealand Journal of Agriculture* 82, 65–66.

Loof, B. and Appleqvist, L. (1972) Plant breeding for improved yield and quality. In: Appleqvist, L. and Ohlson, R. (eds) *Rapeseed*. Elsevier Publishing, Amsterdam, 55 pp.

Mahmud, K.A. (1950) Damping-off of *Brassica juncea* caused by *Pythium aphanidermatum* Fitz.. *Science and Culture* 16, 208.

Malvarez, M., Carbone, I., Grunwald, N.J., Subbarao, K.V., Schafer, M. and Kohn, L.M. (2007) New populations of *Sclerotinia sclerotiorum* from lettuce in California and peas and lentils in Washington. *Phytopathology* 97, 470– 483.

McCartney, H.A. and Lacey, M.E. (1999) Timing and infection of sunflowers by *Sclerotinia sclerotiorum* and disease development. *Aspects of Applied Biology* 56, 151–156.

Meena, P.D., Chattopadhyay, C., Singh, F., Singh, B. and Gupta, A. (2002) Yield loss in Indian mustard due to white rust and effect of some cultural practices on Alternaria blight and white rust severity. *Brassica* 4, 18–24.

Meena, P.D., Meena, R.I., Chattopadhyay, C. and Kumar, A. (2004) Identification of the critical stage for disease development and biocontrol of Alternaria blight of Indian mustard (*Brassica juncea*). *Journal of Phytopathology* 152, 204–209.

Meena, P.D., Sharma, A.K., Jha, S.K. and Chattopadhyay, C. (2006) Impact of fungal diseases on yield of Indian mustard – farmers' perception and the on-farm success of garlic clove extract in the management of Sclerotinia rot. *Indian Journal of Plant Protection* 34, 229–232.

Meena, P.D., Meena, R.L. and Chattopadhyay, C. (2008) Eco-friendly options for management of Alternaria blight in Indian mustard (*Brassica juncea*). *Indian Phytopathology* 61, 65–69.

Meena, P.D., Awasthi, R.P., Chattopadhyay, C., Kolte, S.J. and Kumar, A. (2010a) Alternaria blight: a chronic disease in rapeseed-mustard. *Journal of Oilseed Brassica* 1, 1–11.

Meena, P.D., Mondal, K., Sharma, A.K., Chattopadhyay, C. and Kumar, A. (2010b) Bacterial rot: a new threat for rapeseed-mustard production system in India. *Journal of Oilseed Brassica* 1(1), 39–41.

Meena, P.D., Chattopadhyay, C., Meena, S.S. and Kumar, A. (2011a) Area under disease progress curve and apparent infection rate of Alternaria blight disease of Indian mustard (*Brassica juncea*) at different plant age. *Archives of Phytopatholy – Plant Protection* 44, 684–693.

Meena, P.D., Chattopadhyay, C., Kumar, A., Awasthi, R.P., Singh, R., Kaur, S., Thomas, L., Goyal, P. and Chand, P. (2011b) Comparative study on the effect of chemicals on Alternaria blight in Indian mustard – a multi-location study in India. *Journal of Environmental Biology* 32, 375–379.

Meena, R.L., Meena, P.D. and Chattopadhyay, C. (2003) Potential for biocontrol of white rust of Indian mustard (*Brassica juncea*). *Indian Journal of Plant Protection* 31, 120–123.

Mehta, N., Sangwan, M.S., Srivastava, M.P. and Kumar, R. (2002) Survival of *Alternaria brassicae* causing Alternaria blight in rapeseed-mustard. *Journal of Mycology and Plant Pathology* 32, 64–67.

Mehta, N., Saharan, G.S. and Kaushik, C.D. (1996) Efficacy and economics of fungicidal management of white rust and downy mildew complex in mustard. *Indian Journal of Mycology and Plant Pathology* 26, 243–247.

Mishra, K.K., Kolte, S.J., Nashaat, N.I. and Awasthi, R.P. (2009) Pathological and biochemical changes in *Brassica juncea* (mustard) infected with *Albugo candida* (white rust). *Plant Pathology* 58, 80–86.

Mora, A.A. and Earle, E.D. (2001) Resistance to *Alternaria brassicicola* in transgenic broccoli expressing a *Trichoderma harzianum* endochitinase gene. *Molecular Breeding* 8, 1–9.

Morrall, R.A.A. and Dueck, J. (1983) Sclerotinia stem rot of spring rapeseed in Western Canada. In: *Proceedings of the 6th International Rapeseed Congress,* Paris, France, 17–19 May 1983, pp. 181.

Morrall, R.A.A., Duczek, L.J. and Sheard, J.W. (1972) Variations and correlations within and between morphology, pathogenicity and pectolytic enzyme activity in *Sclerotinia* from Saskatchewan. *Canadian Journal of Botany* 50, 767–786.

Mukherjee, A.K., Mohapatra, T., Varshney, A., Sharma, R. and Sharma, R.P. (2001) Molecular mapping of a locus controlling resistance to *Albugo candida* in Indian mustard. *Plant Breeding* 120, 483–487.

Mundkur, B.B. (1959) *Fungi and Plant Diseases*. MacMillan, London.

Napper, M.E. (1933) Observations on spore germination and specialization of parasitism in *Cystopus candidus*. *Journal of Pomology and Horticulture Science* 11, 81–100.

Narain, A. and Siddiqui, J.A. (1965) Field reaction of species of *Brassica* to *Erysiphe polygoni*. *Oilseed Journal* 9, 153.

Nashaat, N.I., Heran, A., Awasthi, R.P. and Kolte, S.J. (2004) Differential response and genes for resistance to *Peronospora parasitica* (downy mildew) in *Brassica juncea* (mustard). *Plant Breeding* 123, 512–515.

Neergaard, P. (1958) Mycelial seed infection of certain crucifers by *Sclerotinia sclerotiorum* (Lib.) de Bary. *Plant Disease Reporter* 42, 1105–1106.

Nehemiah, K.M.A. and Deshpande, K.B. (1976) Detection of cell wall breaking ability by cellulase of *Alternaria brassicae* (Berk.) Sacc. *Current Science* 45, 28–29.

Neto, J.P. (1955) Occurrence and apothecial state of *Sclerotinia sclerotiorum* (Lib.) de. in the Rio Grande do Sui, Brazil. *Review of Agronomy Porto-Alegre* 17, 109.

Newton, H.C. and Sequeria, L. (1972) Ascospores as the primary infective propagule of *Sclerotinia sclerotiorum* in Wisconsin. *Plant Disease Reporter* 56, 798–802.

Pal, S.S., Gupta, T.R., Kumar, V., Dhaliwal, H.S. and Kumar, V. (1999) Transfer of white rust resistance from *Brassica napus* to *B. juncea* cv. RLM 198. *Crop Improvement* 26, 249–251.

Panjabi-Massand, P., Yadava, S.K., Sharma, P., Kaur, A., Kumar, A., Arumugam, N., Sodhi, Y.S., Mukhopadhyay, A., Gupta, V., Pradhan, A.K. and Pental, D. (2010) Molecular mapping reveals two independent loci conferring resistance to *Albugo*. *Theoretical and Applied Genetics* 121, 137–145.

Parui, N.R. and Bandyopadhyay, D.C. (1973) A note on screening of rai [*Brassica junca* (Coss.)]. *Current Science* 42, 798–799.

Patel, M.K., Abhyankar, S.G. and Kulkarni, Y.S. (1949) Black rot of cabbage. *Indian Phytopathology* 2, 58.

Pedras, M.S. and Smith, K.C. (1997) Sinalexin, a phytoalexin from white mustard elicited by destruxin B and *Alternaria brassicae*. *Phytochemistry* 46, 833–837.

Pedras, M.S. and Kallaf, I. (2012) Molecular interactions of the phytotoxins destruxin B and sirodesmin PL with crucifers and cereals: *metabolism and elicitation of plant defenses*. *Phytochemistry* 77, 129–139.

Penaud, A., Huguet, B., Wilson, V. and Leroux, P. (2003) Fungicide resistance of *Sclerotinia sclerotiorum* in French oilseed rape crops. In: *Abstracts, 11th International Rapeseed Congress*, 6–10 July 2003, Copenhagen, Denmark.

Petri, G.A. (1973a) Diseases of *Brassica* species in Saskatchewan, 1970-72. III. Stem and root rots. *Canadian Plant Disease Survey* 53, 88–92.

Petrie, G.A. (1973b) Diseases of *Brassica* species in Saskatchewan, 1970-72. I. Staghead and aster yellows. *Canadian Plant Disease Survey* 53, 19–25.

Petrie, G.A. and Verma, P.R. (1974) A simple method for germinating oospores of *Albugo candida*. *Canadian Journal of Plant Science* 54, 595–596.

Porter, R.H. (1926) A preliminary report of surveys for plant diseases in East China. *Plant Disease Reporter* 46(Suppl.), 153–166.

Pound, G.S. and Williams, P.H. (1963) Biological races of *Albugo candida*. *Phytopathology* 53, 1146–1149.

Price, K. and Colhoun, J. (1975) Pathogenicity of isolates of *Sclerotinia sclerotiorum* (Lib.) de Bary to hosts. *Journal of Phytopathology* 83, 232–238.

Rai, J.N. and Dhawan, S. (1976a) Studies on purification and identification of toxic metabolite produced by *Sclerotinia sclerotiorum* causing white rot disease of Crucifers. *Indian Phytopathology* 29, 407–411.

Rai, J.N. and Dhawan, S. (1976b) Production of polymethyl galacturonase and cellulase and its relationship with virulence in isolates of *Sclerotinia sclerotiorum* (Lib.) de Bary. *Indian Journal of Experimental Biology* 14, 197.

Rai, J.N. and Singh, R.P. (1973) Fusarial wilt of *Brassica juncea*. *Indian Phytopathology* 26, 225–232.

Rajappan, K., Ramaraj, B. and Natarajan, S. (1999) Knol-khol – a new host for club-root disease in the Nilgiris. *Indian Phytopatholology* 52, 328.

Rao, B.R. (1977) Species of *Alternaria* on some *Cruciferae*. *Geobios* 4, 163–166.

Rao, D.V., Hiruki, C. and Chen, M.H. (1977) Mosaic disease of rape in Alberta caused by turnip virus. *Plant Disease Reporter* 61, 1074.

Rawlinson, C.J. and Muthyalu, G. (1975) *Report of the Rothamsted Experimental Station for 1975*, Rothamstead, UK, pp. 264.

Rawlinson, C.J. and Muthyalu, G. (1979) Diseases of winter oilseed rape: occurrence, effects and control. *Journal of Agriculture Science Cambridge* 93, 593–606.

Richardson, L.T. and Bond, E.J. (1978) Methyl bromide fumigation of rapeseed infested with sclerotia of *Sclerotinia sclerotiorum*. *Canadian Journal of Plant Science* 58, 267–268.

Roy, A.K. and Saikia, U.N. (1976) White blight of mustard and its control. *Indian Journal of Agricultural Science* 46, 274–277.

Ruaro, L., Lima Neto, V.-da, C. and Ribeiro Junior, P.J. (2009) Influence of boron, nitrogen sources and soil pH on the control of club root of crucifers caused by *Plasmodiophora brassicae*. *Tropical Plant Pathology* 34, 231–238.

Saharan, G.S. (1992) Disease resistance. In: Labana, K.S., Banga, S.S. and Banga, S.K. (eds) *Breeding Oilseed Brassicas*. Narosa Publishing House, New Delhi, India, pp. 181–205.

Saharan, G.S. and Kadian, A.K. (1983) Analysis of components of horizontal resistance in rapeseed and mustard cultivars against *Alternaria brassicae*. *Indian Phytopatholology* 36, 503–507.

Saharan, G.S. and Kadian, A.K. (1984) Epidemiology of Alternaria blight of rapeseed and mustard. *Cruciferae Newsletter* 9, 84–86.

Saharan, G.S. and Kaushik, J.C. (1981) Occurrence and epidemiology of powdery mildew of *Brassica*. *Indian Phytopathology* 34, 54–57.

Saharan, G.S., Verma, P.R., Meena, P.D. and Kumar, A. (2014) *White Rust of Crucifers: Biology, Ecology and Management*. Springer Publications, New Delhi, 244 pp.

Sandhu, K.S., Singh, H. and Kumar, R. (1985) Effect of different nitrogen levels and dates of planting on Alternaria blight and downy mildew diseases of radish seed crop. *Journal of Research Punjab Agricultural University* 22, 285–290.

Sandhu, R.S., Singh, G. and Bhatia, N.L. (1969) Studies on the effect of sowing dates and spacing on incidence of phyllody in Indian rape (*Brassica campestris* L. var. *toria* Duth). *Indian Journal of Agricultural Sciences* 39, 959.

Sankhla, H.C., Singh, H.G., Dalela, G.G. and Mathur, R.L. (1967) Occurrence of perithecial stage of *Erysiphe polygoni* on *Brassica campestris* var. *sarson* and *B. juncea*. *Plant Disease Reporter* 51, 800.

Sanogo, S. and Puppala, N. (2007) Characterization of darkly pigmented mycelial isolates of *Sclerotinia sclerotiorum* on Valencia peanut in New Mexico. *Plant Disease* 91, 1077–1082.

Seppänen, L. and Helenius, J. (2004) Do inspection practices in organic agriculture serve organic values? A case study from Finland. *Agriculture and Human Values* 21, 1–13.

Sexton, A.C., Whitten, A.R. and Howlett, B.J. (2006) Population structure of *Sclerotinia sclerotiorum* in an Australian canola field at flowering and stem-infection stages of the disease cycle. *Genome* 49, 1408–1415.

Sharma, A.K. (1979) Powdery mildew diseases of some crucifers from Jammu and Kashmir state. *Indian Jornal of Mycology – Plant Pathology* 9, 29–32.

Sharma, G., Kumar V.D., Haque, A., Bhat, S.R., Prakash, S. and Chopra, V.L. (2002) *Brassica* coenospecies: a rich reservoir for genetic resistance to leaf spot caused by *Alternaria brassicae*. *Euphytica* 125, 411–419.

Sharma, K.D. (1980) Symptomatology, yield losses and control of downy mildew and white rust of rapeseed and mustard. MSc (Agric.) thesis, G.B. Pant University, Agric. Tech., Pantnagar, India.

Sharma, P., Meena, P.D., Kumar, A., Chattopadhyay, C. and Goyal, P. (2009) Soil and weather parameters influencing Sclerotinia rot of *Brassica juncea*. In: *Souvenir and Abstracts, 5th International Conference IPS on Plant Pathol.* Globalized Era, IARI, New Delhi, 10–13 Nov 2009, pp. 97.

Sharma, P., Rai, P.K., Meena, P.D., Kumar, S. and Siddiqui, S.A. (2010) Relation of petal infestation to incidence of *Sclerotinia sclerotiorum* in *Brassica juncea*. In: *Abstracts, National Conference on Recent Advances Integrated Disease Management Enhancing Food Production*, S.K.R.A.U., Bikaner, 27–28 Oct 2010, pp. 76.

Sharma, S.R. and Kolte, S.J. (1994) Effect of soil-applied NPK fertilizers on severity of black spot disease (*Alternaria brassicae*) and yield of oilseed rape. *Plant and Soil* 167, 313–320.

Shaw, F.J.F. and Ajrekar, S.L. (1915) The genus *Rhizoctonia* in India. *Memoirs of the Department of Agriculture in India Botanical Series* 7, 177.

Shen, S.L. and Pu, Z.C. (1965) A preliminary study of the two strains of turnip mosaic virus on I Kiangsu Province. *Acta Phytophylacica Sinica* 4, 35.

Shivanna, K.R. and Sawhney, V.K. (1993) Pollen selection for *Alternaria* resistance in oilseed brassicas: responses of pollen grains and leaves to a toxin of *A. brassicae*. *Theoretical and Applied Genetics* 86, 339–344.

Sigareva, M.A. and Earle, E.D. (1999) Camalexin induction in intertribal somatic hybrids between *Camelina sativa* and rapid-cycling *Brassica oleracea*. *Theoretical and Applied Genetics* 98, 164–170.

Sigareva, M., Ren, J.-P. and Earle, E.D. (1999) Introgression of resistance to *Alternaria brassicicola* from *Sinapis alba* to *Brassica oleracea* somatic hybridization and backcrosses. *Cruciferae Newsletter* 21, 135–136.

Simmons, E.G. (2007) *Alternaria: An Identification Manual.* CBS Biodiversity Series No. 6, Utrecht, The Netherlands.

Singh, D., Singh, H. and Yadava, T.P. (1988) Performance of white rust (*Albugo candida*) resistant genotypes developed from interspecific crosses of *B. juncea x B. carinata*. *Cruciferae Newsletter* 13, 110–111.

Singh, D.N., Singh, N.K. and Srivastava, S. (1999) Biochemical and morphological characters in relation to Alternaria blight resistance in rapeseed-mustard. *Annals of Agricultural Research* 20, 472–477.

Singh, R.R. and Solanki, J.S. (1974) Fungicidal control of powdery mildew of *Brassica juncea*. *Indian Journal of Mycology and Plant Pathology* 4, 210.

Singh, R., Tripathi, N.N., Kaushik, C.D. and Singh, R. (1994) Management of Sclerotinia rot of Indian mustard [*Brassica juncea* (L.) Czern and Coss.] by fungicides. *Crop Research Hisar* 7, 275–281.

Singh, R., Singh, D., Li, H., Sivasithamparam, S., Yadav, N.R., Salisbury, P. and Barbetti, M.J. (2008) Management of Sclerotinia rot of oilseed Brassica – a focus on India. *Brassica* 10, 1–27.

Singh, U.S., Nashaat, N.I., Doughty, K.J. and Awasthi, R.P. (2002) Altered phenotypic response to *Peronospora parasitica* in *Brassica juncea* seedlings following prior inoculation with an avirulent isolate of *Albugo candida*. *European Journal of Plant Pathology* 108, 555–564.

Sinha, R.K.P., Rai, B. and Sinha, B.B.P. (1992) Epidemiology of leaf spot of rapeseed mustard caused by *Alternaria brassicae*. *Journal of Applied Biology* 2, 70–75.

Skoropad, W.P. and Tewari, J.P. (1977) Field evaluation of the epicuticular wax in rapeseed and mustard in resistance to *Alternaria brassicae*. *Canadian Journal of Plant Science* 57, 1001–1003.

Srivastava, B.S., Shaw, M. and Vanterpool, T.C. (1962) Effect of *Albugo candida* (Pers. ex. Chev.) Kuntz on growth substances in *Brassica napus* L. *Canadian Journal of Botany* 40, 53–59.

Srivastava, O.P. (1968) Dry root and bottom rot of mustard caused by *Rhizoctonia solani*. *Indian Journal of Microbiology* 8, 277–278.

Srivastava, S.K. and Dhawan, S. (1979) Screening of *Brassica juncea* cultivars for resistance against *Macrophomina* stem and root rot disease. *Geobios* 6, 333–334.

Stelfox, D., Williams, J.R., Soehngen, U. and Topping, R.C. (1978) Transport of *Sclerotinia sclerotiorum* ascospores by rapeseed pollen in Alberta. *Plant Disease Reporter* 62, 576–579.

Stout, G.L., Altstatt, G.E. and Nakayama, R.M. (1954) Club root of *Crucifers*. *Bulletin of the Department of Agriculture* 42, 113.

Sylvester, E.S. (1954) Aphid transmission of non-persistent plant viruses with special reference to *Brassica nigra virus*. *Hilgardia* 23, 53–98.

Tewari, J.P. (1986) Subcuticular growth of *Alternaria brassicae* in rapeseed. *Canadian Journal of Botany* 64, 1227–1231.

Tewari, J.P. (1991) Current understanding of resistance to *Alternaria brassicae* in crucifers. In: *Proceedings of the GCIRC 8th International Rapeseed Congress*, 9–11 July 1991, Saskatoon, Canada, Vol. 2, pp. 471–476.

Tewari, J.P. and Conn, K.L. (1993) Reaction of some wild crucifers to *Alternaria brassicae*. *Bull OILB/SROP* 16, 53–58.

Tewari, J.P. and Skoropad, W.P. (1977) Ultra structure of oospore development in *Albugo candida* on rapeseed. *Canadian Journal of Botany* 55, 2348–2357.

Togashi, K. and Shibasaki, Y. (1934) Biometrical and biological studies of *Albugo candida* (Pers.) O. Kuntze in connection with its specialization. *Imperial College of Agriculture & Forest Morioka Bulletin* (Japan) 1–88.

Tommercup, I.C. and Ingram, D.S. (1971) The life cycle of *Plasmodiophora brassicae* Woron. in *Brassica* tissue cultures and in intact roots. *New Phytologist* 70, 327–332.

Tripathi, N.N. and Kaushik, C.D. (1984) Studies on the survival of *Alternaria brassicae* the causal organism of leaf spot of rapeseed and mustard. *Madras Agriculture Journal* 71, 237–241.

Tsuneda, A. and Skoropad, W.P. (1978) Phylloplane fungal flora of rapeseed. *Transactions of British Mycological Society* 70, 329–334.

Upadhyay, R. and Pavgi, M.S. (1967) Some new hosts for *Pellicularia rolfsii* (Sacc.) West. from India. *Science and Culture* 33, 71–73.

Vanev, S.G., Dimitrova, E.G. and Ilieva, E.I. (1993) *Gybitev Bylgariya. 2 tom. Razred Peronosporales*. [*Fungi Bulgarica*, Vol. 2. Ordo *Peronosporales*]. Bulgarian Academy of Sciences, Sofia, Bulgaria.

Vanterpool, T.C. (1959) Oospore germination in *Albugo candida*. *Canadian Journal of Botany* 39, 30–31.

Varshney, A., Mohapatra, T. and Sharma, R.P. (2004) Development and validation of CAPS and AFLP markers for white rust resistance gene in *Brassica juncea*. *Theoretical and Applied Genetics* 109, 153–159.

Vasudeva, R.S. (1958) Diseases of rape and mustard, In: Singh, D.P. (ed.) *Rape and Mustard*. Indian Central Oilseed Committee, Hyderabad, pp. 77–86.

Vasudeva, R.S. and Sahambi, H.S. (1955) Phyllody of Sesamum (*Sesamum orientale* L.). *Indian Phytopathology* 8, 124–129.

Verma, P.R. and Petrie, G.A. (1975) Germination of oospores of *Albugo candida*. *Canadian Journal of Botany* 53, 836–842.

Verma, P.R. and Petrie, G.A. (1979) Effect of fungicides on germination of *Albugo candida* oospores in *vitro* and on the foliar phase of the white rust disease. *Canadian Plant Disease Survey* 59, 53–59.

Verma, P.R. and Petrie, G. (1980) Effect of seed infestation and flower bud inoculation on systemic infection of turnip rape by *Albugo candida*. *Canadian Journal of Plant Science* 60, 267–271.

Verma, P.R. and Saharan, G.S. (1994) *Monograph on Alternaria diseases of crucifers*. Agriculture Canada Research Station, Saskatoon, Technical Bulletin No. 1994-6E, 162 pp.

Verma, P.R., Harding, H., Petrie, G.A. and Williams, P.H. (1975) Infection and temporal development of mycelium of *Albugo candida* in cotyledons of four *Brassica* species. *Canadian Journal of Botany* 53, 1016–1020.

Vir, S., Kaushik, C.D. and Chand, J.N. (1973) The occurrence of bacterial rot of raya (*Brassica juncea* coss) in Haryana. *PANS* 19, 46–47.

Vishwanath, Kolte, S.J., Singh, M.P. and Awasthi, R.P. (1999) Induction of resistance in mustard (*Brassica juncea*) against Alternaria black spot with an avirulent *Alternaria brassicae* isolate-D. *European Journal of Plant Pathology* 105, 217–220.

Voglmayr, H. and Riethmüller, A. (2006) Phylogenetic relationships of *Albugo* species (white blister rusts) based on LSU rDNA sequence and oospore data. *Mycological Research* 110, 75–85.

Walker, J.C. (1942) Physiologic races of *Plasmodiophora brassicae*. Systemic invasion of cabbage by *Plasmodiophora brassicae*. *Phytopathology* 32, 18.

Walker, J.C. (1952) *Diseases of Vegetable Crops*. McGraw-Hill, New York.

Walker, J.C. (1957) *Plant Pathology*, 2nd edn. McGraw-Hill, New York.

Walker, J.C., Morell, S. and Foster, H.H. (1937) Toxicity of mustard oil and related sulphur compounds to certain fungi. *American Journal of Botany* 24, 536–541.

Wallenhammar, A.C. (2010) Monitoring and control of *Plasmodiophora brassicae* in spring oilseed *Brassica* crops. *Acta Horticulturae* 867, 181–190.

Wang, H.Z., Liu, G.H., Zheng, Y.B., Wang, X.F. and Yang, Q. (2003) Breeding of *Brassica napus* cultivar Zhongshuang No. 9 with resistance to *Sclerotinia sclerotiorum* and dynamics of its important defence enzyme activity. In: *Proceedings of the 11th International Rapeseed Congress,* Copenhagen, Denmark, 43 pp.

Wang, Z., Tan, X., Zhang, Z., Gu, S., Li, G. and Shi, H. (2012) Defense to *Sclerotinia sclerotiorum* in oilseed rape is associated with the sequential activations of salicylic acid signaling and jasmonic acid signaling. *Plant Science* 184, 75–82.

Webster, J. (1980) *Introduction to Fungi*, 2nd edn. Cambridge University Press, Cambridge, UK.

Wei, C.T., Shen, S.L., Wang, J.L., Zhang, C.W. and Zhu, Y.G. (1960) Mosaic disease of Chinese rape and other crucifers in eastern China. *Acta Phytopathologia Sinica* 4, 94–112.

Westman, A.L. and Dickson, M.H. (1998) Disease reaction to *Alternaria brassicicola* and *Xanthomonas campestris* pv. *campestris* in *Brassica nigra* and other weedy crucifers. *Cruciferae Newsletter* 20, 87–88.

Willets, H.J. and Wong, A.L. (1971) Ontogenetic diversity of sclerotia of *Sclerotinia sclerotiorum* and related species. *Transactions of the British Mycological Society* 57, 515–524.

Willets, H.J. and Wong, J.A.L. (1980) The biology of *Sclerotinia sclerotiorum*, S. *trifoliorum* and S. *minor* with emphasis on specific nomenclature. *Botanical Review* 46, 101–165.

Williams, J.R. and Stelfox, D. (1979) Dispersal of ascospores of *Sclerotinia sclerotiorum* in relation to Sclerotinia stem rot of rapeseed. *Plant Disease Reporter* 63, 395–399.

Williams, J. R. and Stelfox, D. (1980) Occurrence of ascospores of *Sclerotinia sclerotiorum* in areas of Alberta. *Canadian Plant Disease Survey* 60, 51–53.

Williams, P.H. and Yukawa, Y.B. (1967) Ultrastructural studies on the host parasite relations of *Plasmodiophora brassicae*. *Phytopathology* 57, 682–687.

Woronin, M. (1878) *Plasmodiophora brassicae*. Urheberder der Kohlpflanzen-Hemie. *Jahrbuch Wiss Botany* 11, 548–574.

Yang, S.M. (1959) An investigation on the host range and some ecological aspects of the Sclerotinia disease of rape plant. *Acta Phytopathologia Sinica* 5, 111–122.

Yarwood, C.F. (1949) Effect of soil moisture and nutrient concentrations on the development mildew. *Phytopathology* 39, 780–788.

Zhang, F.-L., Xu, J., Yan, H., Li, M.Y., Zhang, F.L., Xu, J.B., Yan, H. and Li, M.Y. (1997) A study on inheritance of resistance to black leaf spot in seedlings of Chinese cabbage. *Acta Agriculturae Boreali Sinica* 12, 115–119.

Zhu, L., Li, Y., Ara, N., Yang, J. and Zhang, M. (2012) Role of a newly cloned alternative oxidase gene (*BjAOX1a*) in Turnip Mosaic Virus (TuMV) resistance in mustard. *Plant Molecular Biology Reporter* 30, 309–318.

Zhu, Y., Hairu, C., Jinghua, F., Yunyue W., Yan, L., Jianbing, C., Jin, X.F., Shisheng, Y., Lingping, H., Hei, L., Tom, W.M., Teng, P.S. and Zonghua (2000) Genetic diversity and disease control in rice. *Nature* 406, 718–722.

8 *Albugo candida*

P.R. Verma,¹ G.S. Saharan² and P.D. Meena³*

¹*Agriculture & Agri-Food Canada, Saskatoon, Saskatchewan, Canada;*
²*Department of Plant Pathology, CCS Haryana Agricultural University,
Hisar, India;* ³*ICAR-Directorate of Rapeseed-Mustard Research,
Bharatpur, India*

Introduction

Albugo candida (Pers. ex. Lev.) Kuntze. (*A. cruciferarum* S.F. Gray) is an oomycete belonging to the family *Albuginaceae* (*Albugonales, Peronosporomycetes*). It is an obligate parasite responsible for the white rust disease of many cruciferous crops. It causes both local and general infection (Saharan and Verma, 1992). Local infection produces white to cream pustules on the lower (abaxial) surface of leaves and stems or pods, while general, or flower bud infection (Verma and Petrie, 1980) causes extensive distortion, hypertrophy, hyperplasia and sterility of inflorescences generally called 'staghead'. The staghead phase accounts for most of the yield losses attributed to this disease. The combined infection of leaf and inflorescence caused extensive yield losses up to 30–60% in severely affected fields in turnip rape (*Brassica rapa* L.) (Petrie, 1973; Harper and Pittman, 1974; Petrie and Vanterpool, 1974). In India, the yield losses can be 60% or more in Indian mustard (*Brassica juncea* (L.) Czern. & Coss.) (Bains and Jhooty, 1979; Lakra and Saharan, 1989a; Saharan, 2010). Substantial yield losses have been reported from several countries in radish (*Raphanus*

sativus L.) (Kadow and Anderson, 1940; Williams and Pound, 1963) and rapeseed (Barbetti, 1981; Barbetti and Carter, 1986). Although Canadian and European *B. napus* cultivars are not attacked, many cultivars of this species grown in China are susceptible (Fan *et al.*, 1983). The wide range of losses in yield estimated by oilseed brassica workers indicates the presence of an array of disease-tolerant genes in the host. There is a need to assess these disease-tolerant genes in each genotype under disease stress environmental conditions (Saharan, 2010). The present chapter includes taxonomy and nomenclature, host range, geographical distribution, inoculation techniques, epidemiology, association of *Albugo* and *Hyalopernospora*, pathogenic variability, virulence spectrum, host resistance, genetics of host–parasite interactions, slow white rusting in crucifers and chemical control.

Taxonomy and Nomenclature

The first species of *Albugo* was described by Gmelin in 1792 as *Aecidium candidum* (now *Albugo candida*), later placed in genus *Uredo*,

*Corresponding author, e-mail: pdmeena@gmail.com

subgenus *Albugo* (Persoon, 1801). Based on differences in symptom development, Persoon (1801) described two different species of white blister rust, with *Uredo candida* subdivided into three varieties, parasitic to *Brassicaceae* and *Asteraceae*. After a few years, *Albugo* was established as an independent genus by de Roussel (1806), although speciously Gray (1821) is often given as the author for this genus. De Candolle (1806) added the species *Uredo portulaceae* (now *Wilsoniana portulacae*) and *Uredo candida* beta *trogopogi* to species rank (*Uredo tragopogi*, now *Pustula tragopogonis*) and renamed *Uredo candida cruciferarum*. Leveilla (1847) described the genus *Cystopus*, and later de Bary (1863) described the sexual state of *Albugo*, adopting the generic name *Cystopus*. *Albugo* was typified by Kuntze (1891), who gave *Uredo candida* (Pers) Pers. as the type species. Before Biga (1955) pointed out that names of sexual form had no antecedence over anamorphs in oomycetes, many researchers considered white blister rusts to be members of the superfluous genus *Cystopus* (Wakefield, 1927); at the same time the older genus name, *Albugo*, also persisted. In the early 20th century, numerous other species of genus *Albugo* were described. Wilson (1907) recorded 13 species, and Biga (1955) accepted 30 species in this genus about 50 years later (Mukerji, 1975). Only a few new species were described in the key to the genus *Albugo* published 40 years later (Choi and Priest, 1995), a total of ten species being recognized.

Albugo candida (Pers.) Roussel is an obligate parasitic fungus responsible for white rusts in brassicaceous hosts over widely different geographical areas of the world. The *Albuginaceae* contain four distinct lineages (*Albugo s.s.*, parasitic to *Brassicales*; *Albugo s.l.*, parasitic to *Convolvulaceae*; *Pustula*, parasitic to *Asterales s.l.*; and *Wilsoniana*, parasitic to *Caryophyllales*). *Albugo cruciferarum* is regarded as a synonym of *A. candida* (Choi *et al.*, 2007). Till now, the white blister pathogen on oilseed rape has been considered *A. candida* (Farr and Rossman, 2010). The high degree of genetic diversity exhibited within the *A. candida* complex may warrant their division into several distinct species (Choi *et al.*, 2006).

Host Range

The first record of *A. candida* on *Brassicaceae* seems to be by Colmeiro (1867). *Albugo candida* has been reported in brassicaceous hosts over widely different geographical areas of the world with a host range of some 63 genera and 241 species (Biga, 1955; Saharan and Verma, 1992; Choi *et al.*, 2007), including some weedy species (Farr *et al.*, 1989). According to the USDA-ARS Systemic Botany and Mycology Laboratory (Farr *et al.*, 2004), *A. candida* was recorded on more than 300 hosts. Molecular phylogenetic investigations have revealed that *A. candida* has an extraordinarily broad host range, extending from *Brassicaceae* to *Cleomaceae*, *Fabales* to *Capparoceae* (Choi *et al.*, 2006, 2007, 2008, 2009, 2011a). *Albugo candida* and *A. tragopogonis* each may consist of several distinct lineages (Voglmayr and Riethmüller, 2006). The host specificity of *A. candida* has been demonstrated by Hiura (1930) and Eberhardt (1904). These studies, which examined predominately crop species, resulted in *A. candida* being classified into ten races based on their host specificity (Pound and Williams, 1963; Hill *et al.*, 1988). For example, Khunti *et al.* (2000) showed that *A. candida* isolates from *Brassica* can infect *Amaranthus viridis* (Amaranthaceae) and *Cleomev viscosa* (*Capparaceae*, now included in *Brassicaceae*; APG, 2003), as well as *B. rapa*. Saharan (2010) has provided a list of pathotypes reported globally (Table 8.1).

Geographical Distribution

White rust is prevalent worldwide. Although the list is not exhaustive, countries where the disease occurs include the UK (Berkeley, 1848), the USA (Walker, 1957), Brazil (Viegas and Teixeira, 1943), Canada (Greelman, 1963; Petrie, 1973), Germany (Klemm, 1938), India (Chowdhary, 1944), Japan (Hirata, 1954), Pakistan (Perwaiz *et al.*, 1969), Palestine (Rayss, 1938), Romania (Savulescu, 1946), Turkey (Bremer *et al.*, 1947), Fiji (Parham, 1942), New Zealand (Hammett, 1969), China (Zhang *et al.*, 1984) and Korea (Choi *et al.*, 2011a). White rust of sunflower occurs

Table 8.1. Global virulence of *Albugo candida* pathotypes (Saharan, 2010).

Designate pathotype	Country	International primary host	Reference
AC1	North America	*Raphanus sativus*	Pound and Williams, 1963
AC2	North America	*Brassica juncea*	Pound and Williams, 1963
AC2V	North America	*Brassica napus*	Petrie, 1994
AC3	North America	*Armoracia rusticana*	Pound and Williams, 1963
AC4	North America	*Capsella bursa-pastoris*	Pound and Williams, 1963
AC5	North America	*Sisymbrium officinale*	Pound and Williams, 1963
AC6	North America	*Rorippa islandica*	Pound and Williams, 1963
AC7	North America	*Brassica rapa*	Verma *et al.*, 1975
AC7V	North America	*Brassica rapa* cv. Reward	Petrie, 1994
AC8	North America	*Brassica nigra*	Delwiche and Williams, 1977
AC9	North America	*Brassica oleracea*	Williams, 1985
AC10	North America	*Sinapis alba*	Williams, 1985
AC11	North America	*Brassica carinata*	Williams, 1985
AC12	India	*Brassica juncea*	Verma *et al.*, 1999
AC13	India	*Brassica rapa* var. Toria	Verma *et al.*, 1999
AC1 to 9	India	*Brassica* species	Singh and Bhardwaj, 1984
AC1 to 5	India	*Brassica* species	Lakra and Saharan, 1988c
AC14	India	*Brassica juncea* cv. RL 1359	Gupta and Saharan, 2002
AC15	India	*Brassica juncea* cv. Kranti	Gupta and Saharan, 2002
AC16	India	*Brassica juncea* cv. Kranti	Gupta and Saharan, 2002
AC17	India	*Brassica juncea* cv. RH 30	Gupta and Saharan, 2002
AC18 to 34	India	*Brassica juncea* cv. RH 30; EC 182925; DVS 7-3-1	Jat, 1999
AC35 and AC36	India	*Brassica rapa* var. Brown Sarson	Jat, 1999
AC37	India	*Brassica nigra*	Jat, 1999
AC2A	Western Australia	*Brassica juncea* cvs Vulcan, Commercial Brown	Kaur *et al.*, 2008
AC2V	Western Australia	*Raphanus raphanistrum*	Kaur *et al.*, 2008

in Russia (Novotel'Nova, 1962), Uruguay (Sackston, 1957), Argentina (Sarasola, 1942), Australia (Middleton, 1971; Stovold and Moore, 1972) and in many other countries (Kajomchaiyakul and Brown, 1976). White rust of salsify occurs in Australia, Canada, the USA, South America, Europe, Asia and Africa (Wilson, 1907). White rust of water spinach is a serious disease in India and Hong Kong (Safeefulla and Thirumalachar, 1953; Ho and Edie, 1969) and affects spinach in Texas (Wiant, 1937; Williams and Pound, 1963). Countries and provinces where the white rust on crucifers occurs include Belgium, Cyprus, Denmark, Finland, France, Greece, Hungary, Ireland, Italy, Latvia, Malta, the Netherlands, Poland, Portugal, Slovakia, Spain, Canary Islands, mainland Spain, Sweden, Switzerland, Ukraine and Yugoslavia (Fed. Rep.), Bhutan, Anhui, Fujian, Gansu, Guangxi, Guizhou, Hebei, Hubei, Jiangsu, Jiangxi, Jilin, Liaoning, NeiMenggu, Qinghai, Shaanxi, Shandong, Shanxi, Sichuan, Xinjiang, Xizhang, Yunnan, Zhejiang, Iran, Iraq, Israel, Malaysia, Peninsular Malaysia, Sabah, Sarawak, Nepal, the Philippines, Taiwan and Yemen, Egypt, Ethiopia, Kenya, Libya, Malawi, Mauritius, Sierra Leone, South Africa, Sudan and Tanzania, Barbados, Bermuda, Cuba, Dominican Republic, El Salvador, Jamaica, Puerto Rico, Trinidad and Tobago, Parana, Falkland Islands, Guyana, Suriname, Cook Islands, Polynesia, New Caledonia, Papua New Guinea, Samoa and Vanuatu (USDA, 2014).

Inoculation Techniques

Growth chamber inoculation technique

Seeds of susceptible *B. rapa* cv. Torch were planted 2 cm deep in a soil-free growth medium (Stringam, 1971) in 10 cm square plastic pots. Plants were grown in a growth chamber with 18 h photoperiod (312 $\mu E/m^2/s$) and at day-night temperature of 21°C and 16°C, respectively. Pots were placed in metal trays and watered by flooding the trays. To prepare inoculum suspensions, zoosporangia from pustules on fresh or frozen leaves infected with *A. candida* were suspended in deionized, distilled water, filtered through cheese cloth, germinated for 2–3 h at 5°C and adjusted to 75,000–100,000 zoospores/ml. The inoculum was sprayed on to plants with an atomizer until leaf run-off. Control plants were sprayed with distilled water. The plants were placed in a mist chamber (100% relative humidity) in the growth chamber for 72 h at 16°C to promote infection. Disease incidence and severity observations were recorded 10 days after inoculation.

Oospore germination

The most conspicuous symptom of white rust and probably the major sources of yield loss are distortion and hypertrophy of infected inflorescence called 'staghead'. When ripe, stagheads almost entirely comprise brown, thick-walled oospores. This is the form in which the pathogens survive during the off-season, and are also the sources of primary infection. These are also known as resting spores. Despite their importance in the epidemiology, conditions under which the oospores germinate were largely a mystery until 1974 (Petrie and Verma, 1974; Verma and Petrie, 1975). Prior to this work, only De Bary (1866) and Vanterpool (1959) have described the oospore germination in *A. candida*. Vanterpool (1959) reported germination as 'always irregular and uncertain', never exceeding 4% of the spores. In our studies (Petrie and Verma 1974; Verma and Petrie, 1975), three reproducible methods were devised that resulted in a

very high percentage of germination. In the first method, a small amount of finely ground staghead powder consisting largely of oospores was scattered over moist filter paper placed on wet cotton in a Petri dish; the lid of the dish was also lined with moist cotton. The plates were incubated at 10–15°C for a period of up to 3 weeks.

In the second method, sterile deionized water or sterile or non-sterile tap water was allowed to drip slowly on to sintered glass filters of ultrafine porosity where small amounts of oospore powder were scattered. This was done in an attempt to mimic the leaching action that might occur during spring from melting snow or rain. Most of these experiments were run at 10–15°C. In the third method, the one most routinely used, a small amount of oospore powder was placed in 50 ml sterile water in a 125 ml flask and incubated at 200 rpm on a rotary shaker at 18–20°C for a period of 3–4 days. The spore suspension was then poured into a Petri dish and kept stationary at 13°C for 24 h or more. Counts of germinated oospores were made on materials mounted in lactophenol-aniline blue. Washing of oospores on a rotary shaker for 3–4 days followed by 1 day in still culture was the most rapid method, and it gave the highest percentage germination for most of the samples, including 10 years old or more. Oospores required 2 weeks of washing on a sintered glass filter before maximum germination was obtained. On moist filter paper, maximum germination occurred after an incubation period of about 21 days. Substantial numbers of oospores retained their viability for at least 20 years in dry storage in the laboratory. The highest percentage germination was recorded in the sample that had been collected only 2 weeks prior to the test. It is still not known how long oospores can remain viable in soil. However, germination of 43% of oospores from material kept in dry storage for 20 years does indicate their potential longevity (Verma and Petrie, 1975).

Three distinct types of germination were observed. In the most common type, the oospore content was divided into numerous zoospores which were then extruded into a globular, thin-walled vesicle. Zoospores subsequently escaped from the vesicle. Initiation of a vesicle to zoospore escape was completed in 3.0–5.2 min with an average elapsed time of 4.1 min. Between 40 and 60 zoopsores were formed per vesicle (Verma and Petrie, 1975).

In the second less-frequent type of germination, a germ tube was produced from the germinating oospore and zoospores, which then differentiated in the oospore. These were discharged through the tube into a so-called 'terminal vesicle' formed at the end of the tube. Zoospores subsequently escaped the vesicle (Verma and Petrie, 1975). The germ tube may be simple or branched. Occasionally up to three branches were observed on a germ tube (Verma and Petrie, 1975).

Oospores as the primary source of inoculum

In the absence of reliable methods of germination, the role of oospores both as overwintering agent and incitants of primary infection have been speculated. The most likely infection site in the field is the emerging cotyledons. Therefore, for pathogenicity studies, plants of *B. rapa* cv. Torch grown in the growth chamber were kept at cotyledon stage by removing the growing points. Cotyledons of 10-day-old plants were drop-inoculated with zoospore suspension derived from germinating oospores. Plants were kept under a mist for 3 days. At 10 days after inoculation nearly every inoculated plant showed heavy infection in the form of white pustules on the underside of cotyledons. These infection studies suggest that zoospores from germinating oospores are most likely the actual infecting units for initiation of primary infection (Verma *et al.*, 1975).

The importance of the oospore as a source of primary inoculum was also explored in a field experiment conducted under irrigated and dryland conditions (Verma and Petrie, 1980). The treated plots were seeded with seeds of susceptible *B. rapa* cv. Torch mixed with an equal weight of oospore powder. The control plots received no oospore powder. Both number of pustules per infected leaf and the percentage of plants with stagheads were significantly higher in oospore-infected than those in the non-infected plots. These results suggest that oospores overwintered in soil or carried on the seed are most likely the primary source of infection.

Induction of staghead in flower-bud inoculated plants

A number of plant pathologists believe that the hypertrophies or stagheads are produced as a result of early infection of young seedlings and systemic development of the fungus in the plant. Induction of stagheads by artificially inoculating flower buds of plants grown under growth chamber and greenhouse conditions discounted such a possibility (Verma and Petrie, 1980; Goyal *et al.*, 1996b). In addition to these, field experiments (Verma and Petrie, 1979, 1980) conclusively prove that a large percentage of stagheads in the field are produced as a result of secondary infection of flower buds rather than a systemic development of the fungus in the plant. Flower bud inoculation technique is now routinely used at growth stage 3.1 (Goyal *et al.*, 1996b) for screening advanced breeding lines. These studies have also proved useful in determining the time of application of protectant and systemic fungicides.

Detached-leaf culture technique

In the detached-leaf techniques (Verma and Petrie, 1978), healthy leaves from the rosette of 12–14-day-old *B. rapa* seedlings were detached and transferred to Petri dishes containing 20–25 ml of autoclaved medium containing 0.5 ppm benzyl adenine and 0.8% agar. Leaves are placed in the dishes with their lower surface on the medium, usually within 15 min of detachment. Leaves were drop-inoculated with a zoospore suspension (75,000–100,000 zoospore/ml) derived from zoosporangia of *A. candida* race-7. Control leaves were treated with distilled water. Leaves were kept under 100% relative humidity for 72 h with day-night temperatures of 21 and 16°C, respectively. Following an initial 24 h dark period, a photoperiod of 18 h day (312 $\mu E/m^2/s$) regime was maintained for the duration of the experiment. Observations were recorded 14 days after inoculation.

Plant susceptibility ratings of various *Brassica* species and breeding lines on the inoculated detached leaves were essentially the same as when intact plants were used as the host. The method facilitates the establishment and maintenance of single zoospore cultures and should enable almost complete isolation from extraneous inoculum, including other races of *A. candida*. Detached-leaf culture also results in greater uniformity of experimental units, more economical use of growth and mist chamber space and allows greater use of environmental control. From the plant breeder's point of view, the programme efficiency is increased, since the breeder can select resistant material for intercrossing from among a vigorous growing plant population rather than a weak group of resistant plants that have survived the unfavourable environment necessary to obtain differential infection on potted plants.

In vitro callus cultures of *Albugo candida*

The growth of *A. candida* in *B. juncea* (Goyal *et al.*, 1995) and *B. rapa* (Goyal *et al.*, 1996c) callus cultures was achieved on MS medium (Murashige and Skoog, 1962) supplemented with 1.0 mg/l naphthalene acetic acid and 1.0 mg/l benzylamino-purine. It is likely that the growth regulators and sucrose used in the culture medium for host callogenesis played a role in haustorial production. This was the first report of the growth of *A. candida* race 7C on leaf callus tissues of *B. rapa* and of the establishment of haustorial connections between the fungus and host callus cells in this species (Goyal *et al.*, 1996c). Pathogenicity tests with *in vitro*-produced zoospores and oospores confirmed the viability and the virulence of *A. candida* in dual cultures.

The *A. candida*–*B. rapa* dual culture system has potential for sexual studies of the fungus. Because it was possible to trace the development of antheridia and oogonia from the mycelium, our results support the view that isolates of *A. candida* race 7V are homothallic. This dual culture system could also prove useful in *in vitro* selection studies of recovering resistant cells and plants.

Epidemiology

Temperature effects on disease development

The detached-leaf culture technique was used to study the influence of temperature on the temporal progression of white rust, the development of disease on leaves of different ages and the development of disease on leaves detached at the end of light and dark periods (Verma *et al.*, 1983). This information was necessary so that the detached-leaf culture technique could be used in the screening of rapeseed cultivars for resistance against *A. candida*. Temperature, leaf age, time of leaf detachment and the interaction of these factors had a significant effect on the temporal development of *A. candida* race 7 on detached leaves. Of the temperatures tested (3–32°C), 21°C gave the best disease development, with 18.5°C being the calculated optimum. The disease did not develop at 3°, 29° and 32°C and was slow to develop at 9°, 12° and 27°C. There was highly significant (p<0.01) interaction between length of incubation period and temperature. Unlike intact plants, detached leaves developed pustules on both surfaces. Infection occurred on leaves of all ages, but medium aged leaves supported the maximum number of pustules, followed by the younger leaves. Leaves detached at the end of a dark period developed more pustules than those detached at the end of the light period. When using the detached leaf culture technique for screening germplasm for resistance to white rust, we advise an adaxial (upper) surface inoculation of cotyledons, or medium aged leaves, and an incubation temperature of 18–22°C (Chattopadhyay *et al.*, 2011).

Temperature effects on oospore development

Epidemiological studies on *A. candida* have focused on the production, viability and germination of zoosporangia (Melhus, 1911; Endo and Linn, 1960; Lakra *et al.*, 1989), and the influence of host age and time of leaf detachment on development of the disease (Verma *et al.*, 1983). Little is known about the sexual reproduction and genetics of the fungus due to the difficulty in determining the factors responsible for the induction of the sexual reproductive phase. The effect of temperature on *in vitro* germination of oospores has been reported (Verma and Petrie, 1975); however, information on the optimum temperature and the time required for production of oogonia, antheridia and mature oospore in leaf tissue would assist in designing experiments for the study of oogenesis, fertilization and karyogamy. The detached leaf culture technique was used to determine the effect of temperature and incubation period on progressive development of oospores of *A. candida* race 2V in *B. juncea* leaves (Goyal *et al.*, 1996a).

The progressive development of *A. candida* oospores in detached leaves of *B. juncea* was largely dependent on incubation temperature. Oogonia and oospore production occurred over the entire range of incubation temperatures of 10–27°C. The earliest development of oogonia was observed at 25°C, 7 days after inoculation and incubation. The largest number of oogonia for the 21°, 23°, 24° and 25°C treatments was observed 12 days post-inoculation and numbers decreased after that. At lower and higher temperature development of oogonia occurred later. Maximum numbers of oogonia were recorded at 17 days for the 15°C treatment at the end of the experiments. Mature oospores were observed 12 days after incubation at 23° and 24°C. The number of mature oospores was still increasing at 17 days post-inoculation at all temperatures. Mature oospores developed late and more slowly at lower as well as higher incubation temperatures.

The production of *A. candida* oospores in leaf tissues could be important in the disease perpetuation. Hypertrophied tissues (staghead) are quite resistant to decomposition, and the release of oospores could take 3–4 years. Leaf tissues are quick to decompose, and thus oospore release from such material could be expected the following year. In naturally infected leaves, oospores are produced in the later part of the season when temperatures are warm (Verma, 1989). Warm temperatures hasten leaf senescence, which in turn enhances tissue decomposition and early release of oospores. The knowledge of optimum temperature and time for the development of oospores in detached leaves make it possible to compare the sequential events of oogenesis, fertilization and karyogamy in various *Albugo* species at the earliest stages of their development. These comparative investigations in *Albugo* species could also be useful in fungal taxonomy. The detached leaf culture technique could also be used to determine the heterothallic nature of *A. candida*.

Temporal development of *Albugo candida* infection in cotyledons

Progression of white rust infection was studied in cotyledons of susceptible (*B. rapa*, *B. juncea*), moderately resistant (*B. hirta*) and immune (*B. napus*) hosts (Verma *et al.*, 1975). Cotyledons of all the four *Brassica* species were inoculated with zoospores of *A. candida* produced from germinating oospores or zoosporangia. At different times after inoculation, intact cotyledons were fixed in 95% ethanol-acetic acid (v/v) solution, cleaned in 70% lactic acid at 40°C for 3–4 days and stained with cotton blue in lactophenol and examined under a compound microscope.

Generally, the sequence of events from zoospore encystment to formation of the first haustorium was the same in all hosts, although under field conditions *B. hirta* is moderately resistant and *B. napus* is essentially 'immune'. In *B. juncea*, the first haustorium was observed 16–18 h after inoculation, while in *B. rapa*, *B. hirta* and *B. napus* the first haustorium was observed about 48 h after inoculation. In the susceptible hosts, after the formation of the first haustorium the hyphae grew rapidly and produced variable number of haustoria in each cell. The profusely branched, non-septate mycelium appeared to fill all available intercellular spaces. At 5–6 days after inoculation, the club-shaped zoosporangia developed from a dense layer of mycelium. In the immune host, usually only one haustorium was formed, after which the hyphae ceased to elongate. At about 72 h after inoculation, a fairly thick, densely staining encapsulation was usually detected around each haustorium, and later only 'ghost' outlines of hyphae and haustoria were observed.

Encapsulations were not observed around haustoria of susceptible hosts.

From our observations (Verma *et al.*, 1975), it seems probable that zoospores derived from germinating oospores constitute the primary inoculum for infection of cotyledons of susceptible *Brassica* species. There was no evidence of direct infection by the germ tubes (Verma and Petrie, 1975). The establishment and maintenance of a compatible relationship between *A. candida* and its hosts hinges on the successful formation of the first haustorium. A similar sequence of events in both susceptible and immune hosts up to this point suggests absence of morphological barrier to zoospore encystment, germination and subsequent penetration through stomata. For the incompatible combination, it is not clear whether the parasite fails to produce a functional haustorium or a viable haustorium is formed within the host cell that is subsequently killed by the host's defence mechanism. Fairly dense, thick encapsulation observed around haustorium of immune host tissue suggests that the latter may be the case. In any event, it does seem that the decision between compatibility and incompatibility is made within 48 h after inoculation.

Studies using whole mounts (Verma *et al.*, 1975) can provide a rapid and useful quantitative means of measuring fungal development and can be useful in screening for disease resistance or testing the effects of environmental changes or fungicide treatments. Whole mounts may also provide a useful perspective for ultrastructural studies where the total amount of fungal thallus present in a susceptible host is not always appreciated. Certainly, the massive amount of intercellular mycelium, particularly the much-branched sporangiophore 'base', which the host is capable of supporting while still actively photosynthesizing, emphasizes the highly integrated and delicate control occurring in the type of parasitism that has evolved in *A. candida*.

Association of *Albugo* and *Hyaloperonospora*

The association or mixed infection or simultaneous occurrence of *A. candida* and *Hyloperonospora brassicae* on leaves, inflorescence and silique of oilseed brassica in nature is very common (Saharan and Verma, 1992). The intensity of mixed infections varies from 0.5 to 35.0%. It is reported that *A. candida* predisposes the host tissues to infection by *H. brassicae* (Bains and Jhooty, 1985; Saharan and Verma, 1992; Saharan and Mehta, 2002). In a study where *Hyaloperonospora* was inoculated 24 h prior to *Albugo*, the infection by *Hyaloperonospora* was recorded after 7 days and by *Albugo* after 5–6 days. However, when both the pathogens were inoculated simultaneously at equal proportion, there was a delay in the expression of infection by *Hyaloperonospora* for 2–3 days and *Albugo* did not show any variation. A latent period of 12 days has been recorded when both the pathogens were inoculated alone or in combination of both (prior or after or mixed together). These observations have been supported in a sequence of events during pathogenesis of *Albugo* and *Hyaloperonospora* on Indian mustard through histopathological study by Mehta *et al.* (1995). It seems that genes governing the virulence of *A. candida* in the *Brassica* system elude the plant defence mechanism through reduction of phytoalexin biosynthetic pathway and by producing metabolites preferred as food by the pathogen *Hyaloperonospora* for colonization of host tissues. The compatibility genes of both the pathogens may be the same or situated on the same loci or tightly linked or epistatic.

Pathogenic Variability in *Albugo candida*

Eberhardt (1904) recognized two specialized groupings of *Albugo*, one attacking *Capsella*, *Lepidium* and *Arabis* and the other attacking *Brassica*, *Sinopis* and *Diplotaxis*. However, he did not use the phrase biological forms for this fungus. Melhus (1911) also suggested the existence of specialization in *A. candida*. Pape and Rabbas (1920) demonstrated that the fungus on *Capsella bursa-pastoris* should be considered a distinct form. Savulescu and Rayss (1930) had distinguished eight morphological forms with *A. candida*. Later, Savulescu (1946) established ten varieties of *A. candida* based on host specialization and morphology. Hiura (1930) distinguished three biologic forms of

A. candida on *Raphanus sativus, B. juncea* and *B. rapa* subsp. *chinensis.* Napper (1933) described 20 races of *A. candida* in Britain. Togashi and Shibasaki (1934) found that sporangia of *Albugo* from *Brassica* and *Raphanus* were 20 × 18 µm in size while those from *Cardamine, Capsella, Draba* and *Arabis* measured 15.5 × 14.5 µm; these were classified as *macrospora* and *microspora*, respectively. Results of this Japanese study suggested that five distinct biological forms of *Albugo* were present. Subsequently, Ito and Tokunaga (1935) elevated the forms with larger spores to species rank as *A. macrospora* (Togashu) Ito. Biga (1955) recognized two morphological taxa: *A. candida macrospora* and *A. candida microspora*, as proposed by Togashi and Shibasaki (1934), but renamed them *A. candida microspora* and *A. candida candida*, respectively. Biga (1955), on the basis of conidial measurements from 63 species, reported that *A. candida microspora* (15–17.5 µm dia.) was restricted to *Armoracia, Brassica, Erucastrum, Raphanus* and *Rapistrum*, whereas *A. candida candida* (12.5–15 µm dia.) had a wide range of cruciferous hosts. Endo and Linn (1960) reported a race of *Albugo* on *Armoracia rusticana.*

Pound and Williams (1963) identified six races of *A. candida*; race I from *R. sativus* var. Early Scarlet Globe, race 2 from *B. juncea* var. Southern Giant Curled, race 3 from *Armoracia rusticana* var. Common, race 4 from *Capsella bursa-pastoris*, race 5 from *Sisymbrium officinale* and race 6 from *Rorippa islandica.* Verma *et al.* (1975) and Delwiche and Williams (1977) added race 7 from *B. rapa* Turnip or Polish rapeseed and race 8 from *B. nigra*, respectively. Novotel'nova (1968), from former USSR while analysing intra-specific taxa, established that the *A. candida* species consisted of distinct morphological or specialized forms confined to a particular range of host plants, i.e. to plants of certain species or groups of genera and species. Within the morphological forms, races can be differentiated, while, within heterogeneous populations, both races and forms can be differentiated. It was considered that geographic and climatic conditions leave their distinguishing mark on the processes of form development and populations of the fungus encountered by investigators in different countries were not identical. Novotel'nova

and Minasyan (1970) and Burdyukova (1980) studied the biology of *A. candida* and *A. tragopogonis* in former USSR and conducted an in-depth study on the extent of specialization of *A. candida.* In India, Singh and Bhardwaj (1984) tested 12 *Brassica* species and identified nine races from four hosts, *B. juncea, B. rapa* var. Toria, *B. rapa* var. Brown Sarson and *B. rapa* var. *pekinensis.* Lakra and Saharan (1988c) identified five races of *A. candida* on the basis of the reaction on a set of 16 host differentials. They identified two distinct races from *B. juncea* which were different from the previous records. One (race 2) attacked *B. nigra, B. juncea* and *B. rapa* var. Brown Sarson and the other (race 3) infected only *B. juncea* and *B. rapa* var. Toria. Bhardwaj and Sud (1988) tested 26 cultivated and wild cruciferous hosts and identified nine new biological races from nine hosts, viz. *B. rapa* var. Brown Sarson (cv. BSH 1), *B. rapa* var. Toria (cv. OK-I), *B. juncea* (cv. Varuna), *B. chinensis, B. rapa* var. *pekinensis* (cv. Local), *B. rapa* (cv. PTWG), *R. sativus* (cv. Chinese Pink), *Raphanus raphanistrum* (wild radish) and *Lepidium virginicum* (Wild). They reported that the reaction of nine isolates of *A. candida* differed from each other on 26 differential hosts revealing thereby that the monotypic crucifer pathogen, *A. candida*, existed in the form of different biological races designated as new biological races or forms 1 to 9.

The concept of races in *A. candida*, as proposed by Pound and Williams (1963), was based on species relationships. Studies have, however, clearly demonstrated that cultivars of brassica crops must be included in a set of host differentials to distinguish isolates of the pathogen within an accepted race (Burdyukova, 1980; Pidskalny and Rimmer, 1985). There is an urgent need to standardize host differentials keeping in mind the homogeneity and purity of species and varieties. Petrie (1988) using North American race 2 and 7 from *B. juncea* and *B. rapa*, respectively, have screened accessions of several *Brassica* species including *B. rapa* var. Yellow Sarson, *B. rapa* var. Brown Sarson, *B. rapa* var. Toria and *B. juncea* from India; both *B. rapa* Yellow Sarson and *B. rapa* Brown Sarson were equally highly susceptible to both races, toria only to race 7 and *B. juncea* only to race 2. A detailed study is needed to determine whether the races of

A. candida attacking *B. juncea* and several *B. rapa* crops in India are similar to race 2 and 7 from Canada and the USA. Kolte *et al.* (1991) reported that the white rust isolate obtained from *B. rapa* appeared to be distinct in pathogenicity from the one obtained from *B. juncea* under Indian conditions. Petrie (1994) in Saskatchewan and Alberta discovered new race 7v in 1988 and race 2v in 1989. Verma *et al.* (1999) reported two new races of *A. candida* in India, viz. race 12 from *B. juncea* and race 13 from *B. rapa* var. Toria using 14 (including six standard) crucifer host differentials.

Mathur *et al.* (1995) and Rimmer *et al.* (2000) collected isolates of *A. candida* from different geographic locations in western Canada and tested for virulence on a number of cultivars and accessions of *Brassica* species to determine variability and distribution of different races in the area. Most isolates were identified as race 7, which could be subdivided into 7a and 7v on the basis of their virulence on *B. rapa* cv. Reward. Isolates 28-7 and 29-1 were less virulent as they were avirulent to all the differentials except the rapid cycling *B. rapa* CrGCI-I8. 'Tower' isolates, 11-6 and 41-4, which could infect cultivars of both *B. rapa* and *B. juncea*, appeared to be hybrids between race 2 and race 7. Wu *et al.* (1995) studied genetic variation among isolates of *A. candida* using randomly amplified polymorphic DNA (RAPD) with five selected random primers fingerprint patterns generated for each isolate. Most polymorphism was found between different races than among isolates within a single race. Most Canadian field isolates were grouped as race 7 and could be further subdivided into two groups (7a and 7v). Classification of *A. candida* isolates based on the results from the RAPD analysis was identical to the virulence classification on ten *Brassica* differentials. Four distinct and new pathotypes of *A. candida*, viz. ACI4 from RL 1359, AC15 and AC16 from Kranti and AC17 from RH30 cultivars of *B. juncea*, have been identified on the basis of their differential interactions with 11 host differentials by Gupta and Saharan (2002). Jat (1999) identified 20 distinct pathotypes of *A. candida*, 17 from *B. juncea* (AC18 to AC34), two from *B. rapa* var. Brown Sarson (AC35 to AC36) and one from *B. nigra* (AC 37). From Western Australia,

Kaur *et al.* (2008) identified pathotype AC2A from *B. juncea* and pathotype AC2v from *R. raphanistrum*.

The pathogenic variability recorded in *A. candida* in the form of races from all over the world are 2 from Australia, 20 from Britain, 4 from Canada, 2 from Germany, 49 from India, 8 from Japan, 18 from Romania and 7 from the USA. However, nomenclature of *A. candida* races came into practice after the use of host differentials to distinguish races by Pound and Williams (1963). Global virulence of *A. candida* based on a primary host is documented in Table 8.2. In *A. candida*, sexual reproduction in the form of oospores is very common and massive, especially on *B. juncea*. Therefore, numerous races are expected to exist. In addition, other mechanism of variability such as recombination, mutation and heterokaryosis are also in operation in nature. To get the true picture of *A. candida* races and virulence spectrum, there is an urgent need to standardize host differentials for each crucifer species in the form of isogenic lines at international level. Standard nomenclature of the races, viz. Acjun 1, 2- for *B. juncea* isolates; AC rap 1, 2- for *B. rapa* isolates, AC nig 1, 2- for *B. nigra* isolates and ACol 1,2 for *B. oleracea* isolates, and so on, may be a useful proposition.

Virulence Spectrum of *Albugo candida*

As per the gene-for-gene hypothesis, interaction of *Albugo*–crucifers for compatibility and incompatibility phenotype determines the number of virulence genes in the pathotype and resistance genes in the host genotype. It has been observed that pathotypes of *A. candida* from *B. juncea* have a wide range of virulence genes. Pathotypes like AC23, AC24, AC17 infect only one, two and three differential hosts indicating a limited virulence potential. However, pathotypes of wider virulence, viz. AC29, AC27, AC30, AC18 and AC21, infected 21, 18, 16, 12 and 10 host differentials, respectively (Jat, 1999; Gupta and Saharan, 2002).

Availability of virulence variability in pathotypes from *B. juncea* has indicated the possibility of a greater number of resistant genes. In the absence of isogenic lines, it cannot

Table 8.2. Sources of resistance in crucifer crops against *A. candida* (Saharan, 2010).

Crucifer host	Sources	Country	References
Radish	China Rose Winter (CRW), Round Black Spanish (RBS)	USA	Williams and Pound, 1963
	Caudatus	USA	Humaydan and Williams, 1976
	Rubiso 2	USA	Bonnet, 1981
	Burpee white	Canada	Petrie, 1986
Brassica napus	Regent	Canada	Fan *et al.*, 1983
	Gulivar, GSL 1, GSL 1501, HNS 1, HNS 3, Tower, HNS 4, Midas, Regent, GBS 7006, Norin 14, WRG 15, H 715, EC 174243, GBS 101, GSL 706, HNS 8, Tower 60, ABN, Altox, EC 131625, EC 131626, Karat, Mary, Niklas, VR-OLGA, VR-WW-1313	India	Bhardwaj and Sud, 1989, 1993; Gupta and Singh, 1994; Jain *et al.*, 1998; Lakra and Saharan, 1989a, b; Saharan 1996, 1997; Saharan *et al.*, 1988, 2005
Brassica juncea	CSR 142, Domo 4, EC 126741, EC 126745, EC 126746, EC 129126-1, KOS 1, PHR 1, RC 781, SSK 1, T 4, YRT 1, Zem 1, Domo, Lethbridge, DIR 519, DIR 1507, EC 126743-2, DIRA 313-7, GS 7027, RN 246, BEC 115, BEC 138, IC 41729, EC 126126, RC 1401, RC 1499, YS7B, MS 85, MS 98, MS 104, MS 105, IB 2073, EC 126743, EC 126743-1, EC 129121-1, RH 8541-46, RW 81-89, Blaze, Metapolka, Newton, Purbiraya, Stoke, RC 1001, RC 1405, RC 1408, RC 1424, RC 1425, RH 8545, WRR 3-1, Chamba 1, CSR 741, RC 295, RC 398, MLS 7, MLS 10, MLS 13, MLS 16, MLS 17, MLS 18, MLS 29, MLS 30, MLS 31, MLS 32, MLS 35, MLS 39, DIRM 5, DIRM 11, Gonads 3, IB 499-1, Kranti 43, R 71-2, R 75-2, RC 12, RC 43, RC 14-1, RH 861, RH 8121, RH 8176, RH 55, RLC 1015, RLM 39, RS 78, RW 15-6, RW 22, RW 33-2, RW 75-123-2, EC 129126-1, EC 129126-2, JMMW 19, PR 8805, RH 8651, RH 8695, Shiva, WRR 98-01, NDRS 2004, NDRS 2013, NDRS 2007, PBR 181, EC 399300, EC 399301, EC 399299, EC 414299, IC 443623, IC 555891	India	Bhardwaj and Sud, 1989; 1993; Gupta and Singh, 1994; Jain *et al.*, 1998; Saharan, 1996 1997; Saharan *et al.*, 1988, 2005; Yadav and Singh, 1992; Yadav and Sharma, 2004; Sinah and Mall, 2007; AICRPRM, 2010
Brassica juncea	CBJ 001, CBJ 003, CBJ 004,	China	Li *et al.*, 2008
	JR 049	Australia	Li *et al.*, 2007
Brassica rapa var. Yellow Sarson	T 6, Prain, YST 6, Tobin, NDYS 2, YSK 8502	India	Kolte and Tewari, 1980; Lakra and Saharan, 1989a, b
Brassica rapa var. Brown Sarson	BSH 1, BS 15	India	Kolte and Tewari, 1980; Lakra and Saharan, 1989a, b; Saharan *et al.*, 1988, 2005

Continued

Table 8.2. Continued.

Crucifer host	Sources	Country	References
Brassica rapa var. Toria	IB 586, KTC, PT 303, PT 30	India	Kolte, 1985
Brassica carinata	HC 1, HC 2, HC 3, HC 5, HC 7, PC 3, DIR 1510, DIR, 1522, HC 9001, PBC 9221	India	Bhardwaj and Sud, 1993; Gupta and Singh, 1994; Jain *et al.*, 1998; Lakra and Saharan, 1989a, b; Saharan, 1996, 1997; Saharan *et al.*, 2005
Eruca sativa	RTM 314, RTM 1263, RTM 1471	India	Jain *et al.*, 1998
Brassica chinensis	All accessions	India	Gupta *et al.*, 1995; Lakra and Saharan, 1989a, b
Brassica alba	All accessions	India	Saharan *et al.*, 1988
Brassica spinescens	All accessions		
Brassica tenuifolia	All accessions		
Brassica incana	All accessions		

be inferred that the races with wider virulence have impacted the same genes in the differentials or genes for susceptibility are situated on different loci or are tightly linked.

Host resistance

The transfer of resistance from different sources in brassica crops is possible and is being done through conventional modern technologies all over the world (Saharan *et al.*, 2005).

Genetics of host–parasite interactions

Studies on the genetics of host–parasite interactions in white rust disease have concentrated on the level of specificity among races of pathogens and genotypes of related host species. A large number of genotypes has been reported, but very few have been utilized for developing cultivars with white rust resistance (Table 8.2). Even within the confines of race cultivar specificity, the studies have been lopsided in that no genetic information has been generated on *Albugo*, the causal organism. Interest in such studies was

stimulated by Hougas *et al.* (1952), who investigated the genetic control of resistance in white rust of horse radish. The exhaustive work of Pound and Williams (1963) clearly demonstrated that resistance to white rust was controlled by a single dominant gene in radish cvs China Rose Winter (CRW) and Round Black Spanish (RES). Histological studies revealed that the resistance in CRW was manifested as a hypersensitive reaction, which might be modified to a sporulating tolerant reaction by environmentally controlled minor genes. Humaydan and Williams (1976) while studying the inheritance of resistance in radish to *A. candida* race 1, changed the gene designation R into the more descriptive symbol AC-l derived from the initials and race number of *A. candida*. The resistance to *A. candida* race 1 in *R. sativus* cv. Caudatus was controlled by a single dominant gene, *AC-l*. The resistance gene *AC-1* and the gene *Pi*, controlling pink pigmentation, were found to be linked with a recombination value of 3.20%. Bonnet (1981) found that the white rust resistance in radish var. Rubisco-2 was also controlled by one dominant gene. Among *Brassica* species, monogenic dominant resistance to *A. candida* race 2 has been reported in *B. nigra*, *B. rapa*, *B. carinata* and *B. juncea* (Delwiche and Williams, 1974; Ebrahimi *et al.*, 1976; Thukral and Singh, 1986). A single dominant gene, *AC-2*, controlling resistance to *A. candida* race 2 in *B. nigra* was identified by Delwiche and Williams (1981). In a study to select quantitatively inherited resistance to *A. candida* race 2 in *B. rapa*, CGS-l, Edwards and Williams (1982) found that variability in reaction to *A. candida* race 2 among susceptible *B. rapa* strain PHW-Aaa-l was due to quantitative genetic regulation and suggested that rapid progress in resistance breeding could be made via mass selection when starting with a susceptible base population. Canadian cultivars of *B. napus* were resistant to white rust, but many cultivars of this species grown in China were susceptible (Fan *et al.*, 1983). The inheritance of white rust resistance in *B. napus* cv. Regent was conditioned by independent dominant genes at three loci, designated as *AC-7-1*, *AC-7-2* and *AC-7-3*. Resistance was conferred by dominance at any one of these loci, while plants with recessive alleles at all loci were susceptible. Verma and Bhowmik (1989) were in part agreement with Fan *et al.* (1983), who suggested that resistance of BN-Sel (*B. napus*) to the *B. juncea* pathotype of *A. candida* found in India was conditioned by dominant duplicate genes.

In a study of inheritance of resistant to *A. candida* race 2 in mustard, Tiwari *et al.* (1988) found that resistance was monogenic dominant, and could be easily transferred to adapted susceptible genotypes via backcrossing. In a study of the performance of 15 advanced generation (F_6) progenies of two interspecific crosses of *B. juncea* and *B. carinata* against white rust, Singh *et al.* (1988) showed significant differences among the progenies and that all the hybrid progenies gave a resistant reaction. A later study on five interspecific crosses between *B. juncea* and *B. carinata* revealed that the dominant gene which conferred resistance to white rust was located in the C genome of *B. oleracea*, a progenitor of *B. carinata* (Singh and Singh, 1988). Williams and Hill (1986) and Edwards and Williams (1987) have opened unusual potential for resolving many problems relating to host–parasite interactions and breeding for disease resistance through development of rapid cycling *Brassica* populations. Their studies demonstrated considerable isozyme variations among individuals in a population, which when inoculated with several pathogens showed a wide range of plant to plant variation in the levels of resistance and susceptibility. Gene pools of both major and minor genes for resistance to various crucifer pathogens have been constructed, which will be of immense value to plant breeders seeking sources of resistance (Williams and Hill, 1986; Edwards and Williams, 1987; Hill *et al.*, 1988).

Thukral and Singh (1986) studied the inheritance of white rust resistance in two crosses involving resistant (R) and susceptible (S) types of *B. juncea*, namely EC 12749 × Prakash and EC 12749 × Varuna under normal and late sown conditions and found that analysis of six generations revealed the importance of additive, dominant and epistatic effects. Reciprocal recurrent selection was also advocated for the exploitation of additive and non-additive gene effects for resistance to white rust. Singh and Singh (1987) showed that when *A. candida* resistant Ethiopian mustard (*B. carinata*) was crossed with *B. juncea*, the interspecific hybrids showed tolerance to *A. candida*. In a study regarding inheritance to *A. candida* race 7 in *B. napus*, Liu *et al.* (1987) found a digenic model with dominant resistance governed by R_1 and *R2* genes. Presence of a dominant allele at either of the two loci will confer resistance to the plant, whereas homozygous recessive at both loci will result in a susceptible phenotype expression. Liu and Rimmer (1992) studied the inheritance of resistance to an Ethiopian isolate of *A. candida* collected from *B. carinata* using two *B. napus* lines and suggested that resistance to the *B. carinata* isolate was conditioned by a single dominant resistant gene; results of this study were also discussed in the context of the relationship between genotypes of *Brassica* species and races of *A. candida*. Pal *et al.* (1991) evaluated the genetic component of variation for white rust resistance through 12 × 12 diallel crosses involving resistant and susceptible parents of Indian and exotic origin under four sets of environment, viz. normal sown in natural conditions, normal sown in artificially created epiphytotic conditions, late sown in natural conditions and late sown in artificially epiphytotic conditions. Based on these results, they suggested that both additive and non-additive components of variation were significant for white rust resistance in all four sets of environments, but with an over-dominance under late sown environment. Gadewadikar *et al.* (1993) in the study of inheritance of resistance to *A. candida* in crosses involving exotic and national promising varieties suggested that resistance was governed by a single dominant nuclear gene pair and as such resistance can easily be transferred via backcross breeding. Paladhi *et al.* (1993) while conducting the study on inheritance of field reaction to white rust in Indian mustard concluded that PI-15 has been identified as a resistance source in *B. juncea* and the resistance gene can easily be transferred to a susceptible type via backcrossing as resistance was controlled by a single gene.

Bains (1993) reported that resistance in the leaves differed from that of young flowers and this resistance in the leaves was due to the CC genome transferred from *Brassica* and may be used for breeding purposes. Rao and Raut (1994) observed the susceptibility of varuna (*B. juncea*) to the local Delhi pathotype of *A. candida* was conditioned by two genes, with dominant and recessive gene interaction. Interspecific crosses between *B. juncea* and *B. napus* suggested that the resistance in WW 1507 and ISN 114 to *A. candida* was controlled by a single dominant gene (Jat, 1999). In the study of three interspecific crosses made between *B. juncea* and *B. napus*, Subudhi and Raut (1994) revealed digenic control with epistatic interaction for the white rust resistance trait and a close association of parental species and different grades of leaf waxiness with white rust resistance. Sachan *et al.* (1995) hybridized two white-rust-resistant mustard (*B. juncea*) cvs Domo and Cutlass with two susceptible Indian cvs, Kranti and Varuna, in a diallel fashion. They reported that all F_1 hybrids, except susceptible × susceptible, were resistant. Segregation pattern for resistance to white rust in F_2 and test crosses indicated the control of a single dominant gene present in 'Domo' and 'Cutlass'. Liu *et al.* (1996) in Canada developed monogenic lines for resistance to *A. candida* from a Canadian *B. napus* cultivar and suggested that these monogenic lines could be used to study the mechanism of resistance response conditioned by the individual genes. These lines also facilitate molecular mapping of the loci in *B. napus* for resistance

to *A. candida* race 7. In breeding for genetic resistance to white rust in Indian mustard, Mani *et al.* (1996) used an inter-varietal cross of Pusa Bold (S) × DIRA 313 (R) having six generations and suggested that final intensity of rust on plant (FIP), final intensity of rust on leaf (FIL) and area under disease progress curve (AUDPC) showed significant additive × additive interaction along with the association of complimentary epistatic interactions indicating a close association between the nature of inheritance for AUDPC on the one hand and FIP and FIL on the other. This was also substantiated by significant correlation of FIP and FIL with AUDPC suggesting ease in selection for lower AUDPC (slow rusting) through FIP or FIL. Sridhar and Raut (1998) reported a monogenic inheritance showing complete dominance in four crosses and lack of dominance in seven crosses attempted between *B. juncea* and resistance sources derived from different species. According to Jat (1999), the resistance was dominant in all the crosses except susceptible × susceptible where it was recessive. In intra-specific crosses, inheritance of resistance to *A. candida* was governed by one dominant gene or two genes with either dominant–recessive epistatic interaction or complete dominance at both gene pairs but when either gene was dominant epistatic to other. Under controlled conditions, resistant genes identified by inoculating three different races of *A. candida* on F_2 population of crosses from R × R revealed that these genes may be located on the same locus or different loci. In different intra-specific crosses of *B. juncea* and interspecific crosses of *B. juncea* × *B. carinata* to *A. candida*, resistance was dominant in all the crosses. The resistance to *A. candida* was governed by one dominant gene or two genes with either as dominant, recessive or epistatic interaction or complete dominance at both gene pairs (Saharan and Krishnia, 2001). Partial resistance in *B. napus* to *A. candida* was controlled by a single recessive gene designated *wpr* with a variable expression (Bansal *et al.*, 2005). Dominant alleles at three unlinked loci (AC7h, AC7z and AC7$_3$) conferred resistance in *B. napus* cv. Regent to race AC7 of *A. candida* (Fan *et al.*, 1983; Liu *et al.*, 1996). Two loci also controlled resistance in *B. napus* to *A. candida* race AC 2 collected from *B. juncea* (Verma and Bhowmik, 1989). The Chinese *B. napus* accession 2282-9, which is susceptible to AC 7, has one locus controlling resistance to an isolate of *A. candida* collected from *B. carinata* (Liu and Rimmer, 1992). These studies indicated that only one allele for resistance was sufficient to condition an incompatible reaction in this pathosystem (Ferreira *et al.*, 1995). In addition, a single locus controlling resistance to AC 2 in *B. napus* and *B. rapa* was mapped using restriction fragment length polymorphism (RFLP) marker (Ferreira *et al.*, 1995). A dominant allele at a single locus or two tightly linked loci were reported to confer resistance to both races AC 2 and AC 7 of *A. candida* (Kole *et al.*, 2002). According to Borhan *et al.* (2008), a dominant white rust resistant gene, WRR 4, encodes a TIR-NB-LRR protein that confers broad-spectrum resistance in *Arabidopsis thaliana* to four races (AC 2, AC 4, AC 7 and AC 9) of *A. candida*.

Slow white rusting in crucifers

Rate of infection or disease spread is influenced by incubation and latent periods of *A. candida* in its compatible host. In white rust, the sporangia (inoculum) become visible after the host epidermis is ruptured as a white powdery mass which can readily be dispersed by wind to cause the secondary infection. According to Liu *et al.* (1989), white rust pustules become visible 5–6 days after inoculation. However, Coffey (1975) observed symptoms after 8 days on the undersurface of the cabbage leaves. Slow rusting attributes, higher incubation and latent periods have been observed in *B. juncea* cv. Rajat (11/14 days) and RC 781 (11/15 days). Similarly, *B. rapa* cv. Candle (11/115 days), Tobin (15/118 days) and Span (11/18 days) have longer incubation and latent periods (Lakra and Saharan, 1988b; Jat, 1999; Gupta and Saharan, 2002). There is a need to identify genotypes with slow rusting attributes to curb the epidemic development of white rust in field. Partial resistance in crucifer genotypes can be assessed through low infection frequency, low spore production and long latent period and a short infection period of *A. candida*.

Chemical Control

Efficacy of fungicides on germination of *Albugo candida* oospores *in vitro*

Albugo candida oospores occur commonly on *Brassica* seed samples (Petrie, 1975). Such inoculum levels on seeds may be considerably higher than actually required for initiation of infection considering that, upon germination, a single oospore releases 40–60 zoospores (Verma and Petrie, 1975). Germination of oospores following a period of washing in water, infection of *Brassica* cotyledons by zoospores from germinating oospores, and field experiments showing more foliar and staghead infection in oospore-treated plots than in the controls support the view that seed-borne oospores constitute primary inoculum for infection of *Brassica* species. Thus treatment, even by a protectant fungicide, could be important in controlling white rust infections by inhibiting the oospore germination or by killing the zoospores on emergence. An oospore germination technique was used to study the effectiveness of 27 fungicides in inhibiting germination at various stages (Verma and Petrie, 1979). Three mercurial fungicides, mersil, PMA-10 and panogen, were the best inhibitors of the oospore germination. The total inhibition with any of these fungicides at a concentration of 500 ppm active ingredient was about 75%. Among the non-mercurial compounds, mancozeb and ethazol were the most effective giving total inhibition of about 60%. The inhibition provided by bromosan and pyroxychlor was about 50%. Since none of the fungicides tested in this study was 100% effective, the search for a completely effective, preferably systemic, fungicide needs to be continued.

Efficacy of protectant fungicides in controlling both the foliar and staghead phase of white rust disease

Of the nine protectant fungicides tested in the growth chamber, application of either chlorothalonil or mancozeb, at 250 or 500 ppm, respectively, 6 h before inoculation and then 1 week later, controlled the disease effectively (Verma and Petrie, 1979). In view of their mainly protectant action, failure to control white rust by either fungicide applied 24 h and 7 days after inoculation was not surprising, as establishment of *A. candida* infection of rapeseed cotyledons and perhaps leaves would normally be completed within 24 h of inoculation (Verma *et al.*, 1975). Two foliar spraying of chlorothalonil (Bravo) in June under controlled conditions when the plants were 3–4 weeks old significantly reduced both foliar and staghead infections in the field (Verma and Petrie, 1979). However, in view of the growth-room studies on successful initiation of stagheads (Verma and Petrie, 1980), a third application at the time of flowering is also advised. Multiple applications, however, may not be economically feasible under commercial rapeseed production.

Efficacy of metalaxyl in controlling both the foliar and staghead phase of white rust disease

Among the systemic chemicals, metalaxyl is probably the best fungicide currently available to control white rust. Metalaxyl was active against *A. candida* race 7 in *B. rapa* cv. Torch (Stone *et al.*, 1987a, b). Treating the seed with metalaxyl at 5.0 g a.i./kg controlled foliar infection in the growth chamber up to the sixth leaf stage, 22 days after planting. When sprayed on the plants up to 4 days after inoculation, metalaxyl reduced foliar infection by 95%. Foliar infection was also controlled when applied as a soil drench but phytotoxicity was evident. Foliar spray application at 2.0 kg a.i./ha or higher reduced foliar infections in 3 years of field studies. Foliar applications also reduced staghead infections when applied at growth stages 3.2 or 4.1. Studies in growth chamber and field conditions (Stone *et al.*, 1987a) showed that metalaxyl possessed both protective and eradicative activity against *A. candida*. Control of disease in tissues remote from the site of application indicated that the fungicide moved systemically in rape plants. Disease control was obtained on the foliage, either by seed treatment or soil drenching, and disease eradication was successful when the fungicide was sprayed within 4 days of inoculation, a further evidence of systemicity (Stone

et al., 1987b). Seed treatment results have been promising, but in field situation these provided adequate protection only during early stages of plant growth. The decline in the activity of metalaxyl with increasing age of plants in seed treatment experiments may have been the result of fungicide dilution as the volume of plant tissue increased. Accordingly, infection of flower buds by wind-borne zoosporangia was not controlled by seed dressing. In the growth chamber, metalaxyl was active as a foliar eradicant for up to 4 days, but when applied 5 or 6 days after inoculation the fungicide did not prevent sporulation (Stone *et al.*, 1987a). It would appear, therefore, that after 4 days the fungus had reached a stage of development when fungicide treatment could not completely arrest growth, although pustule size and development were restricted with these late applications.

Results of our studies (Verma and Petrie, 1979, 1980; Stone *et al.*, 1987a, b) suggest that *A. candida* does not require early infections to develop systemically but can produce stagheads from infections of young flower buds by zoospores arising from wind-borne zoosporangia after plant growth stage 2.6. Successful disease control with metalaxyl, therefore, requires that a sufficient quantity of the fungicide be available well into the growing season. Seed dressings only provide protection for a limited period of time and if conditions favour disease development throughout the season, staghead development will not be controlled. By providing early disease control, however, seed treatment could reduce the secondary inoculum potential in the crop and thereby limit the initiation of stagheads from newly infected flower buds.

Bioassay and gas chromatographic analysis of plant tissue extract confirmed the presence of metalaxyl in tissue remote from the site of the treatment (Stone *et al.*, 1987b). Bioassay and chemical analysis of plants grown in metalaxyl-drenched soil showed that the fungicide was readily taken up by plants from the soil solution. The greatest accumulation was in the lower leaves, and metalaxyl was found in decreasing amount in leaves furthest from the roots and in only small concentrations in the stem and inflorescence. These results indicate that root absorption is an efficient way of metalaxyl application to a single leaf. It was not detected in the leaves below or above the treated leaf; thus, it is concluded that negligible symplastic translocation occurs.

Conclusions

Mycologists and taxonomists need to consider the division of *A. candida* complex into different species depending on host specificity. Information regarding production of oospores inside the seeds, and their possible importance in the survival of the pathogen is lacking. There is a need to investigate the role of simple or branched germ tube from germinating oospores. Single zoospore cultures from germinating sporangia and germinating oospores must be prepared for comparing pathogenicity. Studies on inheritance of virulence may be integrated with virulence spectrum. Genetics of *Albugo–Hyaloperonospora* association may be determined both at phenotypic and genotypic

levels. A consensus nomenclature of the *A. candida* races should be standardized, viz. AC jun I, 2- for *B. juncea* isolates, AC rap 1,2- for *B. rapa* isolates, AC nig I, 2- for *B. nigra* isolates, A Col 1,2, etc. for *B. oleracea* isolates, respectively. This, coupled with the uniform set of host differentials in each crucifer species, if possible, in the form of isogenic lines, can be especially useful for obtaining a true picture of *A. candida* races. Emphasis needs to be given to screen genotypes showing resistance to foliar infection for production of stagheads using flower-bud inoculation technique. Sources of resistance can be characterized on a broad spectrum effectiveness of a genotype against specific races. Genotypes exhibiting attributes of slow white rusting, disease tolerance and partial resistance may be very useful. Strong and weak genes for resistance in the host with their suitable combinations for durable resistance may be looked into. Mapping and cloning of genes for resistance and virulence at molecular level can be effectively used for marker assisted breeding.

References

AICRPRM (2010) All India Co-ordinated Research Project on Rapeseed-Mustard. Directorate of Rapeseed-Mustard Research, Bharatpur, India.

APG (Angiosperm Phylogeny Group) (2003) An update of the Angiosperm Phylogeny Group classification for the orders and families of flowering plants: APG II. *Botanical Journal of the Linnean Society* 141, 399–436.

Bains, S.S. (1993) Differential reaction of leaves and young flowers of different cruciferous crops to *A. candida*. *Plant Disease Research* 8, 70–72.

Bains, S.S. and Jhooty, J.S. (1979) Mixed infections by *Albugo candida* and *Peronospora parasitica* on *Brassica juncea* inflorescence and their control. *Indian Phytopathology* 32, 268–271.

Bains, S.S. and Jhooty, J.S. (1985) Association of *Peronospora parasitica* with *Albugo candida* on *B. juncea* leaves. *Phytopathology Z* 112, 28–31.

Bansal, V.K., Tewari, J.P., Stringam, G.R. and Thiagarajah, M.R. (2005) Histological and inheritance studies of partial resistance in the *Brassica napus–Albugo candida* host–pathogen interaction. *Plant Breeding* 124, 27–32.

Barbetti, M.J. (1981) Effects of sowing date and oospore seed contamination upon subsequent crop incidence of white rust (*Albugo candida*) in rapeseed. *Australian Journal of Plant Pathology* 10, 44–46.

Barbetti, M.J. and Carter, E.C. (1986) *Diseases of rapeseed. Rapeseed in Western Australia*. Western Australian Department of Agriculture, Bulletin No. 4105 (ed. Lawson, J.A.), pp. 14–19.

Berkeley, (1848) On the white rust of cabbages. *Journal of Horticulture Society of London* 3, 265–271.

Bhardwaj, C.L. and Sud, A.K. (1988) A study on the variability of *Albugo* candida from Himachal Pradesh. *Journal of Mycology and Plant Pathology* 18, 287–291.

Bhardwaj, C.L. and Sud, A.K. (1989) Reaction of *Brassica* cultivars against *Albugo candida* isolates from Kangra valley. *Indian Phytopathology* 42, 293 (Abstr.).

Bhardwaj, C.L. and Sud, A.K. (1993) Reaction of Brassica cultivars against *Albugo candida* isolates from Kangra valley. *Indian Phytopathology* 46, 258–260.

Biga, M.L.B. (1955) Review of the species of the genus *Albugo* based on the morphology of the conidia. *Sydowia* 9, 339–358 [in Italian].

Bonnet, A. (1981) Resistance to white rust in radish (*Raphanus sativus* L.). *Cruciferae NewsLetter* 6, 60.

Borhan, M.H., Gunn, N., Cooper, A., Gulden, S., Tor, M., Rimmer, S.R. and Holub, E.B. (2008) WRR4 encodes a TIR-NB-LRR protein that confers broad-spectrum white rust resistance in *Arabidopsis thaliana* to four physiological races of *Albugo candida*. *Molecular Plant-Microbe Interactions* 21, 757–768.

Bremer, H., Ismen, H., Karel, G., Ozkan, H. and Ozkan, M. (1947) Contribution to knowledge of the parasitic fungi of Turkey. *Review of the Faculty of Science, University of Istanbul, Series B* 13(2), 122–172.

Burdyukova, L.I. (1980) Albuginacea fungi. Taxonomy, morphology, biology and specialization. *Ukrainskyi Botanichnyi Zhumal* 37, 65–74 [in Russian].

Chattopadhyay, C., Agrawal, R., Kumar, A., Meena, R.L., Faujdar, K., Chakravarthy, N.V.K., Kumar, A., Goyal, P., Meena, P.D. and Shekhar, C. (2011) Epidemiology and development of forecasting models for White rust of *Brassica juncea* in India. *Archives Phytopathology and Plant Protection* 44(8), 751–763.

Choi, D. and Priest, M.J. (1995) A key to the genus *Albugo*. *Mycotaxon* 53, 261–272.

Choi, Y.J., Hong, S.B. and Shin, H.D. (2006) Genetic diversity within the *Albugo candida* complex (Peronosporales, Oomycota) inferred from phylogenetic analysis of ITS rDNA and COX2 mtDNA sequences. *Molecular Phylogenetics Evolution* 40, 400–409.

Choi, Y.J., Shin, H.D., Hong, S.B. and Thines, M. (2007) Morphological and molecular discrimination among *Albugo candida* materials infecting *Capsella bursa-pastoris* world-wide. *Fungal Diversity* 27, 11–34.

Choi, Y.J., Shin, H.D., Ploch, S. and Thines, M. (2008) Evidence for uncharted biodiversity in the *Albugo candida* complex, with the description of a new species. *Mycological Research* 112, 1327–1334.

Choi, Y.J., Shin, H.D., Hong, S.B. and Thines, M. (2009) The host range of *Albugo candida* extends from Brassicaceae through Cleomaceae to Capparaceae. *Mycological Progress* 8, 329–335.

Choi, Y.J., Park, M.J., Park, J.H. and Shin, H.D. (2011a) White blister rust caused by *Albugo candida* on oilseed rape in Korea. *Plant Pathology Journal* 27(2), 192.

Choi, Y.J., Shin, H.D., Ploch, S. and Thines, M. (2011b) Three new phylogenetic lineages are the closest relatives of the widespread species *Albugo candida*. *Fungal Biology* 115(7), 598–607.

Chowdhary, S. (1944) Some fungi from Assam. *Indian Journal of Agriculture Science* 14, 230–233.

Coffey, M.D. (1975) Ultrastructural features of the haustorial apparatus of the white blister fungus *Albugo candida*. *Canadian Journal of Botany* 53, 1285–1299.

Colmeiro, M. (1867) Enumeración de las criptógamas de España y Portugal, II. *Revista de los Progresos de las Ciencias Fis. Nat.* 17–18, 63–164.

De Bary, A. (1863) Recherches sur le development quelques champignons parasites. *Annales des Sciences Naturelles (Botanique) Tome 20, 4th series*, 5–148.

De Bary, A. (1866) *Morphologie und physiologic der Pilze, Flechten und Myxomyceten*. Welhelm Engelmann, Leipzig, pp. 426–439.

de Candolle, A.P. (1806) *Flore Française*. Lyon, France.

de Roussel, H.F.A. (1806) *Flore du Calvados et des terreins adjacens*. IIe Edition, Caen, France.

Delwiche, P.A. and Williams, P.H. (1974) Resistance to *Albugo candida* race 2 in *Brassica* sp. *Proceedings of the American Phytopathological Society* 1, 66 (Abstr.).

Delwiche, P.A. and Williams, P.H. (1977) Genetic studies in *Brassica nigra* (L.) Koch. *Cruciferae NewsLetter* 2, 39.

Delwiche, P.A. and Williams, P.H. (1981) Thirteen marker genes in *Brassica nigra*. *Journal of Heredity* 72, 289–290.

Eberhardt, A. (1904) Contribution a!etude de *Cystopus candidus* Lev. *Zentr. Bakteriol. Parasitenk* 12, 235–249.

Ebrahimi, A.G., Delwiche, P.A. and Williams, P.H. (1976) Resistance in *Brassica juncea* to *Pemnospora parasítica* and *Albugo candida* race 2. *Proceedings of the American Phytopathological Society* 3, 273.

Edwards, M. and Williams, P.H. (1982) Selection for quantitatively inherited resistance to *Albugo candida* race 2 in *B. canpestris*, CGS-1. *Cruciferae NewsLetter* 7, 66–67.

Edwards, M.D. and Williams, P.H. (1987) Selection of minor gene resistance to *A. candida* in rapid-cycling population of *Brassica campestris*. *Phytopathology* 77, 527–532.

Endo, R.M. and Linn, M.B. (1960) The white rust disease of horseradish. *Illinois Agriculture Experiment Station Bulletin* 655, 56.

Fan, Z., Rimmer, S.R. and Stefansson, B.R. (1983) Inheritance of resistance to *Albugo candida* in rape (*Brassica napus* L.). *Canadian Journal of Genetics and Cytology* 25, 420–424.

Farr, D.F. and Rossman, A.Y. (2010) Fungal Databases, Systematic Mycology and Microbiology Laboratory, ARS, USDA. Available at: http://nt.arsgrin.gov/fungaldatabases (accessed 18 May 2010).

Farr, D.F., Bills, G.F., Chamuris, G.P. and Rossman, A.Y. (1989) *Ipomoea*. In: Farr, D.F., Billis, F.G., Chamuris, G.P. and Rossman, A.Y. (eds) *Fungi on Plants and Plant Products in the United States*. APS Press, St Paul, Minnesota, pp. 142–143.

Farr, D.F., Rossman, A.Y., Palm, M.E. and McCray, E.B. (2004) Online Fungal Databases, Systematic Botany & Mycology Laboratory, ARS, USDA. Available at: http://nt.ars-grin.gov/fungaldatabases (accessed 14 March 2014).

Ferreira, M.E., Williams, P.H. and Osborn, T.C. (1995) Mapping of locus controlling resistance to *Albugo candida* in *Brassica napus* using molecular markers. *Phytopathology* 85, 218–220.

Gadewadikar, P.N., Bhadouria, S.S. and Bartaria, A.M. (1993) Inheritance of resistance to white rust (*Albugo candida*) disease in Indian mustard (*Brassica juncea*). In: *National Seminar, Oilseeds Research and Development in India: Status and Strategies,* 2–4 August 1993, ISOR, Hyderabad, India.

Goyal, B.K., Kant, U. and Verma, P.R. (1995) Growth of *Albugo candida* (race unidentified) on *Brassica juncea* callus cultures. *Plant and Soil* 172, 331–337.

Goyal, B.K., Verma, P.R. and Spurr, D.T. (1996a) Temperature effects on oospore development of *Albugo candida* race 2V in detached *Brassica juncea* leaves. *Indian Journal of Mycology and Plant Pathology* 26, 224–228.

Goyal, B.K., Verma, P.R., Spurr, D.T. and Reddy, M.S. (1996b) *Albugo candida* staghead formation in *Brassica juncea* in relation to plant age, inoculation sites and incubation conditions. *Plant Pathology* 45, 787–794.

Goyal, B.K., Verma, P.R., Swartz, G. and Spurr, D.T. (1996c) Growth of *Albugo candida* in leaf callus cultures of *Brassica rapa*. *Canadian Journal of Plant Pathology* 18, 225–232.

Gray, S.F. (1821) *A natural arrangement of British plants: according to their relations to each other as pointed out by Jussieu, De Candolle, Brown: with an introduction to botany*. Baldwin, Cradock and Joy, London,72 pp.

Greelman, D.W. (1963) New and noteworthy diseases. *Canadian Plant Disease Survey* 43, 61–63.

Gupta, K. and Saharan, G.S. (2002) Identification of pathotypes of *Albugo candida* with stable characteristic symptoms on Indian mustard. *Journal of Mycology and Plant Pathology* 32, 46–51.

Gupta, R.B.L. and Singh, M. (1994) Source of resistance to white rust and powdery mildew of mustard. *International Journal of Tropical Plant Diseases* 12, 225–227.

Gupta, S., Sharma, T.R. and Chib, H.S. (1995) Evaluation of wild allies of *Brassica* under natural conditions. *Cruciferae Newsletter* 17, 10–11.

Hammett, K.P.W. (1969) White rust diseases. *New Zealand Gardner* 26, 43.

Harper, F.R. and Pittman, U.J. (1974) Yield loss by *Brassica campestris* and *Brassica napus* from systemic stem infection by *Albugo curciferarum*. *Phytopathology* 64, 408–410.

Hill, C., Crute, I., Sherriff, C. and Williams, P.H. (1988) Specificity of *Albugo candida* and *Peronospora parasitica* pathotypes toward rapid-cycling crucifers. *Cruciferae Newsletter* 13, 112–113.

Hirata, S. (1954) Studies on the phytohormone in the malformed portion of the diseased plants. I. The relation between the growth rate and the amount of free auxin in the fungus galls and virus-infected plants. *Annals of Phytopathological Society Japan* 19, 33–38.

Hiura, M. (1930) Biologic forms of *Albugo candida* (Pers.) Kuntze on some cruciferous plants. *Japan Journal of Botany* 5, 1–20.

Ho, B.W.C. and Edie, H.H. (1969) White rust (*Albugo ipomoeae-aquaticae*) of *Ipomoea aquatic* in Hong Kong. *Plant Disease Reporter* 53, 959–962.

Hougas, R.W., Rieman, G.H. and Stokes, G.W. (1952) Resistance to white rust in horseradish seedlings. *Phytopathology* 42, 109–110.

Humaydan, H.S. and Williams, P.H. (1976) Inheritance of seven characters in *Raphanus sativus*. *HortScience* 11, 146–147.

Ito, S. and Tokunaga, Y. (1935) Notae mycologicae *Asiae orientalis*. I. *Transactions of the Sapporo Natural History Society* 14, 11–33.

Jain, K.L., Gupta, A.K. and Trivedi, A. (1998) Reaction of rapeseed-mustard lines against white rust pathogen *Albugo candida*. *Journal of Mycology Plant Pathology* 28, 72–73.

Jat, R.R. (1999) Pathogenic variability and inheritance of resistance to *Albugo candida* in oilseed *Brassica*. PhD Thesis, CCSHAU, Hisar, 129 pp.

Kadow, K.J. and Anderson, H.W. (1940) A study of horseradish diseases and their control. *University of Illinois Agriculture Experimental Station Bulletin* 469, 531–543.

Kajomchaiyakul, P. and Brown, J.F. (1976) The infection process and factors affecting infection of sunflower by *Albugo tragopogi*. *Transactions of the British Mycological Society* 66, 91–95.

Kaur, P., Sivasithamparam, K. and Barbetti, M.J. (2008) Pathogenic behaviour of strains of *Albugo candida* from *Brassica juncea* (Indian mustard) and *Raphanus raphanistrum* (wild radish) in Western Australia. *Australian Journal of Plant Pathology* 37, 353–356.

Khunti, J.P., Khandar, R.R. and Bhoraniya, M.F. (2000) Studies on host range of *Albugo cuciferarum* the incitant of white rust of mustard. *Agriculture Science Digest* 20, 219–221.

Klemm, M. (1938) The most important diseases and pests of Colza and Rape. *Dtsch. Landw. Pr.* IXV, 19: 239; 20: 251–252 (German).

Kole, C., Williams, P.H., Rimmer, S.R. and Osborn, T.C. (2002) Linkage mapping of genes controlling resistance to white rust (*Albugo candida*) in *Brassica rapa* (syn. *campestris*) and comparative mapping to *Brassica napus* and *Arabidopsis thaliana. Genome* 45(1), 22–27.

Kolte, S.J. (1985) White rust. In: *Diseases of Annual edible oilseed crops*, Vol. II. *Rapeseed-mustard and sesame diseases*. CRC Press, Boca Raton, Florida, pp. 27–35.

Kolte, S.J. and Tewari, A.N. (1980) Note on the susceptibility of certain *oleiferous Brassicae* to downy mildew and white blister diseases. *Indian Journal of Mycology and Plant Pathology* 10, 191–192.

Kolte, S.J., Bordoloi, D.K. and Awasthi, R.P. (1991) The search for resistance to major diseases of rapeseed and mustard in India. *GCIRC 8th International Rapeseed Congress*, Saskatchewan, Canada, pp. 219–225.

Kuntze, O. (1891) *Revisio Generum Plantarum*, Vol. 2. Leipzig, Germany.

Lakra, B.S. and Saharan, G.S. (1988a) Efficacy of fungicides in controlling white rust of mustard through foliar sprays. *Indian Journal of Mycology and Plant Pathology* 18, 157–163.

Lakra, B.S. and Saharan, G.S. (1988b) Influence of host resistance on colonization and incubation period of *Albugo candida* in mustard. *Cruciferae NewsLetter* 13, 108–109.

Lakra, B.S. and Saharan, G.S. (1988c) Morphological and pathological variations in *Albugo candida* associated with *Brassica* species. *Indian Journal of Mycology and Plant Pathology* 18, 149–156.

Lakra, B.S. and Saharan, G.S. (1988d) Progression and management of white rust of mustard in relation to planting time, host resistance and fungicidal spray. *Indian Journal of Mycology and Plant Pathology* 18, 112 (Abstr.).

Lakra, B.S. and Saharan, G.S. (1989a) Correlation of leaf and staghead infection intensities of white rust with yield and yield components of mustard. *Indian Journal of Mycology and Plant Pathology* 19, 279–281.

Lakra, B.S. and Saharan, G.S. (1989b) Location and estimation of oospores of *Albugo candida* in infected plant parts of mustard. *Indian Phytopathology* 42, 467.

Lakra, B.S., Saharan, G.S. and Verma, P.R. (1989) Effect of temperature, relative humidity and light on germination of *Albugo candida* sporangia from mustard. *Indian Journal of Mycology and Plant Pathology* 19, 264–267.

Leveilla, J.H. (1847) On the methodical arrangement of the Uredineae. *Annales Sciences Naturelles Series* 3, 8, 371.

Li, C.X., Sivasithamparam, K., Walton, G.W., Salisbury, P., Burton, W., Banga, S.S., Banga, S., Chattopadhyay, C., Kumar, A., Singh, R., Singh, D., Agnihotri, A., Liu, S., Li, Y., Wang, T.F.Y. and Barbetti, M.J. (2007) Identification of resistance to *Albugo candida* in Indian, Australian and Chinese *Brassica juncea* genotypes. In: *Proceedings 10th International Rapeseed Congress,* Wuhan, China, 1 March 2007, 408–410.

Li, C.X., Sivasithamparam, K., Walton, G., Fels, P. and Barbetti, M.J. (2008) Both incidence and severity of white rust disease reflect host resistance in *Brassica juncea* germplasm from Australia, China and India. *Field Crops Research* 106, 1–8.

Liu, J.Q., Parks, P. and Rimmer, S.R. (1996) Development of monogenic lines for resistance to *Albugo candida* from a Canadian *Brassica napus* cultivar. *Phytopathology* 86, 1000–1004.

Liu, Q. and Rimmer, S.R. (1992) Inheritance of resistance in *Brassica napus* to an Ethiopian isolate of *Albugo candida* from *Brassica carinata. Canadian Journal of Plant Pathology* 14, 116–120.

Liu, Q., Rimmer, S.R., Scarth, R. and McVetty, P.B.E. (1987) Confirmation of a digenic model of inheritance of resistance to *Albugo candida* race 7 in *Brassica napus. Proceedings of the 7th International Rapeseed Congress,* Poznam, Poland, pp. 1204–1209.

Liu, Q., Rimmer, S.R. and Scarth, R. (1989) Histopathology of compatibility and incompatibility between oilseed rape and *Albugo candida. Plant Pathology* 38, 176–182.

Mani, N., Gulati, S.C., Raman, R. and Raman, R. (1996) Breeding for genetic resistance to white rust in Indian mustard. *Crop Improvement* 23(1), 75–79.

Mathur, S., Wu, C. and Rimmer, S.R. (1995) Virulence of isolates of *Albugo candida* from western Canada to *Brassica* species. *Proceedings of the 9th International Rapeseed Congress,* Cambridge, UK, 2, 652–654.

Mehta, N., Saharan, G.S. and Babber, S. (1995) Sequence of events in the pathogenesis of *Pernospora* and *Albugo* on mustard. *Journal Indian Botanical Society* 74, 299–303.

Melhus, I.E. (1911) Experiments on spore germination and infection in certain species of Oomycetes. *Wisconsin Agriculture Experimental Station Research Bulletin* 15, 25–91.

Middleton, K.J. (1971) Sunflower diseases in South Queensland. *Queensland Agriculture Journal* 97, 597–600.

Mukerji, K.G. (1975) *Albugo candida. IMI Descriptions of Fungi and Bacteria* 46, 460.

Murashige, T. and Skoog, F. (1962) A revised medium for rapid growth and bioassays with tobacco tissue cultures. *Physiology of Plant* 15, 473–497.

Napper, M.E. (1933) Observations on spore germination and specialization of parasitism in *Cystopus candidus. Journal of Pomology and Horticulture Science* 11, 81–100.

Novotel'Nova, N.S. (1962) White rust on sunflower. *Zashch. Rast. Moskva* 7, 57 [in Russian].

Novotel'Nova, N.S. (1968) Intraspecific taxa of *Cystopus candidus* (Pers.) Lev. *Novosti Sistematiki Nizsh. Rastenii,* 88–96 [in Russian].

Novotel'Nova, N.S. and Minasyan, M.A. (1970) Contribution to the biology of *Cystopus candidus* (Pers.) Lev. and *Cystopus tragopogonis* (Pers.) Schroet. *Trudy vsesoyuznogo Nauchnoissledovaterskogo Instituta Zashchity rastenii* 29, 121–128 [in Russian].

Pal, Y., Singh, H. and Singh, D. (1991) Genetic components of variation for white rust resistance in Indian x Exotic crosses of Indian mustard. *Crop Research* 4, 280–283.

Paladhi, M.M., Prasad, R.C., Dass, B. and Dass, B. (1993) Inheritance of field reaction to white rust in Indian mustard. *Indian Journal of Genetics and Plant Breeding* 53(3), 327–328.

Pape, H. and Rabbas, P. (1920) Inoculation tests with *Cystopus candidus* Pers. *Mitt. Biol. R.-Anst. L. and U. Forstw. U.* 18, 58–59 [in German].

Parham, B.E. (1942) White rust of cruciferae (*Albugo candida*). *Agriculture Journal Fiji* 13, 27–28.

Persoon, C.H. (1801) *Synopsis methodica fungorum,* Part I, II. Gottingen,706 pp.

Perwaiz, M.S., Moghal, S.M. and Kamal, M. (1969) Studies on the chemical control of white rust and downy mildew of rape (*Sarsoon*). *West Pakistan Journal of Agriculture Research* 7, 71–75.

Petrie, G.A. (1973) Diseases of *Brassica* species in Saskatchewan, 1970-72. I. Staghead and aster yellows. *Canadian Plant Disease Survey* 53, 19–25.

Petrie, G.A. (1975) Prevalence of oospores of *Albugo cruciferarum* in *Brassica* seed samples from western Canada, 1967-73. *Canadian Plant Disease Survey* 55, 19–24.

Petrie, G.A. (1986) *A. candida* on *Raphanus sativus* in Saskatchewan. *Canadian Plant Disease Survey* 66, 43–47.

Petrie, G.A. (1988) Races of *Albugo candida* (white rust and staghead) on cultivated cruciferae in Saskatchewan. *Canadian Journal of Plant Pathology* 10, 142–150.

Petrie, G.A. (1994) New races of *Albugo candida* (white rust) in Saskatchewan and Alberta. *Canadian Journal of Plant Pathology* 16, 251–252.

Petrie, G.A. and Vanterpool, T.C. (1974) Fungi associated with hypertrophies caused by infection of Cruciferae by *Albugo cruciferatum. Canadian Plant Disease Survey* 54, 37–42.

Petrie, G.A. and Verma, P.R. (1974) A simple method for germinating oospores of *Albugo candida. Canadian Journal of Plant Science* 54, 595–596.

Pidskalny, R.S. and Rimmer, S.R. (1985) Virulence of *Albugo candida* from turnip rape (*Brassica campestris*) and mustard (*Brassica juncea*) on various crucifers. *Canadian Journal Plant Pathology* 7, 283–286.

Pound, G.S. and Williams, P.H. (1963) Biological races of *Albugo candida. Phytopathology* 53, 1146–1149.

Rao, M.V.B. and Raut, R.N. (1994) Inheritance of resistance to white rust (*Albugo candida*) in an interspecific cross between Indian mustard (*Brassica juncea*) and rapeseed (*Brassica napus*). *Indian Journal of Agricultural Science* 64(4), 249–251.

Rayss, T. (1938) Nouvelle contribution altetude de la mycofbre de Palestine. *Palestian Journal of Botany* 1, 143–160.

Rimmer, S.R., Mathur, S. and Wu, C.R. (2000) Virulence of isolates of *Albugo candida* from western Canada to *Brassica* species. *Canadian Journal Plant Pathology* 22, 229–235.

Sachan, J.N., Kolte, S.J. and Singh, B. (1995) Genetics of resistance to white rust (*Albugo candida* race-2) in mustard (*Brassica juncea*). In: GCIRC 9th International Rapeseed Congress, Cambridge, UK, pp. 1295–1297.

Sackston, W.E. (1957) Diseases of sunflowers in Uruguay. *Plant Disease Reporter* 41, 885–889.

Safeefulla, K.M. and Thirumalachar, M.J. (1953) Morphological and cytological studies in *Albugo species* on *Ipomoea aquatica* and *Merrimia emarginata. Cellule* 55, 225–231.

Saharan, G.S. (1996) Studies on physiologic specialization, host resistance and epidemiology of white rust and downy mildew disease complex in rapeseed-mustard. Final report of adhoc research project ICAR, Dept of Plant Pathology, CCSHAU, Hisar, India, 83.

Saharan, G.S. (1997) Disease resistance. In: Kalia, H.R. and Gupta, S.C. (eds) *Recent Advances in Oilseed Brassicas.* Kalyani Pub, Ludhiana, pp. 233–259.

Saharan, G.S. (2010) Analysis of genetic diversity in *Albugo*-Crucifer system. *Journal of Mycology Plant Pathology* 40(1), 1–13.

Saharan, G.S. and Krishnia, S.K. (2001) Multiple disease resistance in rapeseed and mustard. In: Nagarajan, S. and Singh, D.P. (eds) *Role of Resistance in Intensive Agriculture*. Kalyani Publishers, New Delhi, pp. 98–108.

Saharan, G.S. and Mehta, N. (2002) Fungal diseases of rapeseed-mustard. In: Gupta, V.K. and Paul, Y.S. (eds) *Diseases of Field Crops*. Indus Publishing, New Delhi, pp. 193–228.

Saharan, G.S. and Verma, P.R. (1992) *White Rust. A review of economically important species*. International Development Research Centre, Ottawa, Ontario, IDRC-MR315e.

Saharan, G.S., Kaushik, C.D. and Kaushik, J.C. (1988) Sources of resistance and epidemiology of white rust of mustard. *Indian Phytopathology* 41, 96–99.

Saharan, G.S., Mehta, N. and Sagwan, M.S. (2005) Development of disease resistance in rapeseed-mustard. In: Saharan, G.S., Mehta, N. and Sagwan, MS. (eds) *Diseases of Oilseed Crops*. Indus Publishing Co., New Delhi, pp. 561–617.

Sarasola, A.A. (1942) *Sunflower Disease*. Publication Directorate Agrio., Buenos Aires.

Savulescu, O. (1946) Study of the species of *Cystopus* (Pers.) Lev. Bucharest, 1946. *Anal. Acad. Rous. Mem. Sect. Stimtiface. Soc.* 21, 13.

Savulescu, T. and Rayss, T. (1930) Contribution to the knowledge of the Peronsoporaceae of Romania. *Annals of Mycology* 28, 297–320.

Sinah, P.K. and Mall, A.K. (2007) Screening of rapeseed and mustard genotypes against white rust (*Albugo cruciferarum* SF, Gray). *Vegetos* 20, 59–62.

Singh, B.M. and Bhardwaj, C.L. (1984) Physiologic races of *Albugo candida* on crucifers in Himachal Pradesh. *Indian Journal of Mycology and Plant Pathology* 14, 25 (Abstr.).

Singh, D. and Singh, H. (1987) Genetic analysis of resistance to white rust in Indian mustard. *Proceedings of the 7th International Rapeseed Congress,* Poznan, Poland, 11–14 May 1987, 126.

Singh, D. and Singh, H. (1988) Inheritance of white rust resistance in interspecific crosses of *Brassica juncea L.* x *Brassica carinata L. Crop Research* 1, 189–193.

Singh, D., Singh, H. and Yadava, T.P. (1988) Performance of white rust *(A. candida)* resistance genotypes developed from interspecific crosses of *B. juncea* L. x *B. carinata* L. *Cruciferae NewsLetter* 13, 110.

Sridhar, K. and Raut, R.N. (1998) Differential expression of white rust resistance in Indian mustard (*Brassica juncea*). *Indian Journal of Genetics and Plant Breeding* 58(3), 319–322.

Stone, J.R., Verma, P.R., Dueck, J. and Westcott, N.D. (1987a) Bioactivity of the fungicide metalaxyl in rape plants after seed treatment and soil drench applications. *Canadian Journal of Plant Pathology* 9, 260–264.

Stone, J.R., Verma, P.R., Dueck, J. and Spurr, D.T. (1987b) Control of *Albugo candida* race 7 in *B. campestris* cv. Torch by foliar, seed and soil applications of metalaxyl. *Canadian Journal of Plant Pathology* 9, 137–145.

Stovold, G.E. and Moore, K.J. (1972) Diseases. *Agricultural Gazette of New South Wales* 83, 262–264.

Stringam, G R. (1971) Genetics of four hypocotyl mutants in *Brassica campestris* L. *Heredity* 62, 248–250.

Subudhi, P.K. and Raut, R.N. (1994) White rust resistance and its association with parental species type and leaf waxiness in *Brassica juncea* x *Brassica napus* crosses under the action of EDTA and gamma-ray. *Euphytica* 74(1/2), 1–7.

Thukral, S.K. and Singh, H. (1986) Inheritance of white rust resistance in *Brassica juncea. Plant Breeding* 97, 75–77.

Tiwari, A.S., Petrie, G.A. and Downey, R.K. (1988) Inheritance of resistance to *Albugo candida* race 2 in mustard [*Brassica juncea* (L.) Czern]. *Canadian Journal of Plant Science* 68, 297–300.

Togashi, K. and Shibasaki, Y. (1934) Biometrical and biological studies of *Albugo candida* (Pers.) O. Kuntze in connection with its specialization. *Bulletin of the Imperial College of Agriculture and Forestry* (Morioka, Japan) 18, 88.

USDA (2014) Fungal database. Available at: http://nt.ars-grin.gov/fungaldatabases (accessed 14 March 2014).

Vanterpool, T.C. (1959) Oospore germination in *Albugo candida. Canadian Journal of Botany* 37, 169–172.

Verma, P.R. (1989) Report for the post-doctoral transfer of work on white rust (*Albugo candida*) conducted during Nov 1989 – March 1990 at the Jawaharlal Nehru Krishi Vishwa Vidyalaya, Regional Agricultural Research Station, Morena, M.P., India. Unofficial Report: Project No. 88-1004, IDRC, Ottawa, Canada.

Verma, P.R. and Petrie, G.A. (1975) Germination of oospores of *Albugo candida. Canadian Journal of Botany* 53, 836–842.

Verma, P.R. and Petrie, G.A. (1978) A detached-leaf culture technique for the study of white rust disease of *Brassica* species. *Canadian Journal of Plant Science* 58, 69–73.

Verma, P.R. and Petrie, G.A. (1979) Effect of fungicides on germination of *Albugo candida* oospores *in vitro* and the foliar phase of the white rust disease. *Canadian Plant Disease Survey* 59, 53–59.

Verma, P.R. and Petrie, G.A. (1980) Effect of seed infestation and flower bud inoculation on systemic infection of turnip rape by *Albugo candida*. *Canadian Journal Plant Science* 60, 267–271.

Verma, P.R., Harding, H., Petrie, G.A. and Williams, P.H. (1975) Infection and temporal development of mycelium of *Albugo candida* in cotyledons of four *Brassica spp. Canadian Journal of Botany* 53, 1016–1020.

Verma, P.R., Spurr, D.T. and Petrie, G.A. (1983) Influence of age and time of detachment on development of white rust in detached *Brassica campestris* leaves at different temperatures. *Canadian Journal of Plant Pathology* 5, 154–157.

Verma, P.R., Saharan, G.S., Bartaria, A.M. and Shivpuri, A. (1999) Biological races of *Albugo candida* on *Brassica juncea* and *Brassica rapa* var. Toria in India. *Journal of Mycology and Plant Pathology* 29, 75–82.

Verma, U. and Bhowmik, T.P. (1989) Inheritance of resistance to a *Brassica juncea* pathotype of *Albugo candida* in *Brassica napus. Canadian Journal of Plant Pathology* 11, 443–444.

Viegas, A.P. and Teixeira, A.R. (1943) Alguns fungos do Brasil. *Bragantia, Sao Paulo* 3, 223–269.

Voglmayr, H. and Riethmüller, A. (2006) Phylogenetic relationships of *Albugo* species (white blister rusts) based on LSU rDNA sequence and oospore data. *Mycological Research* 110, 75–85.

Wakefield, E.M. (1927) The genus *Cystopus* in South Africa. *Transactions of the British Mycological Society* 2, 242–246.

Walker, J.C. (1957) *Plant Pathology*. McGraw-Hill. New York, pp. 214–219.

Wiant, J.S. (1937) White rust on Texas spinach. *Plant Disease Reporter* 21, 114–115.

Williams, P.H. (1985) White rust [*Albugo candida* (Pers. ex. Hook.) Kuntze]. In: *Crucifer Genetics Cooperative (CRGC) Resource Book*. University of Wisconsin, USA, pp. 1–7.

Williams, P.H. and Hill, C.B. (1986) Rapid cycling populations of *Brassica. Science* 232, 1385–1389.

Williams, P.H. and Pound, G.S. (1963) Nature and inheritance of resistance to *Albugo candida* in radish. *Phytopathology* 53, 1150–1154.

Wilson, G.W. (1907) Studies in North American Peronasporales. I. The genus *Albugo. Bulletin of the Torrey Botanical Club* 34, 61–84.

Wu, C.R., Mathur, S. and Rimmer, S.R. (1995) Differentiation of races and isolates of *Albugo candida* by random amplification of polymorphic DNA. In: *Proceedings of the 9th International Rapeseed Congress*, GCIRC, Cambridge, UK, pp. 655–657.

Yadav, R. and Sharma, P. (2004) Genetic diversity for white rust (*Albugo candida*) resistance in rapeseed-mustard. *Indian Journal of Agricultural Science* 74, 281–283.

Yadav, Y.P. and Singh, H. (1992) The potential exotic sources for white rust resistance in Indian mustard [*Brassica juncea* (L) Czern & Coss]. *Oil Crops NewsLetter* 9, 18.

Zhang, Z.Y., Wang, Y.X. and Liu, Y.L. (1984) Taxonomic studies of the family Albuginaceae of China. II. A new species of *Albugo* on Acanthaceae and known species of *Albugo* on cruciferae. *Acta Mycologica Sinica* 3(2), 65–71.

9 Pathogenesis of *Alternaria* Species: Physiological, Biochemical and Molecular Characterization

P.D. Meena,[1]* Gohar Taj[2] and C. Chattopadhyay[3]

[1]*ICAR-Directorate of Rapeseed-Mustard Research, Bharatpur;* [2]*Molecular Biology & Genetic Engineering, G.B. Pant University of Agriculture & Technology, Pantnagar;* [3]*ICAR-National Centre on Integrated Pest Management, Pusa Campus, New Delhi, India*

Introduction

Alternaria blight or black leaf, and silique spot is an exceptionally serious disease of oilseed brassica crops worldwide. It is mainly induced by *Alternaria brassicae* (Berk) Sacc., *A. brassicicola* (Schwein) Wiltshire, and is a necrotrophic pathogen that can infect every plant part in every plant growth stage. The symptoms emerge on all aerial parts of the plant, generally resulting in serious damages to yield and quality of the seed. The disease starts as minute dark-brown to light black pustules on the older leaves that spread rapidly on to the above foliar parts of the plant by producing typical centred bands and a yellow circle of discoloration in and surrounding the lesions.

In the case of serious outbreak, the disease on the leaves and stems causes the entire plant to blight prior to silique development or before any seeds are formed. Globally, estimates of economic damages caused by this disease range from 35 to 70% in diverse oilseed brassicas. The percentage of oil yield losses occurred by infected seeds were analysed between the ranges of 15 and 36% by Ansari *et al.* (1988).

A hypersensitive response of plants to counter the pathogen invasion is indicated by the existence of brown-coloured dead cells at the location of infection that limit the development of the pathogen. The resistance reaction involves the formation of conscious oxygen (Lamb and Dixon, 1997), alteration of ion fluxes (Levine *et al.*, 1996) and signalling molecules comprising jasmonic as well as salicylic acid, and protein kinases (Dixon *et al.*, 1994; Dangl *et al.*, 1996) accompanied by plant resistance genes and transcription factors, build-up of pathogenesis related (PR) proteins, degradation of proteins with the polyubiquitin system and planned cell death. Although pathogenic diversity among the rapeseed-mustard cultivars exists in India, very little effort has been made to increase the available degree of tolerance in cultivars due to the non-availability of a reliable basis of transferable resistance. None of the cultivated rapeseed-mustards possess an absolute resistance to the disease. Although some wild crucifers, including *Sinapis alba, Camelina sativa, Eruca sativa, Capsella bursa-pastoris* and *Neslia paniculata*, show resistance (Tewari and Conn, 1993),

---

*Corresponding author, e-mail: pdmeena@gmail.com

© CAB International 2015. *Brassica Oilseeds: Breeding and Management* (eds A. Kumar *et al.*)

manipulation of resistant genes to rapeseed-mustard crops has proven difficult due to sexual incompatibility. Transfer of Alternaria resistance by ovary culture and protoplast fusion from *S. alba* to rapeseed germplasm has been initiated (Chevre *et al.*, 1991), but tangible results are still awaited.

Host Range

Alternaria brassicae and *A. brassicicola* occur globally and infect numerous brassicaceous crops including *Brassica carinata*, black mustard, broccoli, Brussels sprouts, cabbage, candy ruft, cauliflower, Chinese cabbage, crambe, garden cress, hedge mustard, horse radish, kale, kohlrabi, radish, rutabaga, rapeseed-mustard, stinkweed, swedes, *E. sativa*, turnip, wallflower and *S. alba* (Verma and Saharan, 1994). Other species including *Alternaria raphani* and *Alternaria alternata* have also been reported worldwide infecting oilseed brassicas. Furthermore, the information regarding *A. brassicae* or *A. brassicicola* are unreliable due to lack of precise pathogen identification and/or the fulfilment of the Koch's postulate.

The Genus *Alternaria*

The genus *Alternaria* was first described by Nees von Esenbeck in 1816 with *Alternaria tenuis*. Unambiguous conformity was lacking on *Alternaria* taxonomy until Elliott (1917) recommended six groups based on the conidial morphology, with every group being designated by an individual species. The need for added groups, not covered by Elliott, was recognized very early, when Neergaard (1945) projected three sections for the genus based on conidial formation in long chain, short chain, or no chain. Simmons (1967) defined three genera including *Alternaria*, *Stemphylium* and *Ulocladium*, which has helped in the classification of the genus. Simmons' 14 species-group concept was further advanced by using representative species of certain distinct groups including the *A. alternata*, *A. tenuissima*, *A. infectoria*, *A. arborescens*, *A. brassicicola*, *A. porri* and *A. radicina* groups (Simmons, 1995; Roberts

et al., 2000). Though the exercise of species-group description does not decide typical species limits within *Alternaria*, a benefit of its use is that it organizes the morphologically dissimilar grouping of *Alternaria* species at the sub-generic level, and authorized comprehensive argument of morphologically related species without inappropriate over-restriction due to nomenclatural ambiguity. Similarly, the species-group concept provided significant structure for hypothesis testing in sophisticated studies on *Alternaria* phylogeny.

More than 275 *Alternaria* species are available worldwide. However, inaccuracy in the taxonomy of *Alternaria* species occurs, since the inconsistency of its morphological characteristics is affected not only with inherent factors but also the ecological situation. Consequently, a single species could be erroneously placed into several species (Rotem, 1994). Neergaard (1945) developed a classification that was based on conidial chain-formation. Nishimura and Kohmoto (1983) proposed *formae speciales* or 'pathotypes' based on the morphological similarity.

Survival and Epidemiology

Depending on the kind of *Brassica* species, prevailing temperatures and available soil moisture for seed germination, oilseed brassicas in India are sown from late September to November and harvested from February to May. In non-traditional areas, off-season crops grown from May to September, coupled with the presence of the pathogen on infected alternative hosts (*Anagallis arvensis*, *Convolvulus arvensis*) and other vegetable brassica crops, are the most likely method of carryover of *A. brassicae* from one crop-season to another (Verma and Saharan, 1994); airborne spores of *A. brassicae*, therefore, form the crucial cause of inoculum of this polycyclic disease (Kolte, 1985).

Both initiation and development of the Alternaria blight disease are significantly influenced by the prevailing weather conditions. Positive correlation among temperatures, relative humidity (RH) and sunshine hours and incidence of Alternaria blight disease has been recognized.

The available data, however, should provide enough information about the growth stage at inception of the pathogen, expected peak severity throughout the crop period, and growth stage at the highest disease severity (Meena *et al.*, 2004). The precise information about the time that an epidemic of the disease will occur on the crop enables farmers to decide the optimal time for fungicidal spray. Blight severity on leaves and siliquae was most severe in delayed sown crops (Meena *et al.*, 2002). Under Indian conditions, delayed sowing often corresponds with the susceptible developmental phase of plants within the high and minimum temperatures of 18–26°C and 8–12°C, respectively, and high RH >70%. Blight severity on leaf was highest with maximum, minimum and mean temperatures of 18–27°C, 8–12°C and >10°C, and morning, afternoon and mean RH >92%, >40% and >70, respectively, in the preceding week. Similarly, the blight severity on the silique was positively correlated with the maximum and mean temperatures of 20–30°C, and >4°C, morning, and mean RH >90% and >70%, >9 h sunshine and >10 h leaf wetness. This information combined with the growth stage (age) of the crop for inception of disease allows farmers to perform required groundwork for management 1 week in advance of maximum blight severity on leaves and silique (Chattopadhyay *et al.*, 2005).

During the epidemic years (2001/02, 2002/03), *A. brassicae* conidia were found in and around the crop from October to April. The counting of conidia on the spore trap increased slowly in the early (0–6) hours of the day, attained its peak between 2 pm and 3 pm and decreased thereafter (Chattopadhyay *et al.*, 2005). The conidial population was influenced directly with the diurnal difference in temperatures and not influenced by RH, because the conidia produced during night left over on lesions and affixed due to high RH or moisture on leaf, became free with the increase in temperature and decrease in RH and leaf wetness (Humpherson-Jones, 1992). However, Humpherson-Jones and Phelps (1989) reported that darkness increases the sporulation of *A. brassicae*.

Alternaria Variability

Information regarding the pathogen variability is essential for improvement of resistant varieties, gene pyramiding and expansion of pre-breeding populations. Few reports are available on the subsistence of variation among isolates of *A. brassicae*. Isolates of *A. brassicae* A (highly virulent), C (moderately virulent) and D (avirulent) are prevalent in India (Vishwanath and Kolte, 1997). While *A. brassicae* produces both chlamydospores and microsclerotia, *A. raphani* is documented to produce only chlamydospores. Although cellulase enzymes (Nehemiah and Deshpande, 1976, 1977) and toxins (Degenhardt *et al.*, 1975; Durbin and Uchytil, 1977) are also produced by *A. brassicae*, their precise function in the pathogenesis is not clear. *Alternaria longipes* and *A. napiformae* were also noticed on rapeseed-mustard in India (Purkayastha and Mallik, 1976; Rao, 1977).

Morphological characteristics of 30 geographical isolates of *A. brassicae* in India revealed variations in growth, colony pigmentation and conidial length, width and number of septa. Conidial length varied from 106.7 to 285.9 µm, width from 33.5 to 57.0 µm, beak length from 41.4 to 180.0 µm, number of horizontal septa from 3.2 to 8.0, and transverse septa from 0.3 to 1.4. Meena *et al.* (2012) found that Elliott's medium was best among various synthetic media for profuse growth and sporulation of *A. brassicae*.

Pathogenic Variability

Alternaria brassicae isolates grown on cherry agar medium showed differences in both cultural growth and pathogenicity on rapeseed seedlings. Based on their morphology, growth, spore production and cultural characteristics, Vishwanath and Kolte (1997) identified three distinct *A. brassicae* isolates: A (highly virulent), C (moderately virulent) and D (avirulent). Significant tolerance with small size lesions has been reported in *B. alba*, *B. juncea* (EC-399299, PAB 9511), *E. sativa*, *B. carinata* and *B. napus*. Variation in tolerance or susceptibility of the same host, depending on the aggressiveness of the isolates, revealed that the variability exists among 30 *A. brassicae* isolates from different geographical locations in India; different *Brassica* species showed variable reactions against the same isolate (Meena *et al.*, 2012).

Difference at deoxyribonucleic acid (DNA) level between *A. brassicae*, *A. brassicicola*, *A. raphani* and *A. alternata* has been recognized (Jasalavich *et al.*, 1995) by restriction fragment length polymorphism (RFLP) (Botstin *et al.*, 1980) and random amplified polymorphic DNA (RAPD) markers (William *et al.*, 1990). Hong *et al.* (1996) from China reported differences in virulence among 53 *A. brassicae* isolates on four groups of Chinese cabbage.

Some extent of resistance is available in wild crucifers to *A. brassicae*, including *B. alba* (Hansen and Earle, 1997), *C. sativa, C. bursa-pastoris, E. sativa, N. paniculata* (Tewari and Conn, 1993), *Brassica desnottesii, Coincya pseuderucastrum, Diplotaxis berthautii, D. catholica, D. cretacea, D. erucoides* and *Erucastrum gallicum* (Sharma *et al.*, 2002).

Melanin Production

A number of pathogens produce melanin by oxidation of L-3, 4-dihydroxyphenylalanine (DOPA) obtained from the environment (Butler and Day, 1998); most, including *Alternaria*, derive melanin from the monomeric precursor 1, 8-dihydroxynaphthalene (DHN), produced during the pentaketide pathway (Kimura and Tsuge, 1993). Apart from their important function in conidial formation, melanins also have both direct and indirect roles in virulence. Melanins help in increasing both prolonged existence and survival of fungi by protecting them from unfavourable conditions including extreme temperatures, UV radiation and compounds secreted by microbial antagonists (Kawamura *et al.*, 1999). Melanin not only contributes to virulence by formation of functional melanised appressoria, but is able to act as a seeker of complimentary oxygen radicals, important mechanism of host resistance to pathogen entrance (Jacobson *et al.*, 1995). Further, since synthetic melanin is known to have an immune-suppressive effect on mammalian cells (Mohagheghpour *et al.*, 2000),

it is possible that similar inhibitive activities against the fungal pathogens might be demonstrated in host infectivity.

Depending on the moment in time, localization and production of melanin, however, fungal species differ in their necessity of melanin for pathogenicity, the production of melanin during the pentaketide pathway being known to be preserved by *Alternaria*. Tanabe *et al.* (1995) in his studies reported that fungal pathogens do not always require melanin for their virulence, because *A. alternata* mutants absent in melanin production still retained their pathogenicity. The cell wall structural design of the complemented appressoria, however, differed from the wild-type appressoria; complemented appressoria recover invasion of cellulose membranes excluding cucumber cotyledons.

The Infection Process

In general, foliar pathogens like *Alternaria* cause reasonably severe damage to plant tissues by decrease of photosynthesis. An infection causes the creation of a necrotic spot, which is enclosed by an uninfected yellowing tissue zone, a symptom typically known for the disease progression of saprophytic fungi, which is created due to toxins produced during transmission of the pathogen (Agarwal *et al.*, 1997). *Alternaria* species normally cause latent infections in which the pathogen penetrates the host tissues, where it remains inactive awaiting the occurrence of favourable conditions for development. *Alternaria* infections usually do not adversely manipulate food supply in the host, as it does not directly infect root systems.

In the absence of a sexual stage or overwintering spores, *Alternaria* is able to transmit either as mycelium or as conidia on decomposing plant remains for a substantial period. Infection can also occur through its dormant contamination in seeds on the seedlings once the seed has germinated. Otherwise, spores are mostly carrying through wind, lying on the plant surfaces and causing disease. Weakened tissues from stresses, advanced age or injury, are generally more vulnerable to *Alternaria* invasion than strong tissues.

Heavily melanized-walled spores germinate by producing germ tubes, which penetrate the host through pores or injury by producing appressoria. Although highly virulent species generally penetrate directly, the less virulent species target stomata or wounds. The host cuticle, which contains both cutin and wax, provides the initial plant defence against aggressive pathogens. Yao and Köller (1995) reported some cutinases which are mainly expressed during the saprophytic and pathogenic phase of *A. brassicicola*. Fan and Köller (1998) established that various cutinolytic enzymes are successively induced in cabbage, starting from the placement of spores to the invasion of the leaf. Although cutinases are expressed during the commencement of *A. brassicicola* activity on the cuticle, once the pathogen reaches the subcuticular layers, various cutinases and profuse saprophytic growth are stimulated by cutin monomers (Trail and Köller, 1993). Involvement of extracellular hydrolase during fungal pathogenesis has not been demonstrated till now. Cutinases and lipases simultaneously contribute to cause the infection. Cytokinin, especially 6-benzyl amino purine, has been revealed to reduce disease symptoms and mycelia growth of *A.brassicae in vivo* (Sharma *et al.*, 2010).

Toxin Biosynthesis

Most phytotoxins produced by a fungal invader are chemically diverse secondary metabolites, and low molecular weight components that are not required for normal growth or reproduction. Host-specific response of phytotoxin may be categorized as non-host-specific and host-specific toxins. Non-host-specific toxins, in general, have a moderate phytotoxic property, influence the vast range of hosts, and are believed to be a supplementary cause facilitating the penetration process and enzymatic degradation. They are normally a prerequisite for the host to carry out a virulent response increasing disease severity, and are not crucial for establishing infection, as these are extremely toxic to hosts.

Several non-host-specific phytotoxins perform different mechanisms, including brefeldin-A (dehydro-) curvularin, tenuazonic

acid, tentoxin and zinniol, are formed by diverse *Alternaria* species. Brefeldin-A causes breakup of the Golgi complex and performs as an inhibitor of secretion; curvularin inhibits division of host cell by interruption of the microtubule assembly; tenuazonic acid inhibits the protein synthesis; and zinniol disturbs the membrane permeabilization (Fujiwara *et al.*, 1988).

Host-specific toxins usually exhibit serious effects on a very narrow range of host species, and are crucial for disease. About 12 host-specific toxin-producing *Alternaria* species are known, mostly variants of *A. alternata*. AF- and ACT-toxin-producing strains of *A. alternata* also showed a pathogenic response on pear accessions that are sensitive to AK-toxin, but not *vice versa* (Kohmoto *et al.*, 1993). Clustering of genes has no advantage during vertical gene transmission from one generation to another, since the whole genome is transformed as a unit. However, a comparatively small adjacent DNA sequence is normally transferred during horizontal gene transfer. The presence of homologous toxin biosynthesis genes and clustering of toxin genes on a single chromosome strongly suggest that this is the method by which saprophytic *Alternaria* possibly obtains its pathogenic ability (Walton, 2000).

Mechanism of Host-specific Toxins

Alternaria toxins, although they vary considerably in their location of action, all cause host cell death, and their mechanism of action is not well understood. ACR-toxin induces swellings, morphological alterations of mitochondria, enhances NADH oxidation and plasma membrane disorders most important to electrolyte leakage and necrosis (Akimitsu *et al.*, 1989). The barley antifungal genes class II chitinase and type I ribosome inactivating protein co-expressed in *Brassica juncea* via *Agrobacterium*-mediated transformation provided good defence against Alternaria blight (Chhikara *et al.*, 2012).

Surprisingly, tomato genotypes sensitive to AAL-toxin were also selectively sensitive to Fumonisin B1 (Gilchrist *et al.*, 1992). The AAL-toxin-induced cell death mechanism in plant cells mimics the hereditarily controlled cellular suicide, or apoptosis, in animals, suggesting a lively contribution of the host in AAL-toxin mediated cell death (Wang *et al.*, 1996). The cause of host-specificity of the host-specific toxins is understood only for some of them such as destruxin B of *A. brassicae* (Pedras *et al.*, 2002); the selective phytotoxicity becomes visible due to a fast and well-organized detoxification, sequential hydroxylation, and glycosylation in resistant host tissues; these reactions also happen in susceptible species, but at a slower rate, providing a clarification for selective toxicity (Pedras *et al.*, 2001).

Screening Techniques

Effective screening techniques are the lifeline of breeding for resistance against any pathogen. Conventional field screening under artificial conditions, pollen culture for black spot disease resistance in Indian mustard to toxin destruxin B was suggested to be simple and effective (Shivanna and Sawhney, 1993). For developing a reliable and reproducible method for screening *B. rapa* against *A. brassicicola*, the effects of spore concentration, growth stage and incubation temperature were studied under an artificially manipulated environment. The detached-leaf inoculation method proved most suitable for screening *B. rapa* genotypes, because typical symptoms developed on leaves within 24 h, and a positive correlation existed between results of detached leaf technique and seedling inoculation tests. For successful infection, inoculum concentration of 5×10^4 conidia/ml, incubation temperature between 20°C and 25°C, and 3rd/4th true leaves from 30-day old plants were found optimal for inoculation (Doullah *et al.*, 2006).

Mechanism of Resistance

Plants resist pathogen infection by a number of induced mechanisms, together with the hypersensitive response (HR), the oxidative burst (the production of oxidizing compounds and enzymes), cell wall modifications (e.g. lignification), the production of pathogenesis-related proteins (PR) and phytoalexins

(Walton, 1997; Hammerschmidt, 1999). These localized responses at the infection site often initiate chain reactions including enhanced level of salicylic acid (SA), triggering of the systemic acquired resistance (SAR) and ultimately to the appearance of PR genes in the host.

Generally, lower leaves of *B. juncea* are more susceptible to *A. brassicae*. In the sunflower, greater internode length has been correlated with *Alternaria* tolerance, possibly due to difficulty in supplying inoculum from lower to upper leaves. SAR has also been induced in *B. juncea* through inoculation with an avirulent isolate of *A. brassicae* (Vishwanath *et al.*, 1999). High level of polyphenol oxidase, peroxidase and catalase content in leaves, low level content of nitrogen (Gupta *et al.*, 1990) and higher deposits of epicuticular wax have also increased *Alternaria* tolerance. Epicuticular wax creates a physiologic impediment as a hydrophobic covering which reduces retention of water-borne inoculum, conidial germination and germ-tube formation (Skoropad and Tewari, 1977). *Brassica napus* (cv. Tower), *B. carinata* and *B. alba* possess extra wax on the surface of the leaf compared to *B. rapa* (BSH-1, YSPB-24) and *B. juncea* (RH-30) (Tewari, 1986). Wild crucifers were found to elicit phytoalexins on challenge inoculation (Conn *et al.*, 1988).

An antimicrobial peptide, PmAMP1, isolated from western white pine (*Pinus monticola*), protected canola against several phytopathogenic fungi. The cDNA encoding PmAMP1 incorporated into the *B. napus* genome conferred superior defence against *A. brassicae*, *Leptosphaeria maculans* and *Sclerotinia sclerotiorum* (Verma *et al.*, 2012). Compounds such as camalexin ($C_{11}H_8N_2S$) and 6-methoxycamalexin ($C_{12}H_{10}N_2SO$) were found toxic to *A. brassicae* (Dzurilla *et al.*, 1998). Phytotoxin destruxin B elicits phytotoxic response in *B. alba* (Pedras and Smith, 1997). Multi-layered and multi-component resistance, and sensitivity to host-specific toxin destruxin B, hypersensitive reaction, Ca sequestration, etc. determines the fate of host–pathogen interaction (Tewari, 1991).

However, some tolerant donors of cultivated and wild crucifers are known for Alternaria blight disease. In general, *B. carinata* and *B. napus* are more tolerant than susceptible *B. juncea* and *B. rapa* to the disease. Genotype PAB 9511 (IC 546946) is the only registered donor for Indian mustard against this disease in India. Other genotypes found tolerant to Alternaria blight include *B. juncea* genotypes RC 781, PHR-2, PAB-9534, JMM-915, EC-399296, EC-399301 and RN-490, *B. carinata* genotypes HC-1 and PBC 9221, and *B. napus* genotypes PBN-9501, PBN-9502, PBN-2001 and PBN-2002.

Inheritance of *Alternaria* Resistance

Slow blighting was linked with the mechanism of tolerance to black spot disease of *B. juncea*. Some findings on mechanism of tolerance to Alternaria blight have showed the effect of additive genes or polygene or cluster of genes (Krishnia *et al.*, 2000). Resistance showed partial dominance (Zhang *et al.*, 1997), and there is also an indication of significant correlation between slow blighting and dominance (Kumar and Kolte, 2001). Diploid and autotetraploid type (colchicine induced) of *B. juncea* are more resistant to Alternaria blight than *B. rapa* at both ploidy levels, and this resistance factor is linked with the B-genome (Rai *et al.*, 1994). Although Alternaria blight resistance is unstable in *B. rapa* var. Yellow Sarson (Khan *et al.*, 1991), this is the most susceptible among different *B. rapa* varieties.

Incorporation of Resistance

Wide hybridization (*B. alba*; Sigareva *et al.*, 1999), somatic hybridization (*C. sativa*; Sigareva and Earle, 1999), transgenics expressing *Trichoderma harzianum* endo-chitinase gene (Mora and Earle, 2001), pollen culture, and sensitivity test to destruxin B (Shivanna and Sawhney, 1993) have been attempted.

Rapid modern techniques including tissue culture, protoplast fusion, embryo rescue and genetic engineering have made transfer of disease resistance characters across wide crossability barriers possible. Disease-resistant transgenic plants which over-express PR proteins (chitinase, glucanase, and osmotin) and

ribosome-inhibiting proteins (RIPs; thionins, defensins and phytoalexins) to inhibit pathogens growth seem less successful (Zhou *et al.*, 2002). Resistant transgenic plants through the use of resistant (R) genes, signalling NPR genes and two component systems of *R-Avr* and *barstar*, have also been exploited for HR and SAR responses. A cDNA encoding hevein (chitin-binding lectin from *Hevea brasiliensis*) was transferred into *B. juncea* cv. RLM-198. In whole-plant bioassay under artificial controlled conditions, interspecific hybridization was successful between wild allies of sunflower (*Helianthus annuus*), viz. *H. simulans*, *H. maximiliani*, *H. divaricatus*, *H. pauciflorus*, *H. decapetalus*, *H. tuberosus* and *H. resinosus* and cvs Morden 234A, 234B and 6D-1. The hybrids were found highly resistant to Alternaria blight under high spore load of the pathogen (Sujatha *et al.*, 1997). The gamete assortment was made by applying fungal inoculums to stigma and style, 1 h before pollination. The populations improved through gametophytic selection were more promising as the pollen selection could allow high selection intensity and absence of dominance effects. For partial resistance, the gametophytic selection combined with the conventional sporophytic selection can be exploited as an efficient tool in improving host populations.

Host Defence Against Alternaria Blight

A host-specific toxin, destruxin B produced by *Alternaria*, causes selective phytotoxicity since the hydroxylation product is generated during the detoxification in resistant but not in susceptible species. Consequently, phytotoxin detoxification and concurrent phytoalexin elicitation plays an important role in resistance (Pedras *et al.*, 2001). These phytoalexins have also been involved in pathogen inhibition for other *Alternaria*–host interactions. The production of camalexin, a phytoalexin, has been interrelated with resistance against *A. brassicae* (Jejelowo *et al.*, 1991). *Arabidopsis* also produce camalexin leading to pathogen challenge. A mutant *pad3* (phytoalexin deficient) of *Arabidopsis* is lacking camalexin biosynthesis and displays enhanced vulnerability to *A. brassicicola*, demonstrating that camalexin contributes to blight disease resistance, because it directly inhibits the growth of *Alternaria* (Thomma *et al.*, 1999b). Camalexin also contributes indirectly to disease resistance, because it inhibits the production of the destruxin B in *A. brassicae* (Pedras *et al.*, 1998).

Camelexin biosynthesis is neither induced by jasmonate, nor is the addition of jasmonate-insensitivity responsible for the scarcity of camalexin. Therefore, it is suggested that plant resistance against *Alternaria* is mediated by both camalexin and jasmonate through two separate corresponding pathways in *Arabidopsis* (Thomma *et al.*, 1998, 2001). It has still not been identified which jasmonate-inducible effector molecules are accountable for inhibition of *Alternaria* in *Arabidopsis*.

Results of cDNA microarray analysis of 2375 *Arabidopsis* genes showed that about 10% of up-regulated and repressed 207 genes were concerned in plant defence and about one-third in cell maintenance and development (Schenk *et al.*, 2000). The precise role of any of these genes in the *Alternaria* resistance of *Arabidopsis* has yet to be established. Results of experiments using ethylene- and jasmonate-insensitive *Arabidopsis* mutants suggested that a set of PR-genes encoding plant defensin, hevein-like protein and chitinase are not involved in *Alternaria* resistance (Thomma *et al.*, 1998, 1999a).

Constitutive expression of a radish defensin gene and a rubber tree chitin-binding lectin conferred improved resistance against *Alternaria* species in Indian mustard (Kanrar *et al.*, 2002). Constitutive appearance of an endochitinase gene from mycoparasite *Trichoderma harzianum* has also resulted in superior *Alternaria* resistance in transgenic tobacco, potato and broccoli (Lorito *et al.*, 1998; Mora and Earle, 2001).

Biological Control

Although information is available on controlling both seed-borne and foliar Alternaria blight by use of fungal and bacterial bioagents, none is being utilized at a commercial level. Seed dressing with a formulation of *Streptomyces* species provided 80–90% control of seed-borne *A. brassicicola* even after storing

treated seeds in dry place for 35–40 days (Tahvonen and Avikainen, 1987). Foliar application of diffusate of *Streptomyces rochei* also reduced Alternaria blight intensity in *B. rapa* (Sharma and Gupta, 1978). Seed-dressing with *Bacillus* spp. and *T. harzianum* effectively increased survival of healthy plants grown from *A. linicola*-infected linseed seeds; weekly sprays with *T. viride* isolate was as effective as the fungicide iprodione (rovral) in controlling *A. linicola* infestation on linseed capsules (Mercer *et al.*, 1992, 1993). Alternaria blight control was as effective with fungicide mancozeb as with spraying *T. viride* at critical growth stages in Indian mustard (Meena *et al.*, 2004).

Foliar application of *Auvobasidium pullulens* and *Epicoccum nigrum*, 14 h before the pathogen, reduced the infection by *A. brassisicola* (Pace and Campbell, 1974); *B. amyloliquefaciens* applied as foliar spray also effectively reduced Alternaria blight in *B. napus* (Danielssen *et al.*, 2006). Several antagonists of *A. ricini*, viz. *Mucor varians*, *T. viride*, *Rhizopus nigricans*, *Pseudomonas fluovescans*, *Aspergillus niger* and *A. sativus* (Silva *et al.*, 1998), of *A. helianthi*, viz. *B. mycoides* (Kong *et al.*, 1997), *P. fluovescens*, *P. putida*, *P. cepocid* and *B. polymyxa* (Prasad and Kulshrestha, 1999), have been reported and, if tested, some may prove effective in controlling Alternaria blight in *B. juncea*. *Allium sativum* bulb extract was found useful in reducing Alternaria blight disease in Indian mustard (Meena *et al.*, 2004).

Integrated Alternaria Blight Management in Rapeseed-mustard

Early sowing of certified seeds after deep ploughing, timely weeding, maintenance of optimum plant population and avoidance of irrigation at flowering and pod formation stages may help in managing the disease effectively. Spray applications of mancozeb at two critical growth stages of the pathogen (Meena *et al.*, 2004) and iprodione (Rovral) at 2 months after seeding (Cox *et al.*, 1983) have been reported very useful in inhibiting *A. brassicae*-induced Alternaria blight. Reduction of Alternaria blight severity on *B. juncea* leaves and pods by foliar application of *A. sativum* bulb extract at 45 and 75 days after seeding (DAS) was similar with that of foliar application with *T. viride*. Following foliar spray of *T. viride* and *A. sativum* bulb extract, singly and in combination, have proved effective in significantly reducing Alternaria blight and increasing yield in *B. juncea* (Meena *et al.*, 2004, 2008, 2011): (i) *T. viride* 45+75 DAS; (ii) *A. sativum* bulb extract 45+75 DAS; (iii) *A. sativum* bulb extract 45 DAS + *T. viride* 75 DAS; and (iv) *T. viride* 45 DAS + *A. sativum* bulb extract 75 DAS.

Conclusions

Despite a significant body of information regarding Alternaria blight and its pathogen, there is still a need to develop reliable and reproducible screening techniques and precisely define pathogenic variability. Effective identification and genetic characterization of resistance sources across genera and species are needed for transfer of resistance into already licensed susceptible cultivars. Promotion of integrated disease management (IPM) through media and improved awareness programmes will help farmers to understand, use and value IPM programmes. These could be achieved in a phased manner through training programmes and by regular contact using modern information and communication technology (Chambers *et al.*, 1989). Crop protection bulletins could be issued regularly by linking with a crop–weather watch campaign for value-added agro-advisory use. There is also a need to develop intensive crop- and location-specific climate and economic disease management packages. A Web-based Integrated Decision Support System (IDSS) for Crop Protection Services needs to be evolved in a farmer participatory mode (Chattopadhyay *et al.*, 2011; Kumar *et al.*, 2012). A pest risk analysis (including diagnostics) vis-à-vis climate change and creation of the database for pro-active climate-resilient IPM in relation to globalization is also required as preparedness for the future.

References

Agarwal, A., Garg, G.K., Devi, S., Mishra, D.P. and Singh, U.S. (1997) Ultrastructural changes in *Brassica* leaves caused by *Alternaria brassicae* and destruxin B. *Journal of Plant Biochemistry Biotechnology* 6, 25–28.

Akimitsu, K., Kohmoto, K., Otani, H. and Nishimura, S. (1989) Host-specific effects of toxin from the rough lemon pathotype of *Alternaria alternata* on mitochondria. *Plant Physiology* 89, 925–931.

Ansari, A.N., Khan, M.W. and Muheet, A. (1988) Effect of Alternaria blight on oil content of rapeseed–mustard. *Current Science* 57, 1023–1024.

Botstin, D., White, R.L., Skolnick, M.H. and Davis, R.W. (1980) Construction of a genetic linkage map in man using restriction fragment length poly-morphisms. *American Journal of Human Genetics* 32, 314–331.

Butler, M.J. and Day, A.W. (1998) Fungal melanins: a review. *Canadian Journal of Microbiology* 44, 1115–1136.

Chambers, R., Pacey, A. and Thrupp, L.A. (1989) *Farmer First – Farmer Innovation and Agricultural Research.* Intermediate Technology Publications, London, 218 pp.

Chattopadhyay, C., Agrawal, R., Kumar, A., Bhar, L.M., Meena, P.D., Meena, R.L., Khan, S.A., Chattopadhyay, A.K., Awasthi, R.P., Singh, S.N., Chakravarthy, N.V.K., Kumar, A., Singh, R.B. and Bhunia, C.K. (2005) Epidemiology and forecasting of Alternaria blight of oilseed *Brassica* in India – a case study. *Zeitschrift für Pflanzenkrankheiten und Pflanzenschutz (Journal of Plant Disease Protection)* 112, 351–365.

Chattopadhyay, C., Bhattacharya, B.K., Kumar, V., Kumar, A. and Meena, P.D. (2011) Epidemiology and forecasting of diseases for value-added agro-advisory. In: *Plant Pathology in India: Vision 2030.* Indian Phytopathological Society, New Delhi, pp. 132–140.

Chevre, A.M., Eber, F., Brun, H., Plessis, J., Primard, C. and Renard, M. (1991) Cytogenetic studies of *Brassica napus-Sinapis alba* hybrids from ovary culture and protoplast fusion, Attempts of introduction of Alternaria resistance in rapeseed. In: *Proceedings of the GCIRC 8th International Rapeseed Congress*, Saskatoon, Canada, p. 346.

Chhikara, S., Chaudhury, D., Dhankher, O.P. and Jaiwal, P.K. (2012) Combined expression of a barley class II chitinase and type I ribosome inactivating protein in transgenic *Brassica juncea* provides protection against *Alternaria brassicae*. *Plant Cell, Tissue and Organ Culture* 108, 83–89.

Conn, K.L., Tewari, J.P. and Dahiya, J.S. (1988) Resistance to *Alternaria brassicae* and phytoalexin-elicitation in rapeseed and other crucifers. *Plant Science* 55, 21–25.

Cox, T., Souche, J.L. and Grapel, H. (1983) The control of *Sclerotinia, Alternaria* and *Botrytis* on oilseed rape with spray treatments of flowable formulation of iprodione. In: *Abstracts of the 6th International Rapeseed Congress,* Paris, France, 17–19 May 1983, pp. 928–933.

Dangl, J.L., Dietrich, R.A. and Richberg, M.H. (1996) Death don't have no mercy: cell death programs in plant-microbe interactions. *Plant Cell* 8, 1793–1807.

Danielsson, J., Reva, O. and Meijer, J. (2006) Protection of oilseed rape (*Brassica napus*) toward fungal pathogens by strains of plant-associated *Bacillus amyloliquefaciens. Microbial Ecology* 54, 134–140.

Degenhardt, K.J., Petrie, G.A., Dueck, J. and Pelcher, L.E. (1975) Synergistic action of two phytotoxic metabolites produced by *Alternaria brassicae. Proceedings of Canadian Phytopathological Society* 42, 17.

Dixon, R.A., Harrison, M.J. and Lamb, C.J. (1994) Early events in the activation of plant defense responses. *Annual Review of Phytopathology* 32, 479–501.

Doullah, M.A.U., Meah, M.B. and Okazaki, K. (2006) Development of an effective screening method for partial resistance to *Alternaria brassicicola* (dark leaf spot) in *Brassica rapa. European Journal of Plant Pathology* 116, 33–43.

Durbin, R.D. and Uchytil, T.F. (1977) Survey of plant insensitivity to tentoxin. *Phytopathology* 67, 602.

Dzurilla, M., Kutschy, P., Tewari, J.P., Ruzinsky, M., Senvicky, S. and Kovacik, V. (1998) Synthesis and antifungal activity of new indolypthiazinone derivatives. *Collection Czechoslovakia Chemists Community* 63, 94–102.

Elliott, J.A. (1917) Taxonomic characters of the genera *Alternaria* and *Macrosporium. American Journal of Botany* 4, 439–476.

Fan, C.Y. and Köller, W. (1998) Diversity of cutinases from plant pathogenic fungi: differential and sequential expression of cutinolytic esterases by *Alternaria brassicicola. FEMS Microbiological Letters* 158, 33–38.

Fujiwara, T., Oda, K., Yokota, S., Takatsuki, A. and Ikehara, Y. (1988) Brefeldin-A causes disassembly of the Golgi complex and accumulation of secretory proteins in the endoplasmic reticulum. *Journal of Biological Chemistry* 263, 18545–18552.

Gilchrist, D.G., Ward, B., Moussato, V. and Mirocha, C.J. (1992) Genetic and physiological response to fumonisins and AAL-toxin by intact tissue of a higher plant. *Mycopathologia* 117, 57–64.

Gupta, S.K., Gupta, P.P., Yadava, T.P. and Kaushik, C.D. (1990) Metabolic changes in mustard due to Alternaria leaf blight. *Indian Phytopathology* 43, 64–69.

Hammerschmidt, R. (1999) Phytoalexins: What have we learned after 60 years? *Annual Review of Phytopathology* 37, 285–306.

Hansen, L.H. and Earle, E.D. (1997) Somatic hybrids between *Brassica oleracea* (L.) and *Sinapis alba* (L.) with resistance to *Alternaria brassicae. Theoretical Applied Genetics* 94, 1078–1085.

Hong, C.X., Fitt, B.D.L. and Welham, S.J. (1996) Effects of wetness period and temperature on development of dark pod spot (*Alternaria brassicae*) on oilseed rape (*Brassica napus*). *Plant Pathology* 45, 1077–1089.

Humpherson-Jones, F.M. (1992) The development of weather-related disease forecasts for vegetable crops in the UK: problems and prospects. *European Plant Protection Organisation Bulletin* 21, 425–429.

Humpherson-Jones, F.M. and Phelps, K. (1989) Climatic factors influencing spore production in *Alternaria brassicae* and *A. brassicicola. Annals of Applied Biology* 114, 449–458.

Jacobson, E.S., Hove, E. and Emery, H.S. (1995) Antioxidant function of melanin in black fungi. *Infection and Immunity* 63, 4944–4945.

Jasalavich, C.A., Morales, V.M., Pelcher, L.E. and Seguin-Swartz, G. (1995) Comparison of nuclear ribosomal DNA sequences from *Alternaria* species pathogenic to crucifers. *Mycological Research* 99, 604–614.

Jejelowo, O.A., Conn, K.L. and Tewari, J.P. (1991) Relationship between conidial concentration, germling growth and phytoalexin production by *Camelina sativa* leaves inoculated with *Alternaria brassicae. Mycological Research* 95, 928–934.

Kanrar, S., Venkateswari, J.C., Kirti, P.B. and Chopra, V.L. (2002) Transgenic expression of hevein, the rubber tree lectin, in Indian mustard confers protection against *Alternaria brassicae. Plant Science* 162, 441–448.

Kawamura, C., Tsujimoto, T. and Tsuge, T. (1999) Targeted disruption of a melanin biosynthesis gene affects conidial development and UV tolerance in the Japanese pear pathotype of *Alternaria alternata. Molecular Plant-Microbe Interaction* 12, 59–63.

Khan, M.W., Ansari, N.A. and Muheet, A. (1991) Response of some accessions of rapeseed 'yellow sarson' (*Brassica campestris* L. var. *sarson* Prain) against Alternaria blight. *International Journal of Tropical Plant Disease* 9, 111–113.

Kimura, N. and Tsuge, T. (1993) Gene cluster involved in melanin biosynthesis of the filamentous fungus *Alternaria alternata. Journal of Bacteriology* 175, 4427–4435.

Kohmoto, K., Itoh, Y., Shimomura, N., Kondoh, Y., Otani, H., Kodama, M., Nishimura, S. and Nakatsuka, S. (1993) Isolation and biological activities of two host-specific toxins from the tangerine pathotype of *Alternaria alternata. Phytopathology* 83, 495–502.

Kolte, S.J. (1985) *Diseases of Annual Edible Oilseed Crops,* Vol. II, *Rapeseed-Mustard and Sesame Diseases.* CRC Press, Boca Raton, Florida, 135 pp.

Kong, G.A., Kochman, J.K. and Brown, J.F. (1997) Phylloplane bacteria antagonistic to the sunflower pathogen *Alternaria helianthi. Australian Plant Pathology* 26, 85–97.

Krishnia, S.K., Saharan, G.S. and Singh, D. (2000) Genetic variation for multiple disease resistance in the families of interspecific cross of *Brassica juncea x Brassica carinata. Cruciferae Newsletter* 22, 51–53.

Kumar, B. and Kolte, S.J. (2001) Progression of Alternaria blight of mustard in relation to components of resistance. *Indian Phytopathology* 54, 329–331

Kumar, V., Kumar, A. and Chattopadhyay, C. (2012) Design and implementation of web-based aphid (*Lipaphis erysimi*) forecast system for oilseed Brassicas. *The Indian Journal of Agricultural Sciences* 82, 608–614.

Lamb, C. and Dixon, R.A. (1997) The oxidative burst in plant disease resistance. *Annual Review of Plant Physiology and Plant Molecular Biology* 48, 251–275.

Levine, A., Pennell, R.I., Alvarez, M.E., Palmer, R. and Lamb, C. (1996) Calcium-mediated apoptosis in a plant hypersensitive disease resistance response. *Current Biology* 6, 427–437.

Lorito, M., Woo, S.L., Garcia, I., Colucci, G., Harman, G.E., Pintor-Toro, J.A., Filippone, E., Muccifora, S., Lawrence, C.B., Zoina, A., Tuzun, S., Scala, F. and Fernandez, I.G. (1998) Genes from mycoparasitic fungi as a source for improving plant resistance to fungal pathogens. *Proceedings of the National Academy Sciences USA* 95, 7860–7865.

Meena, P.D., Chattopadhyay, C., Singh, F., Singh, B. and Gupta, A. (2002) Yield loss in Indian mustard due to white rust and effect of some cultural practices on Alternaria blight and white rust severity. *Brassica* 4, 18–24.

Meena, P.D., Meena, R.L., Chattopadhyay, C. and Kumar, A. (2004) Identification of critical stage for disease development and biocontrol of Alternaria blight of Indian mustard (*Brassica juncea*). *Journal of Phytopathology* 152, 204–209.

Meena, P.D., Meena, R.L. and Chattopadhyay, C. (2008) Eco-friendly options for management of Alternaria blight in Indian mustard (*Brassica juncea*). *Indian Phytopathology* 62(1), 65–69.

Meena, P.D., Chattopadhyay, C., Kumar, A., Awasthi, R.P., Singh, R., Kaur, S., Thomas, L., Goyal, P. and Chand, R. (2011) Comparative study on the effect of chemicals on Alternaria blight in Indian mustard – A multilocation study in India. *Journal of Environmental Biology* 32, 375–379.

Meena, P.D., Rani, A., Meena, R., Sharma, P., Gupta, R. and Chowdappa, P. (2012) Aggressiveness, diversity and distribution of *Alternaria brassicae* isolates infecting oilseed *Brassica* in India. *African Journal of Microbiological Research* 6(24), 5249–5258.

Mercer, P.C., Ruddock, A. and McGimpsey, H.C. (1992) Evaluation of iprodione and *Trichoderma viride* against *Alternaria linicola*. Tests of Agrochemicals and Cultivars. No. 13. *Annals of Applied Biology* 120, 20–21.

Mercer, P.C., Ruddock, A., Mee, E. and Papadoplous, S. (1993) Biological control of *Alternaria* diseases of linseed and oilseed rape. *Bulletin OILB/SROP* 16, 89–99.

Mohagheghpour, N., Waleh, N., Garger, S.J., Dousman, L., Grill, L.K. and Tuse, D. (2000) Synthetic melanin suppresses production of proinflammatory cytokines. *Cell Immunology* 199, 25–36.

Mora, A.A. and Earle, E.D. (2001) Resistance to *Alternaria brassicicola* in transgenic broccoli expressing a *Trichoderma harzianum* endochitinase gene. *Molecular Breeding* 8, 1–9

Neergaard, P. (1945) *Danish Species of Alternaria and Stemphylium.* Oxford University Press, London, 560 pp.

Nees von Esenbeck, C.G. (1816) *Das system der pilze and schwamme.* Wurzburg, Germany.

Nehemiah, K.M.A. and Deshpande, K.B. (1976) Detection of cell wall breaking ability by cellulase of *Alternaria brassicae* (Berk.) Sacc. *Current Science* 45, 28–29.

Nehemiah, K.M.A. and Deshpande, K.B. (1977) Cellulase production and decomposition of cotton fabric and filter paper by *Alternaria brassicae. Indian Phytopathology* 29, 55.

Nishimura, S. and Kohmoto, K. (1983) Host-specific toxins and chemical structures from *Alternaria* species. *Annual Review of Phytopathology* 21, 87–116.

Pace, M.A. and Campbell, R. (1974) The effect of saprophytes on infection of leaves of *Brassica* spp. by *Alternaria brassicicola. Transactions of the British Mycological Society* 63, 193–196.

Pedras, M.S. and Smith, K.C. (1997) Sinalexin, a phytoalexin from white mustard elicited by destruxin B and *Alternaria brassicae. Phytochemistry* 46, 833–837.

Pedras, M.S.C., Kahn, A.Q. and Taylor, J.L. (1998) The phytoalexin camalexin is not metabolized by *Phoma lingam, Alternaria brassicae,* or phytopathogenic bacteria. *Plant Science* 139, 1–8.

Pedras, M.S.C., Zaharia, I.L., Gai, Y., Zhou, Y. and Ward, D.E. (2001) *In planta* sequential hydroxylation and glycosylation of a fungal phytotoxin: avoiding cell death and overcoming the fungal invader. *Proceedings of the National Academy of Sciences USA* 98, 747–752.

Pedras, M.S.C., Zaharia, I.L. and Ward, D.E. (2002) The destruxins: synthesis, biosynthesis, biotransformation, and biological activity. *Phytochemistry* 59, 579–596.

Prasad, R.D. and Kulshrestha, D.D. (1999) Bacterial antagonists of *Alternaria helianthi* of sunflower. *Journal of Mycology and Plant Pathology* 29, 127–128.

Purkayastha, P.P. and Mallik, F. (1976) Two new species of Hyphomycetes from India. *Nova Hedwigia* 27, 781.

Rai, B., Kumar, H. and Rao, G.P. (1994) Reaction of diploid and autotetraploid forms of oilseed *Brassica* varieties and their hybrids to infection of leaf-blight caused by *Alternaria brassicae*. *Indian Journal of Agricultural Science* 64, 182–183.

Rao, B.R. (1977) Species of *Alternaria* on some *Cruciferae*. *Geobios* 4, 163–166.

Roberts, R.G., Reymond, S.T. and Anderson, B. (2000) RAPD fragment pattern analysis and morphological segregation of small-spored *Alternaria* species and species group. *Mycological Research* 104, 151–160.

Rotem, J. (1994) *The Genus* Alternaria: *Biology, Epidemiology and Pathogenicity*. APS Press, St Paul, Minnesota.

Schenk, P.M., Kazan, K., Wilson, I., Anderson, J.P., Richmond, T., Somerville, S.C. and Manners, J.M. (2000) Coordinated plant defense responses in Arabidopsis revealed by microarray analysis. *Proceedings of the National Academy of Sciences USA* 97, 11655–11660.

Sharma, G., Kumar, D.V., Haque, A., Bhat, S.R., Prakash, S. and Chopra, V.L. (2002) *Brassica* coenospecies: a rich reservoir for genetic resistance to leaf spot caused by *Alternaria brassicae*. *Euphytica* 125, 411–419.

Sharma, N., Rahman, M.H., Liang, Y. and Kav, N.N.V. (2010) Cytokinin inhibits the growth of *Leptosphaeria maculans* and *Alternaria brassicae*. *Canadian Journal of Plant Pathology* 32(3), 306–314.

Sharma, S.K. and Gupta, J.S. (1978) Biological control of leaf blight disease of brown sarson caused by *Alternaria brassicae* and *Alternaria brassicicola*. *Indian Phytopathology* 31, 448–449.

Shivanna, K.R. and Sawhney, V.K. (1993) Pollen selection for *Alternaria* resistance in oilseed brassicas: responses of pollen grains and leaves to a toxin of *A. brassicae*. *Theoretical Applied Genetics* 86, 339–344

Sigareva, M.A. and Earle, E.D. (1999) Camalexin induction in intertribal somatic hybrids between *Camelina sativa* and rapid-cycling *Brassica oleracea*. *Theoretical Applied Genetics* 98, 164–170.

Sigareva, M.A., Ren, J.-P., and Earle, E.D. (1999) Introgression of resistance to *Alternaria brassicicola* from *Sinapis alba* to *Brassica oleracea* somatic hybridization and backcrosses. *Cruciferae Newsletter* 21, 135–136.

Silva, F. de A.G. da, Peixoto, C. de N., Assis, S.M.P. de, and Mariano, R. de L.R. Padovan (1998) Potential of fluorescent *Pseudomonas* spp. for biological control of *Alternaria ricini* on castor bean. *Brazilian Archives of Biology and Technology* 41, 91–102.

Simmons, E.G. (1967) Typification of *Alternaria*, *Stemphylium* and *Ulocladium*. *Mycologia* 59, 67–92.

Simmons, E.G. (1995) *Alternaria* themes and variations (112-144). *Mycotaxon* 55, 55–163.

Skoropad, W.P. and Tewari, J.P. (1977) Field evaluation of the epicuticular wax in rapeseed and mustard in resistance to *Alternaria brassicae*. *Canadian Journal of Plant Science* 57, 1001–1003.

Sujatha, M., Prabakaran, A.J. and Chattopadhyay, C. (1997) Reaction of wild sunflowers and certain interspecific hybrids to *Alternaria helianthi*. *Helia* 20, 15–24

Tahvonen, R. and Avikainen, H. (1987) The biological control of seed-borne *Alternaria brassicicola* of cruciferous plants with a powdery preparation of *Streptomyces* sp. *Journal of Agricultural Science Finland* 59, 199–208.

Tanabe, K., Park, P., Tsuge, T., Kohmoto, K. and Nishimura, S. (1995) Characterization of the mutants of the *Alternaria alternata* Japanese pear pathotype deficient in melanin production and their pathogenicity. *Annals of Phytopathology Society Japan* 61, 27–33.

Tewari, J.P. (1986) Subcuticular growth of *Alternaria brassicae* in rapeseed. *Canadian Journal of Botany* 64, 1227–1231.

Tewari, J.P. (1991) Structural and biochemical basis of the black spot disease of crucifers. *Advances in Structural Biology* 1, 325–349.

Tewari, J.P. and Conn, K.L. (1993) Reaction of some wild crucifers to *Alternaria brassicae*. *Bulletin OILB/SROP* 16, 53–58.

Thomma, B.P.H.J., Eggermont, K., Penninckx, I.A.M.A., Mauch-Mani, B., Vogelsang, R., Cammue, B.P.A. and Broekaert, W.F. (1998) Separate jasmonate-dependent and salicylate-dependent defense response pathways in *Arabidopsis* are essential for resistance to distinct microbial pathogens. *Proceedings of the National Academy of Sciences USA* 95, 15107–15111.

Thomma, B.P.H.J., Eggermont, K., Tierens, F.M.J. and Broekaert, W.F. (1999a) Requirement of functional *EIN2* (ethylene insensitive 2) gene for efficient resistance of *Arabidopsis thaliana* to infection by *Botrytis cinerea*. *Plant Physiology* 121, 1093–1101.

Thomma, B.P.H.J., Nelissen, I., Eggermont, K. and Broekaert, W.F. (1999b) Deficiency in phytoalexin production causes enhanced susceptibility of *Arabidopsis thaliana* to the fungus *Alternaria brassicicola*. *Plant Journal* 19, 163–171.

Thomma, B.P.H.J., Eggermont, K., Broekaert, W.F. and Cammue, B.P.A. (2000) Disease development of several fungi on *Arabidopsis* can be reduced by treatment with methyl jasmonate. *Plant Physiology and Biochemistry* 38, 421–427.

Thomma, B.P.H.J., Penninckx, I.A.M.A., Broekaert, W.F. and Cammue, B.P.A. (2001) The complexity of the disease signaling in *Arabidopsis*. *Current Opinion in Immunology* 13, 63–68.

Trail, F. and Köller, W. (1993) Diversity of cutinases from plant pathogenic fungi: purification and characterization of two cutinases from *Alternaria brassicicola*. *Physiology and Molecular Plant Pathology* 42, 205–220.

Verma, P.R. and Saharan, G.S. (1994) *Monograph on Alternaria diseases of crucifers*. Technical Bulletin No. 1994–6E, Agriculture Canada Research Station, Saskatoon, 162 pp.

Verma, S.S., Yajima, W.R., Rahman, M.H., Shah, S., Liu, J.J., Ekramoddoullah, A.K.M. and Kav, N.N.V. (2012) A cysteine-rich antimicrobial peptide from *Pinus monticola* (PmAMP1) confers resistance to multiple fungal pathogens in canola (*Brassica napus*). *Plant Molecular Biology* 79, 61–74.

Vishwanath, M and Kolte, S.J. (1997) Variability in *Alternaria brassicae*: response to host genotypes, toxin production and fungicides. *Indian Phytopathology* 50, 373–381.

Vishwanath, Kolte, S.J., Singh, M.P. and Awasthi, R.P. (1999) Induction of resistance in mustard (*Brassica juncea*) against Alternaria black spot with an avirulent *Alternaria brassicae* isolate-D. *European Journal of Plant Pathology* 105, 217–220.

Walton, J.D. (1997) Biochemical plant pathology. In: Harborne, J.B. and Dey, P.M. (eds) *Plant Biochemistry*. Academic Press, San Diego, California, pp. 487–502.

Walton, J.D. (2000) Horizontal gene transfer and the evolution of secondary metabolite gene clusters in fungi: an hypothesis. *Fungal Genetics and Biology* 30, 167–171.

Wang, H., Li, J., Bostock, R.M. and Gilchrist, D.G. (1996) Apoptosis: a functional paradigm for programmed plant cell death induced by a host-selective phytotoxin and invoked during development. *Plant Cell* 8, 375–391.

Williams, J.G.K., Kubelik, A.R., Livak, K.J., Rafalski, J.A. and Tingey, S.V. (1990) DNA polymorphisms amplified by arbitrary primers are useful as genetic markers. *Nucleic Acids Research*, 6531–6535.

Yao, C. and Köller, W. (1995) Diversity of cutinases from plant pathogenic fungi: different cutinases are expressed during saprophytic and pathogenic stages of *Alternaria brassicicola*. *Molecular Plant-Microbe Interactions* 8, 122–130.

Zhang, F.-L., Xu, J.-B., Yan, H., Li, M.-Y., Zhang, F.L., Xu, J.B, Yan, H. and Li, M.Y. (1997) A study on inheritance of resistance to black leaf spot in seedlings of Chinese cabbage. *Acta Agriculturae Boreali Sinica* 12, 115–119.

Zhou, X.J., Lu, S., Xu, Y.H, Wang, J.W. and Chen, X.Y. (2002) A cotton cDNA (GaPR10) encoding a pathogenesis-related to protein with *in vitro* ribonuclease activity. *Plant Science*, 629–636.

10 Plant Disease Resistance Genes: Insights and Concepts for Durable Disease Resistance

Lisong Ma and M. Hossein Borhan*

Agriculture and Agri-Food Canada, Saskatoon, Saskatchewan, Canada

Introduction

Plants are continuously exposed to biotic and abiotic stress. Plant pathogenic fungi, oomycetes, bacteria, viruses and nematodes affect various crops and contribute to major yield loss. Aside from their economic importance, plant pathogens of staple crops could have great social impact. The best example is the Irish famine of the 19th century caused by *Phytophthora infestans*, the oomycete agent of potato late blight (Vurro *et al.*, 2010; Fisher *et al.*, 2012). Despite advances of modern agriculture in controlling plant diseases, emerging infectious diseases are still posing a threat to the global crop yield and food security (Fisher *et al.*, 2012; Gawehns *et al.*, 2013). An estimated 15% of crop yield loss globally, is caused by the pre-harvest plant diseases (Dangl *et al.*, 2013). The emerging infectious diseases result from increased virulence of fungi and fungal-like oomycetes, which is caused by the co-evolution of pathogens with their adapted host (Fisher *et al.*, 2012). Control of plant disease relies widely on the use of fungicides and breeding for disease resistance (Gust *et al.*, 2010). The best method for controlling plant disease is developing resistant varieties, and

understanding the innate resistance mechanisms in plants is central to genetic improvement of plant disease resistance (McDowell and Woffenden, 2003).

Plants rely mainly on the innate defence mechanism to resist pathogen infection. This innate defence is orchestrated by a multilayered innate immune system (Segonzac and Zipfel, 2011). The first layer is based on the membrane-localized pattern-recognition receptors (PRRs) that perceive the microbe or pathogen-associated molecular patterns (MAMPs or PAMPs). Recognition of these essential molecules by PRRs initiates PAMP-triggered immunity (PTI) (Ma *et al.*, 2012). Adapted pathogens have evolved to overcome PTI by the production of effector proteins. Effectors are small molecules secreted by the pathogen into the host to manipulate host physiology and suppress host plant defence (Giraldo and Valent, 2013). To counteract the virulence activities of effectors, plants in turn have developed resistance (*R*) genes of which their protein products recognize the pathogen effectors, resulting in the defence responses that often culminate into a hypersensitive response (HR), a local cell death around the infection site (Thomma *et al.*, 2011; Ma *et al.*, 2012). This

*Corresponding author, e-mail: hossein.borhan@agr.gc.ca

resistance initiated by effector recognition is termed effector-triggered immunity (ETI), among which *R* genes represent the key component of the second layer of the innate immune system. So far more than 70 *R* genes from crop and model plant species have been cloned (Liu *et al.*, 2007; Ma *et al.*, 2012). Pathogens rapidly evolve and produce new strains that overcome plant resistance. This could occur by loss of effectors that initiate the same host signalling pathway (effector with functional redundancy) point mutation or truncation of the corresponding effector genes (Gururani *et al.*, 2012; Dangl *et al.*, 2013). Plant breeding for resistance to pathogens has been widely used to develop new resistant cultivars. But this approach is time consuming and could be extremely challenging when introducing resistance from wild relatives into elite cultivars. Thus to overcome these obstacles it is critical to have a deep knowledge of R proteins' function and structure and insight into the recognition mechanism of effectors by R proteins. In this review, we will highlight the recent insights into the classification, recognition and activation of R proteins, genome-wide analysis and evolution of *R* genes in plant species and new strategies to develop enhanced disease resistance.

Classification, Recognition and Activation of R Proteins

Plant R proteins can be classified into different families based on their conserved domains (Liu *et al.*, 2007; Gururani *et al.*, 2012). The large majority of R proteins contain the leucine-rich repeats (LRRs) domains that play a key role in recognition specificity (van Ooijen *et al.*, 2007). Based on their presumed cellular localizations and domain compositions, these LRR-containing R proteins are divided into three major classes (Fig. 10.1). The first class anchored to the membrane all consist of extracellular LRRs and a transmembrane domain (TD). However, the intracellular C-terminal part is diverse. One subclass has a short C terminus, which is termed as receptor-like proteins (RLPs). The most representative examples are from tomato resistance

genes *Cf-2*, *Cf-4* and *Cf-9* against leaf mould caused by the fungal pathogen *Cladosporium fulvum* (Gururani *et al.*, 2012). A more recent example of this class is *LepR3*, the first brassica resistance gene against *Leptosphaeria maculans* that causes blackleg disease of canola (Larkan *et al.*, 2013). Another subclass contains an intracellular kinase domain and is known as receptor-like kinases (RLKs). The rice *Xa21* and *Xa3/Xa26* resistance genes against bacterial blight, caused by *Xanthomonas oryzae* pv. *oryzae*, are examples of this subclass (Song *et al.*, 1995; Sun *et al.*, 2004). The last subclass resistance genes are *Ve1* and *Ve2* that confer resistance against *Verticillium* and contain the intracellular PEST (Pro-Glu-Ser-Thr) domain (found only in *Ve2* not in *Ve1*) for protein degradation and along with a short protein motif (ECS) (Thomma *et al.*, 2011). The two classes of intracellular LRR-containing R proteins have a conserved central nucleotide-biding domain (NB) and a C-terminal LRR (NLR) but differ in their N-terminal domain. Based on the N-terminal domain, NLR is divided into two subclasses: TIR-NB-LRR (TNLs) and non-TNLs (Fig. 10.1). This division is based on either the presence or absence of an N-terminal domain with homology to the *Drosophila* Toll and the human Interlerkin-Receptor protein called the TIR domain (Ma *et al.*, 2012). Examples of TNLs include the flax *L6*, *Arabidopsis WRR4* (Borhan *et al.*, 2010) and *RPP5* and tobacco *N* gene (van Ooijen *et al.*, 2007). The *Arabidopsis* RRS1-R TNLs has a C-terminal WRKY domain; a protein domain found in a subset of plant-specific transcription factors (Lahaye, 2002). By far the largest group of non-TNL R proteins carries a Coiled-Coil (CC) N-terminal domain and this group is collectively referred as to CC-NB-LRRs (CNLs) (Fig. 10.1). The examples of this class of resistance gene include the *Arabidopsis RPS2* and *RPM1* (resistance against *P. syringae*) and *I-2* gene from tomato for resistance against *Fusarium oxysporum* (van Ooijen *et al.*, 2007; Gururani *et al.*, 2012). The CC domain is sometimes joined by additional domains such as a so-called Solanaceae domain (SD) or a BED (BEAF/DREF) zinc finger DNA-binding domain, which is located between the CC and NB domains (Liu *et al.*, 2007).

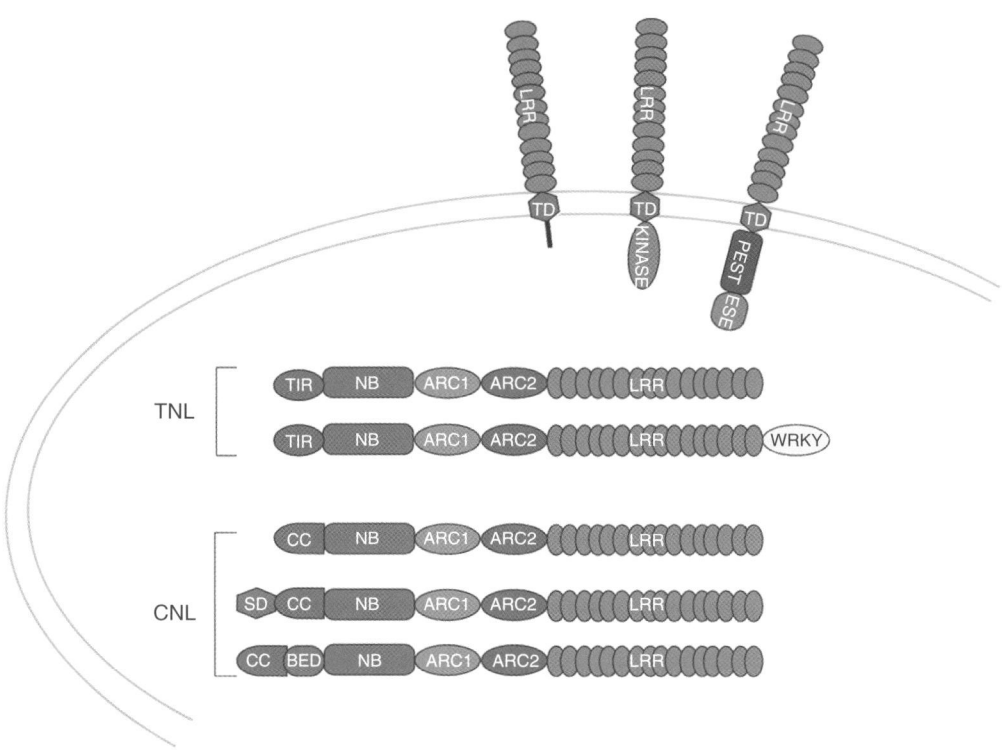

Fig. 10.1. Schematic representation of LRR-containing R proteins. The (sub)domains are represented as different shaded boxes. LRR, leucine-rich repeats; TD, transmembrane domain; KINASE, kinase domain; PEST, protein degradation domain (Pro-Glu-Ser-Thr); ESE, endocytosis cell signalling domain; TIR, toll-interleukin-1 receptor; CC, coiled coil; NB, nucleotide binding; ARC1 and ARC2: APAF1, R protein and CED4; SD, solanaceous domain; BED, BEAF/DREAF zinc finger domain; and WRKY, WRKY transcription factor.

Mechanism of effector recognition by membrane-localized plant resistance RLP or RLK proteins is poorly defined. But in the case of NLRs and based on several examples of effector recognition by NLRs, two models describing direct or indirect perception of pathogen effectors have been proposed. The receptor-ligand model is based on the direct and physical interaction between the NLR and its cognate effector (Ellis *et al.*, 2007). Indirect recognition relies on the host (accessory) protein that interacts with both NLR and the effector. In both cases, the NLRs are kept in a signalling-competent but auto-inhibited state by intra- or inter-molecular interactions in the absence of effector (Takken and Goverse, 2012). Experimental evidence supports that the inter-domain interaction of NLRs, interactions with chaperones, the proteasome and accessory

proteins are required to keep NLRs in a resting state (Ma *et al.*, 2012; Takken and Goverse, 2012). Upon effector recognition by the C-terminal half of LRR, the auto-inhibition state is released; possibly by a conformational change in the interface between the N-terminal half of the LRR and the NB-ARC domain. Release of the auto-inhibition state is thought to trigger a subsequent conformational change in the NB-ARC that allows the exchange of ADP for ATP. Exchange of ADP for ATP is predicted to shift the closed ADP-bound conformation of NLR to an open conformation that results in an activated NLR protein to initiate downstream signalling (Ma *et al.*, 2012; Takken and Goverse, 2012). NLR domain(s) that initiate and transduce the signalling to the downstream component are not well defined, and experimental evidences have shown that different domains from different

NLR proteins are required to initiate down-stream signalling (Takken and Goverse, 2012).

Genome-wide Identification and Evolution of *NLR* Genes

Among cloned *R* genes, the large majority of them are *NLR*s that account for more than half of the *R* genes cloned to date (McHale *et al.*, 2006). With the availability of plant genome sequences, a genome-wide survey of *NLR* genes has been widely conducted. So far, *NLR* repertoires have been characterized in 14 monocot and dicot species from the model plant *Arabidopsis* to crop plants rice and maize (reviewed by Jacob *et al.*, 2013). According to these studies, the majority of plant species surveyed so far have quite a large number of *NLR* genes that account for the approximately 0.3–1.8% of genes encoded by the plant genome (Mun *et al.*, 2009). Interestingly, the numbers of *NLR* genes in the monocot and dicot species is largely variable from 34 in papaya to 459 in wine grape (Yang *et al.*, 2008; Porter *et al.*, 2009). Moreover, Jacob and colleagues suggested that since the variability in the number of *NLR* genes in the plant has no correlation to the phylogeny, expansion and contraction of NLR genes is driven by species-specific mechanisms (Jacob *et al.*, 2013). *NLR* genes are mainly composed of TNL and CNL. CNL are the predominate class in *Brachypodium distachyon*, *Solanum tuberosum* and *Vitis vinifera* (Jacob *et al.*, 2013). TNL are an abundant class in *A. thaliana* (Meyers *et al.*, 2003), *Brassica rapa* (Mun *et al.*, 2009) and *Glycine max* (Kang *et al.*, 2012). However, TNL are absent in the cereal crops *Oryza sativa*, *Sorghum bicolor* and *Zea mays* (Li *et al.*, 2010).

The genomic organization of *NLR* genes shows a clear tendency that they are clustered in specific chromosomal regions (Ameline-Torregrosa *et al.*, 2008; Mun *et al.*, 2009; Jacob *et al.*, 2013). This can be exemplified by the genome-wide identification of *NLR* genes in *Medicago truncatula*, in which 49.5% of *NLR*s are located in clusters, each cluster containing at least three *NLR*s, 34% of TNL encoded by chromosome 6 and 40% of CNL encoded by chromosome 3 (Ameline-Torregrosa *et al.*, 2008). A similar trend is observed in *B. distachyon*; 51%

are clustered (Li *et al.*, 2010). Higher rates are found in *S. tuberosum* and *O. sativa*, where 76% of *NLR*s are located in 44 clusters and 73% of the mapped *NLR* genes are grouped into 63 clusters, respectively (Zhou *et al.*, 2004; Jupe *et al.*, 2012). Moreover, the percentage of gene clustering in the relatively close *Brassicaceae* family is different. In *B. rapa* 92 *NLR* genes are identified and 48% are clustered (Mun *et al.*, 2009). In *A. thaliana* and *Arabidopsis lyrata* 159 and 185 *NLR* genes have been identified where 71.1% and 63.8% form clusters, respectively (Guo *et al.*, 2011). Based on the contents, *NLR* clusters can be classified into two types: homogenous (simple) clusters contain the same type of NLRs (TNL or CNL) and heterogeneous (mixed) clusters contain a mixture of diverse *NLR*s (Jacob *et al.*, 2013). The mixed cluster often contains the majority of the *NLR* gene family members although it varies depending on the species and also contains some phylogenetically distant genes with only nucleotide-binding domain (Marone *et al.*, 2013). As examples, the proportion of heterogeneous *NLR* clusters is similar in *O. sativa* and *Arabidopsis*, with about 25% phylogenetically mixed *NLR* clusters (Ameline-Torregrosa *et al.*, 2008; Marone *et al.*, 2013). A similar percentage of mixed *NLR* clusters have been identified in *S. tuberosum* (Jupe *et al.*, 2012). According to the genome evolution mechanisms, these two types of clusters are evolved differentially. The simple cluster that contains one type of *NLR* gene is derived from tandem duplication and the mixed cluster has undergone ectopic duplication, transposition, or large-scale segmental duplication, with subsequent local rearrangements (Marone *et al.*, 2013). The expansion of the cluster in the chromosome is largely variable and seems to correlate with the density of transposable elements on the same chromosome (Ameline-Torregrosa *et al.*, 2008; Li *et al.*, 2010). Transposable elements have been shown to accelerate *NLR* evolution (Meyers *et al.*, 2005). Interestingly, the recent finding from Tsuchiya and Eulgem demonstrates that transposonal elements can be co-opted to beneficially regulate *NLR* expression (Tsuchiya and Eulgem, 2013).

NLR genes belong to one of the most ancient gene families in plants and their evolution is largely driven by the variable selection pressures from the heterogeneous pathogen population (Marone *et al.*, 2013; Michelmore *et al.*, 2013).

Plants' NLR proteins are composed of TIR, CC, NB and LRR domains, but theses domains are not unique to R proteins and they also exist in various other proteins with diverse functions. During the evolution, fusion of three building blocks (TIR or CC, NB and LRR) acts to generate the ancestral plant NLR proteins, but evolutionary rate and the selection effect on these building blocks are not homogeneous, which leads to each building block (domain) with distinct functional limitations (Yue *et al.*, 2012; Michelmore *et al.*, 2013). The NB domain is highly conserved and the LRR domain is most variable (Zhou *et al.*, 2004). This high variability is consistent with its role to recognize diverse effector gene products from different pathogens. With the availability of whole genome-sequence from higher plants to ancestral taxa common for plants and animals, the comparison of *NLR* repertories among them could further our understanding of *R* gene evolution. Recently, Yue *et al.* (2012) conducted an extensive genome-wide comparison of *NLR* repertories among 38 representative model organisms. They found that the core building blocks of NLRs, such as NB-ARC, TIR and LRR, were very ancient and existed before the split of prokaryotes and eukaryotes (Yue *et al.*, 2012). The origin of NBS domain in plant lineages can be dated back to Coleochaetales and the fusion between NB and LRR or TIR occurred no later than the divergence of bryophytes. Another finding from Yue *et al.* is that TNL probably has an earlier origin than its counterpart non-TNL (CNL), but Cannon *et al.* (2002) and Meyers *et al.* (2003) proposed that non-TIR (CNL) genes are more ancient than TIR-type genes (Cannon *et al.*, 2002; Meyers *et al.*, 2003). Although NLR proteins exhibit homology and similar domain compositions to animal innate immune NOD (nucleotide binding oligomerization domain) protein, the phylogenetic analysis suggested that both have different origins and their striking resemblance was driven by convergent evolution (Yue *et al.*, 2012).

New Strategies for Developing Durable Plant Resistance

Although traditional breeding and transgenic approaches through introducing single *R*

genes into elite cultivars have enabled the provision of effective immunity against infection with pathogens containing the corresponding effector (race-specific resistance), achieving durable resistance remains a problem (Gust *et al.*, 2010). The lack of durability is reflected by failure to confer resistance to even closely related pathogen strains or races lacking the corresponding *Avr* gene. Moreover, race-specific resistance can be overcome by a single mutation in the matching *Avr* gene. Nevertheless, some experimental evidence has shown that some *R* genes can recognize more than one independent effector among pathogens. For example, the extracellular plant immune receptor protein Cf-2 of the red currant tomato can sense perturbations in a common virulence target of two independently evolved effectors of a fungus and a nematode (Lozano-Torres *et al.*, 2012). Introgression of the *Rpi-blb1* gene from *Solanum bulboscastanum* into cultivated potato has provided durable resistance to potato late blight (Song *et al.*, 2003). Similarly, the *Arabidopsis* white rust resistance gene *WRR4* recognizes several races of *Albugo candida*. In addition, transgenic *Brassica* expressing the *Arabidopsis* *WRR4* gene becomes resistant to multiple races of *A. candida* (Borhan *et al.*, 2010). Therefore, *R* genes with the capacity to recognize multiple races/pathogens are candidates for creating more durable resistance. As an alternative to single *R* gene deployment, multiple *R* genes can be incorporated into a single cultivar to achieve durable resistance. This process is referred to as gene pyramiding. In principle, avoidance of multiple *R* gene-mediated resistance requires the pathogen to accumulate mutations in multiple *Avr* genes, which is unlikely to occur because such mutations are likely to have a strong impact on the pathogen virulence (McDowell and Woffenden, 2003). Durable resistance is highly suitable for crop production and yield stability, but breeding multiple *R* genes into an elite cultivar in practice is time-consuming and difficult especially for different genes with similar phenotypes. However, *R* gene pyramiding could be accelerated with the use of marker-assisted selection. The transgenic approach is an alternative tool to generate plants expressing multiple *R* genes. This approach was used to generate the multiple alleles of

Pm3 gene in wheat to confer durable resistance to powdery mildew (Brunner *et al.*, 2012). Another strategy for durable resistance is to understand the polygenic resistance to multiple pathogens (Michelmore *et al.*, 2013). Diverse germplasm collections and highly efficient genotyping of various populations could identify genes for resistance to multiple pathogens. In the near future, genome-wide association studies (GWAS) and meta-QTL (Quantitative Trait Loci) analyses as well as the availability of increasing number of genome sequences will accelerate understanding of polygenic resistance and will assist breeders to incorporate polygenic resistance into new cultivars (Schweizer and Stein, 2011; Michelmore *et al.*, 2013).

In addition to classical *RLP/RLK* and *NLR* genes, another group of plant resistance genes that has evolved to co-opt pathogen virulence function to serve as a trap for transcription activator-like (TAL) effectors from *Xanthomonas* and *Ralstonia* (Dangl *et al.*, 2013). TAL effectors are DNA binding proteins and activate host gene expression to promote pathogen virulence. Resistance genes *Xa27* (from rice) and *Bs3* and *Bs4c* from pepper act as traps for TAL effectors. Presence of TAL binding sites in the promoter of these *R* genes acts as a trap. Binding of TAL effectors to the promoter triggers the expression of the *R* genes and inhibition of pathogen growth (Gu *et al.*, 2005; Strauss *et al.*, 2012). Engineering broad spectrum resistance by adding TAL-effector binding sites against disparate pathogen strains or pathogens to the promoter regions of *R* genes has been demonstrated to confer enhanced and potentially durable disease resistance to pathogen infection.

Another concept to control plant disease resistance is to discover and manipulate host genes that are involved in the pathogen proliferation or infection, which is now termed as susceptibility (*S*) gene-mediated resistance (Dangl *et al.*, 2013; Gawehns *et al.*, 2013). Susceptibility (*S*) genes refer to host genes that are manipulated by a pathogen to facilitate its survival and infection (Dangl *et al.*, 2013; Gawehns *et al.*, 2013). Hence, the loss or inactivation of *S* genes will compromise the pathogen's ability to cause disease (Gawehns *et al.*, 2013). Recently, *S* genes have been proposed as an alternative to *R* genes in resistance breeding as they have the potential to be more durable (Gust *et al.*, 2010; Dangl *et al.*, 2013; Gawehns *et al.*, 2013). Indeed, the well-known recessive disease-resistance genes turned out to be *S* genes. For example, the barley recessive *mlo* gene is required for powdery mildew pathogen invasion and it has been used in practice to provide durable broad-spectrum resistance against powdery mildew for more than 70 years (Humphry *et al.*, 2011). Likewise, loss-of-function mutations of *MLO* genes in pea, *Arabidopsis* and tomato confer highly effective broad-spectrum powdery mildew resistance (Bai *et al.*, 2008; Humphry *et al.*, 2011; Dangl *et al.*, 2013). Based on our understanding of the plant–pathogen interaction mechanism, host proteins that are targeted by pathogen effector proteins might represent *S* genes, as well. For example, the *Pseudomonas syringae* effector HopZ2 target of MLO2 is required for HopZ2 virulence and contributes to resistance against *P. syringae* in *Arabidopsis*, although *mlo2* (T-DNA insertion) knockout lines displayed the variable resistance against *P. syringae* (Lewis *et al.*, 2012). So far, about 30 *S* genes have been identified, among which most come from the genetic screens for recessive resistance in wild germplasm. Nevertheless, only few of them remain the potential to be used in commercial breeding because of their pleiotropic effects (Pavan *et al.*, 2010; Gawehns *et al.*, 2013). Hence, in the future, natural *S*-gene alleles, identified by screening germplasm of certain plant species and retaining their intrinsic function despite insensitivity to effector manipulation, would be the ideal source for a commercial breeding programme (Pavan *et al.*, 2010).

Conclusions

Over the past few decades, much progress has been made in our understanding the molecular mechanism underlying the plant–pathogen interaction. However, there are several aspects of *R*-mediated resistance that still remain fragmentary. Indeed, we need more structural and biochemical evidences to get insights into the NLR activation process. Detailed understanding of the spatial and temporal dynamics of the NLR and the mystery behind its action to trigger different defence responses at cytoplasm or nucleus level remain obscure. Detailed characterization of downstream signalling components of NLR action remains unclear. In the future, the development of new technologies combined with current classical genetics and biochemical tools will accelerate research on NLR structure, recognition, activation and downstream signalling cascade. These fundamental researches will provide mechanistic insights into NLR actions and comprehensive understanding of NLR-mediated resistance. In the long term, a detailed understanding of R proteins will enable us to engineer *R* genes that

efficiently control plant disease. The available strategies for developing durable resistance in the present-day agriculture are numerous and this review covers only a set of the approaches being explored. However, the successful applications of discussed strategies in practice are limited. This is further complicated by the diverse strategies that pathogens exploit and their ability to evolve rapidly. Over the past decade advances in our understanding, the molecular mechanisms of plant immunity and recent progresses in gene editing combined with technological advances in genome sequencing hold great promise for developing durable disease-resistant crops.

References

Ameline-Torregrosa, C., Wang, B.B., O'Bleness, M.S., Deshpande, S., Zhu, H., Roe, B., Young, N.D. and Cannon, S.B. (2008) Identification and characterization of nucleotide-binding site-leucine-rich repeat genes in the model plant *Medicago truncatula*. *Plant Physiology* 146, 5–21.

Bai, Y., Pavan, S., Zheng, Z., Zappel, N.F., Reinstadler, A., Lotti, C., De Giovanni, C., Ricciardi, L., Lindhout, P., Visser, R., Theres, K. and Panstruga, R. (2008) Naturally occurring broad-spectrum powdery mildew resistance in a Central American tomato accession is caused by loss of mlo function. *Molecular Plant-Microbe Interaction* 21, 30–39.

Borhan, M.H., Holub, E.B., Kindrachuk, C., Omidi, M., Bozorgmanesh-Frad, G. and Rimmer, S.R. (2010) WRR4, a broad-spectrum TIR-NB-LRR gene from *Arabidopsis thaliana* that confers white rust resistance in transgenic oilseed Brassica crops. *Molecular Plant Pathology* 11, 283–291.

Brunner, S., Stirnweis, D., Diaz Quijano, C., Buesing, G., Herren, G., Parlange, F., Barret, P., Tassy, C., Sautter, C., Winzeler, M. and Keller, B. (2012) Transgenic Pm3 multilines of wheat show increased powdery mildew resistance in the field. *Plant Biotechnology Journal* 10, 398–409.

Cannon, S.B., Zhu, H., Baumgarten, A.M., Spangler, R., May, G., Cook, D.R. and Young, N.D. (2002) Diversity, distribution and ancient taxonomic relationships within the TIR and non-TIR NBS-LRR resistance gene subfamilies. *Journal of Molecular Evolution* 54, 548–562.

Dangl, J.L., Horvath, D.M. and Staskawicz, B.J. (2013) Pivoting the plant immune system from dissection to deployment. *Science* 341, 746–751.

Ellis, J.G., Dodds, P.N. and Lawrence, G.J. (2007) Flax rust resistance gene specificity is based on direct resistance-avirulence protein interactions. *Annual Review of Phytopathology* 45, 289–306.

Fisher, M.C., Henk, D.A., Briggs, C.J., Brownstein, J.S., Madoff, L.C., McCraw, S.L. and Gurr, S.J. (2012) Emerging fungal threats to animal, plant and ecosystem health. *Nature* 484, 186–194.

Gawehns, F., Cornelissen, B.J. and Takken, F.L. (2013) The potential of effector-target genes in breeding for plant innate immunity. *Microb Biotechnology* 6, 223–229.

Giraldo, M.C. and Valent, B. (2013) Filamentous plant pathogen effectors in action. *Nature Review of Microbiology* 11, 800–814.

Gu, K., Yang, B., Tian, D., Wu, L., Wang, D., Sreekala, C., Yang, F., Chu, Z., Wang, G.L., White, F.F. and Yin, Z. (2005) R gene expression induced by a type-III effector triggers disease resistance in rice. *Nature* 435, 1122–1125.

Guo, Y.L., Fitz, J., Schneeberger, K., Ossowski, S., Cao, J. and Weigel, D. (2011) Genome-wide comparison of nucleotide-binding site-leucine-rich repeat-encoding genes in *Arabidopsis*. *Plant Physiology* 157, 757–769.

Gururani, M.A., Venkatesh, J., Upadhyaya, C.P., Nookaraju, A., Pandey, S.K. and Park, S.W. (2012) Plant disease resistance genes: current status and future directions. *Physiological and Molecular Plant Pathology* 78, 51–65.

Gust, A.A., Brunner, F. and Nurnberger, T. (2010) Biotechnological concepts for improving plant innate immunity. *Current Opinion in Biotechnology* 21, 204–210.

Humphry, M., Reinstadler, A., Ivanov, S., Bisseling, T. and Panstruga, R. (2011) Durable broad-spectrum pow-dery mildew resistance in pea er1 plants is conferred by natural loss-of-function mutations in PsMLO1. *Molecular Plant Pathology* 12, 866–878.

Jacob, F., Vernaldi, S. and Maekawa, T. (2013) Evolution and conservation of plant NLR functions. *Frontier Immunology* 4, 297.

Jupe, F., Pritchard, L., Etherington, G.J., Mackenzie, K., Cock, P.J., Wright, F., Sharma, S.K., Bolser, D., Bryan, G.J., Jones, J.D. and Hein, I. (2012) Identification and localisation of the NB-LRR gene family within the potato genome. *BMC Genomics* 13, 75.

Kang, Y.J., Kim, K.H., Shim, S., Yoon, M.Y., Sun, S., Kim, M.Y., Van, K. and Lee, S.H. (2012) Genome-wide mapping of NBS-LRR genes and their association with disease resistance in soybean. *BMC Plant Biology* 12, 139.

Lahaye, T. (2002) The *Arabidopsis* RRS1-R disease resistance gene – uncovering the plant's nucleus as the new battlefield of plant defense? *Trends in Plant Science* 7, 425–427.

Larkan, N.J., Lydiate, D.J., Parkin, I.A., Nelson, M.N., Epp, D.J., Cowling, W.A., Rimmer, S.R. and Borhan, M.H. (2013) The *Brassica napus* blackleg resistance gene LepR3 encodes a receptor-like protein triggered by the *Leptosphaeria maculans* effector AVRLM1. *New Phytologist* 197, 595–605.

Lewis, J.D., Wan, J., Ford, R., Gong, Y., Fung, P., Nahal, H., Wang, P.W., Desveaux, D. and Guttman, D.S. (2012) Quantitative interactor screening with next-generation sequencing (QIS-Seq) identifies *Arabidopsis thaliana* MLO2 as a target of the *Pseudomonas syringae* type III effector HopZ2. *BMC Genomics* 13, 8.

Li, J., Ding, J., Zhang, W., Zhang, Y., Tang, P., Chen, J.Q., Tian, D. and Yang, S. (2010) Unique evolutionary pattern of numbers of gramineous NBS-LRR genes. *Molecular Genetics Genomics* 283, 427–438.

Liu, J., Liu, X., Dai, L. and Wang, G. (2007) Recent progress in elucidating the structure, function and evolu-tion of disease resistance genes in plants. *Journal of Genetics and Genomics* 34, 765–776.

Lozano-Torres, J.L., Wilbers, R.H., Gawronski, P., Boshoven, J.C., Finkers-Tomczak, A., Cordewener, J.H., America, A.H., Overmars, H.A., Van 't Klooster, J.W., Baranowski, L., Sobczak, M., Ilyas, M., van der Hoorn, R.A., Schots, A., de Wit, P.J., Bakker, J., Goverse, A. and Smant, G. (2012) Dual disease resistance mediated by the immune receptor Cf-2 in tomato requires a common virulence target of a fungus and a nematode. *Proceedings of the National Academy of Sciences USA* 109, 10119–10124.

Ma, L., van den Burg, H.A., Cornelissen, B.J.C. and Takken, F.L.W. (2012) Molecular basis of effector recogni-tion by plant NB-LRR proteins. In: *Molecular Plant Immunity*. Wiley-Blackwell, Oxford, UK, pp. 23–40.

Marone, D., Russo, M.A., Laido, G., De Leonardis, A.M. and Mastrangelo, A.M. (2013) Plant nucleotide bind-ing site leucine-rich repeat (NBS-LRR) genes: active guardians in host defense responses. *International Journal of Molecular Science* 14, 7302–7326.

McDowell, J.M. and Woffenden, B.J. (2003) Plant disease resistance genes: recent insights and potential ap-plications. *Trends in Biotechnology* 21, 178–183.

McHale, L., Tan, X., Koehl, P. and Michelmore, R.W. (2006) Plant NBS-LRR proteins: adaptable guards. *Genome Biology* 7, 212.

Meyers, B.C., Kozik, A., Griego, A., Kuang, H. and Michelmore, R.W. (2003) Genome-wide analysis of NBS-LRR-encoding genes in *Arabidopsis*. *Plant Cell* 15, 809–834.

Meyers, B.C., Kaushik, S. and Nandety, R.S. (2005) Evolving disease resistance genes. *Current Opinion in Plant Biology* 8, 129–134.

Michelmore, R.W., Christopoulou, M. and Caldwell, K.S. (2013) Impacts of resistance gene genetics, function and evolution on a durable future. *Annual Review of Phytopathology* 51, 291–319.

Mun, J.H., Yu, H.J., Park, S. and Park, B.S. (2009) Genome-wide identification of NBS-encoding resistance genes in *Brassica rapa*. *Molecular Genetics Genomics* 282, 617–631.

Pavan, S., Jacobsen, E., Visser, R.G. and Bai, Y. (2010) Loss of susceptibility as a novel breeding strategy for durable and broad-spectrum resistance. *Molecular Breeding* 25, 1–12.

Porter, B.W., Paidi, M., Ming, R., Alam, M., Nishijima, W.T. and Zhu, Y.J. (2009) Genome-wide analysis of *Carica papaya* reveals a small NBS resistance gene family. *Molecular Genetics Genomics* 281, 609–626.

Schweizer, P. and Stein, N. (2011) Large-scale data integration reveals colocalization of gene functional groups with meta-QTL for multiple disease resistance in barley. *Molecular Plant-Microbe Interaction* 24, 1492–1501.

Segonzac, C. and Zipfel, C. (2011) Activation of plant pattern-recognition receptors by bacteria. *Current Opinion in Microbiology* 14, 54–61.

Song, J., Bradeen, J.M., Naess, S.K., Raasch, J.A., Wielgus, S.M., Haberlach, G.T., Liu, J., Kuang, H., Austin-Phillips, S., Buell, C.R., Helgeson, J.P. and Jiang, J. (2003) Gene RB cloned from *Solanum bulbocasta-num* confers broad spectrum resistance to potato late blight. *Proceedings of the National Academy of Sciences USA* 100, 9128–9133.

Song, W.Y., Wang, G.L., Chen, L.L., Kim, H.S., Pi, L.Y., Holsten, T., Gardner, J., Wang, B., Zhai, W.X., Zhu, L.H., Fauquet, C. and Ronald, P. (1995) A receptor kinase-like protein encoded by the rice disease resistance gene, Xa21. *Science* 270, 1804–1806.

Strauss, T., van Poecke, R.M., Strauss, A., Romer, P., Minsavage, G.V., Singh, S., Wolf, C., Kim, S., Lee, H.A., Yeom, S.I., Parniske, M., Stall, R.E., Jones, J.B., Choi, D., Prins, M. and Lahaye, T. (2012) RNA-seq pinpoints a *Xanthomonas* TAL-effector activated resistance gene in a large-crop genome. *Proceedings of the National Academy of Sciences USA* 109, 19480–19485.

Sun, X., Cao, Y., Yang, Z., Xu, C., Li, X., Wang, S. and Zhang, Q. (2004) Xa26, a gene conferring resistance to *Xanthomonas oryzae* pv. *oryzae* in rice, encodes an LRR receptor kinase-like protein. *Plant Journal* 37, 517–527.

Takken, F.L. and Goverse, A. (2012) How to build a pathogen detector: structural basis of NB-LRR function. *Current Opinion in Plant Biology* 15, 375–384.

Thomma, B.P., Nurnberger, T. and Joosten, M.H. (2011) Of PAMPs and effectors: the blurred PTI-ETI dichotomy. *Plant Cell* 23, 4–15.

Tsuchiya, T. and Eulgem, T. (2013) An alternative polyadenylation mechanism coopted to the *Arabidopsis* RPP7 gene through intronic retrotransposon domestication. *Proceedings of the National Academy of Sciences USA* 110, E3535–3543.

van Ooijen, G., van den Burg, H.A., Cornelissen, B.J. and Takken, F.L. (2007) Structure and function of resistance proteins in solanaceous plants. *Annual Review of Phytopathology* 45, 43–72.

Vurro, M., Bonciani, B. and Vannacci, G. (2010) Emerging infectious diseases of crop plants in developing countries; impact on agriculture and socio-economic consequences. *Food Security* 2, 113–132.

Yang, S., Zhang, X., Yue, J.X., Tian, D. and Chen, J.Q. (2008) Recent duplications dominate NBS-encoding gene expansion in two woody species. *Molecular Genetics and Genomics* 280, 187–198.

Yue, J.X., Meyers, B.C., Chen, J.Q., Tian, D. and Yang, S. (2012) Tracing the origin and evolutionary history of plant nucleotide-binding site-leucine-rich repeat (NBS-LRR) genes. *New Phytologist* 193, 1049–1063.

Zhou, T., Wang, Y., Chen, J.Q., Araki, H., Jing, Z., Jiang, K., Shen, J. and Tian, D. (2004) Genome-wide identification of NBS genes in japonica rice reveals significant expansion of divergent non-TIR NBS-LRR genes. *Molecular Genetics and Genomics* 271, 402–415.

11 Insect Pests

Sarwan Kumar[1]* and Y.P. Singh[2]

[1]*Department of Plant Breeding and Genetics, Punjab Agricultural University, Ludhiana, 141004, Punjab, India;* [2]*ICAR-Directorate of Rapeseed Mustard Research, Bharatpur, 321303 Rajasthan, India*

Introduction

Brassicaceae are cultivated the world over under varied agroclimatic conditions (Suwabe *et al.*, 2006; Hong *et al.*, 2008) and for diverse agricultural usage. Oilseed brassicas have become important sources of oil and protein due to the increased oilseed production of *Brassica rapa*, *B. juncea*, *B. napus* and *B. carinata* in the past three decades (Font *et al.*, 2003). *B. napus* alone contributed 58.56 Mt out of total oilseed production of 446.97 Mt amounting to 13.1% during the year 2010/11 (USDA, 2011). Over the course of domestication for thousands of years, domesticated plants have lost many of the genes controlling defence mechanisms employed by their ancestors to ward off herbivores, including insect pests. Very little attention was paid by plant breeders to maintain adequate levels of insect and disease resistance in them as synthetic pesticidal chemicals were available for their management, which at that time were thought to be satisfactory. More emphasis was placed on the selection for crop productivity and desirable quality traits. Further, globalization of the production system and monoculture over large areas allowed insects to multiply with ample food and space available to them.

The Pest Complex

A wide diversity of insects is associated with brassicas in different parts of the world (Lamb, 1989). Even in a specific region, different species of insects attack at different crop growth stages. The insect pests feeding on these crops can be broadly classified as chewing (Coleoptera, Lepidoptera, Hymenoptera), piercing and/or sucking (Heteroptera, Homoptera, Thysanoptera) and putrifying (Diptera). A number of insects reported to be associated with brassicas are listed in Table 11.1.

Aphids Complex

The family Aphididae comprises about 4700 species in the world (Remaudière and Remaudière, 1997), of which 450 species are endemic on crop plants (Blackman and Eastop, 2000), but only 100 have successfully exploited the agricultural environment to the extent that they are of significant economic importance (Blackman and Eastop, 2007). They are the specialized phloem sap feeders resulting in significant yield losses in many crops partly due to their capacity for extremely rapid population

---

*Corresponding author, e-mail: Sarwanent@pau.ent

Table 11.1. Important insect pests associated with brassica crops (modified from Hegedus and Erlandson, 2012).

Order Common name	Scientific name	Family	Distribution
Coleoptera			
Bronzed field beetle	*Adelium brevicorne*	Tenebrionidae	Australia
Black headed pasture cockchafer	*Aphodius tasmaniae*	Scarabaeidae	Australia
Cabbage seed weevil	*Ceutorhyncus assimilis*	Curculionidae	Europe, N. America
Rape stem weevil	*Ceutorhyncus napi*	Curculionidae	Europe, N. America
Cabbage curculio	*Ceutorhyncus rapae*	Curculionidae	Europe, N. America, Russia
Cabbage seed pod weevil	*Ceutorhyncus obstrictus*	Curculionidae	Europe
Cabbage gall weevil	*Ceutorhyncus pleurostigma*	Curculionidae	Europe
Cabbage stem weevil	*Ceutorhyncus quadridens*	Curculionidae	Europe
Desiantha weevil	*Desiantha diversipes*	Curculionidae	Australia
Red turnip beetle	*Entomoscelis americana*	Chrysomelidae	N. America
Grey false wireworm	*Isopteron punctatissimus*	Tenebrionidae	Australia
Vegetable weevil	*Listroderes difficilis*	Curculionidae	Australia
Pollen beetle	*Meligethes aeneus*	Nitidulidae	Asia, Europe, N. America, N. Africa, Russia
Leaf beetle	*Phyllotreta aerea*	Chrysomelidae	Europe, N. America
Crucifer flea beetle	*Phyllotreta cruciferae*	Chrysomelidae	Europe, India, N. America, N. Africa
Yellow striped flea beetle	*Phyllotreta nemorum*	Chrysomelidae	Europe
Striped flea beetle	*Phyllotreta striolata*	Chrysomelidae	Europe, Canada
Turnip flea beetle	*Phyllotreta undulata*	Chrysomelidae	Australia, N. America, Europe
Cabbage stem flea beetle	*Psylliodes chrysocephala*	Chrysomelidae	Asia, Europe, N. America
Sitona weevil	*Sitona discoideus*	Curculionidae	Australia
Diptera			
Cabbage maggot/ swede midge	*Contarinia nasturtii*	Cecidomyiidae	Europe, Canada
Brassica pod midge	*Dasineura brassicae*	Cecidomyiidae	Europe
Turnip maggot	*Delia floralis*	Anthomyiidae	Europe, Canada, China, Japan, Russia
Root maggot	*Delia platura*	Anthomyiidae	Canada
Root maggot	*Delia planipalpis*	Anthomyiidae	Canada
Root maggot	*Delia florilega*	Anthomyiidae	Canada
Cabbage maggot	*Delia radicum*	Anthomyiidae	Canada
Cruciferous leaf miner	*Phytomyza horticola*	Agromyzidae	Asia, Europe, India
Heteroptera			
Bagrada, stink or painted bug	*Bagrada cruciferarum*	Pentatomidae	Africa, Arabia, India, Sri Lanka
Cabbage bug	*Eurydema oleracea*	Pentatomidae	Europe, Russia
Orange stink and shield bugs	*Eurydema* spp.	Pentatomidae	Asia, Australia, Europe, India, N. Africa, Russia, the Philippines
Lygus bug	*Lygus borealis*	Miridae	Canada
Pale legume bug	*Lygus elisus*	Miridae	N. America
Western tarnished plant bug	*Lygus hesperous*	Miridae	N. America
Tarnished plant bug	*Lygus lineolaris*	Miridae	N. America

Continued

Table 11.1. Continued.

Order Common name	Scientific name	Family	Distribution
European tarnished plant bug	*Lygus rugulipennis*	Miridae	Europe, Russia
Harlequin bug	*Murgantia histrionica*	Pentatomidae	USA
Rutherglen bug	*Nysius vinitor*	Lygaeidae	Australia
Homoptera			
Whiteflies	*Aleyrodes* spp.	Aleyrodidae	Europe, N. America, Russia, South Pacific
Cowpea aphid	*Aphis craccivora*	Aphididae	Australia
Mealy cabbage aphid	*Brevicoryne brassicae*	Aphididae	Worldwide
Turnip or mustard aphid	*Lipaphis erysimi*	Aphdidae	Worldwide
Green peach or peach potato aphid	*Myzus persicae*	Aphididae	Worldwide (warmer climates)
Turnip root aphid	*Pemphigus populitransversus*	Aphididae	USA
Hymenoptera			
Mustard sawfly	*Athalia lugens proxima*	Tenthredinidae	India
Turnip sawfly	*Athalia rosae*	Tenthredinidae	Africa, Asia, Europe, India, Japan, N. America
Lepidoptera			
Common cutworm	*Agrotis infusa*	Noctuidae	Australia
Brown cutworm	*Agrotis munda*	Noctuidae	Australia
Pale western cutworm	*Agrotis orthogonia*	Noctuidae	Canada
Alfalfa looper	*Autographa californica*	Noctuidae	N. America
Tobacco loop caterpillar	*Chrysodeixis argentifera*	Noctuidae	Australia
Brown pasture looper	*Ciampa arietaria*	Geometridae	Australia
Cabbage head caterpillar	*Crocidolomia binotalis*	Crambidae	Asia, India
Cabbage cluster caterpillar	*Crocidolomia pavonana*	Crambidae	Africa, India, South Pacific
Bihar/jute hairy caterpillar	*Spilosoma obliqua*	Arctiidae	Asia, India
Clover cutworm	*Discestra trifolii*	Noctuidae	Canada
Redbaked cutworm	*Euxoa ochrogaster*	Noctuidae	Canada
Garden pebble moth	*Evergestis forficalis*	Crambidae	Europe
Purple-backed cabbageworm	*Evergestis pallidata*	Crambidae	Europe, N. America
Cross-striped cabbageworm	*Evergestis rimosalis*	Crambidae	N. America
Native budworm	*Helicoverpa punctigera*	Noctuidae	Australia
Cotton bollworm	*Helicoverpa armigera*	Noctuidae	India
Cabbage centre-grub	*Hellula hydralis*	Crambidae	Australia
Cabbage head borer	*Hellula undalis*	Crambidae	India
Cabbage webworm	*Hellula rogatalis*	Crambidae	USA
Beet webworm	*Loxostege sticticalis*	Crambidae	Canada
Cabbage moth	*Mamestra brassicae*	Noctuidae	Asia, Europe, Japan, Russia
Bertha armyworm	*Mamestra configurata*	Noctuidae	Canada, N. America
Large white butterfly	*Pieris brassicae*	Pieridae	China, Europe, India, Russia, N. Africa
Imported cabbageworm or small white butterfly	*Pieris rapae*	Pieridae	Asia, Australia, Europe, Japan, New Zealand, N. America

Continued

Table 11.1. Continued.

Order Common name	Scientific name	Family	Distribution
Mustard white butterfly	*Pieris napi*	Pieridae	Europe, Japan, N. America
Indian cabbage white	*Pieris canidia*	Pieridae	India and adjoining countries
Diamondback moth or cabbage moth	*Plutella xylostella*	Plutellidae	Worldwide
Beet armyworm	*Spodoptera exigua*	Noctuidae	Worldwide
Tobacco cutworm	*Spodotera litura*	Noctuidae	India, Australia
Cabbage looper	*Trichoplusia ni*	Noctuidae	Worldwide
Painted lady butterfly	*Vanessa cardui*	Nymphalidae	Canada
Thysanoptera			
Plague thrips	*Thrips imaginis*	Thripidae	Australia
Onion thrips	*Thrips tabaci*	Thripidae	Worldwide

growth. Unlike the majority of insects, aphids exhibit parthenogenetic viviparity – a phenomenon that limits the need for males to fertilize females and eliminates the egg stage from the life cycle. Thus aphids produce clonally and give birth to young, and embryonic development of an aphid begins before its mother's birth, leading to telescoping of generations. All these traits help aphids to conserve energy and allow for short generation times; nymphs of certain aphid species can reach maturity in as little as 5 days (Goggin, 2007). Such an enormous propagation rate becomes manifested in abnormally high aphid populations under favourable conditions. Further, to conserve energy and to invest it in maximizing their reproduction and survival, aphid colonies exhibit wing dimorphism to produce highly fecund wingless morphs or less prolific winged progeny that can disperse to new host plant depending on the resource availability. All these strategies contribute to aphids' success and their abundance in temperate zones.

They are the main insect pests in Indian and European agriculture and important pests in horticultural crops (Ahuja *et al.*, 2009). Because of their prolific breeding and short generation time, they multiply very fast and drain large quantities of water and nutrients from plants. In addition to direct feeding damage, they also act as vectors of plant viral diseases (Dawson *et al.*, 1990; Sekhon, 1999). Aphids feed on phloem sap of their host through stylets, the needle-like piercing-sucking mouthparts. Sustained feeding and insertion

of toxic saliva by colonies of nymphs and adults in plant tissue results in the manifestation of yellowing, curling and crumpling of shoots, pods and leaves. This parasitic feeding and resource restriction results in retarded growth, poor seed formation and low oil content (Singhvi *et al.*, 1973; Bakhetia, 1983, 1987; Malik and Anand, 1984; Rohilla *et al.*, 1987). A severe attack during outbreak years can result in up to 75% loss in brassica crops (Sekhon, 1999). *Brevicoryne brassicae*, commonly known as mealy cabbage aphid, a native to Europe, is worldwide in distribution. It is a brassica specialist feeding on the phloem sap of brassica plants (Cole, 1997). Though it is primarily a pest of brassica vegetables, it may also infest other members of the genus *Brassica* (Cole, 1994a, 1997; Kift *et al.*, 2000). An exceptionally high infestation of this species was recorded on oilseed brassicas in research trials at Ludhiana, India (30.91°N, 75.85°E) during the crop season 2013/14, replacing the already prevalent turnip aphid, *Lipaphis erysimi* (Kumar, unpublished). It is a global problem with strong yield-reducing impacts. The turnip aphid, *L. erysimi*, a species of eastern Asian origin (Blackman and Eastop, 2000), is now the most serious aphid pest of brassica worldwide. It is a major productivity limitation factor in India and other subtropical regions of the world. The damage by this insect ranges from as low as 10% to as high as 90% depending on the intensity of population development and crop growth stage (Ahuja *et al.*, 2009). In addition to direct feeding damage, it is a vector of ten non-persistent plant viruses including cabbage black ring spot and mosaic diseases of cauliflower, radish and turnip (Blackman and Eastop, 1984; Rana, 2005). Being a specialist brassica feeder, it can develop on all brassica plants; however, *B. rapa* and *B. juncea* are considered better hosts than other *Brassica* species (*B. napus*, *B. nigra*, *B. carinata*) (Rana, 2005). The period of peak activities also tends to vary with species (Table 11.2). This is mainly because the aphid infestation is associated with the onset of flowering.

Large colonies of nymphs and adults cover the entire surface of shoots, flowers and pods and drain out phloem sap from them. Parthenogenesis and fast growth results in nymphs attaining reproductive maturity in less than 10 days. Since, the generation time is very short, about 45 generations are completed in 1 year. Although most *Lipaphis* populations throughout the world are parthenogenetic, a holocyclic reproduction does occur on cruciferous crops (*B. rapa*, *Raphanus sativus*) in western Honshu, Japan (Kawada and Murai, 1979). These aphids were reported to have 2n=8, while the most anholocyclic parthenogenetic populations have 2n=9, probably derived from eight chromosomes through dissociation of one autosome to produce a small, unpaired element. Though sexual morphs have been reported from north India, populations were mostly anholocyclic (Blackman and Eastop, 2007).

While these two above-mentioned aphid species are specialist feeders, green peach aphid/peach potato aphid *Myzus persicae* is a generalist feeder. The greater economic importance of this species is due to its role as a vector of plant viruses. The species has been reported to transmit more than 100 plant viruses including potato virus Y and potato leaf roll virus to the plants from family *Solanaceae* and various mosaic viruses to many other food crops such as the persistent beet western yellow virus (Ponsen, 1972; Eskanderi *et al.*, 1979; Bwye *et al.*, 1997).

Bioecology of mustard aphid

Landing (1982), Sidhu and Singh (1964) have attempted to find out the weak links in its life history so that they can be exploited in evolving an effective integrated pest management (IPM) programme. Many other workers subsequently supplemented this research. In relation to the changed spectrum of cultivars, cultural practices and the global environment, there are still many gaps in our knowledge. The mustard aphid has been reported to survive on some vegetables and wild crucifers during summer months (Sidhu and Singh, 1964; Agarwala and Bhattacharya, 1999). However, Sachan and Srivastava (1972) could not locate the pest on cabbage in Rajasthan from July to October. Lal (1977) also stated that it was not traceable in the plains of India during the summer. The studies

Table 11.2. Period of peak activity of *Lipaphis erysimi* in India as influenced by different types of cruciferous plants.

State	Peak period	Crop	Reference
Rajasthan	End-January	*B. juncea*	Ahuja, 1990
Punjab	Mid-February	*B. juncea*	Bakhetia and Sidhu, 1983
	Jan–Feb	*B. rapa*	
	Jan–March	*B. juncea*	
	March	*B. napus*	
		B. carinata	Bakhetia *et al.*, 1986
Haryana	Jan–Feb	Brassicas	Bishnoi *et al.*, 1992; Rana *et al.*, 1993
Delhi	February	*B. rapa*	Phadke, 1986
Bihar	Jan–Feb	Rape/mustard	Sinha *et al.*, 1990
Orissa	January	Rape/mustard	Rout and Senapati, 1968
Uttar Pradesh	January	*B. juncea*	Srivastava and Srivastava, 1973

on alate aphids and their movement in the country considerably help in developing a forecasting or early warning system of aphid outbreaks and thus ensure effective and economic management of this serious pest besides preventing prophylactic application of chemical insecticides. The alates of *L. erysimi* appear on rapeseed-mustard in October when these crops are just at the seedling stage. They gradually attain peak population from the end of December to mid-March on different host plants in different agroecological regions (Table 11.2). This is followed by a decline during February–March (Sachan and Srivastava, 1972; Roy, 1975; Ghosh and Mitra, 1983). The aphid population is primarily regulated by the weather parameters, composition and phenology of the host plants. Although a number of natural enemies (coccinellids, chrysopids, syrphids and *Diaeretiella rapae*) are reported to be associated with this pest, there is a lack of synchrony between their peak populations since natural enemies are active late in the season when most of the damage has already occurred (Sarwar, 2009; Kumar, 2014).

Effect of environmental factors on mustard aphid

Multiplication of mustard aphid is favoured by cool, moist and cloudy weather (Hasan *et al.*, 2009). The prevalence of favourable weather conditions for a longer period can result in a severe outbreak of aphids. Various climatic factors like fog, frost, rain, sub-zero and high temperatures have been identified as key mortality factors of mustard aphid.

Temperature

Peak incidence of mustard aphid occurs at an average temperature of 17–18°C (Bishnoi *et al.*, 1992). Severe cold during December and high temperature after March deterred its multiplication. Roy (1975) holds the view that temperature had very little effect on its multiplication during the crop growth season in West Bengal, India. Its appearance on inflorescence of the plants was positively correlated to a maximum temperature of 20–29°C in the preceding week (Chattopadhyay *et al.*, 2005).

The species had the higher intrinsic rate of increase, higher net reproductive rate and longer mean generation time at 25°C compared to a range of other temperatures tested (Hsiao, 1999). Kulat *et al.* (1997), in Nagpur, India, reported that a combination of maximum temperature and minimum temperatures in the range of 26.4–29.0°C and 8.4–12.6°C together with relative humidity (RH) of 75–85% in January provided the favourable conditions for aphid multiplication, while its population started declining at RH ≤65%.

Humidity

Relative humidity in the range of 65–83% favours the multiplication of mustard aphid. However, its influence on aphid population during the crop season, i.e. mid-January to mid-March, was found to be negligible. The appearance of the pest on inflorescence of plants was positively correlated to RH (Samdur *et al.*, 1997; Chattopadhyay *et al.*, 2005; Narjary *et al.*, 2013) with morning RH >92% and daily mean RH >75% congenial for population development.

Rainfall

The population of mustard aphid declines after heavy rainfall but increases within 1 week during the spring season. Bakhetia and Sidhu (1983) reported that the continuous rainfall for 4–5 days towards the end of February results in widespread mortality of mustard aphid populations, which fail to build up to higher proportions in the subsequent weeks. Even slight rainfall was found to have a detrimental effect on population development (Hasan *et al.*, 2009).

Diamondback Moth

Diamondback moth (DBM; *Plutella xylostella*) is the most serious pest of crucifers in many countries. It causes extensive damage to both vegetable and oilseed brassicas with yield losses up to 90% (Charleston and Kfir, 2000). It requires over US$1.0 billion in estimated annual management costs worldwide (Javier, 1990; Talekar and Shelton, 1993) in addition to the crop losses it causes. In certain regions of the world such as Nicaragua and Pakistan, heavy infestations have led farmers to plough down their crops even after repeated insecticide applications (Abro *et al.*, 1994; Pérez *et al.*, 2000). The pest attacks almost all the crop growth stages and economic damage occurs due to larval feeding on leaves. The pest is very difficult to control as it has developed resistance to several classes of insecticides including relatively new chemicals (Mohan and Gujar, 2003; Sayyed *et al.*, 2004) and was the first crop insect found to have developed resistance to DDT in 1953 in Indonesia (Johnson, 1953) and to the biological pesticide, *Bacillus thuringiensis* delta-endotoxin, in the field (Tabashnik *et al.*, 1990). The pest status of this insect can be attributed to some of the inherent characters such as high reproductive potential, short generation time, highly mobile nature of adults, ability to migrate long distances and ability to develop resistance to insecticides, besides the diversity and abundance of its host plants (Shelton, 2001b; Vickers *et al.*, 2004). Damage is caused by larvae feeding on the leaves. While the smaller larvae feed on the green tissue of the leaf leaving the cuticle intact resulting in 'window paned' leaves, the larger ones make irregular holes in the leaves.

Bioecology

The seasonal population dynamics of the pest had been studied by Sachan and Srivastava (1972) and Jayarathnam (1977) in India. The peak larval population on brassica vegetables was recorded during February–March and August–September. The population during July–September (rainy season) was found to be significantly high compared to the other seasons (Nagarkatti and Jayanth, 1982; Ahmad and Ansari, 2010). Rainfall had a negative effect on population development (Harcourt, 1985; Ahmad and Ansari, 2010) while comparatively high temperature and dry conditions were found to be conducive (Shelton, 2001a). Heavy rains lead to dislodgement and drowning of larvae (Talekar and Shelton, 1993; Kobori and Amano, 2003). The first instar larvae are susceptible to drowning even on the leaf surface where they get trapped in water at the leaf axil (Sivapragasam

et al., 1988). Temperatures above 30°C had a negative effect on egg production and larval survival (Yamada and Kawasaki, 1983). However, Kuwahara *et al.* (1996) recorded consistently high larval populations even during the hottest part of the year, i.e. March–May. The population growth in tropical climates is largely dependent on the annual pattern of atmospheric circulation (Campos *et al.*, 2006). Though a number of natural enemies such as *Cotesia plutellae*, *Diadromus collaris*, *D. fenestralis*, *D. insulare*, *Microplitis plutellae* and *Brachymeria excarinata* are present in the field (Chandramohan, 1994; Xu *et al.*, 2001; Chauhan and Sharma, 2002; Mosiane *et al.*, 2003; Navatha and Murthy, 2006), the excessive use of insecticidal chemicals is responsible for the lack of their effectiveness (Talekar and Shelton, 1993; Shelton *et al.*, 2000).

Cabbage Whites

Different species of white butterflies that attacked brassicas are *Pieris brassicae*, *P. rapae*, *P. napi* and *P. canidia*. *P. brassicae*, commonly known as large white butterfly to distinguish it from the closely related small white butterfly/imported cabbage worm, is common throughout its range of distribution in China, Europe, India and Russia. Though *P. brassicae* and *P. rapae* look similar (Capinera, 2004), differences in certain biological attributes exist. While *P. brassicae* lays eggs in masses (Bruinsma *et al.*, 2007) and larvae feed gregariously for the first three instars, *P. rapae* lays eggs singly and larvae are solitary. *P. brassicae* is one among the serious pests of vegetable brassicas and is becoming a threat to oilseed brassicas in certain regions (Kumar, 2011). On the other hand, the damage caused by *P. rapae* is slight, although it can be severe in years with high infestation (Hern *et al.*, 1996). The subspecies *P. rapae rapae* is found in Europe while the Asian populations are placed in subspecies *P. rapae crucivora*. The small species of the white butterfly, *P. rapae*, is commonly known as cabbage moth in the Americas. *P. napi*, commonly known as green-veined white butterfly, is distributed throughout Europe, Japan and North America. The larvae of this species, although they feed primarily on wild *Brassica* species, may also cause significant losses in some vegetable brassicas. The Indian cabbage white, *P. canidia*, is distributed in sub-Himalayan regions of India and Pakistan to the north-east in Assam extending up to China and also in hills of southern India. It is not as serious a pest as the large white butterfly. All these *Pieris* species are brassica specialists and are not reported to feed on plants outside of *Brassicaceae*.

Bioecology

Pieris brassicae is among the most abundant and damaging species on cole crops. In India, it passes the winter season in plains, while in summers it migrates to hilly areas (Gupta, 1984). Since the female lays eggs in masses and larvae are gregarious up to first three instars, the damage caused is more severe compared to *P. rapae*. Mortality among the younger larval instars is higher compared to the grown-up instars (Ahmad *et al.*, 2007).

There is always some natural mortality of young larvae in their life cycle when they grow from first instar to last instar larvae, pupae and finally adults. If a plant is suitable for development, then this natural mortality will be low. There are many reasons for this low mortality: (i) the plant is suitable and will contain more nutrients and less antinutritional factors; (ii) the development time on more suitable plant will be less, therefore an insect will develop quickly on a more suitable host; and (iii) since development time will be less, there will be less exposure of the insect to vagaries of nature such as biotic and abiotic mortality factors: parasitoids, predators, rains etc. Since *P. brassicae* is primarily a pest of vegetable brassicas, the mortality is the lowest on cabbage compared to other *Brassica* species (Gupta, 2002; Ahmad *et al.*, 2007; Hasan and Ansari, 2011) and the cabbage plants with higher content of volatile allyl nitriles are more attractive for *P. brassicae* females to lay eggs (Renwick *et al.*, 1983). The pest is continuously expanding its range from vegetable brassicas to oilseed brassicas (Kumar, 2011). On oilseed brassicas, it is active from January to April till crop maturity. A number of natural

enemies are associated with the pest with eggs and early instar larvae predated upon by generalist predators such as spiders, chrysopids, staphylinids and carabids (Pfiffner *et al.*, 2009), while a range of larval parasitoids result in variable level of larval natural mortality, such as *Cotesia glomerata*, *Brachymeria femorata*, *Aprostocetus taxi*, *Agrothereutes adustus*, *Blapsidotes vicinus*, *Hyposoter clauses*, *Pteromalus puparum*, *Exorista larvarum*, *E. segregata* and *Phryxe vulgaris* (Razmi *et al.*, 2011).

Cabbage Looper

Cabbage looper (*Trichoplusia ni*) is a generalist feeder and feeds on host plants such as cole crops, potato, tomato, cucumber etc. It is distributed throughout the world. In the UK, it is commonly known as Ni moth. This name is derived from the marking on the forewing of adults which resembles the lowercase Greek letter Ni ('*v*') (Waring *et al.*, 2003). Damage results from voracious feeding on plants by pale green larvae sometimes resulting in severe defoliation. On brassicas, it generally results in damage to broccoli, cauliflower, cabbage, kale and mustards. The adult of this species is a nocturnal brown moth and the females have very high fecundity of 300–600 eggs, laying them singly on the underside of leaves. Incubation period is just 3–4 days after which larvae come out of eggs and feed on leaves (Chow *et al.*, 2005; Mossler, 2005).

Cabbage Moth and Bertha Armyworm

Despite its name, cabbage moth (*Mamestra brassicae*) is not completely limited to cabbage and is a generalist feeder with an enormous host range that includes: tomato, potato, onion, beans, peas, lettuce, tobacco and even apple, pear, peaches and grapes (Ovsyannikova and Grichanov, 2010). It is widely distributed in Asia, Japan, Europe and Russia. There are variations in severity of infestations among years in parts of its distribution range, but it is an important pest in central areas of its distribution with up to 80% losses. Members of family *Brassicaceae* and *Chenopodiaceae* are the most preferred host plants (Popova, 1993). Damage results from the defoliation caused by larval feeding. Other species of genus *Mamestra*, i.e. *Mamestra configurata*, commonly known as bertha army worm, is native to North America and is distributed in the western part of North America and Canada (Powell and Opler, 2009). It is also a polyphagous pest and larvae mostly feed on the underside of leaves by making irregular holes, as well as on developing pods of canola. High larval density can result in severe crop losses.

Flea Beetles

Flea beetles are agricultural pests of many crops including cruciferous plants. The different species that infest brassicas include *Phyllotreta nemorum*, *P. undulata*, *P. cruciferae*, *P. striolata*, *P. atra*, *P. nigripes* and *P. armoraciae*. Of these, *P. nemorum*, *P. undulata*, *P. cruciferae* and *P. striolata* are important flea beetle species infesting crucifers (Ester *et al.*, 2003; Ahuja *et al.*, 2009). *Phyllotreta cruciferae* has become the dominant flea beetle pest of canola in America (Ahuja *et al.*, 2009). In northern temperate climates, flea beetles (*Phyllotreta* spp.) are univoltine and adults overwinter and emerge in spring and feed on brassica seedlings. Large number of beetles feed on young seedlings and can very quickly destroy canola fields. Since the attack occurs at the early stage of the crop, timely detection and management is very important to prevent damage. *Psylliodes chrysocephala* and *Phyllotreta nemorum*, common in Europe, bore into the stem and leaf tissue resulting in severe economic losses (Soroka, 2008). Random feeding patterns of beetles result in the formation of pits on cotyledons of *Brassica* species.

Pollen Beetles

Meligethes aeneus is the most common pest, especially on oilseed rape and turnip rape (Alford *et al.*, 2003). Feeding by large number of adults on flower buds can cause them to abort (Williams and Free, 1978). Winter rape is affected less than spring rape in Europe because it usually starts to flower before the beetles are active. Hansen (2004) in Denmark has reported up to 80% yield reduction. He has further stated that economic damage threshold of 5% of yield can be exceeded by 0.1–3.0 beetles/plant.

Weevils

Ceutorhynchus assimilis, the cabbage seed weevil, is native to Europe and is a major pest of both swede rape (*Brassica napus*) and turnip rape (*B. rapa*). It was first reported from North America (British Columbia) in 1931 from where it spread to much of the continental USA (Harmon and McCaffrey, 1997; Brodeur *et al.*, 2001) and Canada (Dolinski, 1979; Dosdall *et al.*, 2002). It is an oligophagous pest of brassica plants and damage is caused by developing larvae inside the pods. Damage from other pests which are linked to this pest may also be considered as symptoms when these other pests are present. For example, holes in the pods by seed weevil either for feeding or oviposition allows the brassica pod midge (*Dasineura brassicae*) to lay eggs in the pods leading to swollen yellow pods ('bladder pod' symptoms) and premature shedding of seeds. Damage to the pods can also lead to dark-edged spots on pods due to infection by canker pathogen, *Phoma lingam* (Newman, 1984). Another species, cabbage seed pod weevil (*C. obsrictus*), is also native to Europe and is found in Europe and North America. Adults feed on the buds causing bud blasting and larvae feed on seed within pods result in their premature shattering. Mature larvae come out from pods through exit holes to pupate in the ground. Adults emerge in July, which diapause in winter till the following spring. Another related species of genus *Ceutorhyncus*, the cabbage stem weevil (*C. pallidactylus*), is the major stem-mining pest of oilseed rape (Barari *et al.*, 2005).

Root Maggot

As the name implies root maggots feed underneath the soil and are difficult to control. *Delia radicum* syn. *brassicae* (the cabbage root fly)

and *D. floralis* (turnip root fly/summer cabbage fly) are the two economically important
root maggots infesting brassica crops. Damage is caused by maggots feeding on roots,
which is only noticed when plants show signs
of wilting and stunted growth (Klingen *et al.*,
2002). Soroka *et al.* (2004) has reported more
than 95% infestation of root maggots on brassicas in Canada. Flies of both species emerge
to lay eggs at the base of plants near soil during early spring. Larval feeding results in
death of the plant and there is very little recovery leading to a significant reduction in
yield (Blossey and Hunt-Joshi, 2003).

Pod Gall Midge or Brassica Pod Midge

The pod gall midge (*Dasineura brassicae*) is an
important pest of oilseed rape in central and
northern Europe (Hughes and Evans, 2003).
Affected plants show poor seed development
and premature pod shattering (Coll, 1991).
Adult flies are very small with a very delicate
ovipositor. Therefore only young and tender
pods are usually available for ovipositing
(Ankersmith, 1956; Doberitz, 1973). The adult
finds pods of brassica plants to be too hard to
puncture for oviposition, therefore first generation flies prefer to lay eggs on pods already
damaged by cabbage seed weevil (*Ceutorhyncus
assimilis*) and the flies of later generations
lay their eggs on pods damaged by wind or
by a fungal disease or by blossom beetles
(Czajkowska and Dmoch, 1975; Hughes and
Evans, 2003). Eggs are laid in clusters, not visible to the naked eye and there may be up to
50 maggots in each pod feeding within the
same pod throughout their larval life.

Sawflies

The turnip sawfly, *Athalia rosae*, is an oligophagous pest of crucifers. The larvae of this insect
are voracious feeders which feed in masses
and devastate the plants. The genus name itself suggests such damage: *athal*, Greek mean-
ing 'not green, withered', *Athalia*, named in
allusion to the devastation produced by its
larvae (Jaeger, 1978). *A. rosae ruficornis* is a
multivoltine species without aestivation and

in Japan it produces from three to seven generations from the late spring to the late au-
tumn depending on the climatic conditions.
During the winter it hibernates as prepupae
(Oishi *et al.*, 1993). Larvae cause damage by
feeding on the plants, while adults damage
leaves with their saw-like ovipositor (AgroAtlas,
2009). In India, the mustard sawfly *Athalia
lugens proxima* is an important pest on oilseed
brassica especially early in the season when
plants are small.

Lygus Bugs

Plant bugs from the genus *Lygus* are polyphagous pests that feed on different crops and
weeds. There are four different species in
Canada, *L. keltoni*, *L. lineolaris*, *L. elisus* and
L. borealis (Kelton, 1975; Young, 1986; Timlick
et al., 1993; Carcamo *et al.*, 2002) and have equal
economic importance. In addition to these,
other species of *Lygus* are also found in North
America, Europe and Russia. Both the nymphs
and adults suck sap from the reproductive
structures of plants. Feeding on buds, flowers
and pods results in bud blasting (buds turn
white and fail to develop), shedding of flowers
and premature shattering of pods (Butts and
Lamb, 1990, 1991a, b; Turnock *et al.*, 1995).
Plant sap oozes out from the feeding punctures
on the pods and stem leading to stickiness of
the plants, which favours infection by various
pathogens. *Lygus* feeding creates small, dark
patches on the pod surface. They are sporadic
pests on canola fields and result in only 3–5%
seed loss which can be higher at times (Turnock
et al., 1995). In addition to these pests, other
pests such as the stink bug (*Bagrada cruciferarum*), cabbage head caterpillar (*Crocidolomia
binotalis*) and crucifer leaf miner (*Chromatomyia
horticola*) are also found damaging crucifer
crops in one or another part of the world.

Plant Resistance in Brassicas:
The Bases of Resistance

Prior to crop domestication, there was strong selection pressure on plants to develop resistance
and only resistant plants survived as a consequence of the laws of adaptation and natural

selection. Unlike herbivorous insects, plants being sessile cannot move to avoid any herbivore attack. To overcome this they have developed different physical/mechanical and biochemical defence strategies. Like other plants, member of family *Brassicaceae* also evolved different strategies such as surface waxes, trichomes, toxic secondary plant metabolites and plant volatiles etc., which provided varying degrees of protection against different herbivorous insects. In the forthcoming sections these are discussed in detail.

Physical/mechanical defence strategies

Epicuticular wax

The plant cuticle is covered with epicuticular waxes. These are complex mixtures of very long chain lipids substituted with primary alcohols, aldehydes, fatty acids and alkyl esters, all of which occur predominantly with even-numbered chain lengths and hydrocarbons, secondary alcohols and ketones with predominance of odd-numbered chain lengths (Walton, 1990). The chemical composition of epicuticular wax is the determining factor in the decision of an insect to feed, probe or oviposit on a plant, since it is the primary site of interaction between insect and its host plant. The neonate larvae of DBM, *P. xylostella*, have been shown to spend more time walking at faster pace and frequency on a waxy resistant line of cabbage than on non-waxy susceptible lines (Eigenbrode *et al.*, 1991). Similarly, Eigenbrode and Shelton (1990) reported that neonate larvae of *P. xylostella* spent more time wandering and palpating before selecting a site for feeding on waxy cabbage cultivars. Waxiness, not hairiness, has been found to hinder *L. erysimi* from reaching the undersurface of leaves, where it normally feeds during the vegetative stage of plants (Åhman, 1990), however, Lamb *et al.* (1993) reported that elevation of leaf wax did not improve resistance of *B. napus* or *B. oleracea* (kale and collard) to *L. erysimi*. The neonate larvae of the mustard beetle, *Phaedon cochleariae*, find it difficult to climb the heavily waxed culm of cabbage on waxy cultivars and do not reach the feeding site, whereas they walked easily on non-waxy cultivars (Stork, 1980). Žnidarčič *et al.* (2008)

reported less damage by flea beetles (*Phyllotreta* spp.), stink bugs (*Eurydema ventrale*) and onion thrips (*Thrips tabaci*) on cabbage varieties with high epicuticular wax with cv. Holandsko Pozno Rdeče to be the most resistant among 12 cultivars evaluated. Introgression of glossy trait for resistance against insect pests is not practical in vegetable brassicas as glossiness is not liked by consumers, but it can be utilized in oilseed brassicas. Genotypes showing glossy leaves in *B. oleracea* and *B. rapa* showed better resistance to *P. xylostella* (Ulmer *et al.*, 2002). Removal of epicuticular wax from leaves of *B. napus* and *B. oleracea* lead to a significant increase in flea beetle, *Phyllotreta cruciferae*, feeding on the leaf area where the wax was removed (Bodnaryk, 1992) and the presence of leaf wax explained the majority of the difference in feeding preference. Similarly, removal of leaf wax also increased *P. cochleariae* activity suggesting that epicuticular wax occludes stimulatory signals such as glucosinolates (Reifenrath *et al.*, 2005) and the resistance was primarily anti-xenosis. The importance of waxes on phylloplane has recently received increased attention due to their close association with polar compounds such as the glucosinolates, which are key host recognition signals for some specialist insects (Städler and Reifenrath, 2009; Badenes-Pérez *et al.*, 2010). Glucosinolates were found to be present on the leaf surface of three *Barbarea* species but not on the surface of the *B. napus* genotype tested (Badenes-Pérez *et al.*, 2010). At the third trophic level, aphid parasitoids' host-recognition behaviour is influenced by aphid cuticular waxes, which in turn are related to plant surface waxes (Muratori *et al.*, 2006). The parasitoid behaviour is also affected via host caterpillar 'footprint' kairomones on waxy leaf surfaces (Rostas *et al.*, 2008).

Trichomes

Trichomes are small, sometimes branched, hair-like structures that are produced from cells of aerial epidermis and are produced by most plant species (Werker, 2000). They provide some degree of protection to the plant from insect herbivory (Levin, 1973). Trichomes can be unicellular, multicellular, straight, spiral, hooked, branched or unbranched (Southwood, 1986; Werker, 2000). Glandular trichomes produce secondary metabolites (e.g. terpenoids, flavonoids, alkaloids) which can either repel or trap insects or can be poisonous (Duffey, 1986). Presence of trichomes has been related with resistance against insect herbivores. The glabrous form of *Arabidopsis lyrata* was reported to be more damaged by insect herbivores than a trichome-producing morphotype (Loe *et al.*, 2007). Non-glandular trichomes mainly function as structural defence against small herbivores. They interfere with insect movement over the plant surface and make it difficult for the insect to access leaf epidermis (Southwood, 1986). The insects feeding on trichome-bearing lines show poor weight gain due to cellulose-rich trichomes, which are of poor nutritive value. This results in increased mortality. High density of trichomes on leaves of *B. nigra* reduced growth of *P. rapae* and caused increased mortality of *P. Cruciferae* (Traw and Dawson, 2002). They are considered relatively as the soft 'weapon' in plant defence, but such defences are associated with less selection pressure and consequently reduced chances of development of counter adaptations in insects (Feeny, 1976). Agrawal (1999) showed that two lepidopteran pests, *Pieris rapae* (specialist) and *Spodoptera exigua* (generalist), were negatively influenced by presence of trichomes on wild radish plants, *Raphanus raphanistrum* and *R. sativus*. Interestingly, there appeared to be an increase in trichome density in response to the insect damage in *R. raphanistrum*. Furthermore, a change in relative proportion of glandular and non-glandular trichomes is also induced by herbivory (Agrawal, 1999). Similarly, an increase in trichome density and glucosinolates level was recorded in black mustard after feeding by *P. rapae* (Traw, 2002). Clauss *et al.* (2006) also observed that resistance of *P. xylostella* (specialist) and *T. ni* (generalist) was due to a combination of glucosinolate levels and trichome density in *A. lyrata*. Furthermore, an increase in glucosinolate levels and trichome density was observed in *Sinapis alba* after herbivore damage (Travers-Martin and Müller, 2008). Trichome-bearing pods of *S. alba* have been reported to be resistant to flea beetle, while glabrous pods of cultivated *Brassica* species are readily attacked (Lamb, 1980). Trichomes on the leaves of

Brassica villosa physically disrupted the movement of flea beetle to settle and feed (Palaniswamy and Bodnaryk, 1994). There have been some attempts to produce altered plant types with increased trichome density. For example, expression of *A. thaliana* myb-like transcription factor GLABRA3 (GL3) in *B. napus* resulted in a dense coat of trichomes on the adaxial leaf surface (Gruber *et al.*, 2006) and larvae of *P. xylostella* had difficulty feeding and grew slower on these lines (Adamson *et al.*, 2008). Despite their negative effects on herbivore insects, trichomes may have their effect at the third trophic level. For example, trichomes on the leaves of a trichome-bearing line of *A. thaliana* affected the movement of aphid predator *Episyrphus balteatus* and resulted in reduced performance (Wietsma, 2010). Further, trichomes play a significant role in the acceptance of host-plants for oviposition (Sadeghi, 2002) and there was comparatively less oviposition on the *A. thaliana* line having higher trichome density (Wietsma, 2010).

Biochemical defences

Glucosinolates and myrosinase–glucosinolate system

Brassicaceae glucosinolates are the most studied class of secondary metabolites. These amino acid-derived secondary plant products containing a sulfate and thioglucose moiety are found almost exclusively in the order *Capparales* (Halkier and Gershenzon, 2006). They are a large group of naturally occurring, non-volatile, sulfur-containing, anionic compounds. To date, almost 140 different structures of glucosinolates are known, 30 of which are present in *Brassica* species (Bellostas *et al.*, 2007). The amount of glucosinolates varies depending on species, plant parts, agronomic and climatic conditions (Sadasivam and Thayumanavan, 2003; Font *et al.*, 2005; Tripathi and Mishra, 2007). Even though the intact glucosinolates may confer resistance to insects feeding on them, their breakdown products released following myrosinase hydrolysis can be more toxic to insects. Myrosinase catalyses the cleavage of glucosinolates to produce isothiocyanates, nitriles and oxazolidiethiones.

The enzymatic cleavage of the glucosinolate molecule leads to the formation of an unstable thiohydroxamate-O-sulfonate and other aglycones (Sadasivam and Thayumanavan, 2003). Under neutral conditions, the bond between sulfur and glucose is cleaved; the aglycone moiety gives rise to sulfate and, by a Loessen-type molecular rearrangement, yields isothiocyanate (Sadasivam and Thayumanavan, 2003). Formation of nitrile (also known as organic cyanide) involves only the sulfur loss from the molecule. At low pH, nitrile formation is favoured over isothiocyante. The presence of Fe^{2+} or thio- compounds and epithiospecifier protein increases the likelihood of a nitrile formation. Epithionitrile formation requires the same conditions as those for nitrile (Sadasivam and Thayumanavan, 2003).

In intact plant tissues, glucosinolates and myrosinase are housed separately; glucosinolates are present in vacuoles of various types of cells while myrosinases are localized in the myrosin cells (Kissen et al., 2009) scattered throughout most plant tissues. This arrangement renders them inactive and protects the plant from self-toxicity (Jones and Vogt, 2001). The myrosin cells contain fewer lipids, high content of endoplasmic reticulum and harbour smooth-looking protein bodies referred to as myrosin grains (Bones et al., 1991; Kissen et al., 2009). Myrosin grains have been shown to form a continuous reticular system known as the myrosin body (Andreasson et al., 2001; Ahuja et al., 2009). Mechanical tissue damage due to insect feeding brings together glucosinolates and myrosinase resulting in rapid release of glucosinolate breakdown products (Bones and Rossiter, 2006). This defensive response (or 'mustard oil bomb') is shown to have multiple roles in plant–insect interactions and insect pest management (Rask et al., 2000; Kissen et al., 2009). The flip side of this system is that while, on one side, it defends against the attacks by generalist feeders (Rask et al., 2000), on the other side it makes plants vulnerable to attack by specialist feeders (Renwick, 2002; Björkman et al., 2011). Glucosinolates are feeding and oviposition stimulants for more than 25 insect species in the orders Coleoptera, Lepidoptera and Diptera that are specialized on these plants (Hopkins et al., 2009). As a consequence of

co-evolution, insects such as Brevicoryne brassicae and L. erysimi (both crucifer specialists) take advantage of glucosinolates by their sequestration from host plant to protect themselves from predators. These insects have developed a mechanism to synthesize their own thioglucosidase endogenously, which is spatially separated in the insect body from sequestered glucosinolates in their non-flight muscles. When an insect is crushed or fed upon by a predator, the enzyme leads to hydrolysis of sequestered glucosinolates (the concentration of glucosinolates in haemolymph is 15–20 times relative to that in leaf tissue) to produce toxic products (Bridges et al., 2002; Rossiter et al., 2003). These crushed insects smell as well as taste badly and release volatiles alarming other insects in the colony. In contrast, the generalist aphid M. persicae rapidly excretes glucosinolates in its honeydew (Hopkins et al., 2009). Diamondback moth P. xylostella (specialist) (Ratzka et al., 2002) and desert locust Schistocerca gregaria (generalist) (Falk and Gershenson, 2007) produce a glucosinolate sulfatase enzyme (GSS) which removes sulfur from glucosinolates to produce desulfoglucosinolates that are not hydrolysed by myrosinase, thus preventing the formation of toxic isothiocyanates. This enables the insects to feed on glucosinolate-containing plants (Ratzka et al., 2002; Falk and Gershenson, 2007). On the other hand, P. rapae redirects the glucosinolates' hydrolysis reaction from the formation of toxic isothiocyanates to the formation of less toxic nitriles through a specific gut protein (nitrile specifier protein) (Wittstock et al., 2004). The glucosinolates are also known to stimulate larval feeding and oviposition by adults of the large white butterfly P. brassicae and small white butterfly P. rapae (David and Gardiner, 1966; Renwick et al., 1992; Smallegange et al., 2007). These are also known to stimulate oviposition by P. xylostella (Renwick et al., 2006). Many insects use isothiocyanates to their advantage for host location. Studies have shown the presence of receptor neurons that can detect isothiocyanates in many specialist insects such as B. brassicae (Nottingham et al., 1991) and P. xylostella (Renwick et al., 2006).

Since glucosinolates play a defensive role in plants against herbivorous insects, there had been doubts that double zero ('00') canola plants, which are exceptionally low in these compounds, might be susceptible to many insects. Such questions may be misplaced because low glucosinolate levels in '00' canola plants were confined primarily to the seeds (Milford *et al.*, 1989) and high and low glucosinolate cultivars did not differ in their susceptibility to pod midge (*Dasineura brassicae*), though the level of glucosinolates in leaf tissue was not determined (Åhman, 1982). Extensive studies in India with both *B. napus* and *B. juncea* canola have shown no reasons to believe that canola-quality cultivars were more susceptible than their non-canola counterparts. In fact, the inheritance mechanism of glucosinolates in *B. juncea* seemed to be different in leaves and seeds. Major QTLs accounting for a large variation in seeds or leaves were not co-localized (Gupta *et al.*, unpublished). Though there are no supporting references, theoretically these low glucosinolate plants may be less attractive to specialist insects for which these compounds serve as attractants and feeding stimuli (Gabrys and Tjallingii, 2002; Mewis *et al.*, 2002). This is again supported by the work of Giamoustaris and Mithen (1995), who reported that increase in the content of glucosinolates in *B. napus* resulted in increased feeding damage by specialist insects, flea beetles (*Psylliodes chrysocephala*) and greater incidence of small white butterfly (*P. rapae*) while the damage by generalist pests, i.e. pigeons and slugs, was reduced. Further, glucosinolate-rich flower tissues are preferred more by *P. brassicae* and sustain higher growth compared to leaf tissues (Smallegange *et al.*, 2007), indicating the selective role of glucosinolates to elicit feeding in this specialist insect and the adaptation of the insect to use these compounds to its advantage.

Other phytoalexins and phytoanticipins

Many different compounds such as polyphenolics, phenolic acids, flavonoids and lignans, terpenoids, phytosterols and alkaloids have been associated with plant defences besides the glucosinolates. Phenolics, especially in the form of condensed tannins, have been reported to be feeding deterrents to several pests on *B. napus* (Meisner and Mitchell, 1984; Muir *et al.*, 1999). They are also reported to bind to and in-

activate digestive enzymes (Nguz *et al.*, 1998) and show antibiotic effects (Duffey and Stout, 1996). Jones *et al.* (1988) reported that a sinapic acid precursor of sinapine deterred the oviposition by *Delia radicum* on cauliflower plants sprayed with water extract of frass of garden pebble moth caterpillars, *Evergestis forficalis*. Both stimulatory as well as deterrent effects of flavonoids have been reported on insects feeding on brassica plants. For example, quercetin and kaempferol found in *Armoracia rusticana* stimulated feeding by *Phyllotreta armoraciae* (Nielsen *et al.*, 1979) and *P. xylostella* (van Loon *et al.*, 2002). On the other hand, flavonoids, isorhamnetin-3-sophoroside-7-glucoside and kaempferol 3,7-diglucoside found in *B. napus* were deterrent to *Mamestra configurata* at levels above those found in vegetative tissues (Onyilagha *et al.*, 2004). The phytosterols, strophanthidin and strophantidol, found in *Cheiranthus* and *Erysimum* species, exhibited feeding deterrent action against different species of flea beetles, *Phyllotreta undulata*, *P. tetrastigma* and *Phaedon cochleariae*, but no effect on *P. nemorum* (Nielsen, 1978). The alkaloid camalexin found in *A. thaliana* has been found to play a role in insect resistance as evident from the camalexin-deficient *A. thaliana* mutants that were more susceptible to the mealy cabbage aphid *B. brassicae* (Kusnierczyk *et al.*, 2008).

Volatile compounds

Brassicas synthesize a wide range of volatile compounds which can play different functions such as adaptation to stresses, plant–insect communication, plant–pathogen communication and plant–plant communication (Baldwin *et al.*, 2002). These volatiles vary from monoterpenes, sesquiterpenes, indole and to 'green leafy volatiles' (Tumlinson *et al.*, 1999). The hydrolysis of glucosinolates leads to the production of volatile thiocyanates, isothiocyanates and nitriles. Cabbage seed weevils, *C. assimilis*, were attracted to 3-butenyl and 4-pentenyl isothiocyanate in *B. napus*, but not to 2-phenylethyl isothiocyanate (Bartlet *et al.*, 1993). Similary, cabbage root fly *D. brassicae* was attracted to 4-methylthio-3-butenyl isothiocyanate and 1-cyano-4-methylthio-3- butene produced after glucosinolate hydrolysis in *R. sativus* (Ellis *et al.*, 1980). Though different herbivore insects use these volatile compounds as cues to locate their hosts, these plant volatiles also serve as a means of indirect defence against these herbivores. Such insect-attacked plants release volatiles to attract natural enemies of insects that keep a check on the herbivore insect population. Volatile z-jasmone not only repels *L. erysimi* but also attracts its parasitoids on brassica plants (Birkett *et al.,* 2000). Blande *et al.* (2007) also reported the attraction of the aphid parasitoid *Diaeretiella rapae* towards semiochemicals produced by turnip plants after feeding by *L. erysimi* (specialist) and *M. persicae* (generalist). Pope *et al.* (2008) studied the orientation response of cabbage aphid *B. brassicae* and its parasitioid *D. rapae* to alkenyl glucosinolate hydrolysis products. The electroantennogram responses indicated peripheral odour perception in *D. rapae* females to all the 3-butenylglucosinolate hydrolysis products.

Lectins

Lectins are found in a wide range of plant, microbial and animal tissues (Nachbar and Oppenheim, 1980; Komath *et al.,* 2006). These are the proteins that selectively bind carbohydrates and more importantly the carbohydrate moieties of glycoproteins that are present on the surface of most animal cells. Lectins incorporated in artificial diets have been shown to reduce performance of several insect pests (Murdock *et al.,* 1990; Powell *et al.,* 1993; Sauvion *et al.,* 2004a). Although the actual mechanism of insecticidal action is not clearly known, those lectins which are not efficiently degraded by digestive enzymes can be poisonous due to their affinity for the surface of gut epithelial cells (Vasconcelos and Oliveira, 2004). They can form complexes with gut proteins (probably glycosylated proteins) with high affinity (Macedo *et al.,* 2004; Sauvion *et al.,* 2004b). Since lectins interact with mono- and oligosaccharides, the insecticidal activity may involve a specific carbohydrate–lectin interaction with glycoconjugates on the surface of digestive tract epithelial cells (Macedo *et al.,* 2004). Acute symptoms following ingestion include nausea, vomiting and diarrhoea. They may lead to membrane disruption of epithelial cell microvilli of insects fed upon diet containing lectin (Hart *et al.,* 1988). Lectins have been reported to show biological

activity against a wide range of insects, especially the sap-sucking insects (Foissac *et al.,* 2000; Powell, 2001). In brassicas, they are of interest as aphids, especially the turnip/mustard aphid *Lipaphis erysimi*, is a limiting factor in the successful cultivation of oilseed brassicas in the Indian subcontinent. *Brassica fruticulosa*, a wild relative of cultivated brassicas, has been reported to possess resistance against cabbage aphid *B. brassicae* (Cole, 1994a, b; Ellis and Farrell, 1995; Ellis *et al.,* 2000) and the high concentration of lectins was reported to be responsible for resistance. Screening of a diverse array of wild *Brassica* species in India resulted in the identification of an accession of *B. fruticulosa* to be resistant to *L. erysimi* (Kumar *et al.,* 2011). The results of feeding preference/choice test revealed that *L. erysimi* showed maximum preference for feeding on *B. rapa* ecotype Brown Sarson cv. BSH 1, while the least preference was shown for *B. fruticulosa*. The antixenosis to feeding in *B. fruticulosa* has earlier been reported for cabbage aphid *B. brassicae*. Monitoring of feeding behaviour of this species electronically by electrical penetration graph (EPG) showed a large reduction in the duration of passive phloem uptake on *B. fruticulosa* compared to the susceptible *B. oleracea* var. *capitata* cv. Offenham Compacta. There was either quick withdrawal of stylets from sieve elements or disrupted phloem uptake (Cole, 1994a). Attempts have been made to introgress the gene of interest from this wild species to the cultivated plants (Kumar *et al.,* 2011; Atri *et al.,* 2012).

Practical Application in Pest Management

Host plant resistance

Due to the presence of many defence-related compounds and their consequent effects on the herbivorous insects, brassica plants can theoretically be tailored to enhance resistance. Genes encoding for these defences may be exploited by breeding methods to reduce insect damage. Most of the plant breeding efforts have been associated with characters other than those related to insect resistance. Although specialist insects have developed ways to counteract the toxic effects of glucosinolates, their higher concentrations, as well as their fission products, can still have adverse effects on these specialists (Agrawal and Kurashige, 2003; Hopkins *et al.*, 2009). Thus, the levels of glucosinolates can be manipulated with breeding methods to enhance the level of resistance. Low damage by *P. xylostella* larvae was observed in *B. juncea* lines with high myrosinase activity compared to those with lower level of activity. Allyl isothiocyanate formed after fission was more toxic than the allyl glucosinolate (Li *et al.*, 2000). However, this trend is not uniform for all the pests. For example, there was no correlation of damage by mealy cabbage aphid *Brevicoryne brassicae* and levels of myrosinase in plant tissue (Barth and Jander, 2006) and to *S. alba* by sawfly *Athalia rosae* (Muller and Sieling, 2006). Giamoustaris and Mithen (1995) observed a positive correlation between total glucosinolate content and herbivore damage for cabbage stem flea beetle *Psylliodes chrysocephala* and small white butterfly *P. rapae* on different *B. napus* lines. Thus, glucosinolates alone may not be responsible for the resistance in brassica plants and a thorough understanding of the interactions of glucosinolates with other phytochemicals may be helpful. For example, polar glucosinolates present on the leaf surface are known to interact with non-polar epicuticular waxes, which can be a research area of increasing interest (Hopkins *et al.*, 1997, 2009; Städler and Reifenrath, 2009).

Since many specialist insects have developed strategies to cope with the toxic effects of glucosinolates and their breakdown products, the level of plant resistance can be further increased by finding ways to counteract these coping strategies. Although there are several reports, there is a general paucity of any significant data showing existence of resistance to various insects in the primary gene pool of crop brassicas (Kumar *et al.*, 2011). Rai and Sehgal (1975) inferred that *B. rapa* varieties having tender and thickly packed inflorescence were more susceptible to mustard aphid *L. erysimi* than *B. juncea* having harder and loose inflorescence. The yellow, bright colour of flower, sparse arrangement of the flower buds on the inflorescence and succulence

of the same contribute to the susceptibility of *Brassica* genotypes to aphid infestation. In general, among the rapeseed-mustard crops, *B. juncea* and *B. carinata* genotypes were observed to be comparatively tolerant to aphid attack compared with *B. rapa* (Teotia and Lal, 1970; Kalra *et al.*, 1987). *B. juncea* strain Laha 101 exhibited all the three types of resistance, i.e. non-preference, antibiosis and tolerance. Slow development of mustard aphid and more alate production on 'T 27' (*Eruca sativa*) and 'Purple Mutant' (*B. juncea*) exhibited antibiosis (Rohilla *et al.*, 1999). Non-waxiness of brassica plants has been found to impart resistance to aphids (Srinivasachar and Malik, 1972; Yadava *et al.*, 1985; Angadi *et al.*, 1987; Chatterjee and Sengupta, 1987). It has also been observed that white petal and non-waxy (glossy surface) mutants had the lower number of initial settlers and subsequently smaller aphid colonies as against normal (yellow petal) and waxy stems (non-glossy). Rich amounts of glucosinolates and phenols have been reported to be negatively correlated, whereas rich amounts of sugars and nitrogen content had a positive correlation with mustard aphid infestation (Gill and Bakhetia, 1985; Singh and Rohilla, 1998). *B. tournefortii* was also found to harbour a relatively low population, indicating tolerance therein. *B. napus* strains and colchicine-induced tetraploid *B. rapa* var. Toria have been shown to possess antibiotic factors (Singh *et al.*, 1965; Jarvis, 1970; Gill and Bakhetia, 1985; Kalra *et al.*, 1987). The importance of host-plant resistance as a tool in IPM has been long documented (Snelling, 1941; Painter, 1951; Chesnokov, 1953; Russel, 1978; Lara, 1979; Panda, 1979; Smith, 1989, 1999; Dhaliwal and Dilwari, 1993; Smith *et al.*, 1994; Panda and Khush, 1995; Dhaliwal and Singh, 2004). Extensive genetic variability for many defensive traits is known to exist in wild and weedy crucifers (Kumar *et al.*, 2011). Attempts have been made to introgress this variation to cultivated brassicas. In one such instance, genetic factors associated with aphid resistance could be successfully introgressed into *B. juncea* through hybridization with a wild crucifer *B. fruticulosa* (Atri *et al.*, 2012). There have also been many attempts at developing *Brassica* transgenics carrying genes for lectin production that offered higher levels of resistance against *L. erysimi* (Hossain *et al.*, 2006). Field testing of such transgenics is still awaited.

Among insect-resistant genes, the efficacy of *Bacillus thuringiensis* (*Bt*) genes encoding *Bt* crystal protein (or δ-endotoxin) have been proven and widely used against a number of insect pests including those of rapeseed (Cheng *et al.*, 1998; Xiang *et al.*, 2000; Cho *et al.*, 2001; Guan *et al.*, 2001). However, some of the pests have developed resistance to *Bt* resulting in control failures with the well known example of the DBM (Shelton *et al.*, 1993; Roush and Shelton 1997; Tang *et al.*, 1997). This has precipitated attempts either at seeking new insect-resistant genes or pyramiding of more than one gene. Wang *et al.* (2005) has made one such attempt leading to transformation of *B. napus* with a simultaneously expressed combination of *chitinase* (*chi*) and *BmkIT* (*Bmk*) containing an insect-specific chitinase gene and a scorpion insect-toxic gene. Both the genes have unique modes of action. Continuous expression of *chitinase* in the plants leads to lysis of insect cuticle fed upon transformed plants and hence can prevent insect damage (Kramer *et al.*, 1997). The other gene codes for an insect-specific neurotoxin – BamKIT – originally derived from the venom of scorpion, *Buthus martensii*, which is a scorpion neurotoxin of contractive paralysis type. The toxin consists of 69 amino acids (Liang *et al.*, 1999) and is toxic to many lepidopteran insects. Bioassays of these transgenic plants on *P. xylostella* exhibited high larval mortality by some of the plants which showed high expression of both the chitinase and scorpion toxin proteins (Wang *et al.*, 2005).

In addition to this, other molecular and genomic approaches such as transcript profiling, mutational analysis, over-expression and gene silencing are also now being considered to develop host-plant resistance to aphids (Goggin, 2007; Bhatia *et al.*, 2011).

Cultural practices

Trap cropping

Allelochemicals endogenous to crop brassicas are known to influence host location for feeding and oviposition by a pest (even by parasitoids). This behaviour of pests can be manipulated to attract them towards a trap crop that is of lower value than the main crop. An ideal trap crop is that which is more attractive to the pest for feeding/oviposition than the main crop.

The use of trap crops is thought to have special importance to subsistence and small-holder farming in developing countries by reducing the reliance on pesticides and reducing production costs (Hokkanen, 1991). It may reduce the prophylactic use of insecticides, which are often applied without regard to the occurrence of insect pests and their associated natural enemies such as parasitoids and predators. Barari *et al.* (2005) used early-flowering *B. rapa* as trap crop to protect *B. napus* from damage by cabbage stem flea beetle, *P. chrysocephala*. The preference of turnip rape over oilseed rape has also been reported by earlier workers (Buchi, 1995; Lambdon *et al.*, 1998). Similarly, Cook *et al.* (2004) used turnip rape, *B. rapa* ssp. *oleifera* as an effective trap crop to protect main-crop oilseed rape, *B. napus* from damage by pollen beetle, *Meligethes aeneus*. A preference of turnip rape over oilseed rape has also been found for other pests of oilseed rape including *M. aeneus* and *C. assimilis* (Hokkanen *et al.*, 1986; Buchi, 1990; Cook *et al.*, 2002). The increased attractiveness of these pests to turnip rape was attributed to various phenological, morphological and chemical factors, for example, turnip rape grows faster, flowers earlier and has a more attractive colour and odour than oilseed rape (Barari *et al.*, 2005). In India, Srinivasan and Krishna Moorthy (1990) have reported that Indian mustard, *B. juncea* can be successfully used as a trap crop for the DBM *P. xylostella* in cabbage. Similarly, Charleston and Kfir (2000) suggested Indian mustard has a potential as trap crop for *P. xylostella* in Africa as female moths significantly preferred Indian mustard for laying their eggs. The pest population on the trap crop can be effectively managed either by the use of insecticides or biocontrol agents of the pests. However, in certain cases the trap crop may act as 'dead end', i.e. it is highly attractive to the adults for oviposition while larvae do not survive. For example, Shelton and Nault (2004) found that the weed winter cress, *Barbarea vulgaris* was highly attractive to *P. xylostella* females for oviposition but larvae failed to survive on it. *B. nigra* was reported to serve as a potential trap crop to protect spring oilseed rape, *B. napus*, before flowering (the most vulnerable stage) from damage by pollen beetle, *Meligethes aeneus*, due to its faster development and acceptability for both feeding and oviposition to over-wintered beetles, while *R. sativus* can serve as a dead-end trap since beetles showed preference for feeding and oviposition but larvae fail to develop in its flowers (Veromann *et al.*, 2012).

Intercropping

The main idea behind intercropping is to disrupt the normal host finding and further colonization in an intercrop compared to that in the monoculture. Further, the intercrop may also provide better agroecosystem diversity to the natural enemies such as parasitoids and predators of the pest for population establishment and subsequent pest suppression (Finch and Collier, 2000; Khan *et al.*, 2000). Intecropping studies have shown varying levels of success in reducing the pest pressure. For example, there was no effect of intercropping on host selection by small white butterfly *P. rapae* due to its superior visual and olfactory senses and active flight ability, while the cabbage aphid *B. brassicae* was affected due to its limited dispersal ability (Banks, 1998). However, a better approach is to mix trap-cropping with intercropping in a 'push-pull' strategy for effective management of different pests.

Push-pull strategy

Merging the trap crop and intercropping together in the form of a 'push-pull' strategy can have greater potential for pest management. A 'push-pull' strategy is a cropping system in which specifically chosen companion plants are grown in between and/or around the main crop. These plants release semiochemicals that either repel insect pests from the main crop using an intercrop, which is the 'push' element, or attract insect pests away from the main crop using a trap crop, which is the 'pull' element (Cook *et al.*, 2007). According to Aiyer (1949), a lesser number of

specialist insects are found on the main crop because the companion plants make the host plant harder to find (disruptive-crop hypothesis), act as alternative host plants (trap crop hypothesis), or serve as a repellent to the pest (see Vandermeer, 1989). Host-plant finding by eight insect-pest species of brassica from four orders was adversely affected when *B. rapa* and *B. oleracea* were surrounded by subterranean clover plants, *Trifolium subterraneum* (Finch and Kienegger, 1997). The pest species monitored were the small white butterfly *P. rapae*, the large white butterfly *P. brassicae*, cabbage root fly *D. radicum*, mustard beetle *Phaedon cochleariae*, DBM *P. xylostella*, garden pebble moth *Evergestis forficalis*, cabbage moth *Mamestra brassicae* and mealy cabbage aphid *B. brassicae*. Björkman *et al.* (2007) demonstrated that intercropping of cabbage with red clover significantly reduced *D. floralis* oviposition and that the cabbage monoculture control had the effect of a trap crop when bordering a cabbage red-clover intercrop plot. The push-pull strategy can also be used to enhance the effectiveness of microbial control agents such as insect pathogenic fungi. Use of traps baited with such fungi such as *Metarhizium anisopliae* and *Beauvaria bassiana* along with ethyl isocyanate (attractant) were very effective for auto-dissemination of the fungi against *Delia radicum* and *D. floralis* in vegetable brassicas (Klingen *et al.*, 2000). Flies were attracted to the ethyl isocyanate, which resulted in their selected mortality. Use of such traps along with Chinese cabbage as intercropping in brassica vegetables can be an interesting component in a push-pull strategy which can enhance the epidemic development of beneficial fungi. Studies have shown that Chinese cabbage is more attractive to *D. radicum* than other brassica vegetables (Rousse *et al.*, 2003). The practical adoption of the strategy can be hampered by varied levels of success obtained. Thus, development of a reliable and sustainable strategy requires a complete scientific understanding of the pest's biology and behavioural and chemical ecology of the pest interactions with host plants, organisms of the same and different species (Khan and Pickett, 2004) to avoid any drift from effectiveness. The push-pull strategy is a powerful and effective IPM tool

(Cook *et al.*, 2007); however, its potential has been largely underexploited.

Time of sowing

Alteration in crop sowing can avoid phenological synchrony between the most susceptible stage of crop with the peak period of pest activity. This asynchrony can also be achieved through the breeding techniques by incorporating genes for earliness and lateness in *Brassica* species. Flowering period (end of December, the first fortnight of January to mid-February) is the crucial period for aphid infestation in India. Therefore, the crop sown early before 20 October mostly escapes aphid infestation (Ghosh and Ghosh, 1981; Yein, 1985; Bhagat and Singh, 1989; Joshi *et al.*, 1989; Kular *et al.*, 2012) since plants become hardy before the peak period of infestation (Singh *et al.*, 1984; Singh and Bakhetia, 1987). However, sowing too early especially in dry regions of the country, such as Rajasthan, can result in more damage by painted bug (Ahuja *et al.*, 2008). Pal *et al.* (1976) also opined that the aphid infestation was a major factor for yield reduction in late-sown crops.

Nutrient application

Balanced use of NPK fertilizers is important because they are an integral part of crop culture. In general, heavily fertilized crops are more prone to the incidence of mustard aphid and many other pests. The population of mustard aphid was 4–8 times more in the mustard crop receiving N at 40 and 60 kg/ha as compared to that in the crop receiving no fertilizer (Rawat *et al.*, 1968; Singh *et al.*, 1990). However, Bakhetia and Sharma (1978) reported that the population of *L. erysimi* was not affected with increase in N level up to 80 kg/ha, while increased sulfur nutrition had a negative correlation with aphid population (Bakhetia *et al.*, 1982). Host plants grown under increased supply of K caused adverse effects on the reproduction and excretion of the mustard aphid (Bhat and Sidhu, 1983). Staley *et al.* (2010) reported that abundance of cabbage aphid *B. brassicae* on organically (green manure and animal manure) fertilized *B. oleracea* was more

compared to that on the synthetically fertilized plants, while the opposite was true for peach potato aphid *M. persicae* and DBM *P. xylostella.* An increase in soil nitrogen level resulted in increase in abundance of *M. persicae*, *B. brassicae*, *Artogeia rapae* and *P. xylostella* on different brassica plants (Letourneau, 1988). Conventionally fertilized broccoli fields harboured a greater population of flea beetles *P. cruciferae* and *M. persicae* than organically fertilized counterparts (Altieri *et al.*, 1998).

Irrigation

Mustard crops facing water stress supported a lower population of *L. erysimi* (Sidhu and Kaur, 1977) while the opposite was true for green peach aphid *M. persicae* (Khan *et al.*, 2010). Significantly large populations of *M. persicae* were recorded on the water-stressed *B. oleracea* var. *italica* plants while there seemed to be no significant effect on mealy cabbage aphid *B. brassicae* (Khan *et al.*, 2010). This was again supported by findings of Mewis *et al.* (2012) that *M. persicae* exhibited higher growth on water-stressed *A. thaliana* plants compared to that on watered plants, while *B. brassicae* performed equally in both the treatments. However, heavy infestation of *B. brassicae* on water-stressed compared to unstressed plants was reported on *B. napus* (Burgess *et al.*, 1994; Popov *et al.*, 2006). There was increase in the levels of sucrose, several amino acids such as glutamic acid, proline, isoleucine and lysine while the levels of 4-methoxyindol-3-yl methyl glucosinolate decreased as a result of water stress (Mewis *et al.*, 2012). In contrast, Chadda and Arora (1982) observed reduction in amino acid concentration in plants kept under dry conditions that resulted in imbalance in excretion of mustard aphid, which was responsible for the reduced fecundity under water stress. Bakhetia and Brar (1988) reported that the rainfed crop had a heavy infestation whereas the irrigated crop maintained a good stand even under heavy aphid attack partly due to differences in plant vigour. Miles *et al.* (1982) reported an increase in amino acid contents of phloem sap in water-stressed rape plants, which enhanced the development of cabbage aphid *B. brassicae.* Joshan *et al.* (1992) also observed a concomitant decrease in net reproduction rate of aphid on potted plants with increasing water stress in the three generations.

Empirical methods

Infestation of mustard aphid occurs in India in the last week of December or first week of January on rapeseed-mustard crops. Initially, the north-east border plants or a few plants at the border of the fields or in some pockets of the crop are found to be infested with this pest. Singh *et al.* (1995) suggested the removal and destruction of infested twigs helps in checking their further development and spread in the field. Similarly, monitoring, collection and destruction of egg masses and gregarious larvae of *P. brassicae* help to reduce pest damage early in the season (Kumar, unpublished).

Microbial pesticides

Microbial pesticides such as *Bt*, fungal bioagents and baculoviruses have been widely used in the management of insect pests as alternatives to chemical insecticides. Entomopathogenic fungi and bacteria paralyse or kill their hosts by adversely affecting their growth and development. These pathogens are now commercially available though their marketability is low. The use of *Verticillium lacanii* against mustard aphid and other insect pests has been found encouraging (Singh and Meghwal, 2009). Entomopathogenic fungi such as *Beauveria bassiana* and *Paecilomyces fumosoroseus* have been reported to be highly pathogenic to *P. xylostella*, but most of these fungi have been evaluated as direct replacements for synthetic insecticides rather than as inoculative agents for classical biological control (Grzywacz *et al.*, 2010). *Metarhizium anisopliae* and *B. bassiana* have given promising results in field trials at the International Institue of Tropical Agriculture (James *et al.*, 2007). A natural colonization of *Bt* on *B. rapa* subsp. *chinensis* resulted in 35% larval mortality of *P. xylostella* (Prabhakar and Bishop, 2009).

Because of their host specificity, virulence and aggressiveness, baculoviruses have great potential in microbial control of lepidopteran pests. They have the advantage over other microbial control agents of both horizontal and vertical transmission. A strain of *P. brassicae* granulovirus (PbGV) isolated from Sangla Valley of Himachal Pradesh, India (31°25′56″ N, 78°15′4″ E, 2590 m a.s.l.) was reported to be highly effective against *P. brassicae* (Sood, 2004) and is known to survive severe winter conditions inside the diapausing insects and get vertically transmitted to the next generation (Sood *et al.*, 2010). Among the several viruses including *Plutella xylostella* granulovirus (PlxyGV), DBM NPV and a DBM cypovirus that infect *P. xylostella*, PlxyGV has been found to be promising in trials conducted in Asia (Su, 1991; Talekar, 1996) and Africa (Grzywacz *et al.*, 2001), but no commercial products have been registered in any country except China (Sun and Peng, 2007).

Use of botanicals

As many as 2121 plant species have been reported to possess pest control properties. Of these, 1005 have insecticidal, 384 antifeedant, 297 repellent, 27 attractant and 31 growth-inhibiting properties (Puri, 1999). India has a vast potential for botanical pesticides, particularly neem. An estimated 18 million neem trees with a potential yield of 0.7 Mt of fruits provides about 142–350 t of azadirachtin (0.2–0.5%) out of 10% seed kernel yield (Narval *et al.*, 1997). Many plant materials as extracts have been evaluated against mustard aphid, which include nicotine sulfate, rotenone and pyrethrins. All these have exhibited variable toxicity. Plant extracts of *Azadirachta indica*, *Lantana camara*, *Ipomoea carnea*, *Acorus* sp., *Solanum xanthocarpum*, *Swertia chirata*, *Melia azedarach* and *Argemone maxicana* proved toxic to mustard aphid (Pandey *et al.*, 1977). In a field evaluation on mustard crop (*B. juncea*), thermo- and photostable tetrahydroazadirachtin-A provided superior control of mustard aphid as compared to azadirachtin, besides being safe to natural predatory arthropods (Dhingra *et al.*, 2006). Singh (2007) reported neem seed kernel extract (5%) and neem leaf extract (5%) effective against mustard aphid. Field trials with neem leaf and seed extract have also shown very promising results against *P. xylostella* in Africa (Schmutterer, 1990).

Integration of botanical extracts with microbial pesticides has been found to have a synergistic action. For example, integration of petroleum ether neem seed kernel extract (NSKE) with PbGV resulted in maximum reduction in LC_{50} value of PbGV (4.39×10^2 occlusion bodies/ml) over PbGV alone (LC_{50}: 1.85×10^4 occlusion bodies/ml) on 2nd instar *P. brassicae* larvae (Bhandari *et al.*, 2009). Botanicals such as neem are of particular interest in organic farming and are being promoted by the organic farming sector (HDRA, 2000). There is much needed to be done for successful commercial exploitation of botanical pesticides. For example, the current application rates of neem seed formulation (0.5–2.0 kg/ha) or fresh leaves (10–20 kg/ha) are very high and hamper their acceptability by growers (Grzywacz *et al.*, 2010).

Behavioural control

Insects react to stimuli from the environment. These reactions are stereotyped and predictable. Chemicals regulating insect behaviour are collectively called semiochemicals, which belong to four categories: pheromones, allomones, kairomones and allomone-kairomones. The pheromones have been successfully isolated and chemically identified for a range of pests. For example, anti-feedants in the clarodane class of diterpenoids such as ajugarins were found to be very effective against mustard beetle, *Phaedon cochleariae* at 0.000001% and against *P. xylostella* at 0.01% applied as foliar treatment (Griffiths *et al.*, 1988). The use of pheromone-baited traps can hold great promise in attraction and management of aphids either early in the season to disguise and disperse the initial settlers or later during crop maturity. Spraying of such chemicals on non-host plants can be very effective in attracting the winged aphids and finally killing them with potent insecticides. It is reported that the alarm pheromone (β-farnesene) is of common occurrence in most of the aphids and thus spraying of formulations based on β-farnesene on the infested plants or leaves

produce peculiar reactions in the aphids and they were not able to settle for feeding on the hosts. The aphid alarm pheromone, E-β-farnesene (Eβf), appears to hold strong potential for controlling a wide variety of aphid pests. Combining imidacloprid with Eβf further reduced numbers of apterous aphids at distances of 5 m from pheromone emitters (Cui *et al.*, 2012). If spraying can be synchronized with the first appearance of aphids in the fields, the early settlers can be prevented from alighting on the hosts and thus build-up of their population can be checked to a greater extent. Thus, there is a good scope for the use of alarm pheromones in the management of aphid pests (Bakhetia *et al.*, 2002). Pheromones can also be used for mating disruption of lepidopteran adults and there have been a number of studies against DBM in many countries (Ando *et al.*, 1979; Chisholm *et al.*, 1984; Ohbayashi *et al.*, 1990). However, such approaches based solely on semiochemicals are insufficiently robust for general farming practice and need to be integrated with other methods in the ambit of IPM.

The use of trap crops and intercropping as discussed earlier in push-pull system are some of the other examples of behavioural manipulations for management of brassica pests.

Use of natural enemies (parasitoids and predators)

The natural enemies play an important role in the reduction of insect populations. In a well-studied fauna, such as in the British Isles, over 26% of the species may be parasitoids and another 4% predators, making 30% of the insect fauna (Price, 1980). Several families of predators and the primary parasitoids were attacked by 28 species of secondary parasitoids (Muller and Godfray, 1998). When natural enemies were employed as biological control agents, mortality of host may reach 90–100% in numerous cases and successful biological control is brought about even when only 40% mortality is achieved in some cases (Hawkins, 1994). The brassica agroecosystem, like other agricultural systems, is thought to be more prone to herbivore outbreaks than natural ecosystems. This has been attributed to monoculture resulting in loss of biodiversity (van Emden and Williams, 1974). However, detailed reviews by some workers such as Hawkins *et al.* (1999) have concluded that perhaps one or two particularly effective natural enemies are all that are needed for effective pest control. *Cotesia glomerata*, a gregarious larval parasitoid of *P. brassicae*, has been reported to result in up to 86% parasitization at Ludhiana, India (Kumar, 2011) and has a great potential in conservation biological control. Kristensen (1994) in Denmark has also recorded up to 82% larval parasitisation of this pest. Though *P. brassicae* is a specialist pest and does not appear to be affected by any chemicals induced in damaged cabbage tissue (Coleman *et al.*, 1996), the wounding of plants as a result of larval feeding results in elevated leaf chemicals (Edwards and Wratten, 1987), which may contribute to herbivore-induced synomones (HIS), constituting a signal to parasitoids (the 'call or cry' for help by host plant as discussed earlier). Bakhetia and Sekhon (1989) in India reported six species of coccinellids, 16 syrphids, one chamaemyiid, hemerobiid (predators), four species of hymenopterous parasitoids, four species of entomogenous fungi and one predatory bird to be associated with mustard aphid as natural enemies. Coccinellids are the important predators of mustard aphid with some species, namely *Coccinella septempunctata*, *C. repanda*, *C. transversalis*, *Brumoides suturalis*, *Menochilus sexmaculatus* and *Hippodamia variegata*, found to be abundant in the brassica agroecosystem. Despite their abundance, these natural enemies fail to provide satisfactory control of mustard aphid. One possible reason for this may be that while aphids flourish at temperatures below 20°C, coccinellids do well above 20°C, resulting in phonological asynchrony in their peak periods of activity. This has been supported by Sarwar (2009), who reported a lack of synchronization between populations of mustard aphid and its predators on canola rape. Singh and Singh (1983) had the view that predators reduce the already declining population of mustard aphid. *Coccinella septempunctata* at 5000 beetles/ha and *Verticillium lecanii* at 10^8 conidial spores/ml were found effective in significantly reducing aphids on Indian mustard 10 days after

release (Singh and Meghwal, 2009). A long list of syrphids has been found to predate on mustard aphid. However, their number in the field is quite low and had a limited scope for the control of mustard aphid. Further, syrphids are reported to oviposit only when their prey population reaches a certain level and do not oviposit below that level. For example, Luna and Jepson (2003) reported that syrphids do not oviposit until aphid infestations exceed 50 aphids per broccoli plant. Some of the important species of syrphids are *Sophaerophoria scutellaris*, *Eristalis obscuritarsis*, *E. tenax*, *Episyphus balteatus*, *Metasyrphus adligatus*, *Metasyrphus corollae*, *Xanthogramma scutellaris*, *Syrphus serarius* and *Syrphus issaci*. In addition to these, the green lacewings *Chrysopa scaslastes* and *Chrysoperla carnea* have been found actively predating upon the mustard aphid. However, their scope in population suppression of pest insects is very limited. A small bird (grey wagtail) has also been found as a predator of mustard aphid.

Among the parasitoids, *Diaeretiella rapae* and *Encyrtus* sp. have also been observed parasitizing the mustard aphid. The parasitoids also appear at the end of the season, i.e. around mid-February. *D. rapae* had been reported to be an important parasitoid of aphids in Punjab, India, which resulted in more than 70% parasitization (Atwal *et al.*, 1969). However, recent study by Kumar (2014) in Punjab has reported maximum parasitization of only 15.6% on *B. rapa* ecotype Yellow Sarson cv. YST 151, while different species of coccinellids were the predominant natural enemies. These temporal differences in parasitization can be attributed to variation in intra-guild competition among different natural enemies, which was high in the recent study. Biological control can be 'elegant, sustainable, non-polluting and inexpensive' (Gurr *et al.*, 2000a) but its success rate has not been particularly high, especially for biological control of arthropods by arthropods (Gurr *et al.*, 2000b). As pest invasions increase with 'globalization' and pesticide use continues to increase (Pimentel, 2004), greater expectations for natural enemies to reduce pest numbers will result (Howarth, 2003).

Chemical control

In the absence of stable and effective resistant cultivars against major brassica pests, insecticides are and will continue to be an integral part of a pest management programme. However, they should be used only as a last option when other methods fail to provide satisfactory pest control owing to a number of problems inherent to their use. Although insecticides are used extensively on brassica crops in India, their timing of application is mostly erratic with those requiring treatment left unsprayed and others sprayed unnecessarily (Chattopadhyay et al., 2005). Even in the UK, routine spraying of vegetable brassicas was widespread in the event of lack of established action thresholds (Blood-Smyth et al., 1992). The newer insecticides such a neonicotinoids have not been introduced on a large scale in brassicas unlike other crops (Dewar, 2007). In India, one spray of imidacloprid provided 99% control of Lipaphis pseudobrassicae (Sreekanth and Babu, 2001) while acetamiprid provided highest mortality of this pest among a range of insecticides tested (Chinnabbai et al., 1999). In glasshouse trials, thiacloprid (50 mg a.i./l) as a foliar treatment on cabbage provided complete control of M. persicae for 18 days (Elbert et al., 2000). Though more than 90% mortality of aphids is possible by use of systemic insecticides, the population attains levels similar to those in untreated fields within a period of 2–3 weeks due to the high rate of multiplication (Singh et al., 1984). Rather than going for prophylactic application, the decision to apply insecticides should be based on economic threshold level. But, unfortunately, economic threshold levels for most of the pests are not available and need to be calculated.

Insect forecasting models

It is a well-known fact that forecasting is a better strategy than either no control or prophylaxis. In a developing country like India, much of the knowledge on pest management lies in the domain of scientists and is inaccessible to farmers. It may be due to many reasons, such as inadequate communication of the results to end-users or because the results are too difficult to simplify for use. Such data can be used to develop the forecasting model(s) that can help farmers in making timely pest-management decisions. This can serve in reducing the unnecessary application of toxic insecticidal chemicals on crops as well as environmental pollution. Weather-based models to forecast mustard aphid outbreaks on oilseed brassicas have been developed for multiple locations across major crop-growing regions (Chattopadhyay et al., 2005; Matis et al., 2008; Rao et al., 2012). These are all based on observations recorded at surface level for pest and meteorological factors. Data recorded at surface meteorological observatories remain valid to a maximum range of 75 km radius. Taking India as an example, weather data recording would be needed at a minimum of ~1200 observatories in order to cover all agroecological zones of India (3,287,240 km^2), as well as multi-year observations on pest outbreaks in those many locations. Remote sensing overcomes such limitations with ability to access all parts of the country and can often achieve high spatial resolution (5 m × 5 m) by multi-spectral LISS IV at 25-day intervals, thus leading to an accurate estimation of area affected. In case of possible failures in forecasts, an accurate assessment of pest damage is possible for providing compensation to farmers. Research efforts have been made to apply or refine ground-based models using satellite-based spatial weather and high-resolution remote sensing (RS) observations for mustard aphid infestation (Bhattacharya et al., 2007; Dutta et al., 2008). With support from ICAR, Web-based forecast software (http://www.drmr.res.in/aphidforecast/index.php) for prediction of mustard aphid (L. erysimi) on oilseed brassicas for different geographic locations has been designed in India (Kumar et al., 2012). Such forecasting systems could bring precision in the management of insect pests on rapeseed-mustard and lead to a reduction in the use of chemical insecticides on standing crops to reduce the pesticide load.

Brassica Pest Management: Lessons from the Past

Grzywacz et al. (2010) provided an extensive review of brassica pest management in Asia and Africa. Farmers in the developing countries are heavily dependent on the use of synthetic insecticides for their management as they find it the most effective and easily available option for pest management. The control failures of DBM have been attributed to the indiscriminate use of insecticides on brassicas, which has disturbed the so-called natural control, and intensive selection pressure, leading to high levels of insecticide resistance to almost all the groups of insecticides including the new chemistries. However, education of the farmers about ill effects of use of insecticides led to a temporary reduction in their use, but their use remains overwhelmingly the most common management strategy. The introduction of parasitoids *Diadegma semiclausum*, *Diadromus collaris* and *Oomyzus sokolowskii* against DBM in Malaysia failed to provide long-term management of this pest since farmers continued spraying synthetic insecticides (AVRDC, 1993). Almost similar results were obtained with the release of *Cotesia plutellae* and *Trichogrammatoidea bactrae* in Bangladesh, Bhutan, India, Nepal, Pakistan and Sri Lanka under a network project, the South Asian Vegetable Research Network (SAVERNET, 1996). There is a common perception among farmers that production is not possible without the use of chemicals and a lack of understanding as to how natural enemies work. Farmers in Malaysia deferred the use of insecticides only when neighbouring countries banned the import of high residue cabbage from Malaysia (Grzywacz *et al.*, 2010). This led to a shift to the alternative control strategies such as *Bt* products. As the use of insecticides declined *D. semiclausum* became well established and the dominant parasitoid (Ooi, 1999). Thus, there is an urgent need to educate the farmers about importance of IPM, beneficial roles the different natural enemies play in agroecosystems and harmful effects of insecticides on human health and the environment, which can be achieved through strengthening of the extension system in each Asian country.

In the developing Asian countries there is a functional extension system to educate growers about importance of IPM, natural enemies of pests and ill effects of insecticides. Farmers mostly do not follow the advice of extension functionaries. It has been observed that growers mostly follow the recommendations if they are made into law.

On the other hand, in developed countries such as the UK, there is a well established pest monitoring and forecasting system. It is supposed that farmers in a developed country will follow the advice of extension functionaries – but this is not so. Despite the advice by extension functionaries about when to apply control applications, most of the farmers do not follow their recommendations and go for prophylactic spraying and a much smaller percentage of them apply control measures based on action threshold (as advised by extension functionaries). Growers select insecticides without giving due consideration to their possible harmful effects on natural enemies of pests with the exception of some supermarket protocols which prefer partially resistant cultivars and selective pesticides to be used (Collier and Finch, 2007). In Lockeyer valley, Southern Queensland, Australia the extensive use of toxic insecticidal chemicals in 1980s resulted in high levels of resistance in DBM to most of the insecticides, leading to control failures. This crisis situation stressed the government and local industry to find alternative strategies, which led to the development of a practical IPM system which included crop scouting, a break in production over the summer (November–February), use of narrow-spectrum insecticides in place of earlier broad-spectrum ones, release of predators and well-targeted spray application practices (Heisswolf *et al.*, 1996; Heisswolf and Bilston, 2001). A local grower group named as the Brassica Improvement Group (BIG) involved the active participation of growers and the industry with group approaches being an important component of public interaction and farmer interface in Queensland. The implementation of an IPM programme resulted in 60–70% farmers practising a voluntary 3 months' production break over the summer (November–February). There was a significant increase in the use of *Bt* products and rotation of insecticides from different groups besides increased use of narrow-spectrum insecticides and increase in the abundance of different natural enemies. The successful implementation of the IPM strategy led to management of brassica pests particularly DBM, which was difficult to manage before 1990.

Conclusions

Though insecticides can provide pest control for a short period, to maintain sustainability of production systems IPM should be seen as a best practice. There is a need for awareness about integrated, broad-scale approaches to ecosystem problems. The pest problems are the result of habitat destruction of insect pests, which can only be managed by following the principles of ecology and regional habitat management rather than going for their eradication. The same principles, though with different relevance, apply to the two contrasting scenarios of the technologically developed and the developing world. In the developed countries, these systems are applied systematically to decrease over-reliance on synthetic insecticides. On the other hand, in developing world there will be increased ecologically compatible use of some conventional insecticides and also bioinsecticides, including transgenics (Way and van Emden, 2000). In both scenarios, the top priority is the relevant application of much-needed applied research This can only be achieved through the participation of policy makers, scientists, growers, industry and the public in general.

References

Abro, G.H., Jayo, A.L. and Syed, T.S. (1994) Ecology of diamondback moth, *Plutella xylostella* (L.) in Pakistan 1. Host plant preference. *Pakistan Journal of Zoology* 26, 35–38.

Adamson, J.B., Soroka, J. and Holowachuk, J. (2008) Feeding and oviposition of diamondback moth *(Plutella xylostella)* on modified 'Hairy' canola. Honours Undergraduate thesis, Department of Plant Science, University of Saskatchewan, Saskatoon, Canada.

Agarwala, B.K. and Bhattacharya, S. (1999) Effective bio-control agents and their use in IPM strategy of the mustard aphid. In: Upadhyay, R.K., Mukerji, K.G. and Rajak, R.L. (eds) *IPM System in Agriculture*, Vol. 5, *Oilseeds*. Aditya Books Pvt. Ltd., New Delhi, India, pp. 77–89.

Agrawal, A.A. (1999) Induced responses to herbivory in wild radish: effects on several herbivores and plant fitness. *Ecology* 80, 1713–1723.

Agrawal, A.A. and Kurashige, N.S. (2003) A role for isothiocyanates in plant resistance against the specialist herbivore *Pieris rapae*. *Journal of Chemical Ecology* 29, 1403–1415.

AgroAtlas (2009) Interactive Agricultural Ecological Atlas of Russia and Neighbouring Countries. Economic Plants and their Diseases, Pests and Weeds, Russia. Available at: http://www.agroatlas.ru (accessed 6 April 2014).

Ahmad, H., Shankar, U., Monobrullah, M., Kaul, V. and Singh, S. (2007) Bionomics of cabbage butterfly, *Pieris brassicae* (Linn.) on cabbage. *Annals of Plant Protection Sciences* 15, 47–52.

Ahmad, T. and Ansari, M.S. (2010) Studies on seasonal abundance of diamondback moth *Plutella xylostella* (Lepidoptera: Yponomeutidae) on cauliflower crop. *Journal of Plant Protection Research* 50, 80–87.

Åhman, I. (1982) A comparison between high and low glucosinolate cultivars of summer oilseed rape (*Brassica napus* L.) with regard to their levels of infestation by the brassica pod midge (*Dasineura brassicae* Winn.). *Zeitschrift fur Angewandte Entomologie (Journal of Applied Entomology)* 94, 103–109.

Åhman, I. (1990) Plant surface characteristics and movements of two *Brassica*-feeding aphids, *Lipaphis erysimi* and *Brevicoryne brassicae*. In: *Symposia Biologica Hungaria No. 39*. Publishing House of Hungarian Academy of Sciences, Budapest, pp. 119–125.

Ahuja, B., Kalyan, R.K., Ahuja, U.R., Singh, S.K., Sundria, M.M. and Dhandapani, A. (2008) Integrated management strategy for painted bug, *Bagrada hilaris* (Burm.) inflicting injury at seedling stage of mustard (*Brassica juncea*) in arid Western Rajasthan. *Pesticide Research Journal* 20, 48–51.

Ahuja, D.B. (1990) Population dynamics of mustard aphid, *Lipaphis erysimi* (Kalt.) on Indian mustard, *Brassica juncea* (sub sp. *juncea*). *Indian Journal of Plant Protection* 18, 233–235.

Ahuja, I., Rohloff, J. and Bones, A.M. (2009) Defence mechanisms of Brassicaceae: implications for plant-insect interactions and potential for integrated pest management – a review. *Agronomy for Sustainable Development*, doi: 10.1051/agro/2009025.

Aiyer, A.K.Y.N. (1949) Mixed cropping in India. *Indian Journal of Agricultural Sciences* 19, 439–453.

Alford, D., Nilsson, C. and Ulber, B. (2003) Insect pests of oilseed rape crops. In: Alford, D.V. (ed.) *Biocontrol of Oilseed Rape Pests*. Blackwell Publishing, Oxford, UK, pp. 9–41.

Altieri, M.A., Schmidt, L.L. and Montalba, R. (1998) Assessing the effects of agroecological soil management practices on broccoli insect pest populations. *Biodynamics* 218, 23–26.

Ando, T., Koshihara, T., Yamada, H., Vu, M.H., Takahashi, N. and Tamaki, Y. (1979) Electroantennogram activities of sex pheromone analogues and their synergist effect on field attraction in the diamondback moth. *Applied Entomology and Zoology* 14, 362–364.

Andreasson, E., Jorgensen, L.B., Hoglund, A.S., Rask, L. and Meijer, J. (2001) Different myrosinase and idioblast distribution in *Arabidopsis* and *Brassica napus*. *Plant Physiology* 127, 1750–1763.

Angadi, S.P., Singh, J.P. and Anand, I.J. (1987) Inheritance of non-waxiness and tolerance to aphids in Indian mustard. *Journal of Oilseeds Research* 4, 265–267.

Ankersmith, G.W. (1956) *De Levenwijze en de Bestrijding van de Koolzaadgalmung* (Dasineura brassicae *Winn.*). Meeded Inst Pl Ziekt Onderz, Wageningen, the Netherlands.

Atri, C., Kumar, B., Kumar, H., Kumar, S., Sharma, S. and Banga, S.S. (2012) Development and characterization of *Brassica juncea-fruticulosa* introgression lines exhibiting resistance to mustard aphid, *Lipaphis erysimi* (Kalt.). *BMC Genetics* 13, 104, doi: 10.1186/1471-2156-13-104.

Atwal, A.S., Chaudhary, J.P. and Ramzan, M. (1969) Some preliminary studies in India on the bionomics and rate of parasitization of *Diaeretiella rapae* Curtis (Braconidae: Hymenoptera) a parasite of aphids. *Punjab Agricultural University Journal of Research* 6, 177–182.

AVRDC (1993) Vegetable Research and Development in South-East Asia. The AVNET Final Report Phase I. AVRDC Publication No. 92-385, 52 pp.

Badenes-Pérez, F.R., Reichelt, M., Gershenzon, J. and Heckel, D.G. (2010) Phylloplane location of glucosinolates in *Barbarea* spp. and misleading assessment of host suitability by a specialist herbivore. *New Phytologist* 189, 549–556. doi: 10.1111/j.1469-8137.2010.03486.x.

Bakhetia, D.R.C. (1983) Losses in rapeseed and mustard due to *Lipaphis erysimi* (Kalt.) in India – a literature study. *Proceedings of the 6th International Rapeseed Conference,* 17–19 May 1983, Paris, France, 6, 1142–1147.

Bakhetia, D.R.C. (1987) Insect pests of rapeseed-mustard and their management. In: Veerbhadra Rao, M. and Sithananthan, S. (eds) *Plant Protection in Field Crops*. Plant Protection Association of India, Hyderabad, pp. 249–259.

Bakhetia, D.R.C. and Brar, K.S. (1988) Effect of water stress in Ethiopian mustard (*Brassica carinata*) and Indian mustard (*B. juncea* sub sp. *juncea*) on infestation by *Lipaphis erysimi*. *Indian Journal of Agricultural Sciences* 58, 67–70.

Bakhetia, D.R.C. and Sekhon, B.S. (1989) Insect pests and their management in rapeseed-mustard. *Journal of Oilseeds Research* 6, 269–299.

Bakhetia, D.R.C. and Sharma, A.K. (1979) Preliminary observations on the aphids infestation on *Eruca sativa* Mill. *Indian Journal of Entomology* 41, 288–289.

Bakhetia, D.R.C. and Sidhu, S.S. (1983) Effect of rainfall and temperature on mustard aphid, *Lipaphis erysimi* (Kalt.). *Indian Journal of Entomology* 45, 202–205.

Bakhetia, D.R.C., Rani, S. and Ahuja, K.L. (1982) Effect of sulphur nutrition of some *Brassica* plants on their resistance response to the mustard aphid, *Lipaphis erysimi* (Kaltenbach). *Calicut University Research Journal* (Special Conference Number), 3–5 May, p. 36.

Bakhetia, D.R.C., Brar, K.S. and Sekhon, B.S. (1986) Seasonal incidence of *Lipaphis erysimi* (Kaltenback) on the *Brassica* species in the Punjab. In: Agarwala, B.K. (ed.) *Aphidology in India*. A.R. Printers, Calcutta, pp. 29–36.

Bakhetia, D.R.C., Singh, H. and Chander, H. (2002) *IPM for Sustainable Production of Oilseeds: Oilseeds and Oils Research and Development Needs*. Indian Society of Oilseeds Research, Hyderabad, pp. 184–218.

Baldwin, I.T., Kessler, A. and Halitschke, R. (2002) Volatile signaling in plant-plant-herbivore interactions: what is real? *Current Opinion in Plant Biology* 5, 351–354.

Banks, J.E. (1998) The scale of landscape fragmentation affects herbivore response to vegetation heterogeneity. *Oecologia* 117, 239–246.

Barari, H., Cook, S.M., Clark, S.J. and Williams, I.H. (2005) Effect of a turnip rape (*Brassica rapa*) trap crop on stem-mining pests and their parasitoids in winter oilseed rape (*Brassica napus*). *BioControl* 50, 69–86.

Barth, C. and Jander, G. (2006) *Arabidopsis* myrosinases TGG1 and TGG2 have redundant function in glucosinolate breakdown and insect defense. *Plant Journal* 46, 549–562.

Bartlet, E., Blight, M.M., Hick, A.J. and Williams, I.H. (1993) The responses of the cabbage seed weevil (*Ceutorhynchus assimilis*) to the odour of oilseed rape (*Brassica napus*) and to some volatile isothiocyanates. *Entomologia Experimentalis et Applicata* 68, 295–302.

Bellostas, N., Sorensen, A.D., Sorensen, J.C., Sorensen, H., Sorensen, M.D., Gupta, S.K. and Kader, J.C. (2007) Genetic variation and metabolism of glucosinolates. *Advances in Botanical Research* 45, 369–415.

Bhagat, D.V. and Singh, S. (1989) Effect of date of sowing and varieties on the seed quality, yield and aphid infestation on mustard. *Bhartiya Krishi Anusandhan Patrika* 4, 179–183.

Bhandari, K., Sood, P., Mehta, P.K., Choudhary, A. and Prabhaker, C.S. (2009) Effect of botanical extracts on the biological activity of granulosis virus against *Pieris brassicae*. *Phytoparasitica* 37, 317–332.

Bhat, N.S. and Sidhu, H.S. (1983) Influence of potassium on amino acids of the host plants and reproduction and excretion of the mustard aphid, *Lipaphis erysimi* (Kalt.). *Proceedings of the 6th International Rapeseed Conference*, 17–19 May 1983, Paris, France, 6, 28.

Bhatia, V., Uniyal, P.L. and Bhattacharya, R.C. (2011) Aphid resistance in *Brassica* crops: challenges, biotechnological progress and emerging possibilities. *Biotechnology Advances* 29, 879–888.

Bhattacharya, B.K., Dutta, S., Dadhwal, V.K., Parihar, J.S., Chattopadhyay, C., Agrawal, R., Kumar, V., Khan, S.A., Roy, S. and Shekhar, C. (2007) Predicting aphid (*Lipaphis erysimi*) growth in oilseed *Brassica* using near surface meteorological data from NOAA TOVS – a case study. *International Journal of Remote Sensing* 28, 3759–3773.

Birkett, M.A., Campbell, C.A.M., Chamberlain, K., Guerrieri, E., Hick, A.J., Martin, J.L., Matthes, M., Napier, J.A., Pettersson, J., Pickett, J.A., Poppy, G.M., Pow, E.M., Pye, B.J., Smart, L.E., Wadhams, G.H., Wadhams, L.J. and Woodcock, C.M. (2000) New roles for cis-jasmone as an insect semiochemical and in plant defense. *Proceedings of National Academy of Sciences USA* 97, 9329–9334.

Bishnoi, O.P., Singh, H. and Singh, R. (1992) Incidence and multiplication of mustard aphid, *Lipaphis erysimi* in relation to meteorological variables. *Indian Journal of Agricultural Sciences* 62, 710–712.

Björkman, M., Hambäck, P.A. and Rämert, B. (2007) Neighbouring monocultures enhance the effect of intercropping on the turnip root fly (*Delia floralis*). *Entomologia Experimentalis et Applicata* 124, 319–326.

Björkman, M., Klingen, I., Birch, A.N.E., Bones, A.M., Bruce, T.J.A., Johansen, T.J., Meadow, R., Molmann, J., Seljasen, R., Smart, L.E. and Stewart, D. (2011) Phytochemicals of Brassicaceae in plant protection and human health – Influences of climate, environment and agronomic practice. *Phytochemistry* 72, 538–556.

Blackman, R.L. and Eastop, V.F. (1984) *Aphids on the World's Crops*. John Wiley, Chichester, UK, 466 pp.

Blackman, R.L. and Eastop, V.F. (2000) *Aphids on the World's Crops: An identification and Information Guide*, 2nd edn., John Wiley & Sons, Chichester, UK, 466 pp.

Blackman, R.L. and Eastop, V.F. (2007) Taxonomic issues. In: van Emden, H.F. and Harrington, R. (eds) *Aphids as Crop Pests*. Cromwell Press, Oxfordshire, UK, pp. 1–29.

Blande, J., Pickett, J. and Poppy, G. (2007) A comparison of semiochemically mediated interactions involving specialist and generalist *Brassica* feeding aphids and the braconid parasitoid *Diaeretiella rapae*. *Journal of Chemical Ecology* 33, 767–779.

Blood-Smyth, J.A., Emmett, B.J. and Mead, A. (1992) Supervised control of foliar pests in brassica crops. *Proceedings of the Brighton Crop Protection Conference. Pests and Diseases* 3, 1015–1070.

Blossey, B. and Hunt-Joshi, T.R. (2003) Belowground herbivory by insects: Influence on plants and aboveground herbivores. *Annual Review of Entomology* 48, 521–547.

Bodnaryk, R.P. (1992) Leaf epicuticular wax as an antixenotic factor in Brassicaceae that affects the rate and pattern of feeding of flea beetles *Phyllotreta cruciferae* (Goeze). *Canadian Journal of Plant Science* 72, 1295–1303.

Bones, A.M. and Rossiter, J.T. (2006) The enzymic and chemically induced decomposition of glucosinolates. *Phytochemistry* 67, 1053–1067.

Bones, A.M., Thangstad, O.P., Haugen, O. and Espevik, T. (1991) Fate of myrosin cells – characterization of monoclonal antibodies against myrosinase. *Journal of Experimental Botany* 42, 1541–1549.

Bridges, M., Jones, A.M.E., Bones, A.M., Hodgson, C., Cole, R., Bartlet, E., Wallsgrove, R., Karapapa, V.K., Watts, N. and Rossiter, J.T. (2002) Spatial organization of the glucosinolate-myrosinase system in *Brassica* specialist aphids is similar to that of the host plant. *Proceedings of Royal Society of London B* 269, 187–191.

Brodeur, J., Leclerc, L.A., Fournier, M. and Roy, M. (2001) Cabbage seedpod weevil (Coleoptera: Curculionidae): new pest of canola in northeastern North America. *Canadian Entomologist* 133, 709–711.

Bruinsma, M., van Dam, N., van Loon, J. and Dicke, M. (2007) Jasmonic acid induced changes in *Brassica oleracea* affect oviposition preference of two specialist herbivores. *Journal of Chemical Ecology* 33, 655–668.

Buchi, R. (1990) Investigations on the use of turnip rape as trap plant to control oilseed rape pests. *Bulletin of IOBC/WPRS* 13, 32–39.

Buchi, R. (1995) Combination of trap plants (*Brassica rapa* var. *silvestris*) and insecticide use to control rape pests. *Bulletin of IOBC/WPRS* 18, 102–121.

Burgess, A.J., Warrington, S. and Allen-Williams, L. (1994) Cabbage aphid (*Brevicoryne brassicae* L.) 'performance' on oilseed rape (*Brassica napus* L.) experiencing water deficiency: roles of temperature and food quality. *Acta Horticulturae* 407: ISHS Brassica Symposium- IX Crucifer Genetics Workshop, 0567-7572, 16–19.

Butts, R.A. and Lamb, R.J. (1990) Injury to oilseed rape caused by mirid bugs (*Lygus*) (Heteroptera: Miridae) and its effect on seed production. *Annals of Applied Biology* 117, 253–266.

Butts, R.A. and Lamb, R.J. (1991a) Pest status of *Lygus* bugs (Hemiptera: Miridae) in oilseed Brassica crops. *Journal of Economic Entomology* 84, 1591–1596.

Butts, R.A. and Lamb, R.J. (1991b) Seasonal abundance of three *Lygus* species (Heteroptera:Miridae) in oilseed rape and alfalfa in Alberta. *Journal of Economic Entomology* 84, 450–456.

Bwye, A.M., Proudlove, W., Berlandier, F.A. and Jones, R.A.C. (1997) Effects of applying insecticides to control aphid vectors and cucumber mosaic virus in narrow leafed lupins (*Lupinus angustifolius*). *Australian Journal of Experimental Agriculture* 37, 93–102.

Campos, W.G., Schoereder, J.H. and Desouza, O.F. (2006) Seasonality in neotropical populations of *Plutella xylostella* (Lepidoptera): resource availability and migration. *Population Ecology* 48, 151–158.

Capinera, J.L. (2004) Cabbageworm, *Pieris rapae* (Linnaeus) (Lepidoptera: Pieridae). In: *Encyclopedia of Entomology*. Kluwer, Dordrecht, the Netherlands, pp. 444–445.

Carcamo, H., Otani, J., Herle, C., Dolinski, M., Dosdall, L., Mason, P., Butts, R., Kaminski, L. and Olfert, O. (2002) Variation of *Lygus* (Hemiptera: Miridae) species assemblages in canola agroecosystems in relation to ecoregion and crop stage. *Canadian Entomologist* 134, 97–111.

Chadda, I.C. and Arora, R. (1982) Influence of water stress in the host plant on the mustard aphid, *Lipaphis erysimi* (Kaltenbach). *Entomon* 7, 75–78.

Chandramohan, N. (1994) Seasonal incidence of diamondback moth, *Plutella xylostella* L. and its parasitoids in Nilgiris. *Journal of Biology and Conservation* 8, 77–80.

Charleston, D.S. and Kfir, R. (2000) The possibility of using Indian mustard, *Brassica juncea*, as a trap crop for the diamondback moth *Plutella xylostella* in South Africa. *Crop Protection* 19, 455–460.

Chatterjee, S.D. and Sengupta, K. (1987) Observations on reaction of mustard aphid to white petal and glossy plants of Indian mustard. *Journal of Oilseeds Research* 4, 125–127.

Chattopadhyay, C., Agrawal, R., Kumar, A., Singh, Y.P., Roy, S.K., Khan, S.A., Bhar, L.M., Chakravarthy, N.V.K., Srivastava, A., Patel, B.S., Srivastava, B., Singh, C.P. and Mehta, S.C. (2005) Forecasting of *Lipaphis erysimi* on oilseed Brassicas in India – a case study. *Crop Protection* 24, 1042–1053.

Chauhan, U. and Sharma, K.C. (2002) Status of biocontrol agents of *Plutella xylostella* (L.) (Lepidoptera: Yponomeutidae) in hilly regions of the north-west Himalayas, India. In: Kirk, A.A. and Bordat, D. (eds) *Proceedings of International Symposium*. CIRAD, Montpellier, France, 21–24 October 2002, pp. 204–211.

Cheng, X.Y., Sardana, R., Kaplan, H. and Altosaar, I. (1998) *Agrobacterium* transformed rice plants expressing synthetic *cry1Ab* and *cry1Ac* genes are highly toxic to striped stem borer and yellow stem borer. *Proceedings of National Academy of Sciences* 95, 2767–2772.

Chesnokov, P.G. (1953) *Methods of Investigating Plant Resistance to Pests*. National Science Foundation, Washington, DC.

Chinnabbai, C., Devi, C.H.R. and Venkataiah, M. (1999) Bio-efficacy of some new insecticides against the mustard aphid, *Lipaphis erysimi* (Kalt.) (Aphididae, Homoptera). *Pest Management and Economic Zoology* 7, 47–50.

Chisholm, M.D., Underhill, E.W., Palaniswamy, P. and Gerwing, V.J. (1984) Orientation disruption of male diamondback moths (Lepidoptera: Plutellidae) to traps baited with synthetic chemicals or female moths in small field plots. *Journal of Economic Entomology* 77, 157–160.

Cho, H.S., Cao, J., Ren, J.P. and Earle, E.D. (2001) Control of Lepidopteran insect pests in transgenic Chinese cabbage (*Brassica rapa* ssp. *pekinensis*) transformed with a synthetic *Bacillus thuringiensis cry1c* gene. *Plant Cell Reports* 20, 1–7.

Chow, J.K., Akhtar, Y. and Isman, M.B. (2005) The effects of larval experience with a complex plant latex on subsequent feeding and oviposition by the cabbage looper moth: *Trichoplusia ni* (Lepidoptera: Noctuidae). *Chemoecology* 15, 129–133.

Clauss, M.J., Dietel, S., Schubert, G. and Michell-Olds, T. (2006) Glucosinolate and trichome defenses in a natural *Arabidopsis lyrata* population. *Journal of Chemical Ecology* 32, 2351–2373.

Cole, R.A. (1994a) Locating a resistance mechanism to the cabbage aphid in two wild *Brassicas*. *Entomologia Experimentalis et Applicata* 71, 23–31.

Cole, R.A. (1994b) Isolation of a chitin binding lectin, with insecticidal activity in chemically defined synthetic diets, from two wild brassica species with resistance to cabbage aphid, *Brevicoryne brassicae*. *Entomologia Experimentalis et Applicata* 72, 181–187.

Cole, R.A. (1997) Comparison of feeding behaviour of two *Brassica* pests *Brevicoryne brassicae* and *Myzus persicae* on wild and cultivated *Brassica* species. *Entomologia Experimentalis et Applicata* 85, 135–143.

Coleman, R.A., Barker, A.M. and Fenner, M. (1996) Cabbage (*Brassica oleracea* var. *capitata*) failed to show wound-induced defence against a specialist and a generalist herbivore. *Oecologia* 108, 105–112.

Coll, C. (1991) Insect pests of oilseed rape in Scotland. Technical Note Scottish Agricultural Colleges, vol. 284. Scottish Agricultural College, Perth, 6pp. English Miscellaneous-19921164002 in CAB Abstracts 1991–1992.

Collier, R.H. and Finch, S. (2007) IPM case studies: Brassicas. In: van Emden, H.F. and Harrington, R. (eds) *Aphids as Crop Pests*. Cromwell Press, Oxfordshire, UK, pp. 549–559.

Cook, S.M., Smart, L.E., Potting, R.J.P., Bartlet, E., Martin, J.L., Murray, D.V., Watts, N.P. and Williams, I.H. (2002) Turnip rape (*Brassica rapa*) as a trap crop to protect oilseed rape (*Brassica napus*) from infestation by insect pests: potential and mechanisms of action. *Proceedings of the BCPC Conference: Pests and Diseases*, 2, 2002, British Crop Protection Council, Farnham, UK, pp. 569–574.

Cook, S.M., Murray, D.A. and Williams, I.H. (2004) Do pollen beetles need pollen? The effect of pollen on oviposition, survival and development of a flower-feeding herbivore. *Ecological Entomology* 29, 164–173.

Cook, S.M., Khan, Z.R. and Pickett, J.A. (2007) The use of push–pull strategies in integrated pest management. *Annual Review of Entomology* 52, 375–400.

Cui, L., Dong, J., Francis, F., Liu, Y., Heuskin, S., Lognay, G., Chen, J., Bragard, C., Tooker, J.F. and Liu, Y. (2012) E-β-farnesene synergizes the influence of an insecticide to improve control of cabbage aphids in China. *Crop Protection* 35, 91–96.

Czajkowska, M. and Dmoch, J. (1975) Badania nad pryszczarkiem Kapustnikiem (*Dasineura brassicae* Winn.) I Observaije nad biologia i ekologia szkodnika na rzepaka ozimym. *Rocz Nauk Rolniczych E* 5, 87–98.

David, W.A.L. and Gardiner, B.O. (1966) Mustard oil glucosides as feeding stimulants for *Pieris brassicae* larvae in semi-synthetic diet. *Entomologia Experimentalis et Applicata* 9, 247–255.

Dawson, G.W., Griffiths, D.C., Merritt, L.A., Mudd, A., Pickett, J.A., Wadhams, L.J. and Woodcock, C.M. (1990) Aphid semiochemicals – a review and recent advances on the sex pheromone. *Journal of Chemical Ecology* 16, 3019–3030.

Dewar, A.M. (2007) Chemical control. In: van Emden, H.F. and Harrington, R. (eds) *Aphids as Crop Pests*. Cromwell Press, Oxfordshire, UK, pp. 435–466.

Dhaliwal, G.S. and Dilawari, V.K. (1993) *Advances in Host Plant Resistance to Insects*. Kalyani Publishers, New Delhi.

Dhaliwal, G.S. and Singh, R.P. (2004) *Host Plant Resistance to Insects: Concepts and Applications*. Panima Publishing Corporation, New Delhi, 578 pp.

Dhingra, S., Sharma, D., Walia, S., Kumar, J., Singh, G., Singh, S., Jayaraman, B. and Parmar, B.S. (2006) Field appraisal of stable neem pesticide tetrahydroazadirachtin-A against mustard aphid (*Lipaphis erysimi*). *Indian Journal of Agricultural Sciences* 76, 111–113.

Doberitz, G. (1973) Untersuchungen über die biologische Abhängigkeit der Kohlschotenmücke (*Dasineura brassicae* Winn.) von Kohlschotenrüssler (*Ceutorhynchus assimilis* Payk.) und Vorschläge zur Verbesserung ihrer Bekampfung. *Nachrichtenblatt für den PXanzenschutzdienst in der DDR* 27, 145–149.

Dolinski, M.G. (1979) The cabbage seedpod weevil, *Ceutorhynchus assimilis* (Payk.) (Coleoptera: Curuliondae), as a potential pest of rape production in Canada. M.P.M. thesis, Simon Fraser University, Burnaby, Canada.

Dosdall, L.M., Weiss, R.M., Olfert, O. and Carcamo, H.A. (2002) Temporal and geographical distribution patterns of cabbage seedpod weevil (Coleoptera: Curculionidae) in canola. *Canadian Entomologist* 134, 403–418.

Duffey, S.S. (1986) Plant glandular trichomes: their partial role in defence against insects. In: Juniper, B. and Southwood, S.R. (eds) *Insects and the Plant Surface*. Arnold, London, pp. 151–172.

Duffey, S.S. and Stout, M.J. (1996) Antinutritive and toxic components of plant defense against insects. *Archives of Insect Biochemistry and Physiology* 32, 3–37.

Dutta, S., Bhattacharya, B.K., Rajak, D.R., Chattopadhyay, C., Dadhwal, V.K., Patel, N.K., Parihar, J.S. and Verma, R.S. (2008) Modelling regional level spatial distribution of aphid (*Lipaphis erysimi*) growth in Indian mustard using satellite-based remote sensing data. *International Journal of Pest Management* 54, 51–62.

Edwards, P.J. and Wratten, S.D. (1987) Ecological significance of wound induced changes in plant chemistry. In: Lebeyrie, V., Farbes, G. and Lachaise, D.D. (eds) *Insect-Plants*. Junk Publishers, Pau, France, pp. 213–218.

Eigenbrode, S.D. and Shelton, A.M. (1990) Behaviour of neonate diamondback moth larvae (Lepidoptera: Plutellidae) on glossy-leaved resistant genotypes of *Brassica oleracea*. *Environmental Entomology* 19, 1566–1571.

Eigenbrode, S.D., Espelie, K.E. and Shelton, A.M. (1991) Behaviour of neonate diamondback moth larvae [*Plutella xylostella* (L.)] on leaves and on extracted leaf waxes of resistant and susceptible cabbages. *Journal of Chemical Ecology* L7, 1691–1704.

Elbert, A., Erdelen, C., Kuhnhold, J., Nauen, R., Schmidt, H.W. and Hattori, Y. (2000) Thiacloprid, a novel neonicotinoid insecticide for foliar application. *Proceedings of the British Crop Protection Council Conference, Pests and Diseases*. Brighton, November 2000 1, 21–26.

Ellis, P.R. and Farrell, J.A. (1995) Resistance to cabbage aphid (*Brevicoryne brassicae*) in six Brassica accessions in New Zealand. *New Zealand Journal of Crop and Horticultural Science* 23, 25–29.

Ellis, P.R., Cole, R.A., Crisp, P. and Hardman, J.A. (1980) The relationship between cabbage root fly egg laying and volatile hydrolysis products of radish. *Annals of Applied Biology* 95, 283–289.

Ellis, P.R., Kiff, N.B., Pink, D.A.C., Jukes, P.L., Lynn, J. and Tatchell, G.M. (2000) Variation in resistance to the cabbage aphid (*Brevicoryne brassicae*) between and within wild and cultivated *Brassica* species. *Genetic Resources and Crop Evolution* 47, 395–401.

Eskanderi, F., Sylvester, E.S. and Richardson, J. (1979) Evidence for lack of propagation of potato leaf roll virus in *Myzus persicae*. *Phytopathology* 68, 45–47.

Ester, A., De Putter, H. and Van Bilsen, J.G.P.M. (2003) Filmcoating the seed of cabbage (*Brassica oleracea* L. convar. *Capitata* L.) and cauliflower (*Brassica oleracea* L. var. *Botrytis* L.) with imidacloprid and spinosad to control insect pests. *Crop Protection* 22, 761–768.

Falk, K.L. and Gershenson, J. (2007) The desert locust, *Schistocerca gregaria*, detoxifies the glucosinolates of *Schowia purpurea* by desulfation. *Journal of Chemical Ecology* 33, 1542–1555.

Feeny, P.P. (1976) Plant apparency and chemical defense. In: Wallace, J.M. and Mansell, R.L. (eds) *Biochemical Interaction Between Plants and Insects*. Plenum Press, New York, pp. 1–40.

Finch, S. and Collier, R.H. (2000) Host-plant selection by insects – a theory based on 'appropriate/inappropriate landings' by pest insects of cruciferous plants. *Entomologia Experimentalis et Applicata* 96, 91–102.

Finch, S. and Kienegger, M. (1997) A behavioural study to help clarify how undersowing with clover affects host-plant selection by pest insects of brassica crops. *Entomologia Experimentalis et Applicata* 84, 165–172.

Foissac, X., Nguyen, T.L., Christou, P., Gatehouse, A.M.R. and Gatehouse, J.A. (2000) Resistance to green leaf hopper (*Nephotettix virescens*) and brown plant hopper (*Nilaparvata lugens*) in transgenic rice expressing snowdrop lectin (*Galanthus nivalis* agglutinin; GNA). *Journal of Insect Physiology* 46, 573–583.

Font, R., Del Rio-Celestino, M., Fernandez, J.M. and De Haro, A. (2003) Acid detergent fiber analysis in oilseed brassicas by near-infrared spectroscopy. *Journal of Agricultural and Food Chemistry* 51, 2917–2922.

Font, R., Del Rio-Celestiono, M., Rosa, E., Aires, A. and De Hardo-Bailon, A. (2005) Glucosinolate assessment in *Brassica oleracea* leaves by near-infrared spectroscopy. *Journal of Agricultural Science* 143, 65–73.

Gabrys, B. and Tjallingii, W.F. (2002) The role of sinigrin in host plant recognition by aphids during initial plant penetration. *Entomologia Experimentalis et Applicata* 104, 89–93.

Ghosh, A.K. and Ghosh, M.R. (1981) Effect of time of sowing and insecticidal treatments on the pests of Indian mustard, *Brassica juncea* L. and on seed yield. *Entomon* 6, 357–362.

Ghosh, M.R. and Mitra, A. (1983) Incidence pattern and population composition of *Lipaphis erysimi* (Kaltenbach) on mustard and radish. In: Behaura, B.K. (ed.) *The Aphids*. The Zoological Society of Orissa, Utkal University, Bhubaneshwar, India, pp. 43–51.

Giamoustaris, A. and Mithen, R. (1995) The effect of modifying the glucosinolate content of leaves of oilseed rape (*Brassica napus* ssp. *oleifera*) on its interaction with specialist and generalist pests. *Annals of Applied Biology* 126, 347–363.

Gill, R.S. and Bakhetia, D.R.C. (1985) Resistance of some *Brassica napus* and *B. campestris* strains to *Lipaphis erysimi* (Kaltenbach). *Journal of Oilseeds Research* 2, 227–239.

Goggin, F.L. (2007) Plant-aphid interactions: molecular and ecological perspectives. *Current Opinion in Plant Biology* 10, 399–408.

Griffiths, D.C., Hassanali, A., Merritt, L.A., Mudd, A., Pickett, J.A., Shah, S.J., Smart, L.E., Wadhams, L.J. and Woodcock, C.M. (1988) Highly active antifeedants against coleopteran pests. In: *Proceedings of the Brighton Crop Protection Conference – Pests and Diseases*. British Crop Protection Council, Thornton Heath, UK, pp. 1041–1046.

Gruber, M.Y., Wang, S., Ethier, S., Holowachuk, J., Bonham-Smith, P.C., Soroka, J.J. and Lloyd, A. (2006) 'HAIRY CANOLA' – *Arabidopsis GL3* induces a dense covering of trichomes on *Brassica napus* seedlings. *Plant Molecular Biology* 60, 679–698.

Grzywacz, D., Parnell, D., Kibata, G., Oduor, G., Ogutu, W., Miano, D. and Winstanley, D. (2001) The development of endemic baculoviruses of *Plutella xylostella* (Diamondback moth, DBM) for control of DBM in East Africa. In: Endersby, N. and Ridland, P.M. (eds) *The Management of Diamondback Moth and Other Crucifer Pests*. Proceedings of the Fourth International Workshop, 26–29 November 2001, Melbourne, Australia, pp. 271–280.

Grzywacz, D., Rossbach, A., Rauf, A., Russell, D.A., Srinivasan, R. and Shelton, A.M. (2010) Current control methods for diamondback moth and other brassica insect pests and the prospects for improved management with lepidopteran-resistant Bt vegetable brassicas in Asia and Africa. *Crop Protection* 29, 68–79.

Guan, C., Wang, G., Chen, S., Li, X. and Lin, L. (2001) Breeding and agronomic characters of *Bt* transgenic insect-resistant *Brassica napus* lines. *Cruciferae Newsletter* 23, 43–44.

Gupta, P.P. (1984) Bionomics of the cabbage butterfly, *Pieris brassicae* (Linn.) in the mid hills of Himachal Pradesh. *Himachal Journal of Agricultural Research* 10, 49–54.

Gupta, R. (2002) Food preference of the 5th instar cabbage white butterfly, *Pieris brassicae* to cole crop. *Pest Management and Economic Zoology* 10, 205–207.

Gurr, G.M., Barlow, N.D., Memmott, J., Wratten, S.D. and Greathead, D.J. (2000a) A history of methodological, theoretical and empirical approaches to biological control. In: Gurr, G.M. and Wratten, S.D. (eds) *Biological Control: Measures of Success*. Kluwer Academic Publishers, Dordrecht, the Netherlands, pp. 3–37.

Gurr, G.M., Wratten, S.D. and Barbosa, P. (2000b) Success in conservation biological control of arthropods. In: Gurr, G.M. and Wratten, S.D. (eds) *Biological Control: Measures of Success*. Kluwer Academic Publishers, Dordrecht, the Netherlands, pp. 105–132.

Halkier, B.A. and Gershenzon, J. (2006) Biology and biochemistry of glucosinolates. *Annual Review of Plant Biology* 57, 303–333.

Hansen, L.M. (2004) Economic damage threshold model for pollen beetles (*Meligethes aeneus* F.) in spring oilseed rape (*Brassica napus* L.) crops. *Crop Protection* 23, 43–46.

Harcourt, D.G. (1985) Population dynamics of the diamondback moth in Southern Ontario. In: Talekar, N.S. (ed.) *The Management of Diamondback Moth and Other Crucifer Pests*. Proceedings of the first International Workshop, 11–15 March 1985, Tainan, Taiwan, pp. 3–15.

Harmon, B.L. and McCaffrey, J.P. (1997) Laboratory bioassay to assess *Brassica* spp. germplasm for resistance to the cabbage seedpod weevil (Coleoptera: Curculionidae). *Journal of Economic Entomology* 90, 1392–1399.

Hart, C.A., Batt, R.M., Saunders, J.R. and Getty, B. (1988) Lectin-induced damage to the enterocyte brush border: an electron-microscopic study in rabbits. *Scandinavian Journal of Gastroenterology* 23, 1153–1159.

Hasan, F.H. and Ansari, M.S. (2011) Population growth of *Pieris brassicae* (L.) (Lepidoptera: Pieridae) on different cole crops under laboratory conditions. *Journal of Pest Science* 84, 179–186.

Hasan, M.R., Ahmad, M., Rahman, M.H. and Haque, M.A. (2009) Aphid incidence and its correlation with different environmental factors. *Journal of Bangladesh Agricultural University* 7, 15–18.

Hawkins, B.A. (1994) *Pattern and Process in Host-Parasitoid Interactions*. Cambridge University Press, Cambridge, UK.

Hawkins, B.A., Mills, N.J., Jervis, M.A. and Price, P.W. (1999) Is the biological control of insects a natural phenomenon? *Oikos* 86, 493–506.

HDRA (2000) *Pest Control TCP3 Diamondback Moth*. Henry Doubleday Tropical Advisory Service, Coventry, UK, 3 pp.

Hegedus, D.D. and Erlandson, M. (2012) Genetics and genomics of insect resistance in Brassicaceae crops. In: Edwards, D., Batley, J., Parkin, I. and Kole, C. (eds) *Genetics, Genomics and Breeding of Oilseed Brassicas*. Science Publishers and CRC Press, Taylor and Francis and Science Publishers, Jersey, UK, pp. 319–372.

Heisswolf, S. and Bilston, L. (2001) Development and implementation of *Brassica* IPM systems in the Lockyer Valley, Queensland, Australia. In: *The Management of Diamondback Moth and Other Crucifer Pests*. Proceedings of the Fourth International Workshop, 26–29 November 2001, Australia, pp. 389–396.

Heisswolf, S., Houlding, B.J. and Deuter, P.L. (1996) A decade of integrated pest management (IPM) in brassica vegetable crops. In: Sivapragasam, A., Loke, W.H., Hussan, A.K. and Lim, G.S. (eds) *The Management of Diamondback Moth and Other Crucifer Pests*. Proceedings of the Third International Workshop, 29 October–1 November 1996, Kuala Lumpur, Malaysia. Malaysian Agricultural Research and Development Institute (MARDI), pp. 228–232.

Hern, A., Edwards-Jones, G. and McKinlay, R.G. (1996) A review of the preoviposition behaviour of small cabbage white butterfly *Pieris rapae* (Lepidoptera: Pieridae). *Annals of Applied Biology* 128, 349–371.

Hokkanen, H.H., Granlund, H., Husberg, G.B. and Markkula, M. (1986) Trap crops used successfully to control *Meligethes aeneus* (Coleoptera: Nitidulidae), the rape blossom beetle. *Annales Entomologici Fennici* 52, 115–120.

Hokkanen, H.M.T. (1991) Trap cropping in pest management. *Annual Review of Entomology* 36, 119–138.

Hong, C.P., Kwon, S.J., Kim, J.S., Yang, T.J., Park, B.S. and Lim, Y.P. (2008) Progress in understanding and sequencing the genome of *Brassica rapa*. *International Journal of Plant Genomics* doi:10.1155/2008/582837.

Hopkins, R.J., Birch, A.N.E., Griffiths, D.W., Baur, R., Städler, E. and McKinlay, R.G. (1997) Leaf surface compounds and oviposition preference of turnip root fly *Delia floralis*: The role of glucosinolate and nonglucosinolate compounds. *Journal of Chemical Ecology* 23, 629–643.

Hopkins, R.J., van Dam, N.M. and van Loon, J.J.A. (2009) Role of glucosinolates in insect plant relationships and multitrophic interactions. *Annual Review of Entomology* 54, 57–83.

Hossain, M.A., Maiti, M.K., Basu, A., Sen, S., Ghosh, A.K. and Sen, S.K. (2006) Transgenic expression of onion leaf lectin in Indian mustard offers protection against aphid colonization. *Crop Science* 46, 2022–2232.

Howarth, F.G. (2003) Globalization and pest invasions: where will we be in five years? *Proceedings of the 1st International Symposium on Biological Control of Arthropods*. USDA Forest Service Health Technology Enterprise Team, FHTET-03-05, pp. 34–39.

Hsiao, W.F. (1999) Developmental biology and population growth of turnip aphid, *Lipaphis erysimi* (Homoptera: Aphididae) fed kale. *Chinese Journal of Entomology* 19, 307–318.

Hughes, J.M. and Evans, K.A. (2003) Lygid bug damage as a pod access mechanism for *Dasineura brassicae* (Diptera: Cecidomyiidae) oviposition. *Journal of Applied Entomology* 127, 116–118.

Jaeger, E.C. (1978) *A Source-Book of Biological Names and Terms*, 3rd edn, 6th printing. Charles C. Thomas, Springfield, Massachusetts.

James, B., Godonou, I., Atcha-Ahowe, C., Glitho, I., Vodouhe, S., Ahanchede, A., Kooyman, C. and Goergen, G. (2007) Extending integrated pest management to indigenous vegetables. *Acta Horticulturae* 752, 89–93.

Jarvis, J.L. (1970) Relative injury to some cruciferous oilseeds by the turnip aphid. *Journal of Economic Entomology* 63, 1498–1502.

Javier, E.Q. (1990) Foreword. In: Talekar, N.S. (ed.) *The Management of Diamondback Moth and Other Crucifer Pests*. Proceedings of the Second International Workshop, 10–14 December 1990, Tainan, Taiwan. Asian Vegetable Research Development Center publication 92-368, Taipei.

Jayarathnam, K. (1977) Studies on the population dynamics of the diamondback moth, *Plutella xylostella* (L.) (Lepidoptera: Yponomeutidae) and crop loss due to the pest in cabbage. PhD thesis, University of Agricultural Sciences, Bangalore.

Johnson, D.R. (1953) *Plutella maculipennis* resistance to DDT in Java. *Journal of Economic Entomology* 46, 176.

Jones, T.H., Cole, R.A. and Finch, S. (1988) A cabbage root fly oviposition deterrent in the frass of garden pebble moth caterpillars. *Entomologia Experimentalis et Applicata* 49, 277–282.

Jones, P. and Vogt, T. (2001) Glycosyltransferases in secondary plant metabolism: tranquilizers and stimulant controllers. *Planta* 213, 164–174.

Joshan, V.S., Narang, D.D. and Dilawari, V.K. (1992) Population build-up of mustard aphid, *Lipaphis erysimi* (Kalt) on *Brassica* crops under water stress conditions. *Proceedings of National Symposim on Recent Advances in IPM*, 12–15 October 1992, PAU, Ludhiana.

Joshi, M.L., Ahuja, D.B. and Mathur, B.N. (1989) Loss in seed yield by insect pests and their occurrence on different dates of sowing in Indian mustard (*Brassica juncea*). *Indian Journal of Agricultural Sciences* 59, 166–168.

Kalra, V.K., Singh, H. and Rohilla, H.R. (1987) Influence of various genotypes of *Brassica juncea* on biology of mustard aphid, *Lipaphis erysimi* (Kalt.). *Indian Journal of Agricultural Sciences* 57, 277–279.

Kawada, K. and Murai, T. (1979) Apterous males and holocyclic reproduction of *Lipaphis erysimi* in Japan. *Entomologia Experimentalis et Applicata* 26, 343–345.

Kelton, L.A. (1975) The Lygus bugs (Genus *Lygus* Hahn) of North America (Heteroptera: Miridae). *Memoirs of the Entomological Society of Canada* 107, 5–107.

Khan, M.A.M., Ulrichs, C. and Mewis, I. (2010) Influence of water stress on the glucosinolate profile of *Brassica oleracea* var. italica and the performance of *Brevicoryne brassicae* and *Myzus persicae*. *Entomologia Experimentalis et Applicata* 137, 229–236.

Khan, Z.R. and Pickett, J.A. (2004) The 'push-pull' strategy for stemborer management: a case study in exploiting biodiversity and chemical ecology. In: Gurr, G.M., Wratten, S.D. and Altieri, M.A. (eds) *Ecological*

Engineering for Pest Management: Advances in Habitat Manipulation for Arthropods. CSIRO Publishing, Australia, pp. 155–164.

Khan, Z.R., Pickett, J.A., Berg, J.V.D., Wadhams, L.J., Woodcock, C.M. and Berg, J.V.D. (2000) Exploiting chemical ecology and species diversity: stem borer and striga control for maize and sorghum in Africa. *Pest Management Science* 56, 957–962.

Kift, N.B., Ellis, P.R., Tatchell, G.M. and Pink, D.A.C. (2000) The influence of genetic background on resistance to the cabbage aphid (*Brevicoryne brassicae*) in kale (*Brassica oleracea* var. *acephala*). *Annals of Applied Biology* 136, 189–195.

Kissen, R., Rossiter, J.T. and Bones, A.M. (2009) The 'mustard oil bomb': not so easy to assemble?! Localization, expression and distribution of the components of the myrosinase enzyme system. *Phytochemistry Reviews* 8, 69–86.

Klingen, I., Meadow, R. and Eilenberg, J. (2000) Prevalence of fungal infections in adult *Delia radicum* and *Delia floralis* captured on the edge of a cabbage field. *Entomologia Experimentalis et Applicata* 97, 265–274.

Klingen, I., Meadow, R. and Aandal, T. (2002) Mortality of *Delia floralis*, *Galleria mellonella* and *Mamestra brassicae* treated with insect pathogenic hyphomycetous fungi. *Journal of Applied Entomology* 126, 231–237.

Kobori, Y. and Amano, H. (2003) Effect of rainfall on a population of the diamondback moth, *Plutella xylostella* (Lepidoptera: Plutellidae). *Japanese Journal of Applied Entomology and Zoology* 38, 249–253.

Komath, S.S., Kavitha, M. and Swamy, M.J. (2006) Beyond carbohydrate binding: new directions in plant lectin research. *Organic and Biomolecular Chemistry* 4, 973–988.

Kramer, K.J., Muthukrishnan, S., John, L. and White, F. (1997) Chitinase for insect control. In: Carozzi, N. and Koziel, M. (eds) *Advances in Insect Control: Transgenic plants for the control of insect pests.* Taylor and Francis, Washington, DC, pp. 185–193.

Kristensen, C.O. (1994) Investigations on the natural mortality of eggs and larvae of the large white *Pieris brassicae* (L.) (Lep., Pieridae). *Journal of Applied Entomology* 117, 92–98.

Kular, J.S., Brar, A.S. and Kumar, S. (2012) Population development of turnip aphid *Lipaphis erysimi* (Kaltenbach, 1843) (Hemiptera: Aphididae) and the associated predator *Coccinella septempunctata* Linnaeus, 1758 as affected by changes in sowing dates of oilseed Brassica. *Entomotropica* 27, 19–25.

Kulat, S.S., Radke, S.G., Tambe, V.J. and Wankhede, D.K. (1997) Role of abiotic components on the development of mustard aphid, *Lipaphis erysimi* Kalt. *PKV Research Journal* 21, 53–56.

Kumar, S. (2011) *Cotesia glomeratus* – a potential biocontrol agent for large white butterfly, *Pieris brassicae* in Indian Punjab. *Proceedings of 13th International Rapeseed Congress*, Prague, Czech Republic, 5–9 June 2011, 13, 1141–1143.

Kumar, S. (2014) Population development of turnip aphid, *Lipaphis erysimi* and the associated resident natural enemies on oilseed *Brassica.* In: Kumar, V., Meena, P.D., Singh, D., Banga, S., Sardana, V. and Banaga S.S. (eds) *Abstracts, 2nd National Brassica Conference on Brassicas for Addressing Edible Oil and Nutritional Security.* organized by Society for Rapeseed Mustard Research at Punjab Agricultrual University, Ludhiana during 14–16 February 2014, p. 80.

Kumar, S., Atri, C., Sangha, M.K. and Banga S.S. (2011) Screening of wild crucifers for resistance to mustard aphid, *Lipaphis erysimi* (Kaltenbach) and attempt at introgression of resistance gene(s) from *Brassica fruticulosa* to *Brassica juncea. Euphytica* 179, 461–470.

Kumar, V., Kumar, A. and Chattopadhyay, C. (2012) Design and implementation of web-based aphid (*Lipaphis erysimi*) forecast system for oilseed Brassicas. *The Indian Journal of Agricultural Sciences* 82, 608–614.

Kusnierczyk, A., Winge, P., Jørstad, T., Troczyńska, J., Rossiter, J.T. and Bones, A.M. (2008) Towards global understanding of plant defence against aphids – timing and dynamics of early *Arabidopsis* defence responses to cabbage aphid (*Brevicoryne brassicae*) attack. *Plant Cell and Environment* 31, 1097–1115.

Kuwahara, M., Keinmeesuke, P. and Shirai, Y. (1996) Seasonal trend in population density and adult body size of the diamondback moth, *Plutella xylostella* (L.) (Lepidoptera: Yponomeutidae), in Central Thailand. *Japanese Journal of Applied Entomology and Zoology* 30, 551–555.

Lal, O.P. (1977) *Lipaphis erysimi* (Kalt.). In: Kranz, J., Schmutterer, H. and Koch, W. (eds) *Diseases, Pests and Weeds in Tropical Crops.* Verlag Paul Parey, Berlin and Hamburg, Germany, pp. 335–336.

Lamb, R.J. (1980) Hairs protect pods of mustard (*Brassica hirta* 'Gisilba') from flea beetle feeding damage. *Journal of Plant Science* 60, 1439–1440.

Lamb, R.J. (1989) Entomology of oilseed *Brassica* crops. *Annual Review of Entomology* 34, 211–229.

Lamb, R.J., Smith, M.A.H. and Bodnaryk, R.P. (1993) Leaf waxiness and the performance of *Lipaphis erysimi* (Kaltenbach) (Homoptera: Aphididae) on three *Brassica* crops. *Canadian Entomologist* 125, 1023–1031.

Lambdon, P.W., Hassall, M. and Mithen, R. (1998) Feeding preferences of wood pigeons and flea-beetles for oilseed rape and turnip rape. *Annals of Applied Biology* 133, 313–328.

Landing, J. (1982) *Life History, Population Dynamics and Dispersal in the Mustard Aphid,* Lipaphis erysimi: *a Literature Study*. Swedish University of Agricultural Research, Research Information Centre, Uppsala, Sweden.

Lara, F.M. (1979) *Principios de Resistancia de Plants a Insects* (in Portugese). Livroceres Ltda, Piracicaba, Brazil.

Letourneau, D.K. (1988) Soil management for pest control: a critical appraisal of the concepts. In: *Proceedings of the 6th International Science Conference of IFOAM on Global Perspectives on Agroecology and Sustainable Agricultural Systems*, Santa Cruz, California, pp. 581–587.

Levin, D.A. (1973) The role of trichomes in plant defence. *Quarterly Review of Biology* 48, 3–15.

Li, Q., Eigenbrode, S.D., Stringam, G.R. and Thiagarajah, M.R. (2000) Feeding and growth of *Plutella xylostella* and *Spodoptera eridania* on *Brassica juncea* with varying glucosinolate concentrations and myrosinase activities. *Journal of Chemical Ecology* 26, 2401–2419.

Liang, A., Su, Z., Li, X. and Wang, W. (1999) Expression of the neurotoxin of scorpion *Buthus martensii* Karsch in *Escherichia coli*. *Chinese Journal of Biochemistry and Molecular Biology* 15, 205–210 (in Chinese).

Loe, G., Torang, P., Gaudeul, M. and Agren, J. (2007) Trichome production and spatiotemporal variation in herbivory in the perennial herb *Arabidopsis lyrata*. *Oikos* 116, 134–142.

Luna, J. and Jepson P. (2003) Enhancing biological conrol with insectary plantings. WSARE, 2003. Available at: http://wsare.usu.edu/projects/2003/SW99-061A.doc (accessed 13 April 2014).

Macedo, M.L.R., de Castro, M.M. and das GraÇas Machado Freire, M. (2004) Mechanisms of the insecticidal action of TEL (*Talisia esculenta* Lectin) against *Callosobruchus maculates* (Coleoptera: Bruchidae). *Archives of Insect Biochemistry and Physiology* 56, 84–96.

Malik, R.S. and Anand, I.J. (1984) Effect of aphid infestation on the oil yielding attributes in Brassica. *Journal of Oilseeds Research* 1, 147–155.

Matis, J.M., Kiffe, T.R., Matis, T.I. and Chattopadhyay, C. (2008) Generalised aphid population growth models with immigration and cumulative-size dependent dynamics. *Mathematical Biosciences* 215, 137–143.

Meisner, J. and Mitchell, B.K. (1984) Phagodeterrency induced by some secondary plant substances in adults of the flea beetle *Phyllotreta striolata*. *Journal of Plant Disease Protection* 91, 301–304.

Mewis, I., Khan, M.A.M., Glawischnig, E., Schreiner, M. and Ulrichs, C. (2012) Water stress and aphid feeding differentially influence metabolite composition in *Arabidopsis thaliana* (L.). *PLoS One* 7(11), e48661.

Mewis, I.Z., Ulrich, C. and Schnitzler, W.H. (2002) The role of glucosinolates and their hydrolysis products in oviposition and host plant finding by cabbage webworm, *Hellula undalis*. *Entomologia Experimentalis et Applicata* 105, 129–139.

Miles, P.W., Aspinall, D. and Rosenberg, L. (1982) Performance of cabbage aphid, *Brevicoryne brassicae* on water stressed rape plants in relation to the changes in their chemical composition. *Australian Journal of Zoology* 30, 337–345.

Milford, G.F.J., Fieldsend, J.K., Porter, A.J.R., Rawlinson, C.J., Evans, E.J. and Bilsborrow, P.E. (1989) Changes in glucosinolate concentrations during the vegetative growth of single- and double-low cultivars of winter oilseed rape. *Aspects of Applied Biology* 23, 83–90.

Mohan, M. and Gujar, G.T. (2003) Local variation in susceptibility of the diamondback moth, *Plutella xylostella* (Linneaus) to insecticides and detoxification enzymes. *Crop Protection* 22, 495–504.

Mosiane, S.M., Kfir, R. and Villet, M.H. (2003) Seasonal phenology of the diamondback moth, *Plutella xylostella* (L.) (Lepidoptera: Plutellidae) and its parasitoids on canola, *Brassica napus* (L.), in Gauteng province, South Africa. *African Entomologist* 11, 277–285.

Mossler, M.A. (2005) *Florida Crop/Pest Management Profile: Specialty Brassicas (Arrugula, Bok Choy, Chinese Broccoli, Chinese Mustard, Napa)*. PI-70, Pesticide Information Office, University of Florida, Florida.

Muir, A.D., Gruber, M.Y., Hinks, C.F., Lees, G.L., Onyilagha, J., Soroka, J. and Erlandson, M. (1999) Effect of condensed tannins in the diets of major crop insects. In: Gross, G., Hemingway, R.W. and Yoshida, T. (eds) *Plant Polyphenols 2: Chemstry, Biology, Pharmacology. Ecology*. Kluwer Academic/Plenum Publishers, New York, pp. 867–881.

Muller, C. and Sieling, N. (2006) Effects of glucosinolate and myrosinase levels in *Brassica juncea* on a glucosinolate-sequestering herbivore – and vice versa. *Chemoecology* 16, 191–201.

Muller, C.B. and Godfray, H.C.J. (1998) Host-parasitoid dynamics. In: Dempster, J.P. and McLean, I.F.G. (eds) *Insect Population in Theory and in Practice*. Kluwer Academic Publishers, Dordrecht, the Netherlands, pp. 135–168.

Muratori, F., Le Ralec, A., Lognay, G. and Hance, T. (2006) Epicuticular factors involved in host recognition for the aphid parasitoid *Aphidius rhopalosiphi*. *Journal of Chemical Ecology* 32, 579–593.

Murdock, L.L., Huesing, J.E., Nielsen, S.S., Pratt, R.C. and Shade, R.E. (1990) Biological effects of plant lectins on the cowpea weevil. *Phytochemistry* 29, 85–89.

Nachbar, M.S. and Oppenheim, J.D. (1980) Lectins in the United States diet: a survey of lectins in commonly consumed foods and a review of the literature. *American Journal of Clinical Nutrition* 33, 2338–2345.

Nagarkatti, S. and Jayanth, K.P. (1982) Population dynamics of major insect pests of cabbage and of their natural enemies in Bangalore district (India). In: *Proceedings of International Conference on Plant Protection in the Tropics*, Malaysian Plant Protection Society of Malaysia, 1–4 March 1982, pp. 325–347.

Narjary, B., Adak, T., Meena, M.D. and Chakravarty, N.V.K. (2013) Population dynamics of mustard aphid in relation to humid thermal ratio and growing degree days. *Journal of Agricultural Physics* 13, 39–47.

Narval, S.S., Tauro, P. and Bisla, S.S. (1997) *Neem in Sustainable Agriculture*. Scientific Publishers, Jodhpur, India.

Navatha, S. and Murthy, K.S. (2006) Host preference for oviposition and feeding by diamondback moth, *Plutella xylostella* (L.). *Annals of Plant Protection Sciences* 14, 283–286.

Newman, P.L. (1984) The effects of insect larval damage upon the incidence of canker in winter oilseed rape. *Proceedings of the 1984 British Crop Protection Conference on Pests and Diseases*, Brighton Metropole, England, 19–22 November 1984, Vol. 2. British Crop Protection Council, Croydon, UK, pp. 815–822.

Nguz, K., van Gaver, D. and Huyghebaert, A. (1998) *In vitro* inhibition of digestive enzymes by sorghum condensed tannins [*Sorghum bicolour* L. (Moench)]. *Sciences des Aliments* 18, 507–514.

Nielsen, J.K. (1978) Host plant discrimination within Cruciferae: feeding responses of four leaf beetles (Coleoptera: Chrysomelidae) to glucosinolates, cucurbitacins and cardenolides. *Entomologia Experimentalis et Applicata* 24, 41–54.

Nielsen, J.K., Larsen, L.M. and Sørensen, H. (1979) Host plant selection of the horseradish flea beetle *Phyllotreta armoraciae* (Coleoptera: Chrysomelidae): identification of two flavonol glycosides stimulating feeding in combination with glucosinolates. *Entomologia Experimentalis et Applicata* 26, 40–48.

Nottingham, S.F., Hardie, J., Dawson, G.W., Hick, A.J., Pickett, J.A., Wadhams, L.J. and Woodcock, C.M. (1991) Behavioral and electrophysiological responses of aphids to host and nonhost plant volatiles. *Journal of Chemical Ecology* 17, 1231–1242.

Ohbayashi, N., Shimizu, K. and Nagata, K. (1990) Control of diamondback moth using synthetic sex pheromones. In: Talekar, N.S. (ed.) *The Management of Diamondback Moth and Other Crucifer Pests*. Proceedings of the Second International Workshop, 10–14 December 1990, Tainan, Taiwan. Asian Vegetable Research Development Centre publication 92-368, Taipei, pp. 99–104.

Oishi, K., Sawa, M., Hatakeyama, M. and Kageyama, Y. (1993) Genetics and biology of the sawfly, *Athalia rosae* (Hymenoptera). *Genetica* 88, 119–127.

Onyilagha, J.C., Lazorko, J., Gruber, M.Y., Soroka, J.J. and Erlandson, M.A. (2004) Effect of flavonoids on feeding preference and development of the crucifer pest *Mamestra configurata* Walker. *Journal of Chemical Ecology* 30, 109–124.

Ooi, P.A.C. (1999) Ensuring successful implementation of biological control in the tropics. *Proceedings of the Symposium on the Biological Control in the Tropics*, 18–19 March 1999, MARDI Training Centre, Serdang, Malaysia, pp. 18–20.

Ovsyannikova, E.I. and Grichanov, I.Y. (2010) Interactive agricultural ecological atlas of Russia and neighboring countries: *Mamestra brassicae* L. – cabbage moth. Available at: http://www.agroatlas.ru/en/content/pests/Mamestra_brassicae (accessed 4 April 2014).

Painter, R.H. (1951) *Insect Resistance in Crop Plants*. Macmillan, New York.

Pal, S.R., Nath, D.K. and Saha, G.N. (1976) Effect of time of sowing and aphid infestation on rai (*Brassica juncea* Coss.). *Indian Agriculturist* 20, 27–34.

Palaniswamy, P. and Bodnaryk, R.P. (1994) A wild *Brassica* from Sicily provides trichome-based resistance against flea beetles, *Phyllotreta cruciferae* (Goeze) (Coleoptera: Chrysomelidae). *Canadian Entomologist* 126, 1119–1130.

Panda, N. (1979) *Principles of Host Plant Resistance to Insect Pests*. Allenheld, Osmun & Universe Books, New York.

Panda, N. and Khush, G.S. (1995) *Host Plants Resistance to Insects*. CAB International, Wallingford, UK, xiii + 431 pp.

Pandey, N.D., Singh, M. and Tiwari, G.C. (1977) Antifeedent, repellant and insecticidal properties of some indigenous plant materials against mustard saw fly, *Athalia lugens proxima*. *Indian Journal of Entomology* 39, 62–64.

Pérez, C.J., Alvarado, P., Narváez, C., Miranda, F., Hernandez, L., Vanegas, H., Hruska, A. and Shelton, A.M. (2000) Assessment of insecticide resistance in five insect pests attacking field and vegetable crops in Nicaragua. *Journal of Economic Entomology* 93, 1779–1787.

Pfiffner, L., Luka, H., Schlatter, C., Juen, A. and Traugott, M. (2009) Impact of wildflower strips on biological control of cabbage lepidopterans. *Agriculture, Ecosystems and Environment* 129, 310–314.

Phadke, K.G. (1986) Ecological factors influencing aphid, *Lipaphis erysimi* (Kaltenback) incidence on mustard crop. In: Agarwala, B.K. (ed.) *Aphidology in India*. A R Printers, Calcutta, pp. 37–42.

Pimentel, D. (2004) *Foreword*. In: *Ecological Engineering for Pest Management: Advances in habitat manipulation for arthropods*. CSIRO Publishing, Collingwood, Victoria, Australia, p. vi.

Ponsen, M.B. (1972) The site of potato leafroll virus multiplication in its vector, *Myzus persicae*. An anatomical study. *Mededlingen Landbouwhogeschool Wageningen* 72–16, 1–147.

Pope, T.W., Kissen, R., Grant, M., Pickett, J.A., Rossiter, J.T. and Powell, G. (2008) Comparative innate responses of the aphid parasitoid *Diaeretiella rapae* to alkenyl glucosinolate derived isothiocyanates, nitriles and epithionitriles. *Journal of Chemical Ecology* 34, 1302–1310.

Popov, C., Trotus, E., Vasilescu, S., Barbulescu, A. and Rasnoveanu, L. (2006) Drought effect on pest attack in field crops. *Romanian Agricultural Research* 23, 43–52.

Popova, T. (1993) A study of antibiotic effects on cabbage cultivars on the cabbage moth *Mamestra brassicae* L. (Lepidoptera: Noctuidae). *Entomological Review* 72, 125–132.

Powell, J.A. and Opler, P.A. (2009) *Moths of Western North America*. University of California Press, California.

Powell, K.S. (2001) Antimetabolic effect of plant lectins towards nymphal stages of the plant hoppers *Tarophagous proserpina* and *Nilaparvata lugens*. *Entomologia Experimentalis et Applicata* 99, 71–77.

Powell, K.S., Gatehouse, A.M.R., Hilder, V.A. and Gatehouse, J.A. (1993) Antimetabolic effects of plant lectins and plant and fungal enzymes on the nymphal stages of two important rice pests, *Nilaparvata lugens* and *Nephotettix cinciteps*. *Entomologia Experimentalis et Applicata* 66, 119–126.

Prabhakar, A. and Bishop, A.H. (2009) Effect of *Bacillus thuringiensis* naturally colonising *Brassica campestris* var. *chinensis* leaves on neonate larvae of *Pieris brassicae*. *Journal of Invertebrate Pathology* 100, 193–194.

Price, P.W. (1980) *Evolutionary Biology of Parasites*. Princeton University Press, Princeton, New Jersey.

Puri, S.N. (1999) Integrated pest management in India. In: Masters trainers training programme on Integrated pest management, 13–17 April 1999. LBS Building, New Delhi.

Rai, B. and Sehgal, B.K. (1975) Field resistance of *Brassica* germplasm to mustard aphid, *Lipaphis erysimi* (Kalt.). *Science and Culture* 4, 444–445.

Rana, J.S. (2005) Performance of *Lipaphis erysimi* (Homoptera: Aphididae) on different *Brassica* species in a tropical environment. *Journal of Pest Science* 78, 155–160.

Rana, J.S., Khokhar, K.S., Singh, H. and Khokhar, S. (1993) Influence of abiotic environment on the population dynamics of mustard aphid, *Lipaphis erysimi* (Kalt.). *Crop Research Hisar* 6, 116–119.

Rao, B.P., Ramaraj, A.P., Chattopadhyay, C., Prasad, Y.G. and Rao, V.U.M. (2012) Predictive model for mustard aphid infestation for eastern plains of Rajasthan. *Journal of Agrometeorology* 14, 60–62.

Rask, L., Andreasson, E., Ekbom, B., Eriksson, S., Pontoppidan, B. and Meijer, J. (2000) Myrosinase: gene family evolution and herbivore defence in Brassicaceae. *Plant Molecular Biology* 42, 93–113.

Ratzka, A., Vogel, H., Kliebenstein, D.J., Mitchell-Olds, T. and Kroymann, J. (2002) Disarming the mustard oil bomb. *Proceedings of National Academy of Sciences USA* 99, 11223–11228.

Rawat, R.R., Misra, U.S., Thakar, A.V. and Dhamdhere, S.V. (1968) Preliminary study on the effect of different doses of nitrogen on the incidence of major pests of mustard. *Madras Agriculture Journal* 55, 363–366.

Razmi, M., Karimpour, Y., Safaralizadeh, M.H. and Safavi, S.A. (2011) Parasitoid complex of cabbage large white butterfly *Pieris brassicae* (L.) (Lepidoptera, Pieridae) in Urmia with new records from Iran. *Journal of Plant Protection Research* 51, 248–251.

Reifenrath, K., Riederer, M. and Müller, C. (2005) Leaf surface wax layers of Brassicaceae lack feeding stimulants for *Phaedon cochleariae*. *Entomologia Experimentalis et Applicata* 115, 41–50.

Remaudière, G. and Remaudière, M. (1997) *Catalogue des Aphididae du Monde*. INRA, Paris.

Renwick, J.A.A. (2002) The chemical world of crucivores: lures, treats and traps. *Entomologia Experimentalis et Applicata* 104, 35–42.

Renwick, J.A., Radke, A. and Celia, D. (1983) Chemical recognition of host plants for oviposition by the cabbage butterfly, *Pieris rapae* (Lepidoptera: Pieridae). *Environmental Entomology* 12, 446–450.

Renwick, J.A.A., Radke, C.D., Sachdev-Gupta, K. and Städler, E. (1992) Leaf surface chemicals stimulating oviposition by *Pieris rapae* (Lepidoptera: Pieridae) on cabbage. *Chemoecology* 3, 33–38.

Renwick, J.A.A., Haribal, M., Gouinguenè, S. and Städler, E. (2006) Isothiocyanates stimulating oviposition by the diamondback moth, *Plutella xylostella*. *Journal of Chemical Ecology* 32, 755–766.

Rohilla, H.R., Singh, H., Kalra, V.K. and Kharub, S.S. (1987) Losses caused by mustard aphid, *Lipaphis erysimi* (Kalt.) on various *Brassica* genotypes. *Proceedings of 7th International Rapeseed Congress*, 11–14 May 1987, Poznan, Poland, 7, 1077–1084.

Rohilla, H.R., Singh, H. and Singh, R. (1999) Evaluation of rapeseed mustard genotypes against mustard aphid, *Lipaphis erysimi* (Kalt.). Test of Agrochemicals and Cultivars No. 20. *Annals of Applied Biology* 134, 42–43.

Rossiter, J.T., Jones, A.M. and Bones, A.M. (2003) A novel myrosinase-glucosinolate defense system in cruciferous specialist aphids. *Recent Advances in Phytochemistry* 37, 127–142.

Rostas, M., Ruf, D., Zabka, V. and Hildebrandt, U. (2008) Plant surface wax affects parasitoid's response to host footprints. *Naturwissenschaften* 95, 997–1002.

Roush, R.T. and Shelton, A.M. (1997) Assessing the odds: the emergence of resistance to *Bt* transgenic plant. *Nature Biotechnology* 15, 816–817.

Rousse, P., Fournet, S., Portenevue, C. and Brunel, E. (2003) Trap cropping to control *Delia radicum* populations in cruciferous crops: first results and future applications. *Entomologia Experimentalis et Applicata* 109, 133–138.

Rout, G. and Senapati, B. (1968) Biology of the mustard aphid, *Lipaphis erysimi* (Kalt.) in India. *Annals of Entomological Society of America* 61, 259–261.

Roy, P. (1975) Population dynamics of mustard aphid, *Lipaphis erysimi* Kaltenbach (Aphididae: Hemiptera) in West Bengal. *Indian Journal of Entomology* 37, 318–321.

Russel, G.E. (1978) *Plant Breeding for Pest and Disease Resistance*. Butterworths, London.

Sachan, J.N. and Srivastava, B.P. (1972) Studies on seasonal incidence of insect pests of cabbage. *Indian Journal of Entomology* 34, 123–129.

Sadasivam, S. and Thayumanavan, B. (2003) *Molecular Host Plant Resistance to Pests*. Marcel Dekker, New York.

Sadeghi, H. (2002) The relationship between oviposition preference and larval performance in an aphidophagous hover fly, *Syrphus ribesii* L. (Diptera: Syrphidae). *Journal of Agricultural Science and Technology* 4, 1–10.

Samdur, M.Y., Gulati, S.C., Raman, R. and Manivel, P. (1997) Effect of environmental factors on mustard aphid (*Lipaphis erysimi* Kalt.) infestation on different germplasm of Indian mustard. *Journal of Oilseeds Research* 14, 278–283.

Sarwar, M. (2009) Populations' synchronization of aphids (Homoptera: Aphididae) and ladybird beetles (Coleoptera: Coccinellidae) and exploitation of food attractants for predator. *Biological Diversity and Conservation* 2, 85–89.

Sauvion, N., Charles, H., Febvay, G. and Rahbé, Y. (2004a) Effects of jackbean lectin (ConA) on the feeding behaviour and kinetics of intoxication of the pea aphid, *Acyrthosiphon pisum*. *Entomologia Experimentalis et Applicata* 110, 31–44.

Sauvion, N., Nardon, C., Febvay, G., Gatehouse, A.M.R. and Rahbé, Y. (2004b) Binding of the insecticidal lectin Concanavalin A in pea aphid, *Acyrthosiphon pisum* (Harris) and induced effects on the structure of midgut epithelial cells. *Journal of Insect Physiology* 50, 1137–1150.

SAVERNET (1996) *Vegetable Research Networking in South East Asia. Final Report Phase I*. Ed. AVRDC Publication No. 96-455, 77 pp.

Sayyed, A.H., Omar, D. and Wright, D.J. (2004) Genetics of spinosad resistance in a multi-resistant field selected population of *Plutella xylostella*. *Pest Management Science* 60, 827–832.

Schmutterer, H. (1990) Control of diamondback moth by application of neem extracts. In: Talekar, N.S. (ed.) *The Management of Diamondback Moth and Other Crucifer Pests*. Proceedings of the second International Workshop, 10–14 December 1990, Tainan, Taiwan. Asian Vegetable Research Development Centre publication 92-368, Taipei, 325–332.

Sekhon, B.S. (1999) Population dynamics of *Lipaphis erysimi* and *Myzus persicae* on different species of *Brassica*. In: *Proceedings of 10th International Rapeseed Congress*, 26–29 September 1999, Canberra, Australia.

Shelton, A.M. (2001a) Regional outbreaks of diamondback moth due to movement of contaminated plants and favourable climatic conditions. In: Endersby, N.M. and Ridland, P.M. (eds) *The Management of Diamondback Moth and Other Crucifer Pests*. Proceedings of the Fourth International Workshop, 26–29 November 2001, Melbourne, Australia, pp. 96–101.

Shelton, A.M. (2001b) Management of the diamondback moth: déja vu all over again? In: Endersby, N.M. and Ridland, P.M. (eds) *The Management of Diamondback Moth and Other Crucifer Pests*. Proceedings of the 4th International Workshop, 26–29 November 2001, Melbourne, Australia, pp. 3–8.

Shelton, A.M. and Nault, B.A. (2004) Dead-end trap cropping: a technique to improve management of the diamondback moth, *Plutella xylostella* (Lepidoptera: Plutellidae). *Crop Protection* 23, 497–503.

Shelton, A.M., Roberton, J.L., Tang, J.D., Perez, C., Eigenbrode, S.D., Preisler, H.K., Wilsey, W.T. and Cooley, R.J. (1993) Resistance of diamondback moth (Lepidoptera: Putellidae) to *Bacillus thuringiensis* subspecies in the field. *Journal of Economic Entomology* 86, 697–705.

Shelton, A.M., Sances, F.V., Hawley, J., Tang, J.D., Bourne, M., Jungers, D., Collins, H.L. and Farias, J. (2000) Assessment of insecticide resistance after the outbreak of diamondback moth in California in 1997. *Journal of Economic Entomology* 93, 931–936.

Sidhu, H.S. and Kaur, P. (1977) Influence of nitrogen applicaion to the host plant on fecundity of mustard aphid, *Lipaphis erysimi* (Kalt.). *Journal of Research Punjab Agricultural University* 14, 445–448.

Sidhu, H.S. and Singh, S. (1964) Control schedule of mustard aphid in Punjab. *Indian Oilseeds Journal* 8, 237–256.

Singh, B. and Bakhetia, D.R.C. (1987) *Screening and breeding techniques for aphid resistance in oleiferous brassicae: a review.* The Oil Crops Network, International Development Research Centre, Canada, 50 pp.

Singh, H. and Rohilla, H.R. (1998) Plant biochemical contents as parameter of resistance in oilseed Brassica to mustard aphid *Lipaphis erysimi* (Kalt). *Journal of Aphidology* 12, 81–84.

Singh, H. and Singh, Z. (1983) New records of insect pests of rapeseed. *Indian Journal of Agricultural Sciences* 53, 970.

Singh, H., Rohilla, H.R., Kalra, V.K. and Yadav, T.P. (1984) Response of *Brassica* varieties sown on different dates to the attack of mustard aphid, *Lipaphis erysimi* Kalt. *Journal of Oilseeds Research* 1, 49–56.

Singh, H., Singh, Z. and Yadava, T.P. (1990) Influence of abiotic factors on alatae production in mustard aphid *Lipaphis erysimi* (Kalt.). *Journal of Aphidology* 4, 71–74.

Singh, H., Rohilla, H.R. and Singh, H. (1995) Empirical approach for the management of mustard aphid, *Lipaphis erysimi* Kalt. *Journal of Insect Science* 8, 203–204.

Singh, S.R., Narain, A., Srivastava, K.P. and Siddique, R.A. (1965) Fecundity of mustard aphid on different rape and mustard species. *Indian Oilseeds Journal* 9, 215–219.

Singh, Y.P. (2007) Efficacy of plant extracts against mustard aphid, *Lipaphis erysimi* on mustard. *Indian Journal of Plant Protection* 35, 116–117.

Singh, Y.P. and Meghwal, H.P. (2009) Evaluation of some bioagents against mustard aphid (*Lipaphis erysimi* Kaltenbach) (Homoptera: Aphididae) on single plant in field conditions. *Journal of Biological Control* 23, 95–97.

Singhvi, S.M., Verma, N.D. and Yadav, T.P. (1973) Estimation of losses in rapeseed (*Brassica campastris* L. var Toria) and mustard, *Brassica juncea* (L.). *Annals of Biology* 16, 145–148.

Sinha, R.P., Yazdani, S.S. and Verma, S.D. (1990) Population dynamics of mustard aphid, *Lipahis erysimi* Kalt. (Homoptera: Aphididae) in relation to ecological parameters. *Indian Journal of Entomology* 52, 387–392.

Sivapragasam, A., Ito, Y. and Saito, T. (1988) Population fluctuations of the diamondback moth, *Plutella xylostella* (L.) on cabbage in *Bacillus thuringiensis* sprayed and non sprayed plots and factors affecting within-generation survival of immatures. *Researches on Population Ecology* 30, 329–342.

Smallegange, R., van Loon, J., Blatt, S., Harvey, J., Agerbirk, N. and Dicke, M. (2007) Flower vs. leaf feeding by *Pieris brassicae*: glucosinolate rich flower tissues are preferred and sustain higher growth rate. *Journal of Chemical Ecology* 33, 1831–1844.

Smith, C.M. (1989) *Plant Resistance to Insects – A Fundamental Approach.* John Wiley & Sons, New York.

Smith, C.M. (1999) Plant Resistance to Insects. In: Rechcigl, J. and Rechcigi, N. (eds) *Biological and Biotechnological Control of Insects.* Lewis Publishers, Boca Raton, Florida, pp. 171–205.

Smith, C.M., Khan, Z.R. and Pathak, M.D. (1994) *Techniques for Evaluating Insect Resistance in Crop Plants.* Lewis Publishers, Boca Raton, Florida.

Snelling, R.O. (1941) Resistance in plants to insect attack. *Botany Reviews* 7, 543–586.

Sood, P. (2004) New record of granulovirus on cabbage white butterfly, *Pieris brassicae* Linn. from dry temperate regions of Himachal Pradesh. *Himachal Journal of Agricultural Research* 30, 146–148.

Sood, P., Mehta, P.K., Bhandari, K. and Prabhakar, C.S. (2010) Transmission and effect of sublethal infection of granulosis virus (PbGV) on *Pieris brassicae* Linn. (Pieridae: Lepidoptera). *Journal of Applied Entomology* 134, 774–780.

Soroka, J.J. (2008) Flea beetles (Coleoptera: Chrysomelidae). In: Capinera, J.L. (ed.) *Encyclopedia of Entomology*, 2nd edn. Springer, Dordrecht, the Netherlands, pp. 43–46.

Soroka, J.J., Dosdall, L.M., Olfert, O.O. and Seidle, E. (2004) Root maggots (*Delia* spp., Diptera: Anthomyiidae) in prairie canola (*Brassica napus* L. and *B. rapa* L.): spatial and temporal surveys of root damage and prediction of damage levels. *Canadian Journal of Plant Science* 84, 1171–1182.

Southwood, S.R. (1986) Plant surfaces and insects – an overview. In: Juniper, B. and Southwood, S.R. (eds) *Insects and the Plant Surface*. Arnold, London, pp. 1–22.

Sreekanth, M. and Babu, T.R. (2001) Evaluation of certain new insecticides against the aphid *Lipaphis erysimi* (Kalt.) on cabbage. *International Pest Control* 43, 242–244.

Srinivasachar, D. and Malik, R.S. (1972) An induced aphid-resistant, non-waxy mutant in turnip, *Brassica rapa. Current Science* 41, 820–821.

Srinivasan, K. and Krishna Moorthy, P.N. (1990) Development and adoption of integrated pest management for major pests of cabbage using Indian mustard as a trap crop. In: Talekar, N.S. (ed.) *The Management of Diamondback Moth and Other Crucifer Pests*. Proceedings of the second International Workshop, Asian Vegetable Research and Development Center, Taipei, Tainan, Taiwan, pp. 511–521.

Srivastava, A.S. and Srivastava, J.L. (1973) Ecological studies on the aphid, painted bug and mustard sawfly attacking mustard and rape in India. *FAO Plant Protection Bulletin* 20, 136–140.

Städler, E. and Reifenrath, K. (2009) Glucosinolates on the leaf surface perceived by insect herbivores: review of ambiguous results and new investigations. *Phytochemistry Reviews* 8, 207–225.

Staley, J.T., Jones, A.S., Pope T.W., Wright, D.J., Leather, S.R., Hadley, P., Rossiter, J.T., van Emden, H.F. and Poppy, G.M. (2010) Varying responses of insect herbivores to altered plant chemistry under organic and conventional treatments. *Proceedings of the Royal Society of London B* 277, 779–786.

Stork, N.E. (1980) Role of waxblooms in preventing attachment to brassicas by the mustard beetle, *Phaedon cochleariae. Entomologia Experimentalis et Applicata* 28, 100–107.

Su, C.Y. (1991) Field trials of granulosis virus and *Bacillus thuringiensis* for control of *Plutella xylostella* and *Artogeia rapae. Chinese Journal of Entomology* 11, 174–178.

Sun, X. and Peng, H. (2007) Recent advances in control of insect pests by using viruses in China. *Virologica Sinica* 22, 158–162.

Suwabe, K., Tsukazaki, H., Iketani, H., Hatakeyama, K., Kondo, M., Fujimura, M., Nunome, T., Fukuoka, H., Hirai, M. and Matsumoto, S. (2006) Simple sequence repeat-based comparative genomics between *Brassica rapa* and *Arabidopsis thaliana*: the genetic origin of clubroot resistance. *Genetics* 173, 309–319.

Tabashnik, B.E., Cushing, N.L., Finson, N. and Johnson, M.W. (1990) Field development of resistance to *Bacillus thuringiensis* in diamondback moth (Lepidoptera: Plutellidae). *Journal of Economic Entomology* 83, 1671–1676.

Talekar, N.S. (1996) Biological control of diamondback moth – a review. *Plant Protectection Research Bulletin (Taipei)* 38, 167–189.

Talekar, N.S. and Shelton, A.M. (1993) Biology, ecology and management of the diamondback moth. *Annual Review of Entomology* 38, 275–301.

Tang, J.D., Gilboa, S., Roush, R.T. and Shelton, A.M. (1997) Inheritance, stability and lack of fitness costs of field-selected resistance to *Bacillus thuringiensis* in diamondback moth (Lepidoptera: Putellidae) from Florida. *Journal of Economic Entomology* 90, 732–741.

Teotia, T.P.S. and Lal, O.P. (1970) Differential reponse of different varieties and strains of oleiferous *brassica* to aphid, *Lipaphis erysimi* (Kalt.). *Labdev Journal of Science and Technology* 8-B(4), 219–226.

Timlick, B.H., Turnock, W.J. and Wise, I. (1993) Distribution and abundance of *Lygus* spp. (Heteroptera: Miridae) on alfalfa and canola in Manitoba. *Canadian Entomologist* 125, 1033–1041.

Travers-Martin, N. and Müller, C. (2008) Specificity of induction responses in *Sinapis alba* L. *Plant Signaling and Behaviour* 3, 311–313.

Traw, B.M. (2002) Is induction response negatively correlated with constitutive resistance in black mustard? *Evolution* 56, 2116–2205.

Traw, M.B. and Dawson, T.E. (2002) Reduced performance of two specialist herbivores (Lepidoptera: Pieridae, Coleoptera: Chrysomelidae) on new leaves of damaged black mustard plants. *Environmental Entomology* 31, 714–722.

Tripathi, M.K. and Mishra, A.S. (2007) Glucosinolates in animal nutrition: a review. *Animal Feed Science and Technology* 132, 1–27.

Tumlinson, J.H., Pare, P. and Lewis, W.J. (1999) Plant production of volatile semiochemicals in response to insect-derived elicitors. In: *Insect-Plant Interactions and Induced Plant Defense*. Wiley, Chichester, UK, pp. 95–109.

Turnock, W.J., Gerber, G.H., Timlick, B.H. and Lamb, R.J. (1995) Losses of canola seeds from feeding by *Lygus* species (Heteroptera: Miridae) in Manitoba. *Canadian Journal of Plant Science* 75, 731–736.

Ulmer, B.J., Gillott, C., Woods, D. and Erlandson, M. (2002) Diamondback moth, *Plutella xylostella* (L.), feeding and oviposition preferences on glossy and waxy *Brassica rapa* (L.) lines. *Crop Protection* 21, 327–331.

USDA (2011) *Oilseeds: World Markets and Trade*. United States Department of Agriculture Foreign Agricultural Service, Circular Series FOP 04-11.

van Emden, H.F. and Williams, G.F. (1974) Insect stability and diversity in agro-ecosystems. *Annual Review of Entomology* 19, 455–475.

van Loon, J.J.A., Chen, Z.W., Nielsen, J.K., Gols, R. and Yu, T.Q. (2002) Flavonoids from cabbage are feeding stimulants for diamondback moth larvae additional to glucosinolates: chemoreception and behaviour. *Entomologia Experimentalis et Applicata* 104, 27–34.

Vandermeer, J. (1989) *The Ecology of Intercropping*. Cambridge University Press, Cambridge, UK.

Vasconcelos, I.M. and Oliveira, J.T. (2004) Antinutritional properties of plant lectins. *Toxicon* 44, 385–403.

Veromann, E., Metspalu, L., Williams, I.H., Hiiesaar, K., Mand, M., Kaasik, R., Kovaes, G., Jogar, K., Svilponis, E., Kivimagi, I., Ploomi, A. and Luik, A. (2012) Relative attractiveness of *Brassica napus*, *Brassica nigra*, *Eruca sativa* and *Raphanus sativus* for pollen beetle (*Meligethes aeneus*) and their potential for use in trap cropping. *Arthropod Plant Interactions* 6, 385–394.

Vickers, R.A., Furlong, M.J., White, A. and Pell, J.K. (2004) Initiation of fungal epizootics in diamondback moth populations within a large field cage: proof of concept of auto-dissimination. *Entomologia Experimentalis et Applicata* 111, 7–17.

Walton, T.J. (1990) Waxes, cutin and suberin. In: Dey, P.M. and Harborne, J.B. (eds) *Methods in Plant Biochemistry*. Academic Press, San Diego, California, pp. 105–158.

Wang, J., Chen, Z., Du, J., Sun, Y. and Liang, A. (2005) Novel insect resistance in *Brassica napus* developed by transformation of chitinase and scorpion toxin genes. *Plant Cell Reports* 24, 549–555.

Waring, P., Townsend, M. and Lewington, R. (2003) *Field Guide to the Moths of Great Britain and Ireland*. British Wildlife Publishing, Rotherwick, UK.

Way, M.J. and van Emden, H.F. (2000) Integrated pest management in practice – pathways towards successful application. *Crop Protection* 19, 81–103.

Werker, E. (2000) Trichome diversity and development. *Advances in Botanical Research* 31, 1–35.

Wietsma, R. (2010) The effect of differences in aliphatic glucosinolate concentrations in *Arabidopsis thaliana* on herbivores of different feeding guilds and different levels of specialization. MSc thesis, Wageningen University and Research Centre, the Netherlands.

Williams, I.H. and Free, J.B. (1978) The feeding and mating behaviour of pollen beetles (*Meligethes aeneus* Fab.) and seed weevils (*Ceuthorhynchus assimilis* Payk.) on oil-seed rape (*Brassica napus* L.). *Journal of Agricultural Science* Cambridge, 91, 453–459.

Wittstock, U., Agerbirk, N., Stauber, E.J., Olsen, C.E., Hippler, M., Mitchell-Olds, T., Gershenzon, J. and Vogel, H. (2004) Successful herbivore attack due to metabolic diversion of a plant chemical defense. *Proceedings of the National Academy of Sciences USA* 101, 4859–4864.

Xiang, Y., Wong, W.K.R., Ma, M.C. and Wong, R.S.C. (2000) *Agrobacterium* mediated transformation of *Brassica campestris* ssp. *parachinensis* with a synthetic *Bacillus thuringiensis* cry1Ab and cry1Ac genes. *Plant Cell Reports* 19, 251–256.

Xu, J., Shelton, A.M. and Xianian, C. (2001) Comparison of *Diadegma insulare* (Hymenoptera: Ichneumonidae) and *Microplitis plutellae* (Hymenoptera: Braconidae) as biological control agents of *Plutella xylostella* (Lepidoptera: Plutellidae): field parasitism, insecticide, susceptibility and host searching. *Journal of Economic Entomology* 94, 14–20.

Yadava, A.K., Singh, H. and Yadava, T.P. (1985) Inheritance of non-waxy trait in Indian mustard and its reaction to aphids. *Journal of Oilseeds Research* 2, 339–342.

Yamada, H. and Kawasaki, K. (1983) The effect of temperature and humidity on the development, fecundity and multiplication of the diamondback moth, *Plutella xylostella* (L.). *Japanese Journal of Applied Entomology and Zoology* 27, 17–21.

Yein, B.R. (1985) Effect of dates of sowing on the incidence of insect pests of toria, *Brassica campestris*. *Journal of Research Assam Agricultural University* 6, 68–70.

Young, O.P. (1986) Host plants of the tarnished plant bug, *Lygus lineolaris* (Heteroptera: Miridae). *Annals of Entomological Society of America* 79, 747–762.

Žnidarčič, D., Valič, N. and Trdan, S. (2008) Epicuticular wax content in the leaves of cabbage (*Brassica oleracea* L. var. *capitata*) as a mechanical barrier against three insect pests. *Acta Agriculturae Slovenica* 91, 361–370.

12 Abiotic Stresses with Emphasis on *Brassica juncea*

D.K. Sharma,[1]* D. Kumar[2] and P.C. Sharma[1]

[1]*ICAR-Central Soil Salinity Research Institute, Karnal, Haryana, India;*
[2]*ICAR-Central Arid Zone Research Institute, Jodhpur, Rajasthan, India*

Introduction

Agricultural productivity is affected by a number of abiotic stresses. These may include deficit or excess water availability, flash floods, high salt levels in soil as well as in irrigation water and extreme temperatures. In addition, mineral deficiency or toxicity is frequently encountered by plants in agricultural systems. In many cases, different abiotic stresses challenge plants in combination. For example, high temperatures and scarcity of water are commonly encountered in periods of drought and can be exacerbated by mineral toxicities that constrain root growth. Further, plants are also exposed to salinity, drought and frost-like conditions in combination in some of the cases. Higher plants have evolved multiple, interconnected strategies that enable them to survive abiotic stresses. However, these strategies are not well developed in most agricultural crops. Across a range of cropping systems around the world, abiotic stresses are estimated to reduce yields to less than half of that possible under ideal growing conditions. Traditional approaches to breeding crop plants with improved stress tolerance have so far met with limited success, in part because of the difficulty of breeding for tolerance traits in traditional breeding programmes. Desired traits can be crossed into crop species from different donors. Since different crops and their genotypes show differential levels of tolerance to many abiotic stresses, efforts are underway to unravel their molecular basis for abiotic stress tolerance to speed up the development of improved genotypes with increased stress tolerance.

In *B. juncea*, drought and salinity are the two major abiotic stresses affecting crop productivity. Most of the drought-prone areas are found in arid, semi-arid and sub-humid regions, which experience less than average annual rainfall. In India, the drought-prone areas have been estimated to be around 51 Mha; the estimates vary with variations in annual rainfall pattern in different parts of the country (Samra *et al.*, 2006). Further, salt-affected areas in India have been assessed to cover 6.73 Mha (Sharma *et al.*, 2004). There are also large areas across the world underlain by brackish groundwater resources. The crop brassicas also experience low temperatures and also frost, though only occasionally, during the growing season. Crops also experience high soil temperature affecting seed germination and crop establishment. The climatic variations further accentuate the stresses

*Corresponding author, e-mail: director@cssri.ernet.in

experienced by the plants. Of late, the climatic variations being observed include frequent variations, both high as well as low and declining number of rainy days but with higher intensity. Germination of seeds and emergence of seedlings is considered to be a very sensitive and critical stage in brassicas, which ultimately influences crop productivity through decreased plant stands. Accordingly, germination, emergence of seedlings and their good growth are pre-requisite attributes for increasing mustard production under rainfed situations as well as in semiarid saline soils and under prevailing high soil temperature during normal sowing time. There is a distinct possibility for the improvement of these traits in brassica oilseeds, as large genetic variation with regards to these traits exists. Response of rapeseed mustard to different stresses and their management is described in following sections. All these environmental factors prevent crop plants reaching their full genetic potential, leading to a decline in crop productivity.

Drought Stress

Rainfed crops may suffer from stringent droughts of fluctuating and unpredictable intensities and drought-induced adverse effects by secondary or even tertiary pathways. These pathways could be through transpirational or heat complexes; drought tolerance is probably the most difficult plant trait to be improved through conventional plant breeding (Ribaut *et al.*, 2002). The challenge is even greater for developing drought-tolerant plants for water-limited environments where occurrence, timing and severity of drought may fluctuate from one zone to the next and also over the years. Furthermore, it induces large impacts on emergence, growth, production and quality of produce through phenological, physiological and biochemical pathways. More precisely, depressed seedling emergence with early seedling mortality may lead to sub-optimal plant stand, delayed growth and removal of sinks in extreme situations and may lead to heavy irreversible production losses. The convenient and traditional solution to drought lies in providing irrigation water to the crop plants experiencing drought. The same, however, appears quite cumbersome and uneconomic due to slow accumulation of salts on the upper surface of the soil following evaporation and sometimes also due to waterlogging. Furthermore, providing irrigation water in arid drylands is too costly to be borne by farmers.

Drought Tolerance

Plant tolerance to soil-moisture deficit is an adaptive feature involving the cell membrane of individual cells and whole plants. Biochemically, specific proteins, their synthesis and accumulation of organic solutes are involved, thus making plant tolerance to drought highly complex. In morphological terms, tolerance can be achieved by substantial and proportional growth reduction and by diversion of carbohydrates to economic parts (sinks). Roots have been observed to play an important role in tolerance of plants to drought, with higher growth and biomass in the root than in the shoot. For instance, in pigeon pea, the taproots were almost 18 cm longer in low rainfall than in a wet situation. A deep rooting pattern makes a major contribution to plant tolerance in soil-moisture deficits. Maize and sorghum roots can grow 2 to 3 m deep into the soil, allowing their survival even when surrounding grasses or crops with shallow roots succumb. The importance of a better-developed root system in *B. juncea* than in *B. carinata*, *B. napus* and *B. campestris* has been related to the greater drought tolerance of that species (Liang *et al.*, 1992). Studies have clearly shown enormous genetic differences in survival under severe dehydration when stomata are closed. Such differences occur between species, for example, between pigeon pea, black gram, soybean and cowpea (Sinclair and Ludlow, 1986). Differences also occur within species such as wheat, oats, maize and soybean (Paje *et al.*, 1988; Rawson and Clarke, 1988). These differences could be used to select genotypes with decreased conductance and, thus, reduced water loss when stomata are closed. Drought tolerance is not simply affected and determined by the depth and pattern of root behaviour and by cuticular resistance. A number of morpho-biochemical factors such as stomatal sensitivity and number, cell turgor, chemical traits, stability of membrane proteins, phospholipids and metabolic activities are also important and actively involved (Table 12.1).

Oilseed brassicas tolerate drought by changing and modifying leaf morphology and structure. For instance, in Poland, swede rape cv. Gorez anksi tolerated drought with leaf expansion due to formation of central vacuoles in mesophyll cells. Thus, desiccation tolerance appeared to be achieved by avoidance of cytoplasm dehydration (Obloj and Kacperska, 1981). Under progressive drought, differential leaf modification was displayed by rape cv. Darmor in France. The older leaves, developed before the onset of water deficit, wilted gradually, whereas the youngest leaves continued to harden when the shoot water potential was below 3 MPa and the leaf water saturation deficit was about 60%. The hardening of leaves was distinguished by leaf turgor and bloom colour was bluish (Reviron

et al., 1992). Morphologically, plants also adapt to water stress by early flowering, decreased plant height, leaf area and leaf dry weight and increased leaf conductance (Yadava and Hari Singh, 1996). Tolerant plants under water stress appear to shift carbohydrate utilization from shoot growth into the production of additional root biomass in the early stages of adaptation. However, at later stages, carbohydrate concentration may shift toward the shoots (Mathur and Wattal, 1996; Wright *et al.*, 1996), accompanied by growth retardation. Physiological changes in water potential and relative water content of the water-stressed leaves through osmoregulation and osmotic adjustment have been observed in *Brassiceae*. In the process of physiological adaptation, maintenance of turgor pressure appears to be the central process. In a study of seven *Brassica* species in India, tolerance was associated with high water use, high stomatal conductance and osmoregulation (Kumar *et al.*, 1987). Net assimilation rates were closely correlated with stomatal conductance. Osmotic adjustment involves the intracellular accumulation of organic solutes, which is believed

Table 12.1. Selection indices for drought tolerance in oilseed brassica.

S. No.	Parameter	Value	Species	Reference
A. Morphological				
(i)	Fresh and dry weight of whole plant	Higher	*B. napus*	Ashraf and Mehmood, 1990
(ii)	Seed germination	High	*B. napus*	Richards and Thurling, 1979
(iii)	Days to flowering	Early	*B. napus*	Richards and Thurling, 1979
(iv)	Leaf expansion	Long	*B. napus*	Yadav and Hari Singh, 1996
(v)	Joint selection (1000 seed weight, seeds per pod, seed yield)	Higher	*B. napus*	Richards and Thurling, 1979
B. Physiological				
(i)	Chlorophyll stability index	0.012–3.16	*B. napus*	Richards and Thurling, 1979
			B. juncea	Chhabra *et al.*, 1981
(ii)	Stomatal conductance		*B. juncea*	Yadav *et al.*, 1990
			B. napus	Kumar and Singh, 1998
(iii)	Osmoregulation	1	*B. juncea*	Singh *et al.*, 1985
				Kumar *et al.*, 1994
(iv)	Turgor pressure	High up to 0.5 Mpa	*B. juncea*	Kumar and Singh, 1998
(v)	Leaf water potential	High	*B. juncea*	Sawhney *et al.*, 1996
(vi)	Transpirational cooling		*B. juncea*	Singh *et al.*, 1985
C. Biochemical				
(i)	Synthesis/accumulation of solutes		*B. napus*	Ashraf and Mehmood, 1990
(ii)	Specific protein	BnD22	*B. napus*	Ilami *et al.*, 1997

to restore turgor pressure. The organic solutes (osmolytes) generally accumulated are polyols and amino acid derivatives (Ashraf and Mehmood, 1990). More than one type of osmolyte may occur in a cell, but specific classes of osmolytes have also been found by many workers.

Solute leakage from the cell membrane of plants exposed to drought stress results from increased membrane permeability due to stress-induced disturbance in its integrity. Accordingly, at the cellular level, integrity and stabilization of cell membrane structure, rather than repair mechanisms, have been observed to operate in rape hypocotyls dehydrated to a water-saturation deficit of 60% (Shcherbakova and Kacperska, 1983). On closure of stomata, plants show various protective mechanisms to reduce photo-inhibition, including thermal dissipation of energy, removal of active oxygen species (AOS) and increase in certain enzyme activities. The AOS system and ascorbate peroxidase (APX) enzyme play key roles in alleviating drought (Holt *et al.*, 2000). An important aspect of severe desiccation is that water content in the cells becomes so scarce that enzyme activities are inhibited (Vertucci and Leopold, 1987a, b). Developing seeds of a range of crops, including rape and mustard, accumulate hydrophilic proteins in the embryo as the desiccation begins (Dure *et al.*, 1989). In certain cases, an alpha helix is present that remains structurally stable during desiccation and this portion of protein has been observed to act as a membrane-stabilizing factor. The mRNAs for these proteins are specifically induced by severe desiccation in very young rape (Harada *et al.*, 1989) and barley (Close and Chandler, 1990) plants. The expression of mRNA was especially increased in shoots that are directly exposed to dehydration under natural conditions. This cellular response suggests that the dehydrin-Em-LEA protein plays an active role in desiccation tolerance of rape seedlings. Drought-stress treatment was found to induce change in the protein pattern in the root system of rape cv. Darmor of 13 new polypeptides of low molecular weight identified in the tap root. Of these, 12 were present in short, tuber-sized roots, a specific root type induced in response to soil moisture deficit

(SMD) situations. Surprisingly, however, reversal of these induced proteins in just 3 days of rehydration suggested involvement of specific proteins in drought tolerance (Vartanian *et al.*, 1987). Similarly, a specific drought-induced protein, BnD22, was shown to accumulate in *B. napus* leaves adapted to progressive osmotic stresses induced by salt and water stresses. The protein might be involved in decreased protease activity in the drought-adapted leaves, thus contributing to delayed leaf senescence (Ilami *et al.*, 1997).

Genetic Management

Adaptation

Rapeseed and mustard present a wide range of adaptation to drought, wherein *B. carinata* and *E. sativa* with high drought tolerance potential represent one extreme and *B. nigra*, *B. alba*, *B. rapa* (Yellow Sarson) and *B. rapa* var. Toria with poor adaptation form the other extreme. Contrary to these extremes, *B. rapa* var. Brown Sarson and *B. tournefortii* fall in an intermediate group (Kumar, 1995). Such differences are due to genetic variations for temperatures, which regulate the phenologic development in rapeseed. However, in moisture stress situations, swede rape is characterized with higher photosynthetic efficiency than in the *B. rapa* group. Lower mortality of *B. napus* plants than with *B. juncea* under water stress conditions is known in advantageous positions, because the *B. napus* plant has faster recovery under moisture stress due to its longer duration than *B. juncea* due to the shorter duration of the crop (Kumar *et al.*, 1994). However, such differences may not lead to yield differences. Amongst the most cultivated and potential amphitetraploids (*B. juncea*, *B. napus* and *B. carinata*), *B. juncea* is the better amphitetraploid (Woods *et al.*, 1991). Indian mustard yielded almost 12.5% more seed yield and 36% more dry matter than canola when grown in limited moisture conditions in Australia. Similarly, Oram (1987) observed from 12 to 67% greater seed yield of Indian mustard (cv. 81792) compared to canola. Various studies under limited soil moisture situations point out distinguished adaptive advantages in Indian mustard than in other diploid and amphitetraploid oilseed brassicas (Lewis, 1992; Mathur and Wattal, 1996).

Drought effects at different growth stages

Soil moisture availability at seedling emergence, anthesis and post-anthesis is crucial for brassicas, which may subsequently determine the extent of production under a soil moisture deficit situation. However, reports indicate that moisture stress is most damaging during the reproductive phase and a delayed irrigation could bring only slight recovery in seed yield. In a similar confirming study, the period of post-flowering recovery was reported to be detrimental for *Brassica* species under limited soil moisture situations on sandy loam soil of Delhi (Mathur and Wattal, 1996). A quantitative assessment study of withholding water at different growth stages of *Brassica* species on sandy loam soils of Delhi revealed that seed yield depressed by from 22.1 to 36.5% and 17.9 to 32.4% when the irrigations were withheld at emergence and flowering stages, respectively. However, yield was slightly reduced (1.5–3.4%) on withholding water at the seed development stage compared to the irrigations applied at all these stages (Singh *et al.*, 1997). Susceptibility of brassica to drought at the reproductive phase may be assigned to two reasons: dehydration losses may prevent cell enlargement and the inadequate water availability at microsporogenesis leading to pollen sterility. Yield differences in brassica are generally ascribed to the extent dry matter is accumulated before the peak anthesis, particularly in a high soil moisture deficit situation (Mendhan *et al.*, 1981; Wright *et al.*, 1996). For instance, canola and rape attained higher leaf area and accumulated maximum total dry weight at the flowering stage, whereas short growth cycle species *B. juncea* and *B. rapa* exhibited maximum total dry weight by the beginning of this stage making the pre-anthesis stage more sensitive in *B. juncea* and *B. rapa*, whereas anthesis to post-anthesis stage appeared more sensitive in Canadian rape (Mathur and Wattal, 1996).

Breeding options

Breeding for drought tolerance may be useful under sub-optimal dry conditions but not in extreme drought situations as in the catastrophic nature of African Sahara. However, breeding for drought tolerance may be undertaken to achieve the following objectives.

1. To reduce the risk of crop failure and maximize yields by manipulating sowing dates and curtailing the maturity period in accordance with the rainfall pattern of the region concerned (drought escape).
2. To maximize water use when soil moisture is available by incorporating important traits such as fast growth and well-developed root systems, rapid leaf-area expansion and physiological efficiency (drought avoidance).
3. To incorporate plant traits responsible for increasing and stabilizing yields and providing resistance to catastrophic events during water-stress periods (drought resistance in complete sense).

Our experience with a variety of crops to date indicates that selection for drought escape followed by drought avoidance has been by far the most successful and effective and this approach is likely to be continued because it favours higher yield vis-à-vis tolerance mechanisms (resistance) under water stress. A number of early maturing varieties in several agricultural crops have been developed and yield potentials have increased under soil-moisture deficit situations in spite of curtailment in the growth period. Selection for earliness behaviour till maturity is complex, due to continued formation of pods. However, selection for drought tolerance is associated with early flowering and partitioning of dry matter to reproductive parts. Partitioning differences to reproductive parts may be more in rainfed rather than in irrigated conditions. However, anthesis, partitioning, harvest index and maturity are all influenced by drought; therefore drought-induced differences make it difficult to detect differences due to genotype. Therefore it is proposed that selecting plants which bear more mature pods early in the season may be useful in improving brassicas in drought situations. Selection based on whole plant maturity may not be useful, as maturity period is strongly influenced by the dry environment. The success of breeding for drought tolerance lies in screening of high yielding and drought-tolerant genotypes separately and hybridizing them, selecting drought-tolerant lines in drought conditions (F_1 to F_2) and finally predicting yields in targeted environments. For this approach, simulation of drought conditions and appropriate drought-tolerance parameters/methodologies are required to be specifically considered. In oilseed brassica, terminal drought is most disastrous due to depletion of soil moisture and increased canopy transpiration towards maturity, ending in forced maturity, shrivelled grain and poor harvest. Thus, an escape mechanism leading to early flowering and partitioning of dry matter towards the sink would be the most practical approach for combating drought. The earliness in reference to the requirement of specific zone, soil type and agroclimatic conditions need to be set out. A number of varieties suitable for rainfed conditions have been bred in India (Table 12.2).

Drought Management

Substantial information on water stress tolerance resulting from tissue pre-hardening treatment at the early stage, its mechanism and impact on growth have been generated in swede rape. It is now evident that the water stress hardening effect of tissue dehydration appears related to the increase in the stability of cell membrane. Dehydration pre-treatment given to these tissues followed by restoration of its turgidity caused decrease in leakage of electrolytes, ultraviolet absorbing material and ninhydrin reactive substance from the desiccated hypocotyl tissues (Lee *et al.*, 1981). Moreover, desiccation or immersion-induced injuries were highly correlated with K effect and the injuries were reversible if tissue damage was not severe. Resultant leakages might result from increased membrane permeability from the stress-induced disturbance in its integrity. Conversely, the increased water stress tolerance observed in hypocotyls pre-hydrated to a water saturation deficit of 60% seems to be due to stabilization of the membrane structure rather to the promotion of a repair mechanism (Shcherbakova and Kacperska, 1983). In an interesting study in Poland with swede

rape cv. Gorczanski, the drought tolerance was found associated with the beginning of leaf expansion. Such increase in drought tolerance was due to the formation of central vacuoles in the mesophyll cells. Thus, high desiccation tolerance of vacuolized tissues may be due to their ability to avoid cytoplasm dehydration (Obloj and Kacperska, 1981). The role of anti-transpirants in checking transpiration rate is well documented. *Brassica napus* cultivars possess a higher assimilative ability than species of the *B. rapa* group under rainfed situations of Saskatoon, Canada (Clark and Thomas, 1982), supporting the view that higher yield potential of swede rape under soil-moisture stress (SMS) conditions is related to its high photosynthetic efficiency. Thus, research results available on this aspect reveal that early imposition of drought at the seedling stage leads to differences in the lipid levels but relative distribution was, in general, independent of SMS. Drought stress-induced new proteins in the taproot system are understood to be involved in drought-tolerance potential. Early establishment and better radicle growth under soil moisture deficit situations ensures moisture extraction from the deeper horizons and the potential to withstand early osmotic shocks. Researches from Hisar revealed that pre-sowing treatment of seeds with Phosphon-D or chloromequat ensures increased seedling emergence rate, radicle and plumule length under drought (Pandya and Khan, 1973; Sheoran and Khan, 1987). At Anand, a single spray of 500 ppm succinic acid at 43 days after sowing (DAS) along with the first and third irrigations increased yield by 11.4 and 14.1%, respectively (Patel and Mehta, 1984). On Diara lands of Faizabad, a mustard trial with anti-transpirants and mulch treatments indicated maximum yield with two foliar applications of 100 ppm cycocel at 30 and 60 DAS (1.03 t/ha) compared to two foliar spray of 6% kaolin (0.90 t/ha), mulch (0.82 t/ha), dust mulch (0.75 t/ha) and control (0.69 t/ha).

Agronomic management

Fertilizer management

Optimum fertility inputs in the form of basal application are known to boost root and

Table 12.2. Improved varieties of *Brassica juncea* for rainfed and drought-affected areas in India.

Variety	Year of Release	Area of adoption
Aravali Mustard (RN 393)	2001	Rainfed areas of Haryana, Punjab and Rajasthan
Durgamani	1974	Rainfed and irrigated areas of Rajasthan
GM1 (Gujarat Mustard 1)	1990	Dry sandy to medium black soils of Gujarat
GM 2 (Gujarat Mustard 2)	1997	North Gujarat areas
GM 3 (Gujarat Mustard 3)	2006	Gujarat
Geeta (RB 9901)	2003	Haryana, Punjab, Western Rajasthan, Delhi
Kranti	1983	Bihar, Delhi, Gujarat, Haryana, Orrisa, Punjab, Rajasthan, Uttar Pradesh, West Bengal
Patan Mustard 67	1984	Rainfed areas of Gujarat
PBR 378	2012	Rainfed areas of Haryana, Punjab, Jammu and Northern Rajasthan
PBR 97	1997	Rainfed areas of Punjab
Pusa Bahar	1991	Rainfed areas of Assam, Bihar, Orissa and West Bengal
RB 50	2009	Haryana, Rajasthan, Delhi and Punjab
RGN 48	2006	Rainfed areas of Haryana, Rajasthan and Punjab
RH 30	1985	Rainfed areas of Haryana, North Rajasthan, Punjab and Western Uttar Pradesh
RH 819	1991	Rainfed areas of Haryana, Rajasthan and Punjab
Rajat (PCR 7)	1997	Rainfed and irrigated areas of Gujarat, Maharashtra and Western Rajasthan
Sanjuncta Asech	1988	Rainfed and irrigated areas of Assam, Bihar, Orissa and West Bengal
Seeta	1982	Rainfed and irrigated areas of West Bengal
TM 2	1993	Rainfed and irrigated areas of Assam (Brahampura valley)
TM 4	1993	Rainfed and irrigated areas of Assam (Brahampura valley)
Vaibhav	1985	Rainfed areas of Madhya Pradesh and Uttar Pradesh
Varuna	1976	Entire mustard-growing areas of country

plumule growth and exert drought resistance leading to improved production during limited soil moisture situations. Hence, fertility management in reference to soil type, soil moisture availability and crop growth is crucial in drought-prone situations. Crop responded favourably up to 40 kg N/ha under unirrigated conditions and up to 80 kg N/ha under 0.4:0.6 irrigation water/cumulative pan evaporation (IW/CPE) ratio (Sharma and Kumar, 1988).

Seed yield of Indian mustard is reported to be increased in limited soil moisture on application of N, P, S and B nutrients. Interestingly, the zinc (Zn) uptake increases with applied phosphorus (P) under SMS situations but decreases in sufficient moisture availability. Thus, Zn and P show synergistic effects in limited moisture availability. Moreover, water use efficiency (WUE) was increased with S application under rainfed conditions

(Chhabra *et al.*, 1981). Nitrogen at 80 kg/ha increased the seed and oil yield under rainfed conditions and the optimum N dose was quantified as 70 kg/ha (Garai *et al.*, 1989).

Irrigation management

Supplementary irrigation at the critical growth stage has been observed crucial in brassica crops leading to enhanced physiological efficiency and yield response in limited soil moisture situations. For instance, one supplementary irrigation at flowering stage increased stomatal conductance, canopy temperature and seed yield; however, the latter was not increased with two irrigations at Hisar (Yadav *et al.*, 1990). At the same location, out of four irrigation methods (furrow, check basin, border, sprinkler) the latter required less water, gave maximum yield (2.7 t/ha) and WUE. The crop also extracted more water from the deeper horizons following sprinkler irrigation (Kochhar *et al.*, 1990). In general, irrigation frequency varies with soil type and agroclimatic conditions, but one supplementary irrigation at 25–30 DAS and/or a second irrigation at 50–60 DAS may be effective in rainfed conditions. Irrigation experiments conducted at Agra and Hisar on toria (*B. rapa* var. Toria) revealed that root volume and root dry weight increased, whereas taproot and lateral root length decreased with irrigation; and with more frequent irrigations leaf relative water content (RWC) increased but there was no effect on seed N, S and oil content (Raja and Bishnoi, 1988, 1990). In respect of taramira (*Eruca sativa*), an irrigation of 50 mm at flowering increased yield by 121.6 and 114.9 kg/ha over 125 and 160 mm soil moisture in 100 mm soil profile, respectively (Singh and Sharma, 1981). The straw yield, seed oil content and seed weight decreased following soil moisture deficit before flower initiation, fruit set and seed maturation stages, respectively (Brzostowicz, 1988). In general, 50–60 kg N + 30–35 kg P_2O_5 + 20–30 kg S + 1 kg B per hectare with sparse plant stand (30–40 × 5–10 cm inter- and intra-row spacings, respectively) have been observed optimum to combat drought and harvesting maximum production of *B. juncea* in rainfed conditions. However, soil properties, winter rainfall pattern, cropping pattern and agroclimatic conditions may specify their applications.

Salt Stress

Salt-affected soils comprise saline and alkali soils. Saline soils contain chlorides and sulfates of Na and K (electrical conductivity >4 dS/m, pH 7.5–8.0), whereas alkali soils comprise carbonates and bicarbonates of Na (electrical conductivity <2 dS/m, pH 7.5–14.0). Reclamation of these salt-affected areas is of paramount importance to bring more areas under cultivation, to enhance the food availability. Generally, three approaches are being followed for their reclamation: (i) the engineering solution for the reclamation of saline soils is beyond the reach of resource-poor farmers due to its prohibitive cost and com-

munity-based application; (ii) the chemical amendment approach, for the reclamation of alkali soils, is generally followed by farmers; and (iii) the biological reclamation approach, by developing saline- and alkaline-tolerant crops, is cost effective and is also economically feasible. Vast literature is available on the effects of salinity on crop plants. Maas (1986) documented salt tolerance of different crops but to a relatively constant salinity in the root zone. However, the exposure of plants to varying salinity levels at different growth stages would change the response of crop plants. Various experiments reveal that the salt tolerance observed during germination and emergence stages does not correlate with later growth stages. In Indian mustard, crop behaviour to salinity stress changes as the crop matures. Indian mustard is sensitive at germination and seedling emergence stages whereas its tolerance to salinity stress increases at a later developmental stage. At EC 15.5 dS/m, seed yield declined by 84, 68 and 56% upon saline irrigation at germination, stem elongation and flower initiation growth stages (Gill and Sharma, 1999) in Indian mustard. Plant growth measured as biomass production is in reality the integration of net photosynthesis over time; therefore factors limiting plant growth are also the factors that limit net photosynthesis. Plant growth is affected by salinity stress, which is a consequence of several physiological processes including photosynthesis. Short-term effects of salinity imposition on photosynthesis were studied at 2, 24 and 120 h of salinity imposition. Though the water stress symptoms were observed immediately after the salinity imposition, the effects on transpiration rate (E), stomatal conductance (gs), assimilation rate (Pn) and internal CO_2 were not observed even after 2 h of salinity treatment. The deteriorating effects of salinity were observed at 24 h of saline irrigation with respect to the above-mentioned parameters. Further, the decline in assimilation rate was also observed at 24 h after salinity imposition with an average decline of about 40–50% compared to control. Even under high salinity, the plant tried to maintain its photosynthetic activity as the effects varied greatly with respect to different leaves. The assimilation rate declined drastically in lower leaves approximately up to position 5 from base where about 60–80% of the effect was observed. The upper leaves were still maintaining higher assimilation rate under salinity compared to lower leaves (Sharma, 2003). Further, the long-term effects of salinity on the above-mentioned parameters were studied in genotypes Varuna, CS 330, CS 609, ST 63 and CS 33, with respect to their 1000 seed weight and grain yield under salinity and irrigated with saline water of EC 12 and 15 dS/m for 45 days. The plants generally adjust to long-term salinity application as evidenced by their lower transpiration rates (36 and 41%), stomatal conductance (45 and 59%) and in turn their effects on lowering the assimilation rates by 25 and 35% in leaves under 12 and 15 dS/m salinity respectively, compared to control. This results in reduced photosynthesis leading to reduction in grain yield under salinity.

Evaluation in salt stress environments

In general, selection for salt tolerance is made by evaluating large number of germplasm lines in germination trays, pots, microplots and field situations. For large-scale screening of varieties at germination and seedling stage, shallow-depth germination trays provided with a polythene sheet lining on the inner face are used. Further, earthern, glazed or plastic pots of different capacities, filled with coarse sand or representative soil are used to evaluate germplasm lines at different growth stages. Microplots pertains to a series of dug-out cavity structures made of brick-mortar-concrete materials measuring 2×2 m or 6×3 m with a depth of about 0.8 m, each filled with artificially prepared soil or original soil of different grades brought from salt-affected fields, so that soil is uniform all through the profile. It is possible to create and maintain desired levels of sodicity and salinity in these microplots in a manner very much comparable to field conditions minus the soil heterogeneity. Generally, the evaluation under microplots was done in a single row for each variety. Data obtained from microplots containing desired grades of saline or alkali soils have been found to be well correlated with those collected from satisfactorily conducted field experiments. The field gradient of soil salinity is determined by soil tests at small intervals of space and a long strip running full length across the salinity/sodicity gradient is allotted to each genotype. Field screening for different germplasm lines was generally restricted to two to three rows of each variety, 20–30 m long. A set of check varieties representing resistant and sensitive types, replicated many times (after every 20 rows) along with genotypes to be evaluated, were planted to take note of general growth conditions. Further, irrigation with saline waters of predetermined composition is also practised to establish desired soil salinity levels, particularly when relative sensitivity of different growth stages are sought to be compared.

When plant breeders are faced with a task of breeding crop varieties which are to be used under specific problem conditions, the criteria of selection is essential to any advancement which may be possible. In case of salt resistance, it would seem that it is essential to work hand to hand with the plant physiologists and soil scientists, in conditions which would make reliable selection possible and to determine if parameters can be developed which can make selection possible and effective. Further, without a concerted research effort, problems such as breeding for salt tolerance cannot be effectively pursued. The conventional methods of improving plant salt tolerance generally employ selection for seed yield following the pedigree method. The advancement of generations was made following pedigree selection simultaneously in moderate-stress and high-stress sodicity and salinity stress environments, known as 'Parallel Pedigree Method', for the development of salt-tolerant varieties in problem soils. Backcross breeding has been used to induce salt tolerance in the prevailing genotypes.

Screening methods

An appropriate and reliable methodology was developed for screening of a large number of mustard genotypes for salt tolerance during germination and seedling emergence growth stages under laboratory conditions (Sinha *et al.*, 2003). Different mustard genotypes were evaluated under solution, sand and soil culture to arrive at a consensus salinity level in solution culture, which is a true representative of soil salinity conditions in the field. Seedling emergence in solution, sand and soil cultures were observed to be significantly correlated. Significant positive correlation ($r=0.92$) was recorded between seedling emergence at 26 dS/m in solution culture and 12.8 dS/m in soil culture. In conclusion, screening of Indian mustard genotypes for salt tolerance at seed germination and seedling emergence stages can be done rapidly in solution culture in the laboratory. This will help in accelerating the progress towards improvement of salt tolerance in Indian mustard. Amongst the six species of *Brassica* genus, evaluated for their performance and ionic accumulation under salinity, *B. juncea* genotypes recorded maximum mean seed yield and accumulated minimum mean Na. *B. nigra* genotypes were recorded to be sensitive to salinity stress. Significantly higher shoot and root weight besides higher seed yield was demonstrated in amphidiplods compared to diploids under salinity stress (Ashraf *et al.*, 2001). The amphidiploids accumulated lower Na^+ and higher K^+ in shoots and roots than in those of diploids. They have assessed that salt tolerance has likely come from the A and C genomes. Kumar *et al.* (2009) have also evaluated the performance of six *Brassica* species under salinity stress and shown *B. juncea* to be more tolerant compared to other *Brassica* species. Evaluation of 158 genotypes, collected from different sources, under a range of salinity stress conditions up to 22 dS/m, in mixed salt solution containing NaCl, $CaCl_2$ and Na_2SO_4 in Hoagland solution, showed decline in seed germination by 51 and 82% at EC 18 and 22 dS/m respectively. Significant differences were observed amongst different genotypes for seed germination under different salinity stress levels. The regression equation for the pooled data of 158 genotypes was calculated to be $y = -3.8544x + 114.17$, ($R^2 = 0.87$) where y and x represent seed germination and salinity level respectively. Further, alkalinity stress showed higher yield reduction compared to salinity and also imposes additional stress when present in conjunction with mild salinity (Javid

et al., 2012). It was also demonstrated that alkaline salinity reduced uptake of essential nutrients and Na^+ exclusion that resulted on more deleterious effects on growth and development compared to salinity alone. The better performing lines under field situations were again tested for their salt tolerance potential under microplots and in pots under sand culture conditions. Different salinity levels were applied at germination stage and maintained throughout the experiment. It was shown that the better performance of a genotype was associated with higher shoot and root fresh and dry weight at seedling stage, minimum percentage reduction in grain yield under salinity, maximum mean susceptibility index values, higher number of pods per plant at higher salinity levels, lower accumulation of Na in shoot and higher in root, higher K levels both in shoot and root and lower shoot Na:K ratio. Recently, Chakraborty *et al.* (2012) have also shown lower Na and higher K accumulation in leaves, stem and roots of salt-tolerant varieties CS 52 and CS 54 compared to salt-sensitive varieties Varuna and T 9 both at flowering and post-flowering stages.

Genetic options

Studies have revealed the complex and polygenic nature of plant salt tolerance. The potential of a genetic approach towards solving the problems of soil salinity and alkalinity is now widely recognized. There is also an overwhelming view that this approach is even more relevant to developing countries often facing severe constraints regarding availability of resources. Genetic adaptation of crops to salinity requires that sufficient heritable variability exists within species to permit selection of salt-tolerant strains and that those plant characteristics that confer salt tolerance be identified. Modern varieties have a relatively narrow genetic base and are poorly adapted to adverse environments such as salinity. However, endemic genotypes from problem environments may provide the basic germplasm for breeding salt-tolerant varieties with acceptable yield potentials. Further, the pool of variability in a crop can also be enhanced by subjecting them to mutagenic agents, which can further be screened for the desired characters. Screening whole plants and the large amount of germplasm available for a particular crop for salinity tolerance in the field situations is time consuming and labour intensive. Keeping these factors in view, rapid screening methodologies have also been developed for screening large amounts of germplasm for salinity tolerance under solution culture in laboratory conditions. There is meagre information available on salt tolerance of Indian mustard at germination and seedling emergence stages under different kinds of stresses. Different genotypes of Indian mustard were shown to differ in their tolerance to soil salinity and alkalinity at seed germination and seedling growth (Sharma and Gill, 1995; Gill and Sharma, 1999) and also at whole plant level (Sharma and Gill, 1994). Different genotypes of Indian mustard showed differential tolerance to saline and alkaline stresses. Genetic variability is the key to any crop improvement programme and the extent to which the desirable characters are heritable is also important. For improving yield and yield component characters, information on their genetic variability and their interrelationships in different characters is necessary. Partitioning the genotypic correlation coefficients of yield components into direct and indirect effects may help to estimate the actual contribution of an attribute and its influence through other characters. Sixty genotypes were evaluated for their adaptation under semi-arid saline soil conditions (Sinha, 1991). High variability was recorded for secondary branches per plant, pods per plant, 1000 seed weight, seed yield per plant and seed yield per pod. Further low variability was recorded for seeds per pod and primary branches per plant. Seeds per plant were least affected by salinity. Seed bearing pods per plant and seed yield per plot showed high genetic coefficients of variation. Heritability was very high for number of seeds per pod (99.7) and 1000 seed weight (61.4) and moderate for pod length (39.7). Genetic advance was high for seed-bearing pods per plant (15.0) and plant height (11.9), followed by seeds per pod (5.7) and sterile pods per plant (8.3). Seedling emergence and

plant height exhibited maximum direct, positive effects on seed yield on saline soil. Further, genetic parameters for variability were also studied under sodic stress conditions in 19 genotypes of Indian mustard (Sinha *et al.*, 2002; Singh and Sharma, 2012, personal communication). The path analysis showed that secondary branches per plant had maximum direct effect on seed yield under sodic stress conditions. These studies point out that the framing of selection criteria should be based on number of primary branches per plant, main shoot length and 1000 seed weight for normal conditions and secondary branches per plant, number of pods per main shoot length and seed yield per plant under sodic stress conditions. Sustained breeding efforts have led to the development and release of two salt-tolerant genotypes CS 52 and CS 54 in the years 1997 and 2005, respectively. Another genotype 'CS 56' has been developed and released in the year 2008 with the characteristics of late sown conditions as well as tolerance to salinity stress conditions. The salt-tolerant variety 'CS 52' yields 20% higher in salt-affected soils compared to the high-yielding released varieties of Indian mustard at the national level. However, its maturity is longer by 1 week compared to the well-known high-yielding released varieties of Indian mustard, i.e. Varuna, Kranti and Pusa Bold. These salt-tolerant varieties perform better under salinity and alkalinity stress conditions by accumulating and compartmentalizing toxic ions in the root part, hence restricting the accumulation of toxic ions in leaves and stem. Further, the toxic ions get accumulated in lower leaves, which shed, ultimately reducing the effects of toxic ions on the plant. These processes help the plant to survive better under salt stress conditions. Another characteristic of these salt-tolerant varieties is their better adaptability under high temperature conditions during germination and seedling emergence stages. Till now, the variability already present in the system was exploited for the development of high yielding salt-tolerant varieties at CSSRI and other places. Variability was also generated further by subjecting the selected lines to different mutagenic agents. Though the success of producing salt-tolerant varieties through mutagenic approach is limited, this process cannot be ignored completely. A large number of mutants have been developed which have shown sensitiveness to salinity and have become an important tool for studying the mechanism of salt tolerance.

Biotechnological approaches

Different research groups are undertaking studies on elucidating salt tolerance mechanisms following molecular and biotechnological approaches. Efforts for the sequencing of the *Brassica* genome are underway at different locations and will take a long time. In the meantime, *Arabidopsis* was preferred as a model system for molecular analysis by different workers. Mutant studies in *Arabidopsis* have shown the involvement of *SOS1*, *SOS2* and *SOS3* genes in salinity tolerance (Hasegawa *et al.*, 2000; Qiu *et al.*, 2002; Zhu, 2003). *SOS* genes (*SOS1*, *SOS2* and *SOS3*) in *Arabidopsis* were genetically confirmed to function in a common pathway of salt tolerance. In response to the sensing of salt stress by the roots, the plasma membrane sensor elicits cytoplasmic Ca^{++} perturbations. In response to Ca^{++} perturbations in the cell, the *SOS3* gene changes its conformation and interacts with an effector kinase to relay the signal downstream. Lower Ca^{++} binding ability was recorded in the *Arabidopsis* plant with mutation in *SOS3* impairing cellular ionic equilibrium and rendering the plant hypersensitive to salt stress. Further, the *SOS2* gene was isolated in *Arabidopsis* mutants oversensitive to salt stress. The mRNA level of *SOS2* was upregulated in response to salt stress in roots. This gene encodes a novel serine/threonine protein kinase. *SOS3* activates *SOS2* protein kinsae activity in a calcium dependent manner, where *SOS2* phosphorylates and activates *SOS1* (a plasma membrane Na^+/H^+ antiporter). Genetic analysis of *SOS1* mutants further helps in identifying the *SOS1* gene as the target of the *SOS3*–*SOS2* pathway. The plasma membrane Na^+/H^+ antiporter helps in excluding excess Na ions through the *SOS* pathway to maintain cellular ion homeostasis. To further elucidate the salinity tolerance mechanisms in *B. juncea*, Kumar *et al.* (2009)

cloned the orthologues of *SOS1*, *SOS2* and *SOS3* from salt-tolerant genotype CS 52 and carried out RNA abundance analysis of root and shoot tissue. It was shown that the tissue specific expression of these *SOS* genes plays a key role in Na^+ regulation. A strong relationship was developed between *SOS* transcript accumulation and physiological parameters across different cultivars of *Brassica* in response to salinity stress. Control of salinity tolerance mechanism in *B. juncea* genotypes through the *SOS* pathway was further elaborated by more recent work from ICAR-Indian Agricultural Research Institute and ICAR-National Research Centre of Plant Biotechnology, New Delhi, India. Gene expression studies revealed the existence of a more efficient *SOS* pathway composed of *SOS1*, *SOS2* and *SOS3* and vacuolar Na^+/K^+ antiporter in salt-tolerant CS 52 and CS 54 compared to salt-sensitive Varuna and T 9 (Chakraborty *et al.*, 2012). Sequence analysis of partial cDNAs showed the conserved nature of these genes and their intra- and intergenic relatedness. The existence of an efficient *SOS* pathway, resulting in higher K/Na ration, could be one of the major factors determining salinity tolerance of *B. juncea* genotypes CS 52 and CS 54.

Agronomic Management

Alkali soils are characterized by poor soil water–plant relations. The capacity of these soils to absorb and supply water is restricted due to poor hydraulic conductivity resulting in slow movement of water from lower to upper layers to meet the evapotranspiration demand of the crop. Since the lower layers in soil are highly sodic, the depletion of water is confined mainly to the upper 15 cm soil layer. Under such situations, the crop requires a pre-sowing irrigation and one post-sowing irrigation, preferably at rosette stage, about 29–30 DAS for significant increase in seed yield (Sharma and Singh, 1993). Irrigation at this stage stimulates root extension by wetting soil layers below the shallow root system. Application of nitrogen at 60 kg/ha recorded optimum yield under normal situations, however, addition of 25% higher nitrogen under salt stress would further increase seed yield.

Frost Stress

In general, different crops have unique day and night temperature requirements for their growth and development. *Brassica* species are grown during the winter season and its reproductive phase, generally, coincides with the chilling low temperatures. In north India, the crop is also exposed to sub-zero temperatures once or twice in its growing season causing widespread damage to seed development. A single exposure of mustard to frost causes yield losses up to 70% in some of the species. Various phenotypic symptoms in response to chilling stress include reduced leaf expansion, wilting, chlorosis and necrosis, as well as severe membrane damage, mainly due to acute dehydration associated with freezing. Chilling-sensitive plants have a higher proportion of saturated fatty acids in their membranes leading to higher transition temperature that affects the fluidity of the membranes under stress conditions. Ice formation takes place in the intercellular spaces of young leaves and developing seeds, which causes mechanical strain on the cell wall and membrane leading to cell rupture and death of the tissues (Olien and Smith, 1997). These factors ultimately cause reduction in yield. In *B. juncea*, both very young and nearly mature seeds are less affected by frost than seeds at an intermediate stage of development. Thus, developing seeds in the lower siliquae of plants sown at normal sowing date and in the upper siliquae of early sown plants were largely affected by frost. Susceptibility of developing seeds was recorded to be maximum at 50% siliquae formation stage in the plant (Dhawan, 1985). Accordingly, growth stage has its clear importance in escaping the frost damage by different varieties.

Cold acclimation enhances the plant's tolerance to cold freezing conditions. It results in physical and biochemical restructuring of cell membranes through changes in the lipid composition and induction of other non-enzymatic proteins that alter the freezing point of water, leading to prevention of ice formation in the tissues. Different species of *Brassica* respond differently to frost. *Brassica rapa* genotypes Span, Torch and Bell were shown to be more tolerant to frost compared to Indian *Brassica* species (Ohlsson, 1983). It has been recorded that the salt-tolerant varieties developed by CSSRI showed tolerance to low temperature and cold conditions as evidenced in the year 2008, when all the germplasm was exposed to frost during January 2008 (Sharma, 2012, personal communication). Dhawan *et al.* (1983) did not observe any effect of amount of N applied on the frost tolerance of *Brassica* species under irrigated or un-irrigated conditions. Further, the un-irrigated crop showed more frost injury than the irrigated crop in *B. rapa* (cv. BSH 1), *B. juncea* (cv. RH 30) and *B. chinensis* (cv. Local). Accordingly, a light irrigation should be applied to the crop when night temperature falls to near 0°C to escape chilling or probable frost stress conditions. Based on a pot-culture experiment on *B. juncea* genotype Prakash, in which plants were exposed to field capacity first and then to varying water stress levels prior to their exposure to frost, it was concluded that short term water stress prior to freezing in an irrigated crop may enhance frost tolerance, a crop raised under un-irrigated conditions is less likely to survive freezing. Further, Dhawan *et al.* (1986) have also shown freeze-hardening responses in *B. juncea* cv. Prakash by short term exposure of plants to low temperature, water stress, short days and dimethyl sulfoxide.

Improved genotypes

Improved genotypes have been developed. Genotypes such as RGN 48, RGN 13, RH 781, RH 7361 and RH 9801 can tolerate low temperature and frost-stress situations that should be preferred over other high-yielding varieties in the areas experiencing frost stress to minimize the loss. Frost tolerance is a multigenic trait. A concerted effort through physiology, molecular biology and biochemistry is required to understand the complex quantitative trait of low temperature and freezing tolerance. Analysis of gene expression in plants during cold acclimation has revealed the existence of low-temperature-responsive genes with complex regulatory mechanisms, pathways and products that assist plant cells to resist and survive freezing (Gamboa *et al.*, 2007). Hadi *et al.* (2011) investigated frost-resistant mutants of *B. oleracea* var. *botrytis* for the presence of CBF/DREB1 and COR15a gene products and induced frost resistance. First *CBF* gene and COR15a protein was detected after cold acclimation for 14 days in *B. oleracea*. The results confirmed the first report of the presence of BoCBF/DREB1 in *B. oleracea* and this only appeared under cold acclimation and their amino acid sequence analysis revealed a very high homology (90%) with CBF sequences of other *Brassica* species (BnCBF5/DREB1, BrDREB1 and BjDREB1B). The genotypes showed positive significant correlation between BoCBF/DREB1 expression and frost resistance. The proline level under acclimation increased about eight-fold and demonstrated positive and significant correlation with BoCBF/DREB1 expression.

High Temperatures

Crop plants depict high productivity when grown in ambient temperature, optimum for various growth and physio-metabolic processes. However, temperature higher than optimum (5–7°C above optimum for the concerned crop) decreases rate and duration of metabolic process and, therefore, decreases yield levels. High temperature is required and is an important factor for realization of optimum growth, development and economic produce in arid and semi-arid regions. *Brassica* species are known to be heat susceptible towards seedling emergence if planted at greater than 30°C mean day/night temperature. Light textured arid soils show very high rate of seedling mortality leading to sub-optimum plant stands. Hence, tolerance to heat at the seedling emergence stage in *Brassica* will facilitate an extended planting period, increasing the cropped area and utilization of conserved soil moisture received due to late rains. Thus, under resource-constraint farming systems, it is desirable to breed for heat tolerance towards critical sensitive growth stages. High-temperature

stresses causes accelerated plant development and consequently reduces both growth and yield of plants (Blum, 1988). Heat-exposed rape plants had distinct morphological and biochemical changes. For instance, heat-exposed seeds had low viability (2%), a tobacco like colour, very high electrolyte leakage, extremely high free fatty acids and a high proportion of jet black seeds (98%) compared to sound rape seeds in Canadian situations. High temperature (32–37°C) under short day (9 h) conditions markedly reduced plant height, produced longer leaves and suppressed flowering in *Brassica* species. Similarly, in a phytotron study in Germany, *Brassica* species at 16.5–24°C showed a decrease by 1.0–1.5% in fatty acids for each rise in 1°C (Marquard, 1985). Heat-stressed rape (cv. 601) had lower endogenous GA_3, IAA and zeatin contents than in control, while ABA and ethylene contents were increased significantly; heat shock may also increase lipid peroxidation and membrane damage (Zhou *et al.*, 1999). High-temperature-induced yield losses could be due to accelerated senescence, reduction in photosynthetic rates but increased respiration and inhibition of metabolic process of grain development such as starch synthesis. Crop species including *Brassica* adapt to heat stress through membrane stability and accumulation of proline content in the anthers. For instance, there was accumulation of free proline in the leaves of *Brassica* vegetables in response to high temperature (35/30°C) as a possible mechanism to suppress heat stress (Takeda *et al.*, 1999). Similarly, in tolerant genotypes there was a surge in the translocations of proline to pollen from anther tissue just before 6 days before anthesis (Talwar *et al.*, 1999). On the other hand, in a susceptible genotype, the proline concentration in anther decreased at higher temperature and there was no increase in the translocation of proline to pollen grain. Thus, high proline concentrations help pollen germination and fast pollen tube growth capable of fertilization and maintaining high pollen and ovule fertility. Morphologically, in *B. chinensis* the heat-tolerant cultivars had longer, thicker leaves with thicker palisade tissue, more leaves per plant, and formed a taller, narrower plant than susceptible lines. Plants generally tolerate heat by way of avoidance and tolerance mechanism. Avoidance keeps the canopy temperature lower than the ambient temperature through transpirational cooling, reflection of solar radiation through leaf hairs and wax deposition. However, the tolerance mechanism operates through bio-membrane saturation, synthesis of isozymes (peroxidase, catalase, super-oxide dismutase) finally stabilizing the bio-membranes. Induction of heat-shock proteins (HSPs) may also play a very important role. Desiccation tolerance may also be achieved through expression of low molecular weight (9–15.5 kDa) heat-stable polypeptides in groundnuts and 80–90 kDa in *Brassica* leading to HSPs inducing thermo-tolerance of required magnitude (Viswanathan and Khanna-Chopra, 1996). Improved genotypes have been developed in *B. juncea* that can tolerate high temperature during germination and later growth stages and should be preferred over other high-yielding varieties. Some of the improved varieties tolerant to high temperature stress include CS 52, RGN 13, Pusa Agrani, Urvashi, Pusa Mustard 25 and Pusa Mustard 27.

Conclusions

Oilseed brassica crops are cultivated under wide ecological diversities that include: irrigated/ rainfed and salt-affected soils experiencing a vast diversity of agroclimatic conditions. The crops are also exposed to extreme variations in temperatures and climatic conditions. High soil temperature at sowing time and low, chilling temperatures, even frost, during the reproductive phase greatly affects the crop production and productivity. A large chunk of area comes under arid and semi-arid regions, which are affected by drought and salt-stress situations. In India, the crop is grown under residual moisture conditions with lower productivity. It is also irrigated with brackish water in the absence of good quality irrigation water at critical crop growth stages. All these factors decline the crop and oil yield. These abiotic stresses can be managed by exploiting genetic resources and agronomic factors. Traditional approaches to breeding crop plants with improved stress tolerance have so far met with limited success. A number of improved varieties have been developed which perform better under different kinds of stresses, recording lower percentage reduction, upon exposure to stress, compared to high-yielding varieties. Newer molecular approaches with marker assisted selection, identification of QTLs for drought, salt stress and frost besides transgenic approaches would help improve the tolerance potential of rapeseed mustard group.

References

Ashraf, M. and Mehmood, S. (1990) Response of four *Brassica* species to drought stress. *Environmental and Experimental Botany* 30, 93–100.

Ashraf, M., Nazir, N. and McNeilly, T. (2001) Comparative salt tolerance of amphidiploids and diploid *Brassica* species. *Plant Science* 160, 683–689.

Blum, A. (1988) *Plant Breeding for Stress Environments*. CRC Press, Boca Rato, Florida.

Brzostowicz, A. (1988) Use of delayed luminescence method for evaluation of frost resistance in *Brassica* plants. *Cruciferae Newsletter* 13, 118.

Chakraborty, K., Sairam, R.K. and Bhattacharya, R.C. (2012) Differential expression of salt overly sensitive pathway genes determines salinity stress tolerance in *Brassica* genotypes. *Plant Physiology and Biochemistry* 51, 90–101.

Chhabra, M.L., Dhingra, H.R. and Yadara, T.P. (1981) Screening of Indian mustard varieties for drought resistance. *Indian Journal of Plant Physiology* 24, 8–11.

Clark, J.M. and Thomas, N.M. (1982) Leaf diffusive resistance, surface temperature, osmotic potential and CO_2 assimilation capability as indicators of drought intensity in rape. *Canadian Journal of Plant Science* 62, 785–789.

Close, T.J. and Chandler, P.M. (1990) Cereal dehydration: serology gene mapping and potential functional roles. *Australian Journal of Plant Physiology* 17, 333–334.

Dhawan, A.K. (1985) Freezing in oil-seed *Brassica* spp.: some factors affecting injury. *Journal of Agricultural Science* (Cambridge) 104, 513–518.

Dhawan, A.K., Chhabra, M.L. and Yadava, T.P. (1983) Freezing injury in oilseed *Brassica* species. *Annals of Botany* 51, 673–677.

Dhawan, A.K., Hooda, A. and Goyal, R.K. (1986) Effect of low temperature, short days, water stress and dimethyl-sulphoxide on frost tolerance of *Brassica juncea* (L.) Coss and Czern var. Prakash. *Annals of Botany* 58, 267–271.

Dure, L., Crouch, M., Harda, J., Ho, T.H.D., Humdy, J., Quatrano, R., Thomas, T. and Sung, Z.R. (1989) Common amino acids sequence domains among the LEA proteins of higher plants. *Plant Molecular Biology* 12, 475–486.

Gamboa, M.C., Rasmussen-Poblete, S., Pablo, D.T. and Valenzuela, E.K. (2007) Isolation and characterization of a cDNA encoding a CBF transcription factor from *E. globules*. *Plant Physiology and Biochemistry* 45, 1–5.

Garai, A.K., Jana, P.K., Mandal, B.B. and Barik, A. (1989) Effect of nitrogen on yield, consumptive use and water-use efficiency of toria (*B. napus* var. *napus*) and Indian mustard (*B. juncea* L.) under rainfed conditions. *Indian Journal of Agricultural Sciences* 59, 791–794.

Gill, K.S. and Sharma, P.C. (1999) Growth, yield and physiological responses to saline water application at various growth stages in *Brassica juncea*. *Indian Journal of Environment and Ecoplanning* 2, 11–17.

Hadi, F., Gilpin, M. and Fuller, M.P. (2011) Identification and expression analysis of CBF/DREB1 and COR15 genes in mutants of *Brassica oleracea* var. *botrytis* with enhanced proline production and frost resistance. *Plant Physiology and Biochemistry* 49, 1323–1332.

Harada, J.J., Delisle, A.J., Baden, C.S. and Crouch, M.L. (1989) Unusual sequence of an abscisic acid inducible mRNA which accumulates late in *Brassica napus* seed development. *Plant Molecular Biology* 12, 395–401.

Hasegawa, P.M., Bressan, R.A., Zhu, J.K. and Bonhert, H.J. (2000) Plant cellular and molecular responses to high salinity. *Annual Review of Plant Physiology and Plant Molecular Biology* 51, 463–499.

Holt, D.F., Kawano, N. and Ito, O. (2000) Involvement of active oxygen scavenging system in drought tolerance. *Philippines Journal of Crop Science* 25 (Suppl.), 50.

Ilami, G., Nespoulous, C., Huet, J.C., Vartanian, N. and Pemollet, J.C. (1997) Characterization of BND22, a drought-induced protein expressed in *Brassica napus* leaves. *Phytochemistry* 45, 1–8.

Javid, M., Ford, R. and Nicolas, E. (2012) Tolerance responses of *Brassica juncea* to salinity, alkalinity and alkaline salinity. *Functional Plant Biology* 39, 699–707.

Kochhar, S., Chaudhary, R.G. and Gautam, J.K.S. (1990) Mustard varieties suitable to stress conditions of Arunachal Pradesh, *National Seminar Gent. Brassicas*. Durgapura, Jaipur, 8–9 August.

Kumar, A. and Singh, D.P. (1998) Use of physiological indices as a screening technique for drought tolerance in oilseed *Brassica* species. *Annals of Botany* 81, 413–420.

Kumar, A., Singh, D.P. and Singh, P. (1987) Genotypic variation in the response of *Brassica* sp. to water deficit. *Journal of Agriculture Science* (Cambridge) 109, 615–618.

Kumar, A., Elston, J. and Singh, V.P. (1994) Leaf area growth of two *Brassica* species in response to water stress. *Crop Research* (Hisar) 8, 594–602.

Kumar, D. (1995) Salt tolerance in oilseed Brassicas – present status and future prospects. *Plant Breeding Abstracts* (Cambridge) 65, 1439–1447.

Kumar, G., Purty, R.S., Sharma, M.P., Singla-Pareek, S.L. and Pareek, A. (2009) Physiological responses among *Brassica* species under salinity stress show strong correlation with transcript abundance for SOS pathway related genes. *Journal of Plant Physiology* 166, 507–520.

Lee, J.I., Kwon, B.S., Kim, I.H. and Ham, Y.S. (1981) The high yielding, cold tolerant rape cultivar Naehan Yuohae with high quality of oil and cake. *Research Report Office Rural Development (Crops)* 23, 188–192.

Lewis, G.J. (1992) Strategies for improvement of canola (*B. napus*) yields in Mediterranean environments. PhD thesis, University of Western Australia, Perth.

Liang, Z.S., Ding, Z.R. and Want, S.T.R. (1992) Study on type of water stress adaptation in both *Brassica napus* and *B. juncea* L. species. *Acta Botanika* 12, 38–45.

Maas, E.V. (1986) Salt tolerance of plants. *Applied Agricultural Research* 1, 12–26.

Marquard, R. (1985) The influence of temperature and photoperiod of fat content, fatty acid composition and tocopherols of rapeseeds (*B. napus*) and mustard species (*Sinopsis alba*, *B. juncea* and *B. nigra*). *Agrochimica* 29, 145–153.

Mathur, D. and Wattal, P.N. (1996) Physiological analysis of growth and development in three species of rapeseed-mustard (*Brassica juncea*, *B. compestris* and *B. napus*) under irrigated and unirrigated conditions. *Indian Journal of Plant Physiology* 1, 171–174.

Mendhan, N.J., Shipway, P.A. and Scott, R.K. (1981) The effect of delayed sowing and water stress on growth, development and yield of winter oilseed rape (*B. napus*). *Journal of Agriculture Science* (Cambridge) 96, 389–416.

Obloj, H. and Kacperska, A.A. (1981) Desiccation tolerance changes in winter rape leaves grown under different experimental conditions. *Biologia Plantarum* 23, 209–213.

Ohlsson, I. (1983) Indian *Brassica* material tested in Sweden for cold tolerance. In: *Proceedings of the 6th International Rapeseed Conference*, Vol. I, May 1983, Paris, pp. 504–511.

Olien, C.R. and Smith, M.N. (1997) Ice adhesions in relation to freeze stress. *Plant Physiology* 60, 499–503.

Oram, R.N. (1987) Adapting *B. juncea* to southern Australia. In: *Proceedings of the 4th Australian Agronomy Conference*, The Australian Society of Agronomy, Parkville, Victoria, pp. 227.

Paje, M.C.M., Ludlow, M.M. and Lawn, R.J. (1988) Variation among soybean (*Glycine max* L.) accessions in epidermal conductance of leaves. *Australian Journal of Agricultural Research* 39, 363–373.

Pandya, R.B. and Khan, M.I. (1973) Enhancement of raya (*B. juncea* L.) germination under simulated drought by seed treatment with Cycocel trimethyl ammonium chloride. *Biochemistry and Physiology Pflanzen* 164, 112–115.

Patel, M.M. and Mehta, H.M. (1984) Effect of growth regulators on mustard (*B. juncea* L.) variety Varuna under stress and non-stress conditions of soil moisture. *Annals of Arid Zone* 23, 240–254.

Qiu, Q.S., Guo, Y., Dietrich, M.A., Schumaker, K.S. and Zhu, J.K. (2002) Regulation of SOS1, a plasma membrane Na^+/K^+ exchanger in *Arabidopsis thaliana* by SOS2 and SOS3. *Proceedings of National Academy of Sciences USA* 99, 8436–8441.

Raja, V. and Bishnoi, K.C. (1988) Studies on the seed quality and oil yield of toria genotypes under varying irrigation schedules. *Indian Journal of Agronomy* 33, 77–83.

Raja, V. and Bishnoi, K.C. (1990) Soil-water-and dry conditions. *Indian Journal of Agronomy* 35, 91–98.

Rawson, H.M. and Clarke, J.M. (1988) Nocturnal transpiration in wheat. *Australian Journal of Plant Physiology* 15, 397–406.

Reviron, M.P., Vartanian, N., Sallantin, M., Huet, J.C., Pernollet, T.C. and de Vienne, D.E. (1992) Characterization of a novel protein induced by progressive on rapid drought and salinity in *Brassica napus* leaves. *Plant Physiology* 100, 1486–1493.

Ribaut, J.M., Banziger, M. and Hoisington, D. (2002) Genetic engineering of crop plants for abiotic stress. *Proceedings of APECJIRAC Joint Symposium and Workshop*, Bangkok, 3–7 September 2001, JIRAC Working Report 23, 85–92.

Richards, R.A. and Thurling, N. (1979) Genetic analysis of drought stress response in rapeseeds (*Brassica campestris* and *B. napus*). II. Yield improvement and application of selection indices. *Euphytica* 28, 169–177.

Samra, J.S., Singh, G. and Dagar, J.C. (2006) *Drought Management Strategies in India*. Indian Council of Agricultural Research, New Delhi.

Sawhney, V., Singal, H.R., Singh Phool, Sawhney, S.K. and Singh, D.P. (1996) Pattern of plant relations and carbon dioxide exchange rates in contrasting genotypes of *Brassica juncea* under water deficit conditions. *Indian Journal of Plant Physiology* 1, 203–206.

Shcherbakova, A. and Kacperska, A. (1983) Water stress injuries and tolerance as related to potassium efflux from winter rape hypocotyls. *Plant Physiology* 57, 296–300.

Sharma, D.K. and Kumar, A. (1988) Effect of irrigation scheduling and nitrogen on yield and N uptake of mustard. *Indian Journal of Agronomy* 33, 436–441.

Sharma, D.K. and Singh, K.N. (1993) Effect of irrigation on growth, yield and evapotranspiration of mustard (*Brassica juncea*) in partially reclaimed sodic soils. *Agricultural Water Management* 23, 225–232.

Sharma, P.C. (2003) Salt tolerance of Indian mustard (*Brassica juncea* L.): factors affecting growth and yield. *Indian Journal of Plant Physiology* 8, 368–372.

Sharma, P.C. and Gill, K.S. (1994) Salinity induced effects on biomass, yield, yield attributing characters and ionic contents in genotypes of Indian mustard (*Brassica juncea*). *Indian Journal of Agricultural Sciences* 64, 785–788.

Sharma, P.C. and Gill, K.S. (1995) Performance and ionic accumulation in *Brassica juncea* and *B. carinata* genotypes under salinity. *Plant Physiology and Biochemistry* 22, 154–158.

Sharma, R.C., Rao, B.R.M. and Saxena, R.K. (2004) Salt affected soils in India – Current asessment. In: *Advances in Sodic Land Reclamation. International Conference on Sustainable Management of Sodic Lands*, 9–14 February at Lucknow, India, pp. 1–26.

Sheoran, I.S. and Khan, M.I. (1987) Induction of drought tolerance in Indian mustard during germination by pre-sowing seed treatment with growth regulators. *Indian Journal of Agricultural Sciences* 57, 54–56.

Sinclair, T.R. and Ludlow, M.M. (1986) Influence of soil water supply on the plant water balance of four tropical grain legumes. *Australian Journal of Plant Physiology* 13, 329–341.

Singh, G., Thakur, P.S. and Rai, V.K. (1985) Free amino acid pattern in stressed leaves of two contrasting resistant and susceptible cultivars of chickpea. *Experimental Agriculture* 41, 40–41.

Singh, S. and Sharma, H.C. (1981) Effect of soil moisture status and fertility levels on the yield of taramira. *Indian Journal of Agricultural Science* 51, 875–880.

Singh, S., Singh, N.P. and Bandyopadhyay, S.K. (1997) Effect of limited irrigation on seed production, oil yield and water use by Indian mustard. *Annals of Agriculture Research* 18, 265–269.

Sinha, T.S. (1991) Genetic adoption of Indian mustard (*Brassica juncea*) to semi-arid saline soil conditions. *Indian Journal of Agricultural Science* 61, 251–254.

Sinha, T.S., Singh, D., Sharma, P.C. and Sharma, H.B. (2002) Genetic variability, correlation and path coefficient studies and their implications of selections of high yielding genotypes in Indian mustard (*Brassica juncea* L.) under normal and sodic soil conditions. *Indian Journal of Coastal Agricultural Research* 20, 31–36.

Sinha, T.S., Singh, D., Sharma, P.C. and Sharma, H.B. (2003) Rapid screening methodology for salt tolerance during germination and seedling emergence in Indian mustard (*Brassica juncea* L.). *Indian Journal of Plant Physiology* 8, 363–367.

Takeda, H., Cenpukedee, U., Chauhan, Y.S., Srinivasan, A., Hossain, M.M., Rashad, M.H., Lin Bai Quin, Talwar, H.S., Senboku, T., Yashima, S., Shono, M., Ancha, S., Lin, B. Q., Yajima, M. and Hayashi, T. (1999) Studies in heat tolerance of *Brassica* vegetables and legumes at the International Collaboration Research Section from 1992. *Proceedings of Workshop on Heat Tolerance of Crops*, Okinawa, Japan 7–9 October 1997. JIRCAS Working Report 14, 17–29.

Talwar, H.S., Takeda, H., Yashima, S. and Senboku, T. (1999) Growth and photosynthetic responses of groundnut genotypes to high temperature. *Crop Science* 39, 460–466.

Vartanian, N., Damernval, C. and de Vienne, D. (1987) Drought induced changes in protein patterns of *Brassica napus* var. *oleifera* roots. *Plant Physiology* 84, 989–992.

Vertucci, C.W. and Leopold, A.C. (1987a) The relationship between water binding and desiccation tolerance in tissues. *Plant Physiology* 85, 232–238.

Vertucci, C.W. and Leopold, A.C. (1987b) Water binding in legume seeds. *Plant Physiology* 85, 224–231.

Viswanathan, C. and Khanna-Chopra, R. (1996) Heat shock proteins – role in thermotolerance of crop plants. *Current Science* 71, 275–284.

Woods, D.L., Capraca, J.J. and Downey, R.K. (1991) The potential of mustard (*B. juncea* L.) as an edible crop on the Canadian priorities. *Canadian Journal of Plant Science* 71, 195–198.

Wright, P.R., Morgan, J.M. and Jessop, R.S. (1996) Comparative adaptation of canola (*B. napus*) and Indian mustard (*B. juncea*) to soil water deficit. *Field Crops Research* 49, 51–64.

Yadav, S.K., Kumar, A. and Singh, D.P. (1990) Micro-environment, water relations and yield of mustard cultivars in unirrigated and irrigated soils. *Beitrage Zur Tropischen Landwirtschaft and Verterinamedizin* 28, 399–404.

Yadava, T.P. and Hari Singh (1996) Morpho-physiological determinants of yield under water stress conditions in Indian mustard (*Brassica juncea*). *Acta Horticulturae* 407, 155–160.

Zhou, W., Leul, M. and Zhou, W.J. (1999) Uniconazole-induced tolerance of rape plants to heat stress in relation to change in hormonal levels, enzyme activities and lipid peroxidase. *Plant Growth Regulation* 27, 99–104.

Zhu, K.K. (2003) Regulation of ion homeostasis under salt stress. *Current Opinion in Plant Biology* 6, 441–445.

Index

Page numbers in **bold** refer to illustrations and tables